TOYOTA TACOMA
2005-09 REPAIR MANUAL

CHILTON'S

Covers all U.S. and Canadian models of Toyota Tacoma 2WD and 4WD 2005 through 2009

by Joe L Hamilton

CHILTON Automotive Books

PUBLISHED BY HAYNES NORTH AMERICA, Inc.

Haynes®

APA AUTOMOTIVE PARTS & ACCESSORIES ASSOCIATION MEMBER

Manufactured in USA

ISBN-13: 978-1-56392-780-5
ISBN-10: 1-56392-780-2
Library of Congress Control Number 2009931667

ABCDE
FGHIJ
KLMNO
PQRST

Haynes Publishing Group
Sparkford Nr Yeovil
Somerset BA22 7JJ England

Haynes North America, Inc
861 Lawrence Drive
Newbury Park
California 91320 USA

Contents

Mechanic and photographer with a 2006 Toyota Tacoma

ACKNOWLEDGEMENTS

Technical writers who contributed to this project include Mike Stubblefield, Rob Maddox, Tim Imhoff and Jamie Sarté. Wiring diagrams provided exclusively for Haynes North America, Inc. by Solution Builders.

About this manual

ITS PURPOSE

The purpose of this manual is to help you get the best value from your vehicle. It can do so in several ways. It can help you decide what work must be done, even if you choose to have it done by a dealer service department or a repair shop; it provides information and procedures for routine maintenance and servicing; and it offers diagnostic and repair procedures to follow when trouble occurs.

We hope you use the manual to tackle the work yourself. For many simpler jobs, doing it yourself may be quicker than arranging an appointment to get the vehicle into a shop and making the trips to leave it and pick it up. More importantly, a lot of money can be saved by avoiding the expense the shop must pass on to you to cover its labor and overhead costs. An added benefit is the sense of satisfaction and accomplishment that you feel after doing the job yourself.

USING THE MANUAL

The manual is divided into Chapters. Each Chapter is divided into numbered Sections, which are headed in bold type between horizontal lines. Each Section consists of consecutively numbered paragraphs.

At the beginning of each numbered Section you will be referred to any illustrations which apply to the procedures in that Section. The reference numbers used in illustration captions pinpoint the pertinent Section and the Step within that Section. That is, illustration 3.2 means the illustration refers to Section 3 and Step (or paragraph) 2 within that Section.

Procedures, once described in the text, are not normally repeated. When it's necessary to refer to another Chapter, the reference will be given as Chapter and Section number. Cross references given without use of the word "Chapter" apply to Sections and/or paragraphs in the same Chapter. For example, "see Section 8" means in the same Chapter.

References to the left or right side of the vehicle assume you are sitting in the driver's seat, facing forward.

Even though we have prepared this manual with extreme care, neither the publisher nor the author can accept responsibility for any errors in, or omissions from, the information given.

➡NOTE

A *Note* provides information necessary to properly complete a procedure or information which will make the procedure easier to understand.

⁂ CAUTION

A *Caution* provides a special procedure or special steps which must be taken while completing the procedure where the Caution is found. Not heeding a Caution can result in damage to the assembly being worked on.

⁂ WARNING

A *Warning* provides a special procedure or special steps which must be taken while completing the procedure where the Warning is found. Not heeding a Warning can result in personal injury.

Introduction

The Tacoma is equipped with either a 2.7L four-cylinder engine or a 4.0L V6 engine. All engines are equipped with an Electronic Fuel Injection (EFI) system. The engine drives the rear wheels through either a manual or automatic transmission via a driveshaft and solid rear axle. A transfer case and driveshaft are used to drive the front axle on 4WD models.

The front suspension is fully independent; it consists of upper and lower control arms, a stabilizer bar, and integral coil spring/shock absorber assemblies. A solid axle at the rear is suspended by leaf springs and shock absorbers. The steering gear is a rack-and-pinion type and is connected to the steering knuckles by tie-rods. The front brakes are disc type, while the rear brakes are drum type. Power assist and ABS is standard on all models.

Vehicle identification numbers

Modifications are a continuing and unpublicized process in vehicle manufacturing. Since spare parts manuals and lists are compiled on a numerical basis, the individual vehicle numbers are essential to correctly identify the component required.

VEHICLE IDENTIFICATION NUMBER (VIN)

This very important identification number is stamped on a plate attached to the left side of the dashboard just inside the windshield on the driver's side of the vehicle (see illustration). The VIN also appears on the Vehicle Certificate of Title and Registration. It contains information such as where and when the vehicle was manufactured, the model year and the body style.

Counting from the left, the engine code is the eighth digit and the model year code is the tenth digit.

Model year codes:

5	2005
6	2006
7	2007
8	2008
9	2009

ENGINE SERIAL NUMBER

On V6 engines, the engine serial number is located on the right side of the engine block, at the rear of the engine (see illustration). On four-cylinder engines it's located on the left side, at the rear of the engine, near the starter (see illustration).

MANUFACTURER CERTIFICATION LABEL

The Manufacturer Certification label is affixed to the front door pillar. The plate contains the name of the manufacturer, the month and year of production, the Gross Vehicle Weight Rating (GVWR) and the certification statement (see illustration).

The Vehicle Identification Number (VIN) is visible through the driver's side of the windshield

The engine serial number on the four-cylinder engine is located at the left rear of the engine block, below the starter

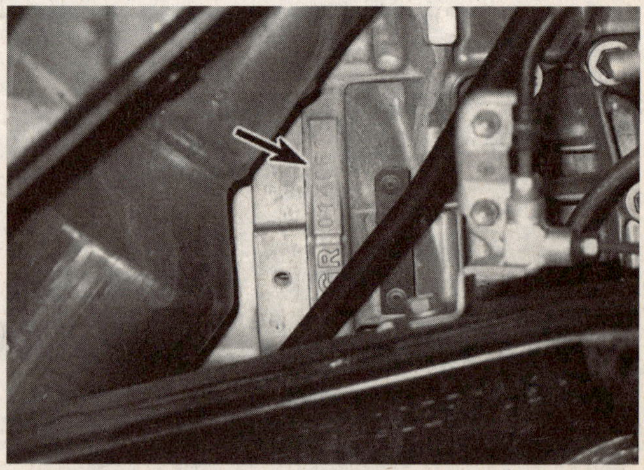

The engine serial number on the V6 engine is stamped onto a pad on the right side of the engine block, at the rear of the engine

VEHICLE EMISSIONS CONTROL INFORMATION (VECI) LABEL

The emissions control information label is found under the hood. This label contains information on the emissions control equipment installed on the vehicle, as well as tune-up specifications (see illustration).

TRANSFER CASE AND TRANSMISSION IDENTIFICATION NUMBER

The transfer case and manual transmission identification number is stamped into the case of the component. Automatic transmission numbers are stamped onto ID plates (see illustration).

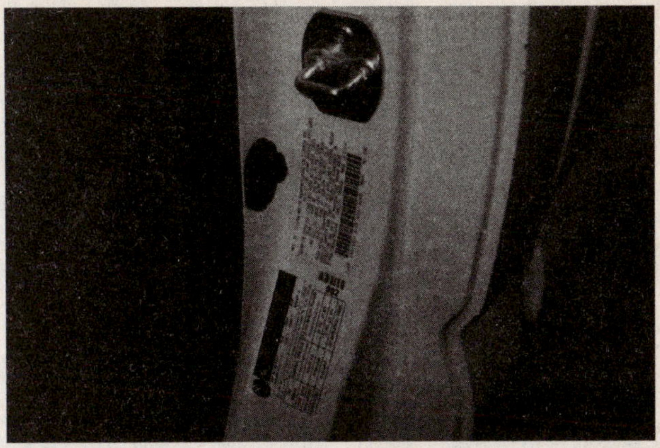

The manufacturer's certification label is affixed to the driver's side door jamb

The Vehicle Emissions Control Label (VECI) is located on the underside of the hood

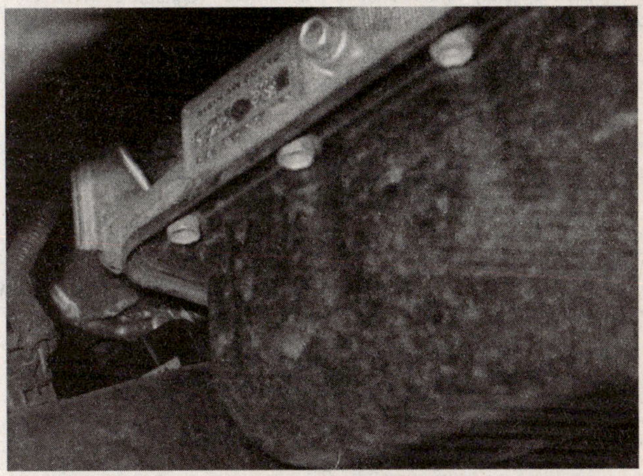

The identification number on automatic transmissions is stamped into a plate located on the side of the transmission

Recall information

Vehicle recalls are carried out by the manufacturer in the rare event of a possible safety-related defect. The vehicle's registered owner is contacted at the address on file at the Department of Motor Vehicles and given the details of the recall. Remedial work is carried out free of charge at a dealer service department.

If you are the new owner of a used vehicle which was subject to a recall and you want to be sure that the work has been carried out, it's best to contact a dealer service department and ask about your individual vehicle - you'll need to furnish them your Vehicle Identification Number (VIN).

The table below is based on information provided by the National Highway Traffic Safety Administration (NHTSA), the body which oversees vehicle recalls in the United States. The recall database is updated constantly. For the latest information on vehicle recalls, check the NHTSA website at www.nhtsa.gov, www.safercar.gov, or call the NHTSA hotline at 1-888-327-4236.

Recall date	Recall campaign number	Model(s) affected	Concern
Feb 10, 2005	05V050000	2005 Tacoma	On certain pickup trucks equipped with an automatic transmission, there is a possibility that the parking brake pedal cable lock nut may not have been tightened to the proper torque specification. In this condition, the lock nut may loosen and come off, which will reduce the effectiveness of the parking brake. This condition could allow the vehicle to roll if the transmission is not placed into the "PARK" position and the vehicle is parked on a slope, thus raising the possibility of a crash.
Jun 27, 2005	05V302000	2005 Tacoma	On certain regular cab pickup trucks equipped with a bench seat, the seat position and seat belt buckle sensor connector pins are incorrectly positioned due to a wire harness manufacturing process error. In this condition, the seat position and seat belt buckle sensor may not function as designed, affecting the adaptive airbag deployment, which, if the vehicle is involved in a crash, could result in improper occupant restraint.
Feb 27, 2006	06V061000	2006 Tacoma	On certain pickup trucks, the bead of the tire may be damaged due to improper assembly of the tires onto the wheels. If the vehicle is operated in this condition, there is a possibility that a bulge may be formed on the sidewall and air may leak from the area of the damaged bead and a crash could occur.

Recall date	Recall campaign number	Model(s) affected	Concern
Mar 28, 2006	06V096000	2005 and 2006 Tacoma	On certain vehicles, due to improper assembly of the airbag inflator, which is used in the side airbag, the curtain shield airbag, and the knee airbag assembly, some inflators were produced with an insufficient amount of the heating agents necessary for proper airbag deployment. In this condition, the expansion force of the gas may be insufficient to properly inflate the airbag when the SRS system is activated during a crash. This may increase the risk of injury to the occupant in the involved seating position in the event of a crash.
Jul 24, 2007	07V324000	2007 Tacoma	On certain pickup trucks equipped with four wheel drive, a section of the rear propeller shaft may have been improperly cast. The front side of the rear propeller shaft could separate at the joint and come into contact with the road surface, which could result in a loss of vehicle control, increasing the possibility of a crash.

Buying parts

Replacement parts are available from many sources, which generally fall into one of two categories - authorized dealer parts departments and independent retail auto parts stores. Our advice concerning these parts is as follows:

Retail auto parts stores: Good auto parts stores will stock frequently needed components which wear out relatively fast, such as clutch components, exhaust systems, brake parts, tune-up parts, etc. These stores often supply new or reconditioned parts on an exchange basis, which can save a considerable amount of money. Discount auto parts stores are often very good places to buy materials and parts needed for general vehicle maintenance such as oil, grease, filters, spark plugs, belts, touch-up paint, bulbs, etc. They also usually sell tools and general accessories, have convenient hours, charge lower prices and can often be found not far from home.

Authorized dealer parts department: This is the best source for parts which are unique to the vehicle and not generally available elsewhere (such as major engine parts, transmission parts, trim pieces, etc.).

Warranty information: If the vehicle is still covered under warranty, be sure that any replacement parts purchased - regardless of the source - do not invalidate the warranty!

To be sure of obtaining the correct parts, have engine and chassis numbers available and, if possible, take the old parts along for positive identification.

Maintenance techniques, tools and working facilities

MAINTENANCE TECHNIQUES

There are a number of techniques involved in maintenance and repair that will be referred to throughout this manual. Application of these techniques will enable the home mechanic to be more efficient, better organized and capable of performing the various tasks properly, which will ensure that the repair job is thorough and complete.

Fasteners

Fasteners are nuts, bolts, studs and screws used to hold two or more parts together. There are a few things to keep in mind when working with fasteners. Almost all of them use a locking device of some type, either a lockwasher, locknut, locking tab or thread adhesive. All threaded fasteners should be clean and straight, with undamaged threads and undamaged corners on the hex head where the wrench fits. Develop the habit of replacing all damaged nuts and bolts with new ones. Special locknuts with nylon or fiber inserts can only be used once. If they are removed, they lose their locking ability and must be replaced with new ones.

Rusted nuts and bolts should be treated with a penetrating fluid to ease removal and prevent breakage. Some mechanics use turpentine in a spout-type oil can, which works quite well. After applying the rust penetrant, let it work for a few minutes before trying to loosen the nut or bolt. Badly rusted fasteners may have to be chiseled or sawed off or removed with a special nut breaker, available at tool stores.

If a bolt or stud breaks off in an assembly, it can be drilled and removed with a special tool commonly available for this purpose. Most automotive machine shops can perform this task, as well as other repair procedures, such as the repair of threaded holes that have been stripped out.

Flat washers and lockwashers, when removed from an assembly, should always be replaced exactly as removed. Replace any damaged washers with new ones. Never use a lockwasher on any soft metal surface (such as aluminum), thin sheet metal or plastic.

Fastener sizes

For a number of reasons, automobile manufacturers are making wider and wider use of metric fasteners. Therefore, it is important to be able to tell the difference between standard (sometimes called U.S. or SAE) and metric hardware, since they cannot be interchanged.

All bolts, whether standard or metric, are sized according to diameter, thread pitch and length. For example, a standard 1/2 - 13 x 1 bolt is 1/2 inch in diameter, has 13 threads per inch and is 1 inch long. An M12 - 1.75 x 25 metric bolt is 12 mm in diameter, has a thread pitch of 1.75 mm (the distance between threads) and is 25 mm long. The two bolts are nearly identical, and easily confused, but they are not interchangeable.

In addition to the differences in diameter, thread pitch and length, metric and standard bolts can also be distinguished by examining the bolt heads. To begin with, the distance across the flats on a standard bolt head is measured in inches, while the same dimension on a metric bolt is sized in millimeters (the same is true for nuts). As a result, a standard wrench should not be used on a metric bolt and a metric wrench should not be used on a standard bolt. Also, most standard bolts have slashes radiating out from the center of the head to denote the grade or strength of the bolt, which is an indication of the amount of torque that can be applied to it. The greater the number of slashes, the greater the strength of the bolt. Grades 0 through 5 are commonly used on automobiles. Metric bolts have a property class (grade) number, rather than a slash, molded into their heads to indicate bolt strength. In this case, the higher the number, the stronger the bolt. Property class numbers 8.8, 9.8 and 10.9 are commonly used on automobiles.

Strength markings can also be used to distinguish standard hex nuts from metric hex nuts. Many standard nuts have dots stamped into one side, while metric nuts are marked with a number. The greater the number of dots, or the higher the number, the greater the strength of the nut.

Metric studs are also marked on their ends according to property class (grade). Larger studs are numbered (the same as metric bolts), while smaller studs carry a geometric code to denote grade.

It should be noted that many fasteners, especially Grades 0 through 2, have no distinguishing marks on them. When such is the case, the only way to determine whether it is standard or metric is to measure the thread pitch or compare it to a known fastener of the same size.

Standard fasteners are often referred to as SAE, as opposed to metric. However, it should be noted that SAE technically refers to a non-metric fine thread fastener only. Coarse thread non-metric fasteners are referred to as USS sizes.

Since fasteners of the same size (both standard and metric) may have different strength ratings, be sure to reinstall any bolts, studs or nuts removed from your vehicle in their original locations. Also, when replacing a fastener with a new one, make sure that the new one has a strength rating equal to or greater than the original.

Tightening sequences and procedures

Most threaded fasteners should be tightened to a specific torque value (torque is the twisting force applied to a threaded component such as a nut or bolt). Overtightening the fastener can weaken it and cause it to break, while undertightening can cause it to eventually come loose. Bolts, screws and studs, depending on the material they are made of and their thread diameters, have specific torque values, many of which are noted in the Specifications at the end of each Chapter. Be sure to follow the torque recommendations closely. For fasteners not assigned a specific torque, a general torque value chart is presented here as a guide. These torque values are for dry (unlubricated) fasteners threaded into steel or cast iron (not aluminum). As was previously mentioned, the size and grade of a fastener determine the amount of torque that can safely be applied to it. The figures listed here are approximate for Grade 2 and Grade 3 fasteners. Higher grades can tolerate higher torque values.

Fasteners laid out in a pattern, such as cylinder head bolts, oil pan bolts, differential cover bolts, etc., must be loosened or tightened in sequence to avoid warping the component. This sequence will normally be shown in the appropriate Chapter. If a specific pattern is not given, the following procedures can be used to prevent warping.

Initially, the bolts or nuts should be assembled finger-tight only. Next, they should be tightened one full turn each, in a criss-cross or diagonal pattern. After each one has been tightened one full turn, return to the first one and tighten them all one-half turn, following the same

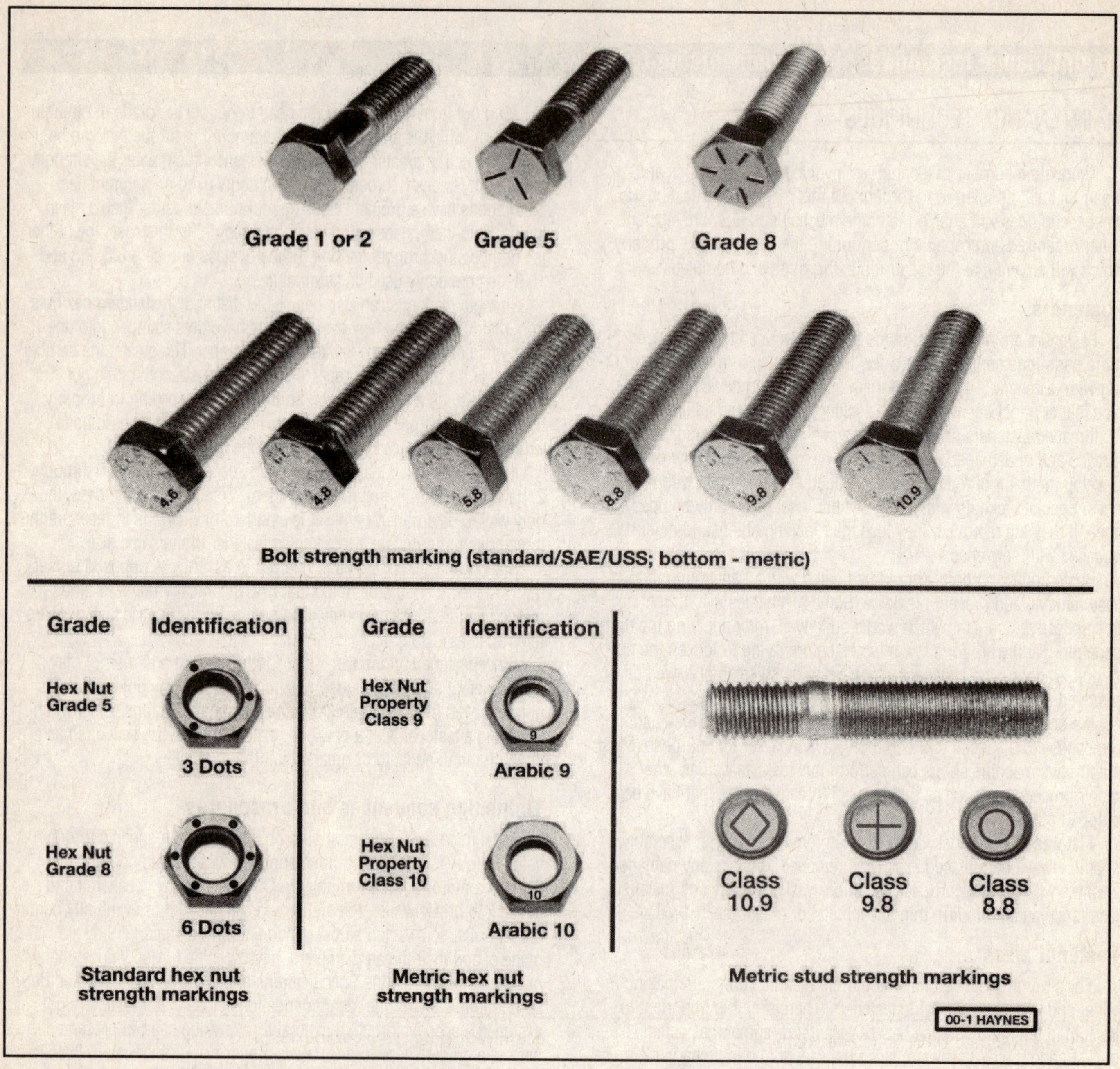

Bolt strength marking (standard/SAE/USS; bottom - metric)

Standard hex nut strength markings

Metric hex nut strength markings

Metric stud strength markings

pattern. Finally, tighten each of them one-quarter turn at a time until each fastener has been tightened to the proper torque. To loosen and remove the fasteners, the procedure would be reversed.

Component disassembly

Component disassembly should be done with care and purpose to help ensure that the parts go back together properly. Always keep track of the sequence in which parts are removed. Make note of special characteristics or marks on parts that can be installed more than one way, such as a grooved thrust washer on a shaft. It is a good idea to lay the disassembled parts out on a clean surface in the order that they were removed. It may also be helpful to make sketches or take instant photos of components before removal.

When removing fasteners from a component, keep track of their locations. Sometimes threading a bolt back in a part, or putting the washers and nut back on a stud, can prevent mix-ups later. If nuts and bolts cannot be returned to their original locations, they should be kept in a compartmented box or a series of small boxes. A cupcake or muffin tin is ideal for this purpose, since each cavity can hold the bolts and nuts from a particular area (i.e. oil pan bolts, valve cover bolts, engine

	Ft-lbs	Nm
Metric thread sizes		
M-6	6 to 9	9 to 12
M-8	14 to 21	19 to 28
M-10	28 to 40	38 to 54
M-12	50 to 71	68 to 96
M-14	80 to 140	109 to 154
Pipe thread sizes		
1/8	5 to 8	7 to 10
1/4	12 to 18	17 to 24
3/8	22 to 33	30 to 44
1/2	25 to 35	34 to 47
U.S. thread sizes		
1/4 - 20	6 to 9	9 to 12
5/16 - 18	12 to 18	17 to 24
5/16 - 24	14 to 20	19 to 27
3/8 - 16	22 to 32	30 to 43
3/8 - 24	27 to 38	37 to 51
7/16 - 14	40 to 55	55 to 74
7/16 - 20	40 to 60	55 to 81
1/2 - 13	55 to 80	75 to 108

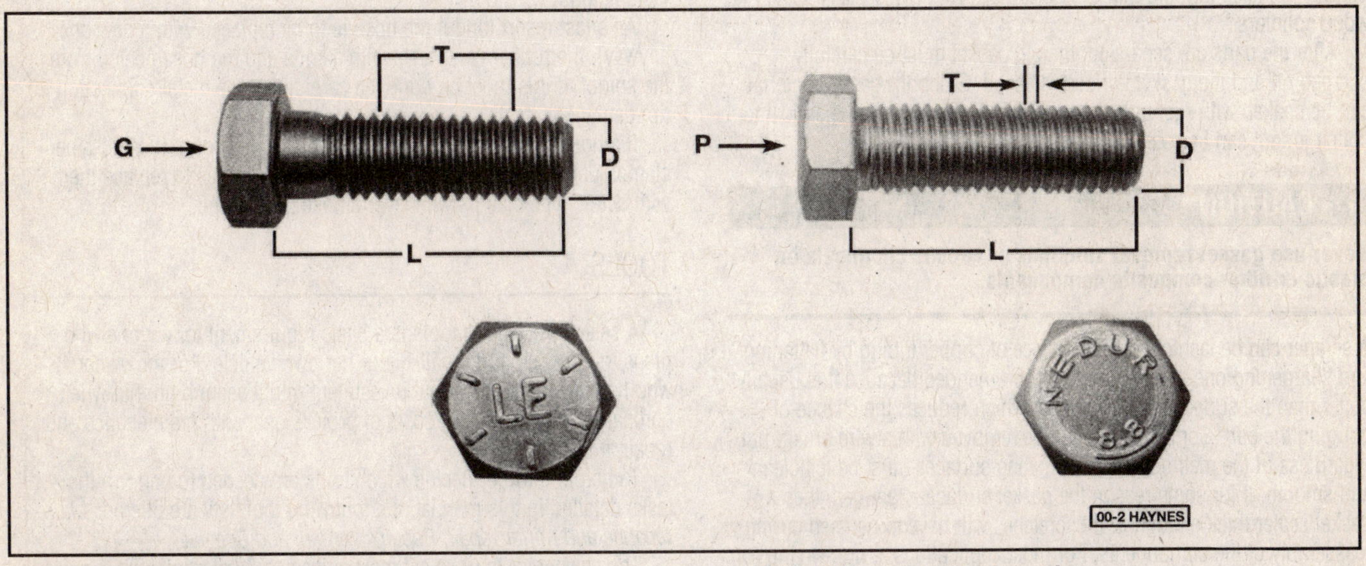

Standard (SAE and USS) bolt dimensions/grade marks

G *Grade marks (bolt strength)*
L *Length (in inches)*
T *Thread pitch (number of threads per inch)*
D *Nominal diameter (in inches)*

Metric bolt dimensions/grade marks

P *Property class (bolt strength)*
L *Length (in millimeters)*
T *Thread pitch (distance between threads in millimeters)*
D *Diameter*

mount bolts, etc.). A pan of this type is especially helpful when working on assemblies with very small parts, such as the carburetor, alternator, valve train or interior dash and trim pieces. The cavities can be marked with paint or tape to identify the contents.

Whenever wiring looms, harnesses or connectors are separated, it is a good idea to identify the two halves with numbered pieces of masking tape so they can be easily reconnected.

Gasket sealing surfaces

Throughout any vehicle, gaskets are used to seal the mating surfaces between two parts and keep lubricants, fluids, vacuum or pressure contained in an assembly.

Many times these gaskets are coated with a liquid or paste-type gasket sealing compound before assembly. Age, heat and pressure can sometimes cause the two parts to stick together so tightly that they are very difficult to separate. Often, the assembly can be loosened by striking it with a soft-face hammer near the mating surfaces. A regular hammer can be used if a block of wood is placed between the hammer and the part. Do not hammer on cast parts or parts that could be easily damaged. With any particularly stubborn part, always recheck to make sure that every fastener has been removed.

Avoid using a screwdriver or bar to pry apart an assembly, as they

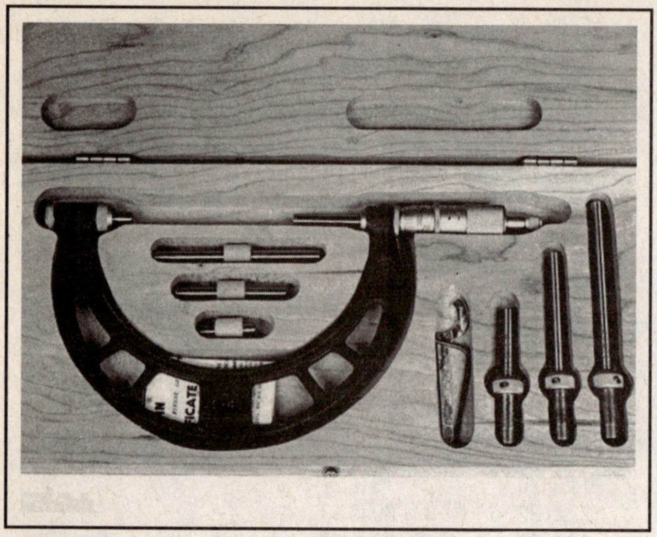

Micrometer set

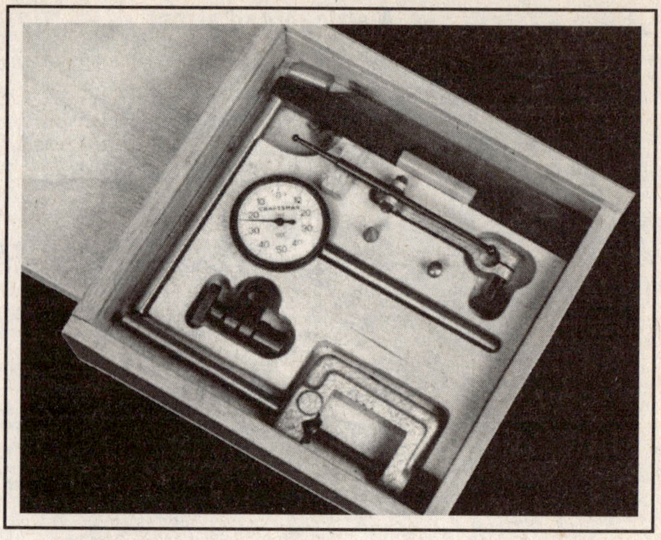

Dial indicator set

can easily mar the gasket sealing surfaces of the parts, which must remain smooth. If prying is absolutely necessary, use an old broom handle, but keep in mind that extra clean up will be necessary if the wood splinters.

After the parts are separated, the old gasket must be carefully scraped off and the gasket surfaces cleaned. Stubborn gasket material can be soaked with rust penetrant or treated with a special chemical to soften it so it can be easily scraped off.

✻✻ CAUTION:

Never use gasket removal solutions or caustic chemicals on plastic or other composite components.

A scraper can be fashioned from a piece of copper tubing by flattening and sharpening one end. Copper is recommended because it is usually softer than the surfaces to be scraped, which reduces the chance of gouging the part. Some gaskets can be removed with a wire brush, but regardless of the method used, the mating surfaces must be left clean and smooth. If for some reason the gasket surface is gouged, then a gasket sealer thick enough to fill scratches will have to be used during reassembly of the components. For most applications, a non-drying (or semi-drying) gasket sealer should be used.

Hose removal tips

✻✻ WARNING:

If the vehicle is equipped with air conditioning, do not disconnect any of the A/C hoses without first having the system depressurized by a dealer service department or a service station.

Hose removal precautions closely parallel gasket removal precautions. Avoid scratching or gouging the surface that the hose mates against or the connection may leak. This is especially true for radiator hoses. Because of various chemical reactions, the rubber in hoses can bond itself to the metal spigot that the hose fits over. To remove a hose, first loosen the hose clamps that secure it to the spigot. Then, with slip-joint pliers, grab the hose at the clamp and rotate it around the spigot. Work it back and forth until it is completely free, then pull it off. Silicone or other lubricants will ease removal if they can be applied between the hose and the outside of the spigot. Apply the same lubricant to the inside of the hose and the outside of the spigot to simplify installation.

As a last resort (and if the hose is to be replaced with a new one anyway), the rubber can be slit with a knife and the hose peeled from the spigot. If this must be done, be careful that the metal connection is not damaged.

If a hose clamp is broken or damaged, do not reuse it. Wire-type clamps usually weaken with age, so it is a good idea to replace them with screw-type clamps whenever a hose is removed.

TOOLS

A selection of good tools is a basic requirement for anyone who plans to maintain and repair his or her own vehicle. For the owner who has few tools, the initial investment might seem high, but when compared to the spiraling costs of professional auto maintenance and repair, it is a wise one.

To help the owner decide which tools are needed to perform the tasks detailed in this manual, the following tool lists are offered: *Maintenance and minor repair, Repair/overhaul and Special.*

The newcomer to practical mechanics should start off with the *maintenance and minor repair* tool kit, which is adequate for the simpler jobs performed on a vehicle. Then, as confidence and experience grow, the owner can tackle more difficult tasks, buying additional tools as they are needed. Eventually the basic kit will be expanded into the *repair and overhaul* tool set. Over a period of time, the experienced do-it-yourselfer will assemble a tool set complete enough for most repair and overhaul procedures and will add tools from the special category when it is felt that the expense is justified by the frequency of use.

Maintenance and minor repair tool kit

The tools in this list should be considered the minimum required for performance of routine maintenance, servicing and minor repair work. We recommend the purchase of combination wrenches (box-end and open-end combined in one wrench). While more expensive than open end wrenches, they offer the advantages of both types of wrench.

Combination wrench set (1/4-inch to 1 inch or 6 mm to 19 mm)
Adjustable wrench, 8 inch
Spark plug wrench with rubber insert

Spark plug gap adjusting tool
Feeler gauge set
Brake bleeder wrench
Standard screwdriver (5/16-inch x 6 inch)
Phillips screwdriver (No. 2 x 6 inch)
Combination pliers - 6 inch
Hacksaw and assortment of blades
Tire pressure gauge
Grease gun
Oil can
Fine emery cloth
Wire brush
Battery post and cable cleaning tool
Oil filter wrench
Funnel (medium size)
Safety goggles
Jackstands (2)
Drain pan

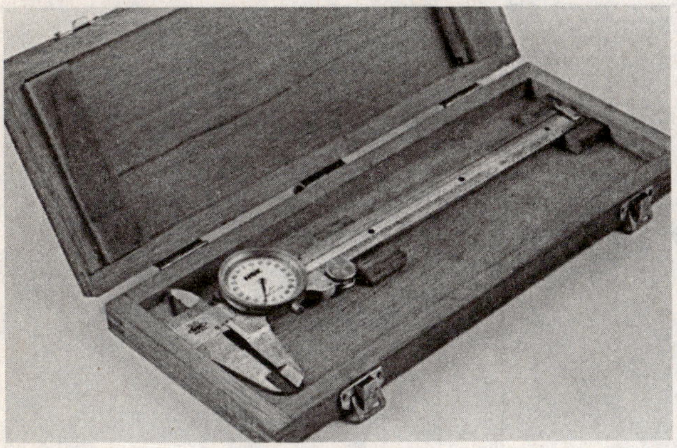

Dial caliper

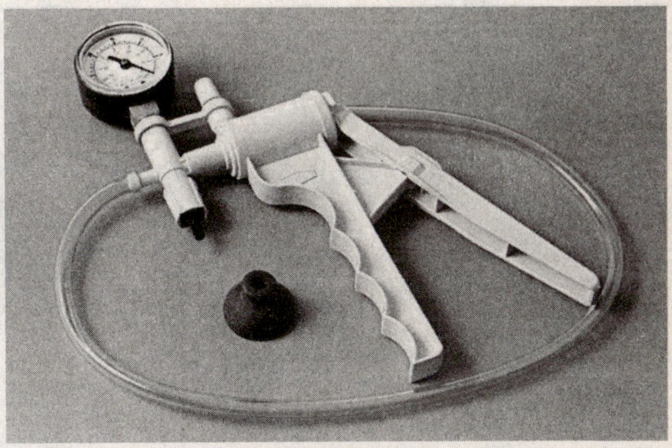

Hand-operated vacuum pump

Timing light

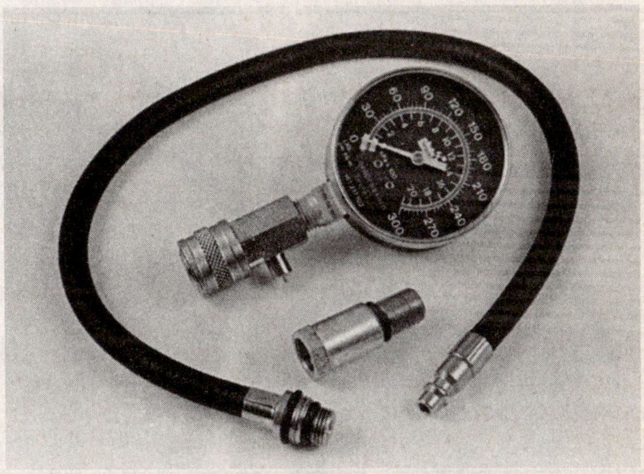

Compression gauge with spark plug hole adapter

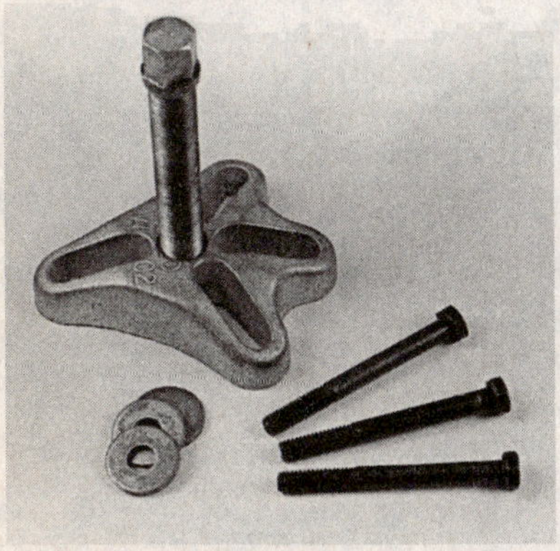

Damper/steering wheel puller

General purpose puller

Hydraulic lifter removal tool

➡Note: If basic tune-ups are going to be part of routine maintenance, it will be necessary to purchase a good quality stroboscopic timing light and combination tachometer/dwell meter. Although they are included in the list of special tools, it is mentioned here because they are absolutely necessary for tuning most vehicles properly.

Repair and overhaul tool set

These tools are essential for anyone who plans to perform major repairs and are in addition to those in the maintenance and minor repair tool kit. Included is a comprehensive set of sockets which, though expensive, are invaluable because of their versatility, especially when various extensions and drives are available. We recommend the 1/2-inch drive over the 3/8-inch drive. Although the larger drive is bulky and more expensive, it has the capacity of accepting a very wide range of large sockets. Ideally, however, the mechanic should have a 3/8-inch drive set and a 1/2-inch drive set.

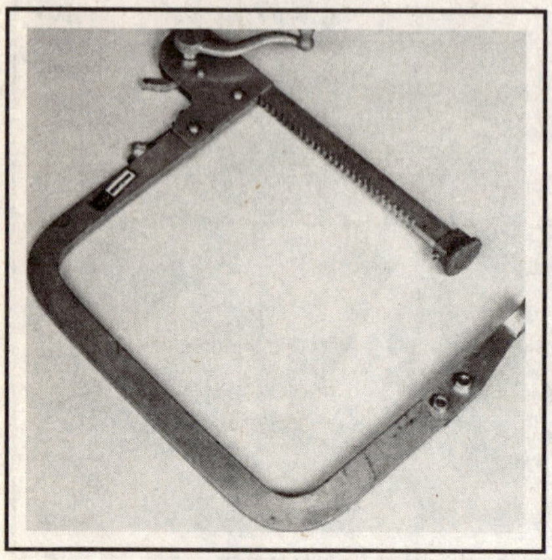

Valve spring compressor

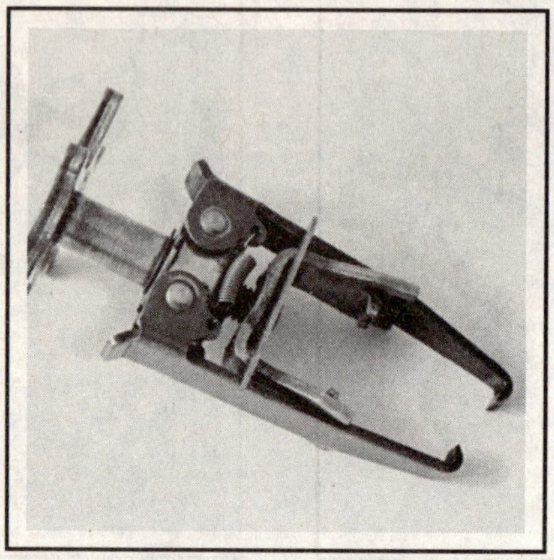

Valve spring compressor

Ridge reamer

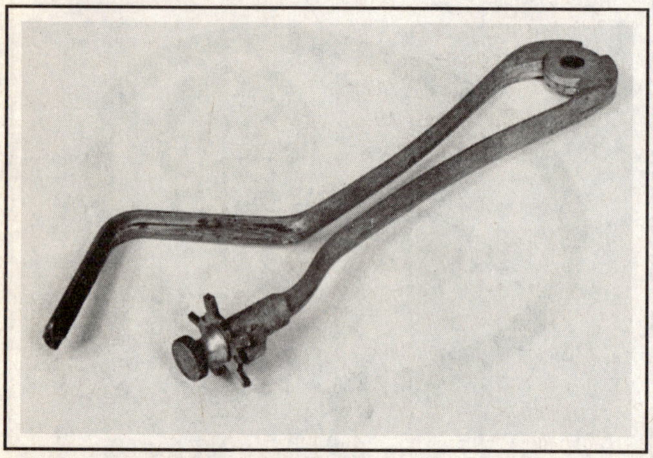

Piston ring groove cleaning tool

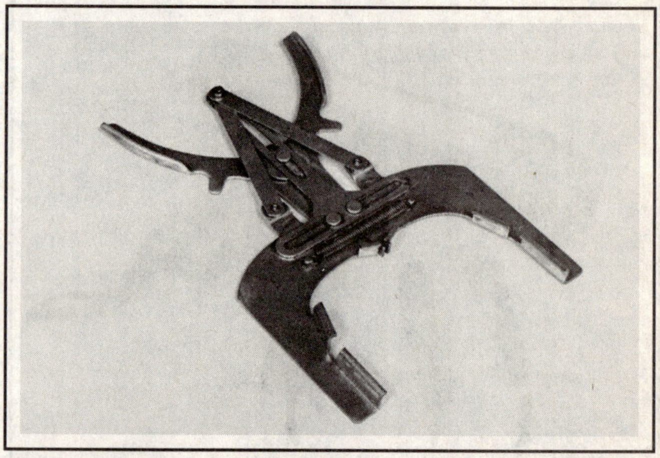

Ring removal/installation tool

Ring compressor

Cylinder hone

Brake hold-down spring tool

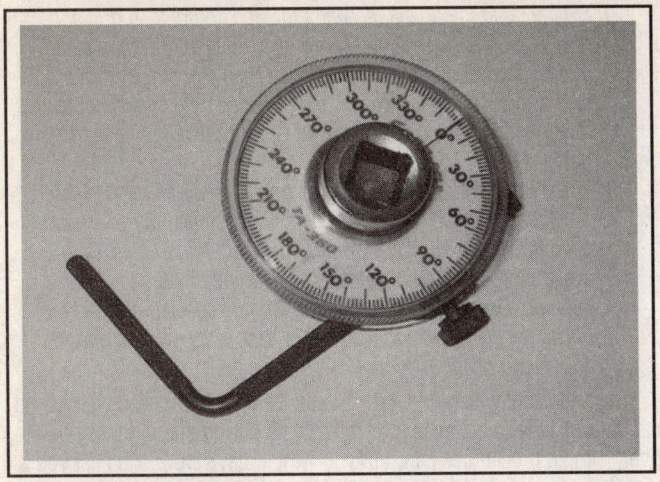

Torque angle gauge

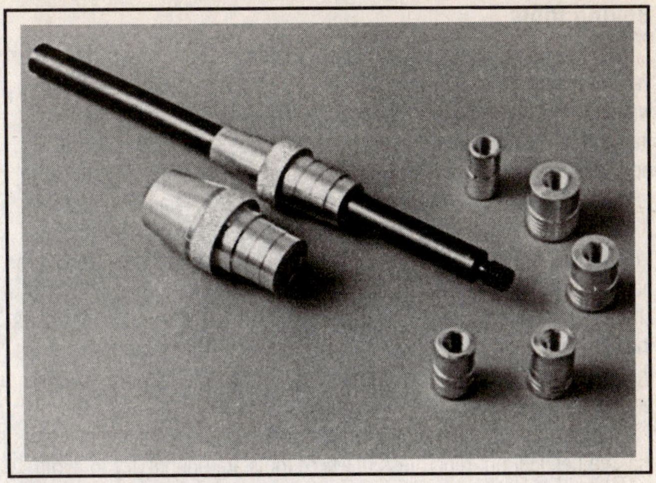

Clutch plate alignment tool

Socket set(s)
Reversible ratchet
Extension - 10 inch
Universal joint
Torque wrench (same size drive as sockets)
Ball peen hammer - 8 ounce
Soft-face hammer (plastic/rubber)
Standard screwdriver (1/4-inch x 6 inch)
Standard screwdriver (stubby - 5/16-inch)
Phillips screwdriver (No. 3 x 8 inch)
Phillips screwdriver (stubby - No. 2)
Pliers - vise grip
Pliers - lineman's
Pliers - needle nose
Pliers - snap-ring (internal and external)
Cold chisel - 1/2-inch
Scribe
Scraper (made from flattened copper tubing)
Centerpunch
Pin punches (1/16, 1/8, 3/16-inch)
Steel rule/straightedge - 12 inch
Allen wrench set (1/8 to 3/8-inch or 4 mm to 10 mm)
A selection of files
Wire brush (large)
Jackstands (second set)
Jack (scissor or hydraulic type)

➡**Note: Another tool which is often useful is an electric drill with a chuck capacity of 3/8-inch and a set of good quality drill bits.**

Special tools

The tools in this list include those which are not used regularly, are expensive to buy, or which need to be used in accordance with their manufacturer's instructions. Unless these tools will be used frequently, it is not very economical to purchase many of them. A consideration would be to split the cost and use between yourself and a friend or friends. In addition, most of these tools can be obtained from a tool rental shop on a temporary basis.

This list primarily contains only those tools and instruments widely available to the public, and not those special tools produced by the vehicle manufacturer for distribution to dealer service departments. Occasionally, references to the manufacturer's special tools are included in the text of this manual. Generally, an alternative method of doing the job without the special tool is offered. However, sometimes there is no alternative to their use. Where this is the case, and the tool cannot be purchased or borrowed, the work should be turned over to the dealer service department or an automotive repair shop.

Valve spring compressor
Piston ring groove cleaning tool
Piston ring compressor
Piston ring installation tool
Cylinder compression gauge
Cylinder ridge reamer
Cylinder surfacing hone
Cylinder bore gauge
Micrometers and/or dial calipers
Hydraulic lifter removal tool
Balljoint separator
Universal-type puller
Impact screwdriver
Dial indicator set
Stroboscopic timing light (inductive pick-up)
Hand operated vacuum/pressure pump
Tachometer/dwell meter
Universal electrical multimeter
Cable hoist
Brake spring removal and installation tools
Floor jack

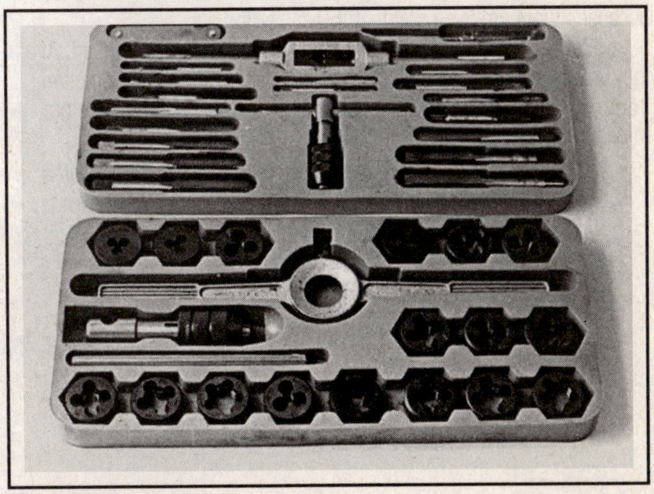

Tap and die set

Buying tools

For the do-it-yourselfer who is just starting to get involved in vehicle maintenance and repair, there are a number of options available when purchasing tools. If maintenance and minor repair is the extent of the work to be done, the purchase of individual tools is satisfactory. If, on the other hand, extensive work is planned, it would be a good idea to purchase a modest tool set from one of the large retail chain stores. A set can usually be bought at a substantial savings over the individual tool prices, and they often come with a tool box. As additional tools are needed, add-on sets, individual tools and a larger tool box can be purchased to expand the tool selection. Building a tool set gradually allows the cost of the tools to be spread over a longer period of time and gives the mechanic the freedom to choose only those tools that will actually be used.

Tool stores will often be the only source of some of the special tools that are needed, but regardless of where tools are bought, try to avoid cheap ones, especially when buying screwdrivers and sockets, because they won't last very long. The expense involved in replacing cheap tools will eventually be greater than the initial cost of quality tools.

Care and maintenance of tools

Good tools are expensive, so it makes sense to treat them with respect. Keep them clean and in usable condition and store them properly when not in use. Always wipe off any dirt, grease or metal chips before putting them away. Never leave tools lying around in the work area. Upon completion of a job, always check closely under the hood for tools that may have been left there so they won't get lost during a test drive.

Some tools, such as screwdrivers, pliers, wrenches and sockets, can be hung on a panel mounted on the garage or workshop wall, while others should be kept in a tool box or tray. Measuring instruments, gauges, meters, etc. must be carefully stored where they cannot be damaged by weather or impact from other tools.

When tools are used with care and stored properly, they will last a very long time. Even with the best of care, though, tools will wear out if used frequently. When a tool is damaged or worn out, replace it. Subsequent jobs will be safer and more enjoyable if you do.

HOW TO REPAIR DAMAGED THREADS

Sometimes, the internal threads of a nut or bolt hole can become stripped, usually from overtightening. Stripping threads is an all-too-common occurrence, especially when working with aluminum parts, because aluminum is so soft that it easily strips out.

Usually, external or internal threads are only partially stripped. After they've been cleaned up with a tap or die, they'll still work. Sometimes, however, threads are badly damaged. When this happens, you've got three choices:

1) *Drill and tap the hole to the next suitable oversize and install a larger diameter bolt, screw or stud.*
2) *Drill and tap the hole to accept a threaded plug, then drill and tap the plug to the original screw size. You can also buy a plug already threaded to the original size. Then you simply drill a hole to the specified size, then run the threaded plug into the hole with a bolt and jam nut. Once the plug is fully seated, remove the jam nut and bolt.*
3) *The third method uses a patented thread repair kit like Heli-Coil or Slimsert. These easy-to-use kits are designed to repair damaged threads in straight-through holes and blind holes. Both are available as kits which can handle a variety of sizes and thread patterns. Drill the hole, then tap it with the special included tap. Install the Heli-Coil and the hole is back to its original diameter and thread pitch.*

Regardless of which method you use, be sure to proceed calmly and carefully. A little impatience or carelessness during one of these relatively simple procedures can ruin your whole day's work and cost you a bundle if you wreck an expensive part.

WORKING FACILITIES

Not to be overlooked when discussing tools is the workshop. If anything more than routine maintenance is to be carried out, some sort of suitable work area is essential.

It is understood, and appreciated, that many home mechanics do not have a good workshop or garage available, and end up removing an engine or doing major repairs outside. It is recommended, however, that the overhaul or repair be completed under the cover of a roof.

A clean, flat workbench or table of comfortable working height is an absolute necessity. The workbench should be equipped with a vise that has a jaw opening of at least four inches.

As mentioned previously, some clean, dry storage space is also required for tools, as well as the lubricants, fluids, cleaning solvents, etc. which soon become necessary.

Sometimes waste oil and fluids, drained from the engine or cooling system during normal maintenance or repairs, present a disposal problem. To avoid pouring them on the ground or into a sewage system, pour the used fluids into large containers, seal them with caps and take them to an authorized disposal site or recycling center. Plastic jugs, such as old antifreeze containers, are ideal for this purpose.

Always keep a supply of old newspapers and clean rags available. Old towels are excellent for mopping up spills. Many mechanics use rolls of paper towels for most work because they are readily available and disposable. To help keep the area under the vehicle clean, a large cardboard box can be cut open and flattened to protect the garage or shop floor.

Whenever working over a painted surface, such as when leaning over a fender to service something under the hood, always cover it with an old blanket or bedspread to protect the finish. Vinyl covered pads, made especially for this purpose, are available at auto parts stores.

Jacking and towing

JACKING

The jack supplied with the vehicle should only be used for raising the vehicle when changing a tire or placing jackstands under the frame.

✲✲ WARNING:

Never work under the vehicle or start the engine while this jack is being used as the only means of support.

The vehicle should be on level ground with the hazard flashers on, the wheels blocked, the parking brake applied and the transmission in Park (automatic) or Reverse (manual). If a tire is being changed, loosen the lug nuts one-half turn and leave them in place until the wheel is raised off the ground.

Place the jack under the vehicle (see illustrations):

Front: Under the frame rail, where the crossmember is attached.

Rear: Under the axle tube nearest the wheel being removed.

Operate the jack with a slow, smooth motion until the wheel is raised off the ground. Remove the lug nuts, pull off the wheel, install the spare and thread the lug nuts back on with the beveled sides facing in. Tighten them snugly, but wait until the vehicle is lowered to tighten them completely.

Lower the vehicle, remove the jack and tighten the nuts (if loosened or removed) in a criss-cross pattern.

TOWING

As a general rule, the vehicle should be towed with professional towing equipment. If towed from the front, the rear wheels should be placed on a towing dolly. If a 4WD model is towed from the rear, the front wheels should be placed on a towing dolly; if a dolly is not available, turn the ignition key to the ACC position, place the transfer case in the 2H position and the transmission in Neutral.

When a vehicle is towed with the rear wheels raised, the steering wheel must be clamped in the straight ahead position with a special device designed for use during towing. The ignition key must not be in the LOCK position, since the steering lock mechanism isn't strong enough to hold the front wheels straight while towing.

Equipment specifically designed for towing should be used. It should be attached to the main structural members of the vehicle, not the bumpers or brackets. Safety is a major consideration when towing and all applicable state and local laws must be obeyed. A safety chain system must be used at all times. Remember that power steering and power brakes will not work with the engine off.

Front jacking point

Rear jacking point

Booster battery (jump) starting

Observe these precautions when using a booster battery to start a vehicle:

a) Before connecting the booster battery, make sure the ignition switch is in the Off position.
b) Turn off the lights, heater and other electrical loads.
c) Your eyes should be shielded. Safety goggles are a good idea.
d) Make sure the booster battery is the same voltage as the dead one in the vehicle.
e) The two vehicles MUST NOT TOUCH each other!
f) Make sure the transaxle is in Neutral (manual) or Park (auto matic).
g) If the booster battery is not a maintenance-free type, remove the vent caps and lay a cloth over the vent holes.

Connect the red jumper cable to the positive (+) terminals of each battery (see illustration).

Connect one end of the black jumper cable to the negative (-) terminal of the booster battery. The other end of this cable should be connected to a good ground on the vehicle to be started, such as a bolt or bracket on the body.

Start the engine using the booster battery, then, with the engine running at idle speed, disconnect the jumper cables in the reverse order of connection.

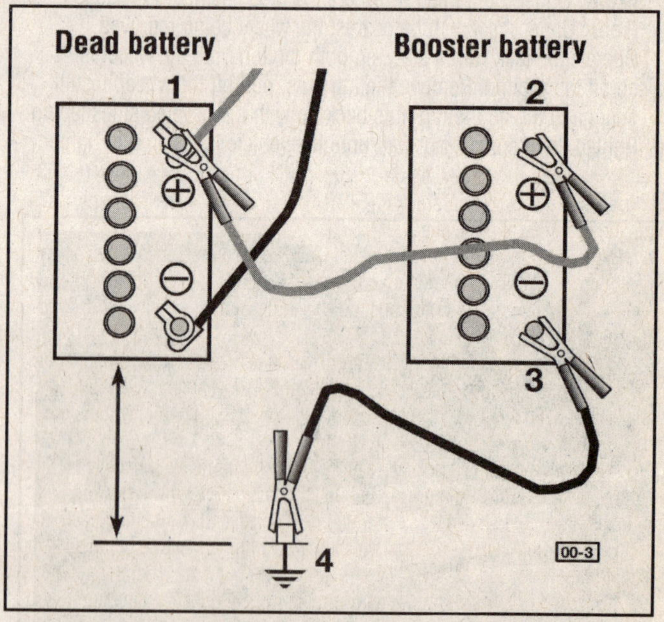

Make the booster battery cable connections in the numerical order shown (note that the negative cable of the booster battery is NOT attached to the negative terminal of the dead battery)

Automotive chemicals and lubricants

A number of automotive chemicals and lubricants are available for use during vehicle maintenance and repair. They include a wide variety of products ranging from cleaning solvents and degreasers to lubricants and protective sprays for rubber, plastic and vinyl.

CLEANERS

Carburetor cleaner and choke cleaner is a strong solvent for gum, varnish and carbon. Most carburetor cleaners leave a dry-type lubricant film which will not harden or gum up. Because of this film it is not recommended for use on electrical components.

Brake system cleaner is used to remove brake dust, grease and brake fluid from the brake system, where clean surfaces are absolutely necessary. It leaves no residue and often eliminates brake squeal caused by contaminants.

Electrical cleaner removes oxidation, corrosion and carbon deposits from electrical contacts, restoring full current flow. It can also be used to clean spark plugs, carburetor jets, voltage regulators and other parts where an oil-free surface is desired.

Demoisturants remove water and moisture from electrical components such as alternators, voltage regulators, electrical connectors and fuse blocks. They are non-conductive and non-corrosive.

Degreasers are heavy-duty solvents used to remove grease from the outside of the engine and from chassis components. They can be sprayed or brushed on and, depending on the type, are rinsed off either with water or solvent.

LUBRICANTS

Motor oil is the lubricant formulated for use in engines. It normally contains a wide variety of additives to prevent corrosion and reduce foaming and wear. Motor oil comes in various weights (viscosity ratings) from 0 to 50. The recommended weight of the oil depends on the season, temperature and the demands on the engine. Light oil is used in cold climates and under light load conditions. Heavy oil is used in hot climates and where high loads are encountered. Multi-viscosity oils are designed to have characteristics of both light and heavy oils and are available in a number of weights from 0W-20 to 20W-50.

Gear oil is designed to be used in differentials, manual transmissions and other areas where high-temperature lubrication is required.

Chassis and wheel bearing grease is a heavy grease used where increased loads and friction are encountered, such as for wheel bearings, balljoints, tie-rod ends and universal joints.

High-temperature wheel bearing grease is designed to withstand the extreme temperatures encountered by wheel bearings in disc brake equipped vehicles. It usually contains molybdenum disulfide (moly), which is a dry-type lubricant.

White grease is a heavy grease for metal-to-metal applications where water is a problem. White grease stays soft under both low and high temperatures (usually from -100 to +190-degrees F), and will not wash off or dilute in the presence of water.

Assembly lube is a special extreme pressure lubricant, usually containing moly, used to lubricate high-load parts (such as main and rod bearings and cam lobes) for initial start-up of a new engine. The assembly lube lubricates the parts without being squeezed out or washed away until the engine oiling system begins to function.

Silicone lubricants are used to protect rubber, plastic, vinyl and nylon parts.

Graphite lubricants are used where oils cannot be used due to contamination problems, such as in locks. The dry graphite will lubricate metal parts while remaining uncontaminated by dirt, water, oil or acids. It is electrically conductive and will not foul electrical contacts in locks such as the ignition switch.

Moly penetrants loosen and lubricate frozen, rusted and corroded fasteners and prevent future rusting or freezing.

Heat-sink grease is a special electrically non-conductive grease that is used for mounting electronic ignition modules where it is essential that heat is transferred away from the module.

SEALANTS

RTV sealant is one of the most widely used gasket compounds. Made from silicone, RTV is air curing, it seals, bonds, waterproofs, fills surface irregularities, remains flexible, doesn't shrink, is relatively easy to remove, and is used as a supplementary sealer with almost all low and medium temperature gaskets.

Anaerobic sealant is much like RTV in that it can be used either to seal gaskets or to form gaskets by itself. It remains flexible, is solvent resistant and fills surface imperfections. The difference between an anaerobic sealant and an RTV-type sealant is in the curing. RTV cures when exposed to air, while an anaerobic sealant cures only in the absence of air. This means that an anaerobic sealant cures only after the assembly of parts, sealing them together.

Thread and pipe sealant is used for sealing hydraulic and pneumatic fittings and vacuum lines. It is usually made from a Teflon compound, and comes in a spray, a paint-on liquid and as a wrap-around tape.

CHEMICALS

Anti-seize compound prevents seizing, galling, cold welding, rust and corrosion in fasteners. High-temperature anti-seize, usually made with copper and graphite lubricants, is used for exhaust system and exhaust manifold bolts.

Anaerobic locking compounds are used to keep fasteners from vibrating or working loose and cure only after installation, in the absence of air. Medium strength locking compound is used for small nuts, bolts and screws that may be removed later. High-strength locking compound is for large nuts, bolts and studs which aren't removed on a regular basis.

Oil additives range from viscosity index improvers to chemical treatments that claim to reduce internal engine friction. It should be noted that most oil manufacturers caution against using additives with their oils.

Gas additives perform several functions, depending on their chemical makeup. They usually contain solvents that help dissolve gum and varnish that build up on carburetor, fuel injection and intake parts. They also serve to break down carbon deposits that form on the inside surfaces of the combustion chambers. Some additives contain upper cylinder lubricants for valves and piston rings, and others contain chemicals to remove condensation from the gas tank.

MISCELLANEOUS

Brake fluid is specially formulated hydraulic fluid that can withstand the heat and pressure encountered in brake systems. Care must be taken so this fluid does not come in contact with painted surfaces or plastics. An opened container should always be resealed to prevent contamination by water or dirt.

Weatherstrip adhesive is used to bond weatherstripping around doors, windows and trunk lids. It is sometimes used to attach trim pieces.

Undercoating is a petroleum-based, tar-like substance that is designed to protect metal surfaces on the underside of the vehicle from corrosion. It also acts as a sound-deadening agent by insulating the bottom of the vehicle.

Waxes and polishes are used to help protect painted and plated surfaces from the weather. Different types of paint may require the use of different types of wax and polish. Some polishes utilize a chemical or abrasive cleaner to help remove the top layer of oxidized (dull) paint on older vehicles. In recent years many non-wax polishes that contain a wide variety of chemicals such as polymers and silicones have been introduced. These non-wax polishes are usually easier to apply and last longer than conventional waxes and polishes.

CONVERSION FACTORS

LENGTH (distance)

Inches (in)	X 25.4	= Millimeters (mm)	X 0.0394	= Inches (in)
Feet (ft)	X 0.305	= Meters (m)	X 3.281	= Feet (ft)
Miles	X 1.609	= Kilometers (km)	X 0.621	= Miles

VOLUME (capacity)

Cubic inches (cu in; in³)	X 16.387	= Cubic centimeters (cc; cm³)	X 0.061	= Cubic inches (cu in; in³)
Imperial pints (Imp pt)	X 0.568	= Liters (l)	X 1.76	= Imperial pints (Imp pt)
Imperial quarts (Imp qt)	X 1.137	= Liters (l)	X 0.88	= Imperial quarts (Imp qt)
Imperial quarts (Imp qt)	X 1.201	= US quarts (US qt)	X 0.833	= Imperial quarts (Imp qt)
US quarts (US qt)	X 0.946	= Liters (l)	X 1.057	= US quarts (US qt)
Imperial gallons (Imp gal)	X 4.546	= Liters (l)	X 0.22	= Imperial gallons (Imp gal)
Imperial gallons (Imp gal)	X 1.201	= US gallons (US gal)	X 0.833	= Imperial gallons (Imp gal)
US gallons (US gal)	X 3.785	= Liters (l)	X 0.264	= US gallons (US gal)

MASS (weight)

Ounces (oz)	X 28.35	= Grams (g)	X 0.035	= Ounces (oz)
Pounds (lb)	X 0.454	= Kilograms (kg)	X 2.205	= Pounds (lb)

FORCE

Ounces-force (ozf; oz)	X 0.278	= Newtons (N)	X 3.6	= Ounces-force (ozf; oz)
Pounds-force (lbf; lb)	X 4.448	= Newtons (N)	X 0.225	= Pounds-force (lbf; lb)
Newtons (N)	X 0.1	= Kilograms-force (kgf; kg)	X 9.81	= Newtons (N)

PRESSURE

Pounds-force per square inch (psi; lbf/in²; lb/in²)	X 0.070	= Kilograms-force per square centimeter (kgf/cm²; kg/cm²)	X 14.223	= Pounds-force per square inch (psi; lbf/in²; lb/in²)
Pounds-force per square inch (psi; lbf/in²; lb/in²)	X 0.068	= Atmospheres (atm)	X 14.696	= Pounds-force per square inch (psi; lbf/in²; lb/in²)
Pounds-force per square inch (psi; lbf/in²; lb/in²)	X 0.069	= Bars	X 14.5	= Pounds-force per square inch (psi; lbf/in²; lb/in²)
Pounds-force per square inch (psi; lbf/in²; lb/in²)	X 6.895	= Kilopascals (kPa)	X 0.145	= Pounds-force per square inch (psi; lbf/in²; lb/in²)
Kilopascals (kPa)	X 0.01	= Kilograms-force per square centimeter (kgf/cm²; kg/cm²)	X 98.1	= Kilopascals (kPa)

TORQUE (moment of force)

Pounds-force inches (lbf in; lb in)	X 1.152	= Kilograms-force centimeter (kgf cm; kg cm)	X 0.868	= Pounds-force inches (lbf in; lb in)
Pounds-force inches (lbf in; lb in)	X 0.113	= Newton meters (Nm)	X 8.85	= Pounds-force inches (lbf in; lb in)
Pounds-force inches (lbf in; lb in)	X 0.083	= Pounds-force feet (lbf ft; lb ft)	X 12	= Pounds-force inches (lbf in; lb in)
Pounds-force feet (lbf ft; lb ft)	X 0.138	= Kilograms-force meters (kgf m; kg m)	X 7.233	= Pounds-force feet (lbf ft; lb ft)
Pounds-force feet (lbf ft; lb ft)	X 1.356	= Newton meters (Nm)	X 0.738	= Pounds-force feet (lbf ft; lb ft)
Newton meters (Nm)	X 0.102	= Kilograms-force meters (kgf m; kg m)	X 9.804	= Newton meters (Nm)

VACUUM

Inches mercury (in. Hg)	X 3.377	= Kilopascals (kPa)	X 0.2961	= Inches mercury
Inches mercury (in. Hg)	X 25.4	= Millimeters mercury (mm Hg)	X 0.0394	= Inches mercury

POWER

Horsepower (hp)	X 745.7	= Watts (W)	X 0.0013	= Horsepower (hp)

VELOCITY (speed)

Miles per hour (miles/hr; mph)	X 1.609	= Kilometers per hour (km/hr; kph)	X 0.621	= Miles per hour (miles/hr; mph)

FUEL CONSUMPTION *

Miles per gallon, Imperial (mpg)	X 0.354	= Kilometers per liter (km/l)	X 2.825	= Miles per gallon, Imperial (mpg)
Miles per gallon, US (mpg)	X 0.425	= Kilometers per liter (km/l)	X 2.352	= Miles per gallon, US (mpg)

TEMPERATURE

Degrees Fahrenheit = (°C x 1.8) + 32

Degrees Celsius (Degrees Centigrade; °C) = (°F - 32) x 0.56

***It is common practice to convert from miles per gallon (mpg) to liters/100 kilometers (l/100km), where mpg (Imperial) x l/100 km = 282 and mpg (US) x l/100 km = 235**

FRACTION/DECIMAL/MILLIMETER EQUIVALENTS

DECIMALS TO MILLIMETERS

Decimal	mm	Decimal	mm
0.001	0.0254	0.500	12.7000
0.002	0.0508	0.510	12.9540
0.003	0.0762	0.520	13.2080
0.004	0.1016	0.530	13.4620
0.005	0.1270	0.540	13.7160
0.006	0.1524	0.550	13.9700
0.007	0.1778	0.560	14.2240
0.008	0.2032	0.570	14.4780
0.009	0.2286	0.580	14.7320
0.010	0.2540	0.590	14.9860
0.020	0.5080	0.600	15.2400
0.030	0.7620	0.610	15.4940
0.040	1.0160	0.620	15.7480
0.050	1.2700	0.630	16.0020
0.060	1.5240	0.640	16.2560
0.070	1.7780	0.650	16.5100
0.080	2.0320	0.660	16.7640
0.090	2.2860	0.670	17.0180
0.100	2.5400	0.680	17.2720
0.110	2.7940	0.690	17.5260
0.120	3.0480	0.700	17.7800
0.130	3.3020	0.710	18.0340
0.140	3.5560	0.720	18.2880
0.150	3.8100	0.730	18.5420
0.160	4.0640	0.740	18.7960
0.170	4.3180	0.750	19.0500
0.180	4.5720	0.760	19.3040
0.190	4.8260	0.770	19.5580
0.200	5.0800	0.780	19.8120
0.210	5.3340	0.790	20.0660
0.220	5.5880	0.800	20.3200
0.230	5.8420	0.810	20.5740
0.240	6.0960	0.820	21.8280
0.250	6.3500	0.830	21.0820
0.260	6.6040	0.840	21.3360
0.270	6.8580	0.850	21.5900
0.280	7.1120	0.860	21.8440
0.290	7.3660	0.870	22.0980
0.300	7.6200	0.880	22.3520
0.310	7.8740	0.890	22.6060
0.320	8.1280	0.900	22.8600
0.330	8.3820	0.910	23.1140
0.340	8.6360	0.920	23.3680
0.350	8.8900	0.930	23.6220
0.360	9.1440	0.940	23.8760
0.370	9.3980	0.950	24.1300
0.380	9.6520	0.960	24.3840
0.390	9.9060	0.970	24.6380
0.400	10.1600	0.980	24.8920
0.410	10.4140	0.990	25.1460
0.420	10.6680	1.000	25.4000
0.430	10.9220		
0.440	11.1760		
0.450	11.4300		
0.460	11.6840		
0.470	11.9380		
0.480	12.1920		
0.490	12.4460		

FRACTIONS TO DECIMALS TO MILLIMETERS

Fraction	Decimal	mm	Fraction	Decimal	mm
1/64	0.0156	0.3969	33/64	0.5156	13.0969
1/32	0.0312	0.7938	17/32	0.5312	13.4938
3/64	0.0469	1.1906	35/64	0.5469	13.8906
1/16	0.0625	1.5875	9/16	0.5625	14.2875
5/64	0.0781	1.9844	37/64	0.5781	14.6844
3/32	0.0938	2.3812	19/32	0.5938	15.0812
7/64	0.1094	2.7781	39/64	0.6094	15.4781
1/8	0.1250	3.1750	5/8	0.6250	15.8750
9/64	0.1406	3.5719	41/64	0.6406	16.2719
5/32	0.1562	3.9688	21/32	0.6562	16.6688
11/64	0.1719	4.3656	43/64	0.6719	17.0656
3/16	0.1875	4.7625	11/16	0.6875	17.4625
13/64	0.2031	5.1594	45/64	0.7031	17.8594
7/32	0.2188	5.5562	23/32	0.7188	18.2562
15/64	0.2344	5.9531	47/64	0.7344	18.6531
1/4	0.2500	6.3500	3/4	0.7500	19.0500
17/64	0.2656	6.7469	49/64	0.7656	19.4469
9/32	0.2812	7.1438	25/32	0.7812	19.8438
19/64	0.2969	7.5406	51/64	0.7969	20.2406
5/16	0.3125	7.9375	13/16	0.8125	20.6375
21/64	0.3281	8.3344	53/64	0.8281	21.0344
11/32	0.3438	8.7312	27/32	0.8438	21.4312
23/64	0.3594	9.1281	55/64	0.8594	21.8281
3/8	0.3750	9.5250	7/8	0.8750	22.2250
25/64	0.3906	9.9219	57/64	0.8906	22.6219
13/32	0.4062	10.3188	29/32	0.9062	23.0188
27/64	0.4219	10.7156	59/64	0.9219	23.4156
7/16	0.4375	11.1125	15/16	0.9375	23.8125
29/64	0.4531	11.5094	61/64	0.9531	24.2094
15/32	0.4688	11.9062	31/32	0.9688	24.6062
31/64	0.4844	12.3031	63/64	0.9844	25.0031
1/2	0.5000	12.7000	1	1.0000	25.4000

Safety first!

Regardless of how enthusiastic you may be about getting on with the job at hand, take the time to ensure that your safety is not jeopardized. A moment's lack of attention can result in an accident, as can failure to observe certain simple safety precautions. The possibility of an accident will always exist, and the following points should not be considered a comprehensive list of all dangers. Rather, they are intended to make you aware of the risks and to encourage a safety conscious approach to all work you carry out on your vehicle.

ESSENTIAL DOS AND DON'TS

DON'T rely on a jack when working under the vehicle. Always use approved jackstands to support the weight of the vehicle and place them under the recommended lift or support points.
DON'T attempt to loosen extremely tight fasteners (i.e. wheel lug nuts) while the vehicle is on a jack - it may fall.
DON'T start the engine without first making sure that the transmission is in Neutral (or Park where applicable) and the parking brake is set.
DON'T remove the radiator cap from a hot cooling system - let it cool or cover it with a cloth and release the pressure gradually.
DON'T attempt to drain the engine oil until you are sure it has cooled to the point that it will not burn you.
DON'T touch any part of the engine or exhaust system until it has cooled sufficiently to avoid burns.
DON'T siphon toxic liquids such as gasoline, antifreeze and brake fluid by mouth, or allow them to remain on your skin.
DON'T inhale brake lining dust - it is potentially hazardous (see Asbestos below).
DON'T allow spilled oil or grease to remain on the floor - wipe it up before someone slips on it.
DON'T use loose fitting wrenches or other tools which may slip and cause injury.
DON'T push on wrenches when loosening or tightening nuts or bolts. Always try to pull the wrench toward you. If the situation calls for pushing the wrench away, push with an open hand to avoid scraped knuckles if the wrench should slip.
DON'T attempt to lift a heavy component alone - get someone to help you.
DON'T rush or take unsafe shortcuts to finish a job.
DON'T allow children or animals in or around the vehicle while you are working on it.
DO wear eye protection when using power tools such as a drill, sander, bench grinder, etc. and when working under a vehicle.
DO keep loose clothing and long hair well out of the way of moving parts.
DO make sure that any hoist used has a safe working load rating adequate for the job.
DO get someone to check on you periodically when working alone on a vehicle.
DO carry out work in a logical sequence and make sure that everything is correctly assembled and tightened.
DO keep chemicals and fluids tightly capped and out of the reach of children and pets.
DO remember that your vehicle's safety affects that of yourself and others. If in doubt on any point, get professional advice.

STEERING, SUSPENSION AND BRAKES

These systems are essential to driving safety, so make sure you have a qualified shop or individual check your work. Also, compressed suspension springs can cause injury if released suddenly - be sure to use a spring compressor.

AIRBAGS

Airbags are explosive devices that can CAUSE injury if they deploy while you're working on the vehicle. Follow the manufacturer's instructions to disable the airbag whenever you're working in the vicinity of airbag components.

ASBESTOS

Certain friction, insulating, sealing, and other products - such as brake linings, brake bands, clutch linings, torque converters, gaskets, etc. - may contain asbestos or other hazardous friction material. Extreme care must be taken to avoid inhalation of dust from such products, since it is hazardous to health. If in doubt, assume that they do contain asbestos.

FIRE

Remember at all times that gasoline is highly flammable. Never smoke or have any kind of open flame around when working on a vehicle. But the risk does not end there. A spark caused by an electrical short circuit, by two metal surfaces contacting each other, or even by static electricity built up in your body under certain conditions, can ignite gasoline vapors, which in a confined space are highly explosive. Do not, under any circumstances, use gasoline for cleaning parts. Use an approved safety solvent.

Always disconnect the battery ground (-) cable at the battery before working on any part of the fuel system or electrical system. Never risk spilling fuel on a hot engine or exhaust component. It is strongly recommended that a fire extinguisher suitable for use on fuel and electrical fires be kept handy in the garage or workshop at all times. Never try to extinguish a fuel or electrical fire with water.

FUMES

Certain fumes are highly toxic and can quickly cause unconsciousness and even death if inhaled to any extent. Gasoline vapor falls into this category, as do the vapors from some cleaning solvents. Any draining or pouring of such volatile fluids should be done in a well ventilated area.

When using cleaning fluids and solvents, read the instructions on the container carefully. Never use materials from unmarked containers.

Never run the engine in an enclosed space, such as a garage. Exhaust fumes contain carbon monoxide, which is extremely poisonous. If you need to run the engine, always do so in the open air, or at least have the rear of the vehicle outside the work area.

THE BATTERY

Never create a spark or allow a bare light bulb near a battery. They normally give off a certain amount of hydrogen gas, which is highly explosive.

Always disconnect the battery ground (-) cable at the battery before working on the fuel or electrical systems.

If possible, loosen the filler caps or cover when charging the battery from an external source (this does not apply to sealed or maintenance-free batteries). Do not charge at an excessive rate or the battery may burst.

Take care when adding water to a non maintenance-free battery and when carrying a battery. The electrolyte, even when diluted, is very corrosive and should not be allowed to contact clothing or skin.

Always wear eye protection when cleaning the battery to prevent the caustic deposits from entering your eyes.

HOUSEHOLD CURRENT

When using an electric power tool, inspection light, etc., which operates on household current, always make sure that the tool is correctly connected to its plug and that, where necessary, it is properly grounded. Do not use such items in damp conditions and, again, do not create a spark or apply excessive heat in the vicinity of fuel or fuel vapor.

SECONDARY IGNITION SYSTEM VOLTAGE

A severe electric shock can result from touching certain parts of the ignition system (such as the spark plug wires) when the engine is running or being cranked, particularly if components are damp or the insulation is defective. In the case of an electronic ignition system, the secondary system voltage is much higher and could prove fatal.

HYDROFLUORIC ACID

This extremely corrosive acid is formed when certain types of synthetic rubber, found in some O-rings, oil seals, fuel hoses, etc. are exposed to temperatures above 750-degrees F (400-degrees C). The rubber changes into a charred or sticky substance containing the acid. *Once formed, the acid remains dangerous for years. If it gets onto the skin, it may be necessary to amputate the limb concerned.*

When dealing with a vehicle which has suffered a fire, or with components salvaged from such a vehicle, wear protective gloves and discard them after use.

Troubleshooting

CONTENTS

This section provides an easy reference guide to the more common problems which may occur during the operation of your vehicle. These problems and their possible causes are grouped under headings denoting various components or systems, such as Engine, Cooling system, etc. They also refer you to the chapter and/or section which deals with the problem.

Remember that successful troubleshooting is not a mysterious "black art" practiced only by professional mechanics. It is simply the result of the right knowledge combined with an intelligent, systematic approach to the problem. Always work by process of elimination, starting with the simplest solution and working through to the most complex - and never overlook the obvious. Anyone can run the gas tank dry or leave the lights on overnight, so don't assume that you are exempt from such oversights.

Finally, always establish a clear idea of why a problem has occurred and take steps to ensure that it doesn't happen again. If the electrical system fails because of a poor connection, check the other connections in the system to make sure that they don't fail as well. If a particular fuse continues to blow, find out why - don't just replace one fuse after another. Remember, failure of a small component can often be indicative of potential failure or incorrect functioning of a more important component or system.

ENGINE AND PERFORMANCE

1 Engine will not rotate when attempting to start

1 Battery terminal connections loose or corroded. Check the cable terminals at the battery; tighten cable clamp and/or clean off corrosion as necessary (see Chapter 1).

2 Battery discharged or faulty. If the cable ends are clean and tight on the battery posts, turn the key to the On position and switch on the headlights or windshield wipers. If they won't run, the battery is discharged.

3 Automatic transmission not engaged in park (P) or Neutral (N).

4 Broken, loose or disconnected wires in the starting circuit. Inspect all wires and connectors at the battery, starter solenoid and ignition switch (on steering column).

5 Starter motor pinion jammed in flywheel ring gear. If manual transmission, place transmission in gear and rock the vehicle to manually turn the engine. Remove starter (Chapter 5) and inspect pinion and flywheel (Chapter 2) at earliest convenience.

6 Starter solenoid faulty (Chapter 5).

7 Starter motor faulty (Chapter 5).

8 Ignition switch faulty (Chapter 12).

9 Engine seized. Try to turn the crankshaft with a large socket and breaker bar on the pulley bolt.

2 Engine rotates but will not start

1 Fuel tank empty.

2 Battery discharged (engine rotates slowly). Check the operation of electrical components as described in previous Section.

3 Battery terminal connections loose or corroded. See previous Section.

4 Fuel not reaching the fuel injectors. Check for clogged fuel filter or lines and defective fuel pump. Also make sure the tank vent lines aren't clogged (Chapter 4).

5 Low cylinder compression. Check as described in Chapter 2.

6 Water in fuel. Drain tank and fill with new fuel.

7 Dirty fuel injector(s) (Chapter 4).

8 Fault with the fuel injection or engine management system (Chapter 4 or 6).

9 Timing chain broken (Chapter 2B).

3 Starter motor operates without turning engine

1 Starter pinion sticking. Remove the starter (Chapter 5) and inspect.

2 Starter pinion or flywheel/driveplate teeth worn or broken. Remove the inspection cover and inspect.

4 Engine hard to start when cold

1 Battery discharged or low. Check as described in Chapter 1.

2 Fuel not reaching the fuel injectors. Check the fuel filter, lines and fuel pump (Chapters 1 and 4).

3 Defective spark plugs (Chapter 1).

4 Fault with the fuel injection or engine management system (Chapter 4 or 6).

5 Engine hard to start when hot

1 Air filter dirty (Chapter 1).

2 Fuel not reaching fuel injectors (see Chapter 4).

3 Fault with the fuel injection or engine management system (Chapter 4 or 6).

6 Starter motor noisy or engages roughly

1 Pinion or flywheel/driveplate teeth worn or broken. Remove the inspection cover on the left side of the engine and inspect.

2 Starter motor mounting bolts loose or missing.

7 Engine starts but stops immediately

1 Loose or damaged wire harness connections in the ignition system or at the alternator.

2 Intake manifold vacuum leaks. Make sure all mounting bolts/nuts are tight and all vacuum hoses connected to the manifold are attached properly and in good condition (Chapters 2A, 2B and 4).

3 Insufficient fuel flow to fuel injectors (Chapter 4).

8 Engine 'lopes' while idling or idles erratically

1 Vacuum leaks. Check mounting bolts at the intake manifold for tightness. Make sure that all vacuum hoses are connected and in good condition. Use a stethoscope or a length of fuel hose held against your ear to listen for vacuum leaks while the engine is running. A hissing sound will be heard. A soapy water solution will also detect leaks. Check the intake manifold gasket surfaces.

2 Plugged PCV valve (see Chapters 1 and 6).

3 Air filter clogged (Chapter 1).

4 Fuel pump not delivering sufficient fuel (Chapter 4).

5 Leaking head gasket. Perform a cylinder compression check (Chapter 2C).

6 Timing chain worn (Chapter 2).

7 Camshaft lobes worn (Chapter 2).

8 Valve clearance out of adjustment (Chapter 1).
9 Valves burned or otherwise leaking (Chapter 2).
10 Ignition system not operating properly (Chapters 1 and 5).
11 Dirty or clogged injector(s) (Chapter 4).
12 Fault with the fuel injection or engine management system (Chapter 4 or 6).

9 Engine misses at idle speed

1 Spark plugs faulty or not gapped properly (Chapter 1).
2 Faulty ignition coil (Chapter 5).
3 Sticking or faulty emissions systems (see Chapter 6).
4 Clogged fuel filter and/or foreign matter in fuel.
5 Vacuum leaks at intake manifold or hose connections. Check as described in Section 8.
6 Low or uneven cylinder compression. Check as described in Chapter 2.
7 Clogged or dirty fuel injectors (Chapter 4).

10 Excessively high idle speed

1 Intake air leak (Chapters 2 and 4).
2 Malfunction in the engine management system (Chapter 6).

11 Battery will not hold a charge

1 Alternator drivebelt defective or not adjusted properly (Chapter 1).
2 Battery cables loose or corroded (Chapter 1).
3 Alternator not charging properly (Chapter 5).
4 Loose, broken or faulty wires in the charging circuit (Chapter 5).
5 Short circuit causing a continuous drain on the battery.
6 Battery defective internally.

12 Alternator light stays on

1 Fault in alternator or charging circuit (Chapter 5).
2 Drivebelt defective or not properly adjusted (Chapter 1).

13 Alternator light fails to come on when key is turned on

1 Faulty bulb (Chapter 12).
2 Defective alternator (Chapter 5).
3 Fault in the printed circuit, dash wiring or bulb holder (Chapter 12).

14 Engine misses throughout driving speed range

1 Fuel filter clogged and/or impurities in the fuel system. Check fuel filter or clean system (Chapter 4).
2 Faulty or incorrectly gapped spark plugs (Chapter 1).
3 Emissions system components faulty (Chapter 6).
4 Low or uneven cylinder compression pressures. Check as described in Chapter 2.
5 Weak or faulty ignition coil(s) (Chapter 5).
6 Vacuum leaks at intake manifold or vacuum hoses (see Section 8).
7 Dirty or clogged fuel injector (Chapter 4).

15 Hesitation or stumble during acceleration

1 Ignition system not operating properly (Chapter 5).
2 Dirty or clogged fuel injector(s) (Chapter 4).
3 Low fuel pressure. Check for proper operation of the fuel pump and for restrictions in the fuel filter and lines (Chapter 4).
4 Fault with the fuel injection or engine management system (Chapter 4 or 6).

16 Engine stalls

1 Fuel filter clogged and/or water and impurities in the fuel system (Chapter 1).
2 Emissions system components faulty (Chapter 6).
3 Faulty or incorrectly gapped spark plugs (Chapter 1).
4 Vacuum leak at the throttle body, the intake manifold or vacuum hoses. Check as described in Section 8.
5 Valve clearances incorrect (Chapter 1).
6 Fault with the fuel injection or engine management system (Chapter 4 or 6).

17 Engine lacks power

1 Faulty or incorrectly gapped spark plugs (Chapter 1).
2 Air filter dirty (Chapter 1).
3 Faulty ignition coil(s) (Chapter 5).
4 Automatic transmission fluid level incorrect, causing slippage (Chapter 1).
5 Clutch slipping (Chapter 8).
6 Fuel filter clogged and/or impurities in the fuel system (Chapters 1 and 4).
7 Use of sub-standard fuel. Fill tank with proper octane fuel.
8 Low or uneven cylinder compression pressures. Check as described in Chapter 2C.
9 Air leak at intake manifold (check as described in Section 8).
10 Fault with the fuel injection or engine management system (Chapter 4 or 6).
11 Brakes binding (Chapters 1 and 10).

18 Engine backfires

1 Thermostatic air cleaner system not operating properly (Chapter 6).
2 Vacuum leak (see Section 8).
3 Valve clearances incorrect (Chapter 1).
4 Damaged valve springs or sticking valves (Chapter 2).
5 Intake air leak (see Section 8).

19 Engine surges while holding accelerator steady

1 Intake air leak (see Section 8).
2 Fuel pump not working properly (Chapter 4).
3 Fault with the fuel injection or engine management system (Chapter 4 or 6).

20 Pinging or knocking engine sounds when engine is under load

1 Incorrect grade of fuel. Fill tank with fuel of the proper octane rating.

2 Carbon build-up in combustion chambers. Remove cylinder heads and clean combustion chambers (Chapter 2).
3 Incorrect spark plugs (Chapter 1).
4 Fault with the fuel injection or engine management system (Chapter 4 or 6).

21 Engine continues to run after being turned off

Faulty ignition switch (Chapter 12).

22 Low oil pressure

1 Improper grade of oil.
2 Oil pump worn or damaged (Chapter 2).
3 Engine overheating (see Section 27).
4 Clogged oil filter (Chapter 1).
5 Clogged oil strainer (Chapter 2).
6 Oil pressure gauge not working properly (Chapter 2C).

23 Excessive oil consumption

1 Loose oil drain plug.
2 Loose bolts or damaged oil pan gasket (Chapter 2).
3 Loose bolts or damaged timing chain cover gasket (Chapter 2).
4 Front or rear crankshaft oil seal leaking (Chapter 2).
5 Loose bolts or damaged valve cover gasket (Chapter 2).
6 Loose oil filter (Chapter 1).
7 Loose or damaged oil pressure switch (Chapter 2).
8 Pistons and cylinders excessively worn (Chapter 2).
9 Piston rings not installed correctly on pistons (Chapter 2).
10 Worn or damaged piston rings (Chapter 2).
11 Intake and/or exhaust valve oil seals worn or damaged (Chapter 2).
12 Worn valve stems.
13 Worn or damaged valves/guides (Chapter 2).

24 Excessive fuel consumption

1 Dirty or clogged air filter element (Chapter 1).
2 Low tire pressure or incorrect tire size (Chapter 11).
3 Fuel leakage. Check all connections, lines and components in the fuel system (Chapter 4).
4 Dirty or clogged fuel injectors (Chapter 4).
5 Fault with the fuel injection or engine management system (Chapter 4 or 6).

25 Fuel odor

1 Fuel leakage. Check all connections, lines and components in the fuel system (Chapter 4).
2 Fuel tank overfilled. Fill only to automatic shut-off.
3 Charcoal canister in Evaporative Emissions Control system defective (Chapter 1).
4 Vapor leaks from Evaporative Emissions Control system lines (Chapter 6).

26 Miscellaneous engine noises

1 A strong dull noise that becomes more rapid as the engine accelerates indicates worn or damaged crankshaft bearings or an unevenly worn crankshaft. To pinpoint the trouble spot, remove the spark plug wire (or ignition coil electrical connector on coil-over-plug systems) from one plug at a time and crank the engine over. If the noise stops, the cylinder with the removed plug wire indicates the problem area. Replace the bearing and/or service or replace the crankshaft (Chapter 2).
2 A similar (yet slightly higher pitched) noise to the crankshaft knocking described in the previous paragraph, that becomes more rapid as the engine accelerates, indicates worn or damaged connecting rod bearings (Chapter 2). The procedure for locating the problem cylinder is the same as described in Paragraph 1.
3 An overlapping metallic noise that increases in intensity as the engine speed increases, yet diminishes as the engine warms up indicates abnormal piston and cylinder wear (Chapter 2). To locate the problem cylinder, use the procedure described in Paragraph 1.
4 A rapid clicking noise that becomes faster as the engine accelerates indicates a worn piston pin or piston pin hole. This sound will happen each time the piston hits the highest and lowest points in the stroke (Chapter 2). The procedure for locating the problem piston is described in Paragraph 1.
5 A metallic clicking noise coming from the water pump indicates worn or damaged water pump bearings or pump. Replace the water pump with a new one (Chapter 3).
6 A rapid tapping sound or clicking sound that becomes faster as the engine speed increases indicates "valve tapping" or improperly adjusted valve clearances. This can be identified by holding one end of a section of hose to your ear and placing the other end at different spots along the valve cover. The point where the sound is loudest indicates the problem valve. Adjust the valve clearance (Chapter 1). If the problem persists, you likely have a collapsed valve lifter or other damaged valve train component. Changing the engine oil and adding a high viscosity oil treatment will sometimes cure a stuck lifter problem. If the problem still persists, the lifters, pushrods and rocker arms must be removed for inspection (see Chapter 2).
7 A steady metallic rattling or rapping sound coming from the area of the timing chain cover indicates a worn, damaged or out-of-adjustment timing chain. Service or replace the chain and related components (Chapter 2).

COOLING SYSTEM

27 Overheating

1 Insufficient coolant in system (Chapter 1).
2 Drivebelt defective or not adjusted properly (Chapter 1).
3 Radiator core blocked or radiator grille dirty or restricted (Chapter 3).
4 Thermostat faulty (Chapter 3).
5 Fan not functioning properly (Chapter 3).
6 Radiator cap not maintaining proper pressure. Have cap pressure tested by gas station or repair shop.
7 Defective water pump (Chapter 3).
8 Improper grade of engine oil.
9 Inaccurate temperature gauge (Chapter 12).

28 Overcooling

1 Thermostat faulty (Chapter 3).
2 Inaccurate temperature gauge (Chapter 12).

29 External coolant leakage

1 Deteriorated or damaged hoses. Loose clamps at hose connections (Chapter 1).
2 Water pump seals defective. If this is the case, water will drip from the weep hole in the water pump body (Chapter 3).
3 Leakage from radiator. This will require the radiator to be professionally repaired (see Chapter 3 for removal procedures).
4 Engine drain plugs or water jacket freeze plugs leaking (see Chapters 1 and 2).
5 Leak from engine coolant temperature switch (Chapter 6).
6 Leak from damaged gaskets or small cracks (Chapter 2).
7 Damaged head gasket. This can be verified by checking the condition of the engine oil as noted in Section 30.

30 Internal coolant leakage

➡Note: Internal coolant leaks can usually be detected by examining the oil. Check the dipstick and inside the valve cover for water deposits and an oil consistency like that of a milkshake.

1 Leaking cylinder head gasket. Have the system pressure tested or remove the cylinder head (Chapter 2) and inspect.
2 Cracked cylinder bore or cylinder head. Dismantle engine and inspect (Chapter 2).

31 Abnormal coolant loss

1 Overfilling system (Chapter 1).
2 Coolant boiling away due to overheating (see causes in Section 27).
3 Internal or external leakage (see Sections 29 and 30).
4 Faulty radiator cap. Have the cap pressure tested.
5 Cooling system being pressurized by engine compression. This could be due to a cracked head or block or leaking head gaskets.

32 Poor coolant circulation

1 Inoperative water pump. A quick test is to pinch the top radiator hose closed with your hand while the engine is idling, then release it. You should feel a surge of coolant if the pump is working properly (Chapter 3).
2 Restriction in cooling system. Drain, flush and refill the system (Chapter 1). If necessary, remove the radiator (Chapter 3) and have it reverse flushed or professionally cleaned.
3 Loose drivebelt (Chapter 1).
4 Thermostat sticking (Chapter 3).
5 Insufficient coolant (Chapter 1).

33 Corrosion

1 Excessive impurities in the water. Soft, clean water is recommended. Distilled or rainwater is satisfactory.
2 Insufficient antifreeze solution (refer to Chapter 1 for the proper ratio of water to antifreeze).
3 Infrequent flushing and draining of system. Regular flushing of the cooling system should be carried out at the specified intervals as described in Chapter 1.

CLUTCH

➡Note: All clutch related service information is located in Chapter 8, unless otherwise noted.

34 Fails to release (pedal pressed to the floor - shift lever does not move freely in and out of Reverse)

1 Freeplay incorrectly adjusted.
2 Clutch contaminated with oil. Remove clutch plate and inspect.
3 Clutch plate warped, distorted or otherwise damaged.
4 Diaphragm spring fatigued. Remove clutch cover/pressure plate assembly and inspect.
5 Leakage of fluid from clutch hydraulic system. Inspect master cylinder, operating cylinder and connecting lines.
6 Air in clutch hydraulic system. Bleed the system.
7 Insufficient pedal stroke. Check and adjust as necessary.
8 Piston seal in master or release cylinder deformed or damaged.
9 Lack of grease on pilot bearing.

35 Clutch slips (engine speed increases with no increase in vehicle speed)

1 Worn or oil-soaked clutch plate.
2 Diaphragm spring weak or damaged. Remove clutch cover/pressure plate assembly and inspect.
3 Clutch hydraulic line damaged internally (not allowing fluid to return to the clutch master cylinder).
4 Binding in the release mechanism.

36 Grabbing (chattering) as clutch is engaged

1 Oil on clutch plate. Remove and inspect. Repair any leaks.
2 Worn or loose engine or transmission mounts. They may move slightly when clutch is released. Inspect mounts and bolts.
3 Worn splines on transmission input shaft. Remove clutch components and inspect.
4 Warped pressure plate or flywheel. Remove clutch components and inspect.
5 Diaphragm spring fatigued. Remove clutch cover/pressure plate assembly and inspect.
6 Clutch linings hardened or warped.
7 Clutch lining rivets loose.

37 Squeal or rumble with clutch engaged (pedal released)

1 Improper pedal adjustment. Adjust pedal freeplay.
2 Release bearing binding on transmission shaft. Remove clutch components and check bearing. Remove any burrs or nicks, clean and relubricate before reinstallation.
4 Clutch rivets loose.
5 Clutch plate cracked.
6 Fatigued clutch plate torsion springs. Replace clutch plate.

38 Squeal or rumble with clutch disengaged (pedal depressed)

1 Worn or damaged release bearing.

2 Worn or broken pressure plate diaphragm fingers.
3 Pilot bearing worn or damaged.

39 Clutch pedal stays on floor when disengaged

Binding release bearing.

MANUAL TRANSMISSION

➡Note: All manual transmission service information is located in Chapter 7A, unless otherwise noted.

40 Noisy in Neutral with engine running

1 Input shaft bearing worn.
2 Damaged main drive gear bearing.
3 Insufficient transmission oil (Chapter 1).
4 Transmission oil in poor condition. Drain and fill with proper grade oil. Check old oil for water and debris (Chapter 1).
5 Noise can be caused by variations in engine torque. Change the idle speed and see if noise disappears.

41 Noisy in all gears

1 Any of the above causes, and/or:
2 Worn or damaged output gear bearings or shaft.

42 Noisy in one particular gear

1 Worn, damaged or chipped gear teeth.
2 Worn or damaged synchronizer.

43 Slips out of gear

1 Stiff shift lever seal.
2 Shift linkage binding.
3 Broken or loose input gear bearing retainer.
4 Dirt between clutch lever and engine housing.
5 Worn linkage.
6 Damaged or worn check balls, fork rod ball grooves or check springs.
7 Worn mainshaft or countershaft bearings.
8 Loose engine mounts (Chapter 2).
9 Excessive gear endplay.
10 Worn synchronizers.

44 Oil leaks

1 Excessive amount of lubricant in transmission (see Chapter 1 for correct checking procedures). Drain lubricant as required.
2 Rear oil seal or speedometer oil seal damaged.
3 To pinpoint a leak, first remove all built-up dirt and grime from the transmission. Degreasing agents and/or steam cleaning will achieve this. With the underside clean, drive the vehicle at low speeds so the air flow will not blow the leak far from its source. Raise the vehicle and determine where the leak is located.

45 Difficulty engaging gears

1 Clutch not releasing completely.
2 Loose or damaged shift linkage. Make a thorough inspection, replacing parts as necessary.
3 Insufficient transmission oil (Chapter 1).
4 Transmission oil in poor condition. Drain and fill with proper grade oil. Check oil for water and debris (Chapter 1).
5 Worn or damaged striking rod.
6 Sticking or jamming gears.

46 Noise occurs while shifting gears

1 Check for proper operation of the clutch (Chapter 8).
2 Faulty synchronizer assemblies.

AUTOMATIC TRANSMISSION

➡Note: Due to the complexity of the automatic transmission, it's difficult for the home mechanic to properly diagnose and service. For problems other than the following, the vehicle should be taken to a reputable mechanic.

47 Fluid leakage

1 Automatic transmission fluid is a deep red color, and fluid leaks should not be confused with engine oil which can easily be blown by air flow to the transmission.
2 To pinpoint a leak, first remove all built-up dirt and grime from the transmission. Degreasing agents and/or steam cleaning will achieve this. With the underside clean, drive the vehicle at low speeds so the air flow will not blow the leak far from its source. Raise the vehicle and determine where the leak is located. Common areas of leakage are:

a) ***Fluid pan:*** *tighten mounting bolts and/or replace pan gasket as necessary (Chapter 1).*
b) ***Rear extension:*** *tighten bolts and/or replace oil seal as necessary.*
c) ***Filler pipe:*** *replace the rubber oil seal where pipe enters transmission case.*
d) ***Transmission oil lines:*** *tighten fittings where lines enter transmission case and/or replace lines.*
e) ***Vent pipe:*** *transmission overfilled and/or water in fluid (see checking procedures, Chapter 1).*
f) ***Speedometer connector:*** *replace the O-ring where speed ometer cable enters transmission case.*

48 General shift mechanism problems

Chapter 7B deals with checking and adjusting the shift cable on automatic transmissions. Common problems which may be caused by out-of-adjustment linkage are:

a) *Engine starting in gears other than P (park) or N (Neutral).*
b) *Indicator pointing to a gear other than the one actually engaged.*
c) *Vehicle moves with transmission in P (Park) position.*

49 Transmission will not downshift with the accelerator pedal pressed to the floor

Since these transmissions are electronically controlled, check for any diagnostic trouble codes stored in the PCM. The actual repair will most likely have to be performed by a qualified repair shop with the proper equipment.

50 Engine will start in gears other than Park or Neutral

Chapter 7B deals with adjusting the Park/Neutral position switch installed on automatic transmissions.

51 Transmission slips, shifts rough, is noisy or has no drive in forward or Reverse gears

1 There are many probable causes for the above problems, but the home mechanic should concern himself only with one possibility: fluid level.

2 Before taking the vehicle to a shop, check the fluid level and condition as described in Chapter 1. Add fluid, if necessary, or change the fluid and filter if needed. If problems persist, have a professional diagnose the transmission.

DRIVESHAFT

➡Note: Refer to Chapter 8, unless otherwise specified, for service information.

52 Leaks at front of driveshaft

Defective transmission rear seal. See Chapter 7 for replacement procedure. As this is done, check the splined yoke for burrs or roughness that could damage the new seal. Remove burrs with a fine file or whetstone.

53 Knock or clunk when transmission is under initial load (just after transmission is put into gear)

1 Loose or disconnected rear suspension components. Check all mounting bolts and bushings (Chapters 7 and 10).

2 Loose driveshaft bolts. Inspect all bolts and nuts and tighten them securely.

3 Worn or damaged universal joint bearings. Inspect the universal joints (Chapter 8).

4 Worn sleeve yoke and mainshaft spline.

54 Metallic grating sound consistent with vehicle speed

Pronounced wear in the universal joint bearings. Replace U-joints or driveshafts, as necessary.

55 Vibration

➡Note: Before blaming the driveshaft, make sure the tires are perfectly balanced and perform the following test.

1 Install a tachometer inside the vehicle to monitor engine speed as the vehicle is driven. Drive the vehicle and note the engine speed at which the vibration (roughness) is most pronounced. Now shift the transmission to a different gear and bring the engine speed to the same point.

2 If the vibration occurs at the same engine speed (rpm) regardless of which gear the transmission is in, the driveshaft is NOT at fault since the driveshaft speed varies.

3 If the vibration decreases or is eliminated when the transmission is in a different gear at the same engine speed, refer to the following probable causes.

4 Bent or dented driveshaft. Inspect and replace as necessary.

5 Undercoating or built-up dirt, etc. on the driveshaft. Clean the shaft thoroughly.

6 Worn universal joint bearings. Replace the U-joints or driveshaft as necessary.

7 Driveshaft and/or companion flange out of balance. Check for missing weights on the shaft. Remove driveshaft and reinstall 180-degrees from original position, then recheck. Have the driveshaft balanced if problem persists.

8 Loose driveshaft mounting bolts/nuts.

9 Defective center bearing, if so equipped.

10 Worn transmission rear bushing (Chapter 7).

56 Scraping noise

Make sure the dust cover on the sleeve yoke isn't rubbing on the transmission extension housing.

57 Whining or whistling noise

Defective center bearing.

AXLES AND DIFFERENTIAL

➡Note: For differential servicing information, refer to Chapter 8, unless otherwise specified.

58 Noise - same when in drive as when vehicle is coasting

1 Road noise. No corrective action available.

2 Tire noise. Inspect tires and check tire pressures (Chapter 1).

3 Front wheel bearings or rear axle bearings, worn or damaged (Chapters 8 and 10).

4 Insufficient differential oil (Chapter 1).

5 Defective differential.

59 Knocking sound when starting or shifting gears

Defective or incorrectly adjusted differential.

60 Noise when turning

Defective differential.

61 Vibration

1 See probable causes under Driveshaft. Proceed under the guidelines listed for the driveshaft. If the problem persists, check the rear wheel bearings by raising the rear of the vehicle and spinning the

wheels by hand. Listen for evidence of rough (noisy) bearings. Remove and inspect (Chapter 8).

2 Worn driveaxle CV joint (Chapter 8).

62 Oil leaks

1 Pinion oil seal damaged (Chapter 8).
2 Axleshaft oil seals damaged (Chapter 8).
3 Loose filler or drain plug on differential (Chapter 1).
4 Clogged or damaged breather on differential.

TRANSFER CASE

➡Note: Unless otherwise specified, refer to Chapter 7C for service and repair information.

63 Gear jumping out of mesh

1 Interference between the control lever and the console.
2 Internal wear or incorrect adjustments.

64 Difficult shifting

1 Lack of oil.
2 Internal wear, damage or incorrect adjustment.

65 Noise

1 Lack of oil in transfer case.

2 Noise in 4H and 4L, but not in 2H indicates cause is in the front differential or front axle.

3 Noise in 2H, 4H and 4L indicates cause is in rear differential or rear axle.

4 Noise in 2H and 4H but not in 4L, or in 4L only, indicates internal wear or damage in transfer case.

BRAKES

➡Note: Before assuming a brake problem exists, make sure the tires are in good condition and inflated properly, the front end alignment is correct and the vehicle is not loaded with weight in an unequal manner. All service procedures for the brakes are included in Chapter 9, unless otherwise noted.

66 Vehicle pulls to one side during braking

1 Defective, damaged or oil contaminated brake pad on one side. Inspect as described in Chapter 1. Refer to Chapter 9 if replacement is required.

2 Excessive wear of brake pad material or disc on one side. Inspect and repair as necessary (Chapter 9).

3 Loose or disconnected front suspension components. Inspect and tighten all bolts securely (Chapters 1 and 11).

4 Defective caliper assembly. Remove caliper and inspect for stuck piston or damage.

5 Scored brake disc (Chapter 9).

67 Noise (grinding or high-pitched squeal)

1 Brake pads or shoes worn out. Replace pads or shoes with new ones immediately!

2 Glazed or contaminated pads.
3 Dirty or scored disc.
4 Bent support plate.

68 Excessive brake pedal travel

1 Partial brake system failure. Inspect entire system (Chapter 1) and correct as required.

2 Insufficient fluid in master cylinder. Check (Chapter 1) and add fluid - bleed system if necessary.

3 Air in system. Bleed system.

4 Drum brakes out of adjustment. Check the operation of the automatic adjusters.

5 Defective proportioning valve. Replace valve and bleed system.

69 Brake pedal feels spongy when depressed

1 Air in brake lines. Bleed the brake system.

2 Deteriorated rubber brake hoses. Inspect all system hoses and lines. Replace parts as necessary.

3 Master cylinder mounting nuts loose. Inspect master cylinder bolts (nuts) and tighten them securely.

4 Master cylinder faulty.
5 Incorrect shoe or pad clearance.
6 Deformed rubber brake lines.
7 Soft or swollen caliper seals.
8 Poor quality brake fluid. Bleed entire system and fill with new approved fluid.

70 Excessive effort required to stop vehicle

1 Power brake booster not operating properly.

2 Excessively worn linings or pads. Check and replace if necessary.

3 One or more caliper or wheel cylinder pistons seized or sticking. Inspect and replace caliper or wheel cylinder as required.

4 Brake pads or linings contaminated with oil or grease. Inspect and replace as required.

5 New pads or linings installed and not yet seated. It'll take a while for the new material to seat against the disc or drum.

6 Worn or damaged master cylinder or caliper assemblies. Check particularly for frozen pistons.

7 Also see causes listed under Section 69.

71 Pedal travels to the floor with little resistance

Little or no fluid in the master cylinder reservoir caused by leaking caliper piston(s) or loose, damaged or disconnected brake lines. Inspect entire system and repair as necessary.

72 Brake pedal pulsates during brake application

1 Wheel bearings damaged or worn (Chapter 10).

2 Disc not within specifications. Check for excessive lateral runout and parallelism. Have the discs resurfaced or replace them with new ones. Also make sure that all discs are the same thickness.

3 Out-of-round rear brake drums. Remove the drums and have them resurfaced or replace them with new ones.

73 Brakes drag (indicated by sluggish engine performance or wheels being very hot after driving)

1 Output rod adjustment incorrect at the brake pedal.
2 Obstructed master cylinder fill port.
3 Master cylinder piston seized in bore.
4 Caliper sticking.
5 Piston cups in master cylinder or caliper assembly deformed.
6 Parking brake assembly will not release.
7 Clogged brake lines.
8 Brake pedal height improperly adjusted.
9 Wheel cylinder sticking.
10 Improper shoe-to-drum clearance. Adjust as necessary.

74 Rear brakes lock up under light brake application

1 Tire pressures too high.
2 Tires excessively worn (Chapter 1).

75 Rear brakes lock up under heavy brake application

1 Tire pressures too high.
2 Tires excessively worn (Chapter 1).
3 Front brake pads contaminated with oil, mud or water. Clean or replace the pads.
4 Front brake pads excessively worn.
5 Defective master cylinder or caliper assembly.

SUSPENSION AND STEERING

➡Note: All service procedures for the suspension and steering systems are included in Chapter 10, unless otherwise noted.

76 Vehicle pulls to one side

1 Tire pressures uneven (Chapter 1).
2 Defective tire (Chapter 1).
3 Excessive wear in suspension or steering components (Chapter 1).
4 Wheel alignment incorrect.
5 Front brakes dragging (Chapter 9).
6 Wheel bearings improperly adjusted (Chapter 1 or 8).
7 Wheel lug nuts loose.

77 Shimmy, shake or vibration

1 Tire or wheel out-of-balance or out-of-round. Have them balanced on the vehicle.
2 Defective wheel bearings (Chapter 10).
3 Suspension components worn or damaged. Check for worn bushings in the upper and lower links.
4 Wheel lug nuts loose.
5 Incorrect tire pressures.
6 Excessively worn or damaged tire.
7 Loosely mounted steering gear housing.
8 Loose, worn or damaged steering components.
9 Worn balljoint.

78 Excessive pitching and/or rolling around corners or during braking

1 Defective shock absorbers. Replace as a set.
2 Broken or weak springs and/or suspension components.
3 Worn or damaged stabilizer bar or bushings.

79 Wandering or general instability

1 Improper tire pressures.
2 Worn or damaged upper and lower control arm bushings.
3 Incorrect front end alignment.
4 Worn or damaged steering gear or suspension components.
5 Out-of-balance wheels.
6 Loose wheel lug nuts.
7 Worn rear shock absorbers.
8 Fatigued or damaged rear leaf springs.

80 Excessively stiff steering

1 Lack of lubricant in power steering fluid reservoir (Chapter 1).
2 Incorrect tire pressures (Chapter 1).
3 Balljoints worn (Chapter 10).
4 Front end out of alignment.
5 Worn or damaged steering gear.

81 Excessive play in steering

1 Excessive wear in suspension bushings (Chapter 1).
2 Steering gear worn.
3 Steering gear mounting bolts loose.
4 Worn tie-rod ends.

82 Lack of power assistance

1 Drivebelt or tensioner faulty (Chapter 1).
2 Fluid level low (Chapter 1).
3 Hoses or pipes restricting the flow. Inspect and replace parts as necessary.
4 Air in power steering system. Bleed system.
5 Defective power steering pump.

83 Steering wheel fails to return to straight-ahead position

1 Incorrect front end alignment.
2 Tire pressures low.
3 Steering gear worn or damaged.
4 Worn or damaged balljoint.
5 Worn or damaged tie-rod end.
6 Lack of fluid in power steering pump.

84 Steering effort not the same in both directions (power system)

1 Leaks in steering gear.
2 Clogged fluid passage in steering gear.

85 Noisy power steering pump

1. Insufficient oil in pump.
2. Clogged hoses or oil filter in pump.
3. Loose pulley.
4. Improperly adjusted drivebelt (Chapter 1).
5. Defective pump.

86 Miscellaneous noises

1. Loose or worn steering gear or suspension components.
2. Defective shock absorber.
3. Defective wheel bearing.
4. Worn or damaged suspension bushings.
5. Damaged leaf spring.
6. Loose wheel lug nuts.

87 Excessive tire wear (not specific to one area)

1. Incorrect tire pressures.
2. Tires out of balance.
3. Wheels damaged. Inspect and replace as necessary.
4. Suspension or steering components worn (Chapter 1).

88 Excessive tire wear on outside edge

1. Incorrect tire pressure.
2. Excessive speed in turns.
3. Front end alignment incorrect (excessive toe-in).

89 Excessive tire wear on inside edge

1. Incorrect tire pressure.
2. Front end alignment incorrect (toe-out).

90 Tire tread worn in one place

1. Tires out of balance. Have them balanced on the vehicle.
2. Damaged or buckled wheel. Inspect and replace if necessary.
3. Defective tire.

1

TUNE-UP AND ROUTINE MAINTENANCE

Engine compartment component locations - four-cylinder engine

1. *Automatic transmission fluid dipstick*
2. *Brake fluid reservoir*
3. *Fuse box*
4. *Battery*
5. *Engine oil dipstick*
6. *Power steering fluid reservoir*
7. *Engine oil filler cap*
8. *Coolant reservoir*
9. *Radiator cap*
10. *Windshield washer fluid reservoir*
11. *Air filter housing*

Engine compartment component locations - V6 engine

1 *Brake fluid reservoir*
2 *Fuse box*
3 *Engine oil filler cap*
4 *Battery*
5 *Oil filter*
6 *Coolant reservoir*
7 *Radiator cap*
8 *Windshield washer fluid reservoir*
9 *Power steering fluid reservoir*
10 *Engine oil dipstick*
11 *Air filter housing*

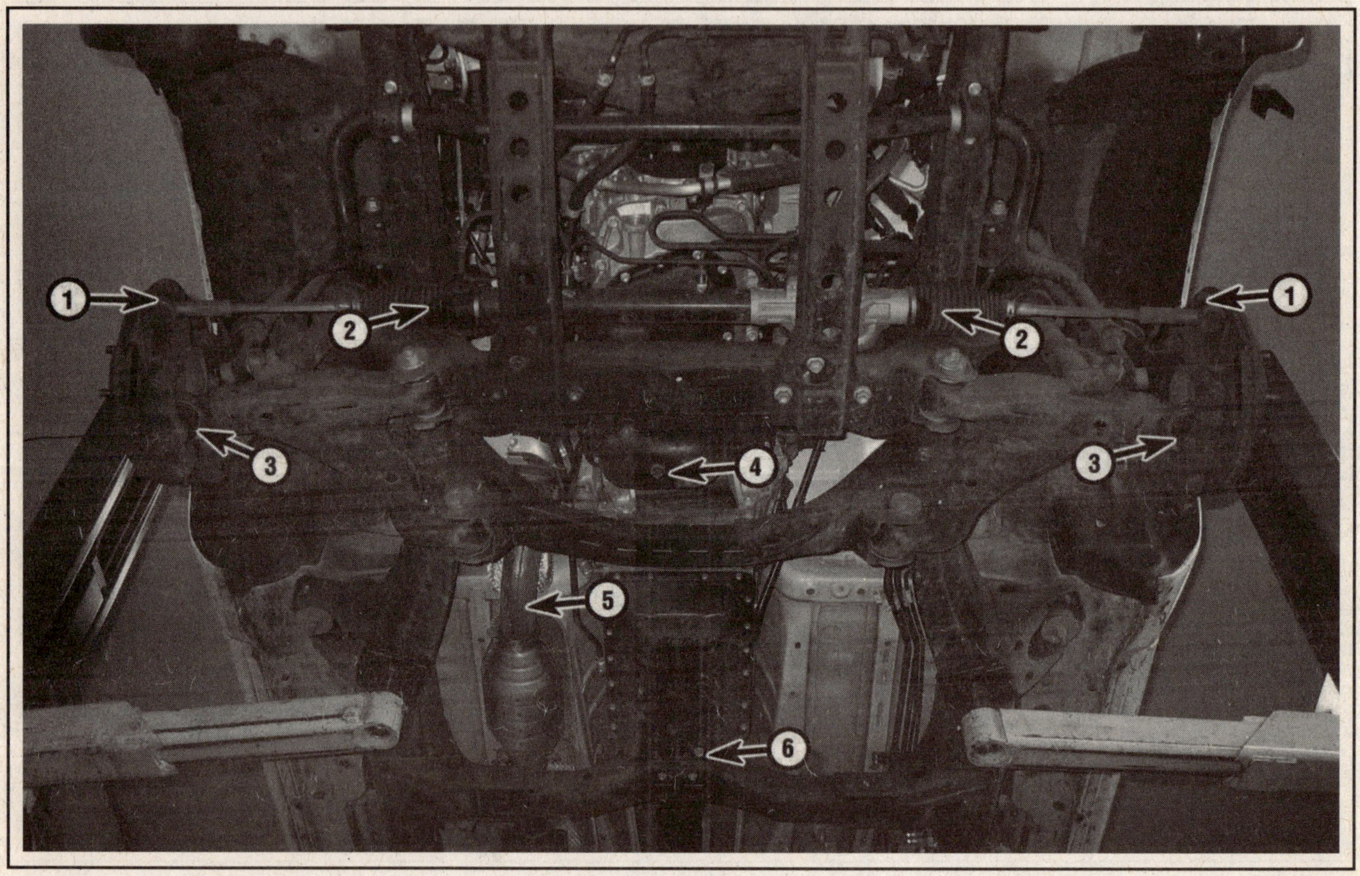

Typical engine compartment underside component locations (2WD model)

1. *Tie-rod end*
2. *Steering gear boot*
3. *Lower balljoint*
4. *Engine oil drain plug*
5. *Exhaust pipe*
6. *Automatic transmission fluid drain plug*

Typical rear underside component locations

1 *Exhaust pipe*
2 *Differential check/fill plug*
3 *Differential drain plug*
4 *Brake drum*
5 *Fuel tank*
6 *Muffler*

1 Toyota Tacoma Maintenance schedule

The maintenance intervals in this manual are provided with the assumption that you, not the dealer, will be doing the work. These are the minimum maintenance intervals recommended by the factory for vehicles that are driven daily. If you wish to keep your vehicle in peak condition at all times, you may wish to perform some of these procedures even more often. Because frequent maintenance enhances the efficiency, performance and resale value of your car, we encourage you to do so. If you drive in dusty areas, tow a trailer, idle or drive at low speeds for extended periods or drive for short distances (less than four miles) in below freezing temperatures, shorter intervals are also recommended.

When your vehicle is new, it should be serviced by a factory authorized dealer service department to protect the factory warranty. In many cases, the initial maintenance check is done at no cost to the owner.

EVERY 250 MILES (400 KM) OR WEEKLY, WHICHEVER COMES FIRST

Check the engine oil level (Section 4)
Check the engine coolant level (Section 4)
Check the windshield washer fluid level (Section 4)
Check the brake fluid level (Section 4)
Check the power steering fluid level (Section 4)
Check the tires and tire pressures (Section 5)

EVERY 3000 MILES (4800 KM) OR 3 MONTHS, WHICHEVER COMES FIRST

All items listed above, plus . . .
Check the automatic transmission fluid level (Section 6)
Change the engine oil and filter (Section 7)

EVERY 7500 MILES (12,000 KM) OR 6 MONTHS, WHICHEVER COMES FIRST

Check and service the battery (Section 8)
Check the cooling system (Section 9)
Inspect and replace, if necessary, all underhood hoses (Section 10)
Inspect and replace, if necessary, the windshield wiper blades (Section 11)
Rotate the tires (Section 12)
Inspect the suspension and steering components (Section 13)
Lubricate the driveshaft and body components (Section 14)*
Inspect the exhaust system (Section 15)
Check the transfer case lubricant level (Section 16)
Check the differential lubricant level (Section 17)
Check the manual transmission lubricant level (Section 17)
Check the seat belts (Section 18)
Re-torque the driveshaft bolts (See Chapter 8)*
Re-torque the front driveshaft bolts and driveaxle/hub nut (4WD) (See Chapter 8)*

EVERY 15,000 MILES (24,000 KM) OR 12 MONTHS, WHICHEVER COMES FIRST

All items listed above, plus . . .
Check the driveaxle boots (4WD) (Section 19)
Replace the air filter (Section 20)*
Check the engine drivebelt(s) (Section 21)
Inspect the fuel system (Section 22)
Check the brakes (Section 23)*

EVERY 30,000 MILES (48,000 KM) OR 24 MONTHS, WHICHEVER COMES FIRST

All items listed above, plus . . .
Replace the spark plugs (non-platinum or iridium type) (Section 24)
Service the cooling system (drain, flush and refill) (Section 25)
Change the automatic transmission fluid and filter (Section 26)**
Change the cabin air filter (Section 27)
Change the transfer case lubricant (Section 28)
Change the differential lubricant (Section 29)*
Change the manual transmission lubricant (Section 29)*
Change the brake fluid (Section 30)

EVERY 60,000 MILES (96,000 KM) OR 48 MONTHS, WHICHEVER COMES FIRST

Check and adjust if necessary, the valve clearance (V6 engine only) (Section 31)
Inspect the evaporative emissions control system (Section 32)
Re-torque the driveshaft bolts (See Chapter 8)
Re-torque the front driveshaft bolts and axle shaft nut (4WD) (See Chapter 8)

EVERY 90,000 MILES (145,000 KM)

Replace the spark plugs (platinum or iridium type) (Section 24)

**This item is affected by "severe" operating conditions, as described below. If the vehicle is operated under severe conditions, perform all maintenance indicated with an asterisk (*) at 5000 mile/four-month intervals. Severe conditions exist if you mainly operate the vehicle . . .*

in dusty areas
towing a trailer
idling for extended periods and/or driving at low speeds when outside temperatures remain below freezing and most trips are less than four miles long

***If operated under one or more of the following conditions, change the automatic transmission fluid every 15,000 miles:*

in heavy city traffic where the outside temperature regularly reaches 90-degrees F or higher
in hilly or mountainous terrain
frequent trailer pulling

2 Introduction

This Chapter is designed to help the home mechanic maintain the Toyota Tacoma for peak performance, economy, safety and long life.

Included is a master maintenance schedule, followed by Sections dealing specifically with each item on the schedule. Visual checks, adjustments, component replacement and other helpful items are included. Refer to the accompanying illustrations of the engine compartment and the underside of the vehicle for the location of various components.

Servicing your vehicle in accordance with the mileage/time maintenance schedule and the following Sections will provide it with a planned maintenance program that should result in a long and reliable service life. This is a comprehensive plan, so maintaining some items but not others at the specified service intervals will not produce the same results.

As you service your vehicle, you will discover that many of the procedures can, and should, be grouped together because of the nature of the particular procedure you're performing or because of the close proximity of two otherwise unrelated components to one another.

For example, if the vehicle is raised for any reason, you should inspect the exhaust, suspension, steering and fuel systems while you're under the vehicle. When you're rotating the tires, it makes good sense to check the brakes and wheel bearings since the wheels are already removed.

Finally, let's suppose you have to borrow or rent a torque wrench. Even if you only need to tighten the spark plugs, you might as well check the torque of as many critical fasteners as time allows.

The first step of this maintenance program is to prepare yourself before the actual work begins. Read through all Sections pertinent to the procedures you're planning to do, then make a list of and gather together all the parts and tools you will need to do the job. If it looks as if you might run into problems during a particular segment of some procedure, seek advice from your local auto parts store or dealer service department.

OWNER'S MANUAL AND VECI LABEL INFORMATION

Your vehicle Owner's Manual was written for your year and model and contains very specific information on component locations, specifications, fuse ratings, part numbers, etc. The Owner's Manual is an important resource for the do-it-yourselfer to have; if one was not supplied with your vehicle, it can generally be ordered from a dealer parts department.

Among other important information, the Vehicle Emissions Control Information (VECI) label contains specifications and procedures for tune-up adjustments (if applicable) and, in some instances, spark plugs (see Chapter 6 for more information on the VECI label). The information on this label is the exact maintenance data recommended by the manufacturer. This data often varies by intended operating altitude, local emissions regulations, month of manufacture, etc.

This Chapter contains procedural details, safety information and more ambitious maintenance intervals than you might find in the manufacturer's literature. However, you may also find procedures or specifications in your Owner's Manual or VECI label that differ with what's printed here. In these cases, the Owner's Manual or VECI label can be considered correct, since it is specific to your particular vehicle.

3 Tune-up general information

The term tune-up is used in this manual to represent a combination of individual operations rather than one specific procedure.

If, from the time the vehicle is new, the routine maintenance schedule is followed closely and frequent checks are made of fluid levels and high wear items, as suggested throughout this manual, the engine will be kept in relatively good running condition and the need for additional work will be minimized.

More likely than not, however, there will be times when the engine is running poorly due to lack of regular maintenance. This is even more likely if a used vehicle, which has not received regular and frequent maintenance checks, is purchased. In such cases, an engine tune-up will be needed outside of the regular routine maintenance intervals.

The first step in any tune-up or diagnostic procedure to help correct a poor running engine is a cylinder compression check. A compression check (see Chapter 2, Part C) will help determine the condition of internal engine components and should be used as a guide for tune-up and repair procedures. If, for instance, the compression check indicates serious internal engine wear, a conventional tune-up won't improve the performance of the engine and would be a waste of time and money. Because of its importance, the compression check should be done by someone with the right equipment and the knowledge to use it properly.

The following procedures are those most often needed to bring a generally poor running engine back into a proper state of tune.

MINOR TUNE-UP

Check all engine related fluids (Section 4)
Clean, inspect and test the battery (Section 8)
Check the cooling system (Section 9)
Check all underhood hoses (Section 10)
Check the air filter (Section 20)
Replace the spark plugs (Section 24)

MAJOR TUNE-UP

All items listed under Minor tune-up, plus . . .

Replace the air filter (Section 20)
Check the fuel system (Section 22)
Check and adjust the valve clearances (Section 31)
Check the ignition system (Chapter 5)
Check the charging system (Chapter 5)

4 Fluid level checks (every 250 miles [400 km] or weekly)

➡Note: The following are fluid level checks to be done on a 250 mile or weekly basis. Additional fluid level checks can be found in specific maintenance procedures which follow. Regardless of intervals, be alert to fluid leaks under the vehicle which would indicate a fault to be corrected immediately.

1 Fluids are an essential part of the lubrication, cooling, brake and windshield washer systems. Because the fluids gradually become depleted and/or contaminated during normal operation of the vehicle, they must be periodically replenished. See *Recommended lubricants and fluids* in this Chapter's Specifications before adding fluid to any of the following components.

➡Note: The vehicle must be on level ground when fluid levels are checked.

ENGINE OIL

➧ Refer to illustrations 4.2a, 4.2b, 4.4 and 4.6

2 The engine oil level is checked with a dipstick that extends through a tube and into the oil pan at the bottom of the engine (see illustration).

3 The oil level should be checked before the vehicle has been driven, or about 5 minutes after the engine has been shut off. If the oil is checked immediately after driving the vehicle, some of the oil will remain in the upper engine components, resulting in an inaccurate reading on the dipstick.

4 Pull the dipstick out of the tube and wipe all the oil from the end with a clean rag or paper towel. Insert the clean dipstick all the way back into the tube, then pull it out again. Note the oil at the end of the dipstick. Add oil as necessary to keep the level between the upper and lower marks on the dipstick (see illustration).

5 Do not overfill the engine by adding too much oil since this may result in oil-fouled spark plugs, oil leaks or oil seal failures.

6 Oil is added to the engine after removing the threaded cap from the valve cover (see illustration). A funnel may help to reduce spills.

7 Checking the oil level is an important preventive maintenance step. A consistently low oil level indicates oil leakage through damaged seals, defective gaskets or past worn rings or valve guides. If the oil looks milky or has water droplets in it, the cylinder head gasket(s) may be blown or the head(s) or block may be cracked. The engine should be checked immediately. The condition of the oil should also be checked. Whenever you check the oil level, slide your thumb and index finger up the dipstick before wiping off the oil. If you see small dirt or metal particles clinging to the dipstick, the oil should be changed (see Section 7).

ENGINE COOLANT

➧ Refer to illustration 4.8

⁂ WARNING:

Do not allow antifreeze to come in contact with your skin or painted surfaces of the vehicle. Flush contaminated areas immediately with plenty of water. Don't store new coolant or leave old coolant lying around where it's accessible to children or pets - they're attracted by its sweet smell. Ingestion of even a small amount of coolant can be fatal! Wipe up garage floor and drip pan spills immediately. Keep antifreeze containers covered and repair cooling system leaks as soon as they're noticed.

8 All vehicles covered by this manual are equipped with a pressurized coolant recovery system. A coolant reservoir which is located in the right front corner of the engine compartment (next to the radiator) is connected by a hose to the base of the coolant filler cap (see illustration). If the coolant gets too hot during engine operation, coolant can escape through a pressurized filler cap, then through a connecting hose into the reservoir. As the engine cools, the coolant is automatically drawn back into the cooling system to maintain the correct level.

9 The coolant level should be checked regularly. It must be between the FULL and LOW lines on the tank. The level will vary with the temperature of the engine. When the engine is cold, the coolant level should be at or slightly above the LOW mark on the tank. Once the engine has warmed up, the level should be at or near the FULL mark. If it isn't, allow the fluid in the tank to cool, then remove the cap from the reservoir and add coolant to bring the level up to the FULL line. Use only the type of coolant and water in the mixture ratio recommended in

4.2a On four-cylinder models, the oil dipstick is located on the left side of the engine

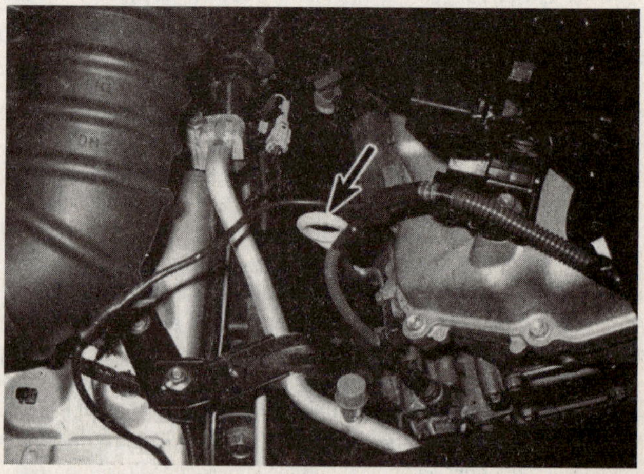

4.2b On V6 models, the oil dipstick is located on the right side of the engine

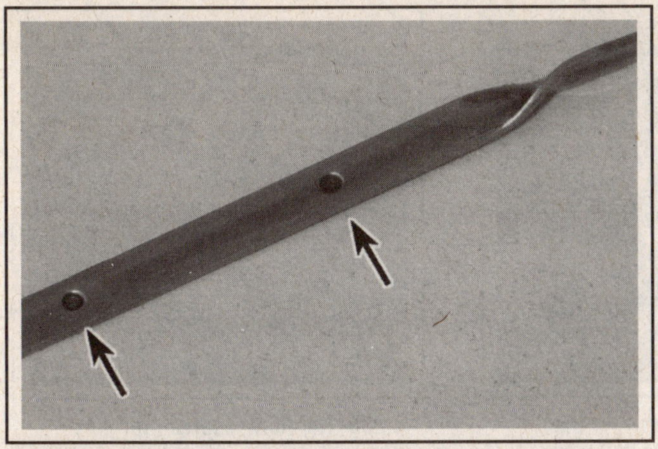

4.4 The oil level must be maintained between the marks at all times - it takes one quart of oil to raise the level from the lower mark to the upper mark

this Chapter's Specifications. Do not use supplemental inhibitor additives. If only a small amount of coolant is required to bring the system up to the proper level, water can be used. However, repeated additions of water will dilute the recommended antifreeze and water solution. In order to maintain the proper ratio of antifreeze and water, it is advisable to top up the coolant level with the correct mixture.

10 If the coolant level drops within a short time after replenishment, there may be a leak in the system. Inspect the radiator, hoses, engine coolant filler cap, drain plugs and water pump. If no leak is evident, have the radiator cap pressure tested by your dealer.

WARNING:

Never remove the radiator cap or the coolant recovery reservoir cap when the engine is running or has just been shut down, because the cooling system is hot. Escaping steam and scalding liquid could cause serious injury.

11 If it is necessary to open the radiator cap, wait until the system has cooled completely, then wrap a thick cloth around the cap and turn it to the first stop. If any steam escapes, wait until the system has cooled further, then remove the cap.

4.8 Check the coolant level in the reservoir with the engine hot - it should be visible through the translucent reservoir

4.6 Oil is added to the engine after removing the twist-off cap located on the valve cover (four-cylinder model shown, V6 models similar)

12 When checking the coolant level, always note its condition. It should be relatively clear. If it is brown or rust colored, the system should be drained, flushed and refilled. Even if the coolant appears to be normal, the corrosion inhibitors wear out with use, so it must be replaced at the specified intervals.

13 Do not allow antifreeze to come in contact with your skin or painted surfaces of the vehicle. Flush contacted areas immediately with plenty of water.

WINDSHIELD WASHER FLUID

Refer to illustration 4.14

14 Fluid for the windshield washer system is located on the right side of the engine compartment (see illustration). In milder climates, plain water can be used to top up the reservoir, but the reservoir should be kept no more than two-thirds full to allow for expansion should the water freeze. In colder climates, the use of a specially designed windshield washer fluid, available at your dealer and any auto parts store, will help lower the freezing point of the fluid. Mix the solution with water in accordance with the manufacturer's directions on the container. Do not use regular antifreeze. It will damage the vehicle's paint.

4.14 The windshield washer fluid reservoir is located in the right front corner of the engine compartment

4.16 To check the fluid level, simply look at the MAX and MIN marks on the reservoir

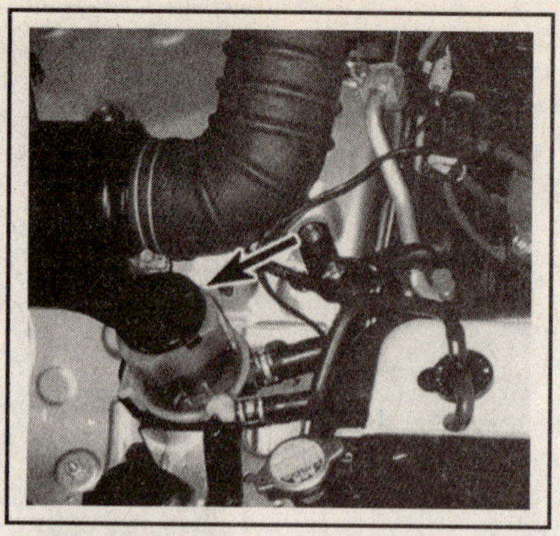

4.23a The fluid reservoir for the power steering pump is located at the right front corner of the engine compartment on V6 models

4.23b The fluid reservoir for the power steering pump is located at the left front corner of the engine on four-cylinder models

BRAKE AND CLUTCH FLUID

➧ **Refer to illustration 4.16**

15 The brake master cylinder is mounted on the front of the power booster unit in the engine compartment. The clutch master cylinder on vehicles with manual transaxles is connected by a hose to the brake master cylinder.

16 To check the fluid level, simply look at the MAX and MIN marks on the reservoir (see illustration). The level should be at or near the maximum fill line.

17 If the level is low, wipe the top of the reservoir cover with a clean rag to prevent contamination of the brake or clutch system before lifting the cover.

18 Add only the specified brake fluid to the reservoir (refer to *Recommended lubricants and fluids* in this Chapter's Specifications or to your owner's manual). Mixing different types of brake fluid can damage the system.

⁂ WARNING:

Use caution when filling the reservoir - brake fluid can harm your eyes and damage painted surfaces. Do not use brake fluid that has been opened for more than one year (even if the cap has been on) or has been left open. Brake fluid absorbs moisture from the air. Excess moisture can cause a dangerous loss of braking.

19 While the reservoir cap is removed, inspect the master cylinder reservoir for contamination. If deposits, dirt particles or water droplets are present, the fluid in the brake system should be changed (see Section 30 for the brake fluid replacement procedure, or Chapter 8 for the clutch hydraulic system bleeding procedure).

20 After filling the reservoir to the proper level, make sure the lid is properly seated to prevent fluid leakage and/or system pressure loss.

21 The brake fluid in the master cylinder will drop slightly as the brake pads at each wheel wear down during normal operation. If the master cylinder requires repeated replenishing to keep it at the proper level, this is an indication of leakage in the brake system, which should be corrected immediately. Check all brake lines and connections, along with the calipers, wheel cylinders and booster (see Section 23 for more information).

22 If, upon checking the master cylinder fluid level, you discover the reservoir empty or nearly empty, the brake and clutch systems must be diagnosed immediately (see Chapters 8 and 9).

POWER STEERING FLUID

➧ **Refer to illustrations 4.23a, 4.23b, 4.26 and 4.27**

23 The fluid reservoir for the power steering pump is located at the right front corner of the engine compartment on V6 models, and it's located on the left front corner of the engine on four-cylinder models (see illustrations).

24 For the check, the front wheels should be pointed ahead and the engine should be off.

Four-cylinder models

25 Use a rag and clean off the area around the cap. This will help prevent foreign material from falling in the reservoir when the cap is removed.

26 Turn and pull out the reservoir cap, which has a dipstick attached to it. Wipe the fluid at the bottom of the dipstick with a clean rag. Reinstall the cap to get a fluid level reading. Remove the cap again and note the fluid level. It should be at the appropriate mark on the dipstick in relation to the engine temperature (see illustration).

4.26 Make sure to use the proper dipstick markings according to fluid temperature (four-cylinder models)

V6 models

27 The reservoir on the V6 model is translucent plastic and the fluid level can be checked visually without removing the cap (see illustration).

All models

28 If additional fluid is required, pour the specified type fluid (see *Recommended lubricants and fluids* in this Chapter's Specifications or your owner's manual) directly into the reservoir using a funnel to prevent spills.

29 If the reservoir requires frequent topping up, all power steering hoses, hose connections, the power steering pump and the steering gear should be carefully examined for leaks.

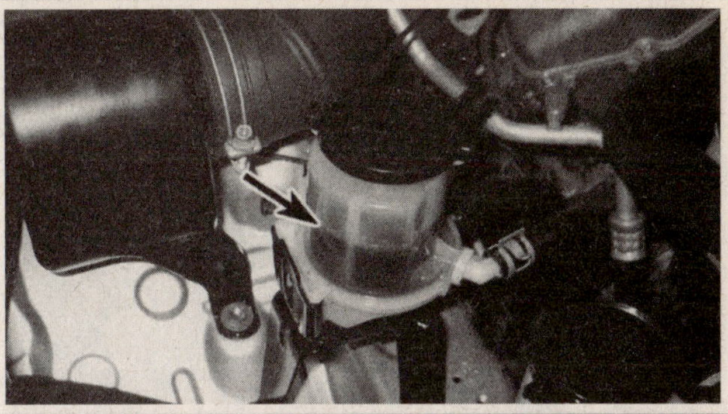

4.27 The fluid level inside the power steering fluid reservoir can be checked by observing the level from the outside (V6 models)

5 Tire and tire pressure checks (every 250 miles [400 km] or weekly)

➧ Refer to illustrations 5.2, 5.3, 5.4a, 5.4b and 5.8

1 Periodic inspection of the tires may spare you the inconvenience of being stranded with a flat tire. It can also provide you with vital information regarding possible problems in the steering and suspension systems before major damage occurs.

2 The original tires on this vehicle are equipped with 1/2-inch wide wear bands that will appear when tread depth reaches 1/16-inch, at which point the tires can be considered worn out. Tread wear can be monitored with a simple, inexpensive device known as a tread depth gauge (see illustration).

3 Note any abnormal tread wear (see illustration). Tread pattern irregularities such as cupping, flat spots and more wear on one side than the other are indications of front end alignment and/or balance problems. If any of these conditions are noted, take the vehicle to a tire shop or service station to correct the problem.

5.2 Use a tire tread depth gauge to monitor tire wear - they are available at auto parts stores and service stations and cost very little

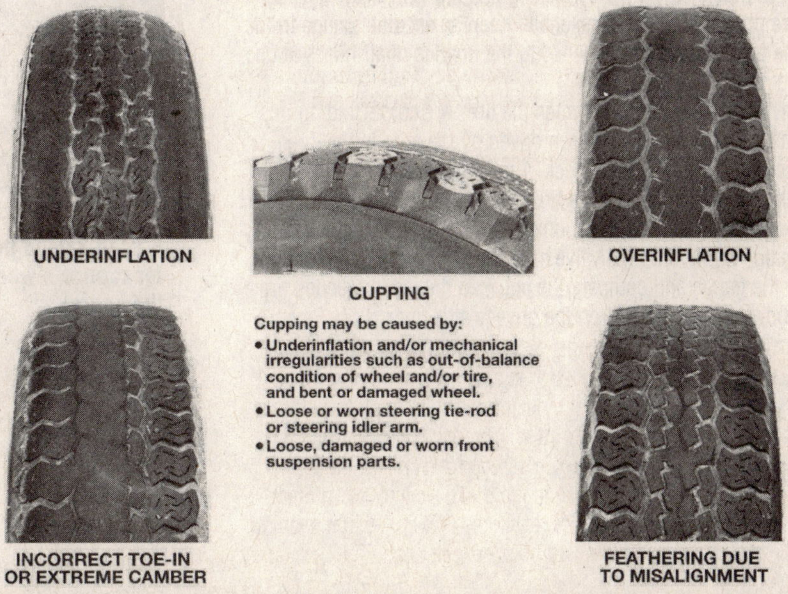

5.3 This chart will help you determine the condition of the tires, the probable cause(s) of abnormal wear and the corrective action necessary

5.4a If a tire loses air on a steady basis, check the valve core first to make sure it's snug (special inexpensive wrenches are commonly available at auto parts stores)

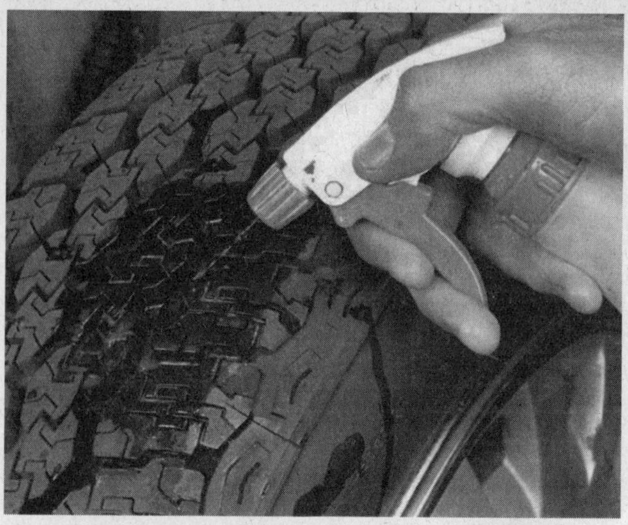

5.4b If the valve core is tight, raise the corner of the vehicle with the low tire and spray a soapy water solution onto the tread as the tire is turned slowly - leaks will cause small bubbles to appear

4 Look closely for cuts, punctures and embedded nails or tacks. Sometimes a tire will hold air pressure for a short time or leak down very slowly after a nail has embedded itself in the tread. If a slow leak persists, check the valve stem core to make sure it's tight (see illustration). Examine the tread for an object that may have embedded itself in the tire or for a plug that may have begun to leak (radial tire punctures are repaired with a rubber plug that's installed in the hole). If a puncture is suspected, it can be easily verified by spraying a solution of soapy water onto the puncture area (see illustration). The soapy solution will bubble if there's a leak. Unless the puncture is unusually large, a tire shop or service station can usually repair the tire.

5 Carefully inspect the inner sidewall of each tire for evidence of brake fluid leakage. If you see any, inspect the brakes immediately.

6 Correct air pressure adds miles to the lifespan of the tires, improves mileage and enhances overall ride quality. Tire pressure cannot be accurately estimated by looking at a tire, especially if it's a radial. A tire pressure gauge is essential. Keep an accurate gauge in the vehicle. The pressure gauges attached to the nozzles of air hoses at gas stations are often inaccurate.

7 Always check tire pressure when the tires are cold. Cold, in this case, means the vehicle has not been driven over a mile in the three hours preceding a tire pressure check. A pressure rise of four to eight pounds is not uncommon once the tires are warm.

8 Unscrew the valve cap protruding from the wheel or hubcap and push the gauge firmly onto the valve stem (see illustration). Note the reading on the gauge and compare the figure to the recommended tire pressure shown on the placard on the driver's side door jamb. Be sure to reinstall the valve cap to keep dirt and moisture out of the valve stem mechanism. Check all four tires and, if necessary, add enough air to bring them up to the recommended pressure.

9 Don't forget to keep the spare tire inflated to the specified pressure (consult your owner's manual).

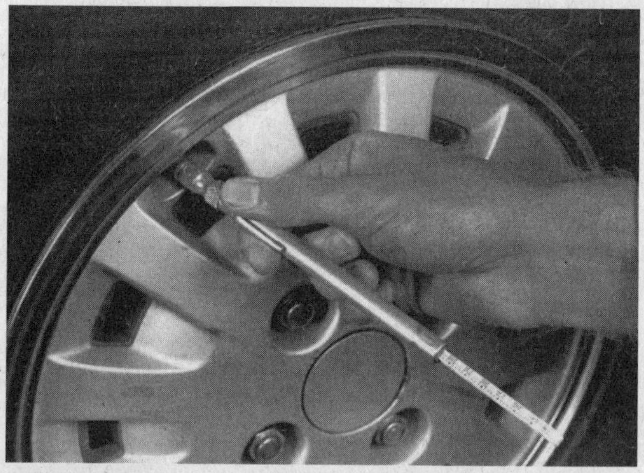

5.8 To extend the life of the tires, check the air pressure at least once a week with an accurate gauge (don't forget the spare)

6 Automatic transmission fluid level check (every 3000 miles [4800 km] or 3 months)

1 The level of the automatic transmission fluid should be carefully maintained. Low fluid level can lead to slipping or loss of drive, while overfilling can cause foaming, loss of fluid and transmission damage.

2 The transmission fluid level should only be checked when the transmission is hot (at its normal operating temperature). If the vehicle has just been driven over 10 miles (15 miles in a frigid climate), and the fluid temperature is 160 to 175-degrees F, the transmission is hot.

⁂ CAUTION:

If the vehicle has just been driven for a long time at high speed or in city traffic in hot weather, or if it has been pulling a trailer, an accurate fluid level reading cannot be obtained. Allow the fluid to cool down for about 30 minutes.

3 If the vehicle has not been driven, park the vehicle on level ground, set the parking brake, then start the engine and bring it to operating temperature. While the engine is idling, depress the brake pedal and move the selector lever through all the gear ranges, beginning and ending in Park.

FOUR-CYLINDER MODELS

➧ Refer to illustrations 6.4 and 6.6

4 With the engine still idling, remove the dipstick from its tube (see illustration). Check the level of the fluid on the dipstick and note its condition. Refer to the underhood photographs at the beginning of this Chapter for the exact location of the automatic transmission dipstick.

5 Wipe the fluid from the dipstick with a clean rag and reinsert it back into the filler tube until the cap seats.

6 Pull the dipstick out again and note the fluid level (see illustration). If the transmission is cold, the level should be in the COOL range on the dipstick. If it is hot, the fluid level should be in the HOT range. If the level is at the low side of either range, add the specified automatic transmission fluid through the dipstick tube with a funnel.

7 Add just enough of the recommended fluid to fill the transmission to the proper level. It takes about one pint to raise the level from the low mark to the high mark when the fluid is hot, so add the fluid a little at a time and keep checking the level until it is correct. Proceed to Step 12.

V6 MODELS

➡Note 1: These models are not equipped with an automatic transmission fluid level dipstick.

➡Note 2: The vehicle must be level for this check. If there is not enough room to crawl under the vehicle, raise both ends and support the vehicle securely on jackstands.

➡Note 3: The fluid temperature must be between 97 degrees F (36 degrees C) and 115 degrees F (46 degrees C) to accurately check the fluid level

8 With the engine idling. Remove the overflow plug from the bottom of the transmission fluid pan.

➡Note: The overflow plug is located along the right side of the fluid pan; the drain plug is located at the back of the pan (don't remove the drain plug).

9 If the fluid runs out of the hole, allow it to drip until it stops. If no fluid comes out of the hole, remove the refill plug, located on the right side of the transmission housing, near the rear.

10 Add the proper type of transmission fluid (see this Chapter's Specifications) until fluid flows from the overflow hole in the bottom of the pan. When the flow of fluid slows to a trickle, install the overflow plug and tighten it to the torque listed in this Chapter's Specifications.

11 Tighten the refill plug to the torque listed in this Chapter's Specifications.

ALL MODELS

12 The condition of the fluid should also be checked along with the level. If the fluid at the end of the dipstick is black or a dark reddish brown color, or if it emits a burned smell, the fluid should be changed (see Section 26). If you are in doubt about the condition of the fluid, purchase some new fluid and compare the two for color and smell.

6.4 The automatic transmission fluid dipstick is located at the rear of the engine compartment (four-cylinder shown)

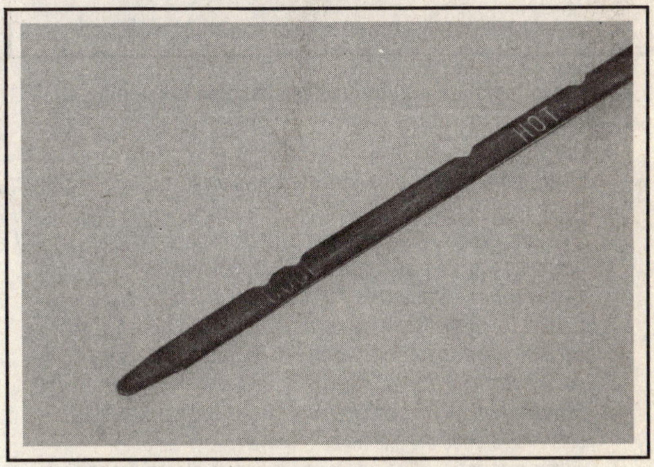

6.6 The automatic transmission fluid level must be maintained between the notches at the indicated operating temperature

7 Engine oil and filter change (every 3000 miles [4800 km] or 3 months)

Refer to illustrations 7.2, 7.7, 7.12a, 7.12b and 7.14

1 Frequent oil changes are the best preventive maintenance the home mechanic can give the engine, because aging oil becomes diluted and contaminated, which leads to premature engine wear.

2 Make sure that you have all the necessary tools before you begin this procedure (see illustration). You should also have plenty of rags or newspapers handy for mopping up any spills.

3 Access to the underside of the vehicle is greatly improved if the vehicle can be lifted on a hoist, driven onto ramps or supported by jackstands.

4 If this is your first oil change, get under the vehicle and familiarize yourself with the location of the oil drain plug. The engine and exhaust components will be warm during the actual work, so try to anticipate any potential problems before the engine and accessories are hot.

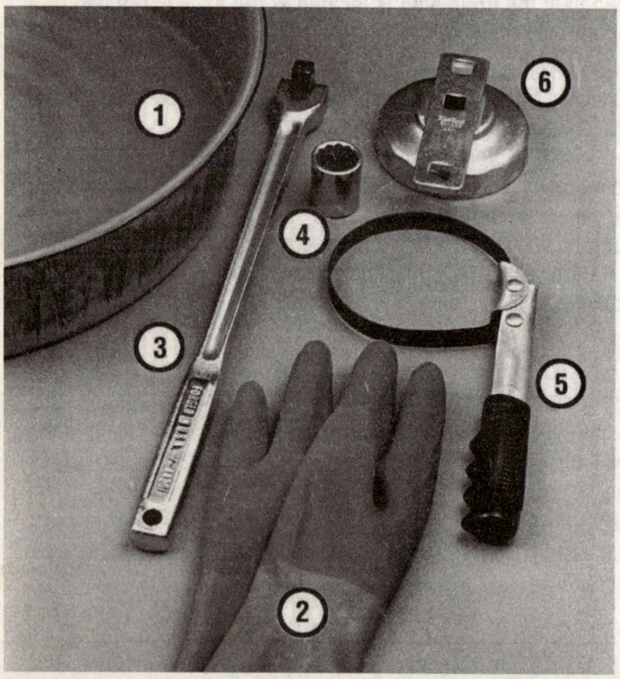

7.2 These tools are required when changing the engine oil and filter

1 ***Drain pan*** - *It should be fairly shallow in depth, but wide to prevent spills*
2 ***Rubber gloves*** - *When removing the drain plug and filter, you will get oil on your hands (the gloves will prevent burns)*
3 ***Breaker bar*** - *Sometimes the oil drain plug is tight, and a long breaker bar is needed to loosen it*
4 ***Socket*** - *To be used with the breaker bar or a ratchet (must be the correct size to fit the drain plug)*
5 ***Filter wrench*** - *This is a metal band-type wrench, which requires clearance around the filter to be effective*
6 ***Filter wrench*** - *This type fits on the bottom of the filter and can be turned with a ratchet or breaker bar (different size wrenches are available for different types of filters)*

5 Park the vehicle on a level spot. Start the engine and allow it to reach its normal operating temperature (the needle on the temperature gauge should be at least above the bottom mark). Warm oil and contaminates will flow out more easily. Turn off the engine when it's warmed up. Remove the filler cap on the valve cover.

6 Raise the front of the vehicle and support it securely on jackstands.

WARNING:

To avoid personal injury, never get beneath the vehicle when it is supported by only by a jack. The jack provided with your vehicle is designed solely for raising the vehicle to remove and replace the wheels. Always use jackstands to support the vehicle when it becomes necessary to place your body underneath the vehicle.

7 Being careful not to touch the hot exhaust components, place the drain pan under the drain plug in the bottom of the pan and remove the plug (see illustration). You may want to wear gloves while unscrewing the plug the final few turns if the engine is really hot.

8 Allow the old oil to drain into the pan. It may be necessary to move the pan farther under the engine as the oil flow slows to a trickle. Inspect the old oil for the presence of metal shavings and chips.

9 After all the oil has drained, wipe off the drain plug with a clean rag. Even minute metal particles clinging to the plug would immediately contaminate the new oil.

10 Clean the area around the drain plug opening, reinstall the plug and tighten it securely, but do not strip the threads.

11 On four-cylinder models, move the drain pan into position under the oil filter.

12 Loosen the oil filter by turning it counterclockwise with the filter wrench (see illustrations). Any standard filter wrench should work. Once the filter is loose, use your hands to unscrew it. Just as the filter is detached, immediately tilt the open end up to prevent the oil inside the filter from spilling out.

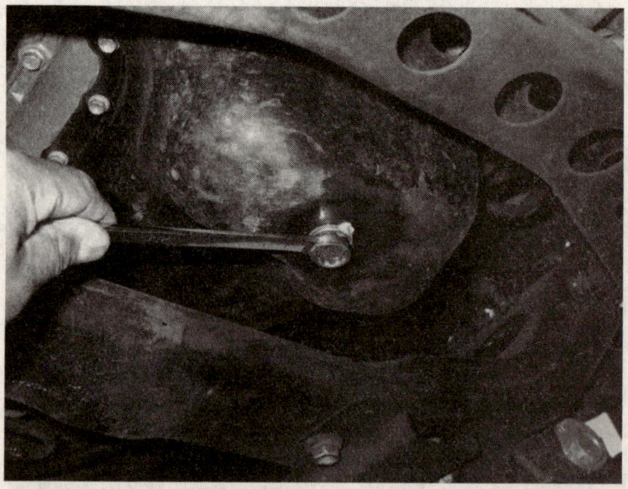

7.7 The engine oil drain plug is located on the bottom of the oil pan - it is usually very tight, so use the proper size box-end wrench or socket to avoid rounding it off

7.12a The oil filter is usually on very tight and will require a special wrench for removal - DO NOT use the wrench to tighten the new filter!

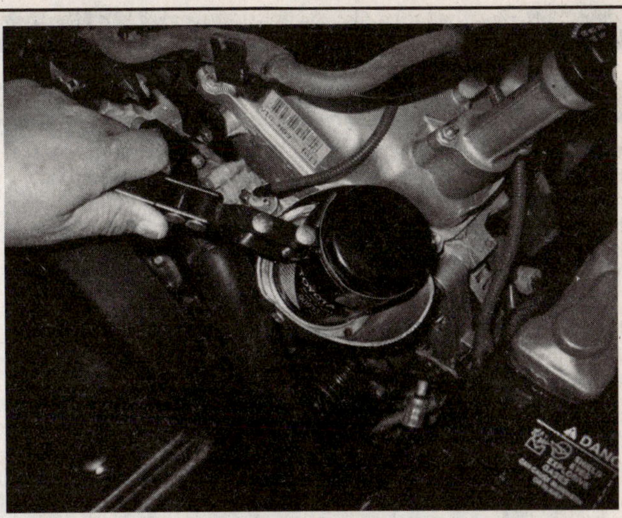

7.12b On V6 models, the oil filter is located on the timing chain cover, so it's easier to access than on other models. But when you unscrew it, some oil will run out before you can tilt the open end up. The shield catches any spilled oil

13 With a clean rag, wipe off the filter mounting surface. If a residue of old oil is allowed to remain, it will smoke when the engine is heated up. It will also prevent the new filter from seating properly. Also make sure that the none of the old gasket remains stuck to the mounting surface. It can be removed with a scraper if necessary.

14 Compare the old filter with the new one to make sure they are the same type. Smear some engine oil on the rubber gasket of the new filter (see illustration). Attach the filter to the engine, following the tightening directions printed on the filter canister or packing box. Most filter manufacturers recommend against using a filter wrench due to the possibility of overtightening and damage to the seal.

15 Remove all tools, rags, etc. from under the vehicle, being careful not to spill the oil in the drain pan, then lower the vehicle.

16 Add new oil to the engine through the oil filler cap in the valve cover. Use a funnel to prevent oil from spilling onto the top of the engine. Pour four quarts of fresh oil into the engine. Wait a few minutes to allow the oil to drain into the pan, then check the level on the oil dipstick (see Section 4 if necessary). If the oil level is at or near the H mark, install the filler cap hand tight, start the engine and allow the new oil to circulate.

17 Allow the engine to run for about a minute. While the engine is running, look under the vehicle and check for leaks at the oil pan drain plug and around the oil filter. If either is leaking, stop the engine and tighten the plug or filter slightly.

18 Wait a few minutes to allow the oil to trickle down into the pan, then recheck the level on the dipstick and, if necessary, add enough oil to bring the level to the upper mark.

19 During the first few trips after an oil change, make it a point to check frequently for leaks and proper oil level.

20 The old oil drained from the engine cannot be reused in its present state and should be disposed of. Check with your local auto parts store, disposal facility or environmental agency to see if they will accept the oil for recycling. After the oil has cooled it can be drained into a container (capped plastic jugs, topped bottles, milk cartons, etc.) for transport to one of these disposal sites. Don't dispose of the oil by pouring it on the ground or down a drain!

7.14 Lubricate the oil filter gasket with clean engine oil before installing the filter on the engine

8 Battery check, maintenance and charging (every 7500 miles [12,000 km] or 6 months)

WARNING:

Certain precautions must be followed when checking and servicing the battery. Hydrogen gas, which is highly flammable, is always present in the battery cells, so keep lighted tobacco and all other open flames and sparks away from the battery. The electrolyte inside the battery is actually dilute sulfuric acid, which will cause injury if splashed on your skin or in your eyes. It will also ruin clothes and painted surfaces. When removing the battery cables, always detach the negative cable first and hook it up last!

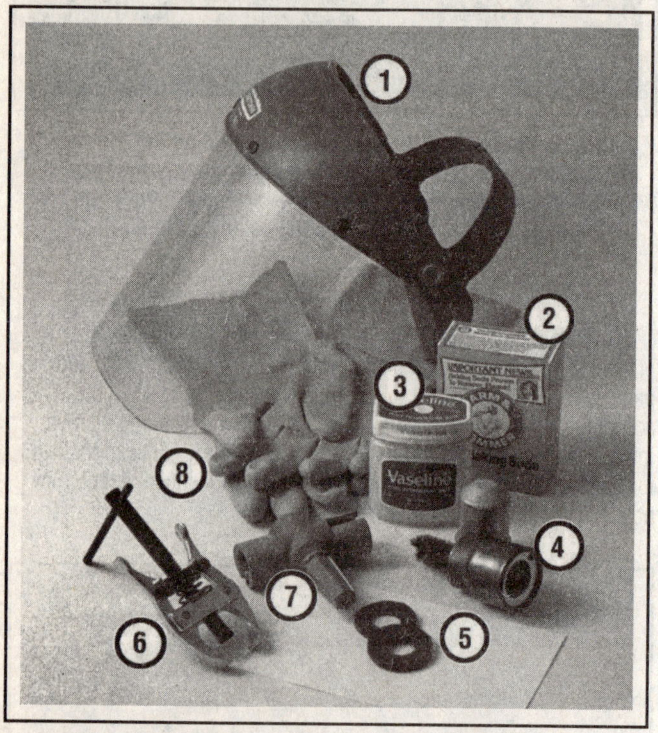

8.1 Tools and materials required for battery maintenance

1 ***Face shield/safety goggles*** - *When removing corrosion with a brush, the acidic particles can easily fly up into your eyes*
2 ***Baking soda*** - *A solution of baking soda and water can be used to neutralize corrosion*
3 ***Petroleum jelly*** - *A layer of this on the battery posts will help prevent corrosion*
4 ***Battery post/cable cleaner*** - *This wire brush cleaning tool will remove all traces of corrosion from the battery posts and cable clamps*
5 ***Treated felt washers*** - *Placing one of these on each post, directly under the cable clamps, will help prevent corrosion*
6 ***Puller*** - *Sometimes the cable clamps are very difficult to pull off the posts, even after the nut/bolt has been completely loosened. This tool pulls the clamp straight up and off the post without damage*
7 ***Battery post/cable cleaner*** - *Here is another cleaning tool which is a slightly different version of Number 4 above, but it does the same thing*
8 ***Rubber gloves*** - *Another safety item to consider when servicing the battery; remember that's acid inside the battery!*

MAINTENANCE

Refer to illustrations 8.1, 8.6a, 8.6b, 8.7a and 8.7b

1 A routine preventive maintenance program for the battery in your vehicle is the only way to ensure quick and reliable starts. But before performing any battery maintenance, make sure that you have the proper equipment necessary to work safely around the battery (see illustration).

2 There are also several precautions that should be taken whenever battery maintenance is performed. Before servicing the battery, always turn the engine and all accessories off and disconnect the cable from the negative terminal of the battery.

3 The battery produces hydrogen gas, which is both flammable and explosive. Never create a spark, smoke or light a match around the battery. Always charge the battery in a ventilated area.

4 Electrolyte contains poisonous and corrosive sulfuric acid. Do not allow it to get in your eyes, on your skin on your clothes. Never ingest it. Wear protective safety glasses when working near the battery. Keep children away from the battery.

5 Note the external condition of the battery. If the positive terminal and cable clamp on your vehicle's battery is equipped with a rubber protector, make sure that it's not torn or damaged. It should completely cover the terminal. Look for any corroded or loose connections, cracks in the case or cover or loose hold-down clamps. Also check the entire length of each cable for cracks and frayed conductors.

6 If corrosion, which looks like white, fluffy deposits (see illustration) is evident, particularly around the terminals, the battery should be removed for cleaning. Loosen the cable clamp bolts with a wrench, being careful to remove the ground cable first, and slide them off the terminals (see illustration). Disconnect the hold-down clamp bolt and nut, remove the clamp and lift the battery from the engine compartment.

7 Clean the cable clamps thoroughly with a battery brush or a terminal cleaner and a solution of warm water and baking soda (see illustration). Wash the terminals and the top of the battery case with the same solution but make sure that the solution doesn't get into the battery. When cleaning the cables, terminals and battery top, wear safety goggles and rubber gloves to prevent any solution from coming in contact with your eyes or hands. Wear old clothes too - even diluted, sulfuric acid splashed onto clothes will burn holes in them. If the terminals

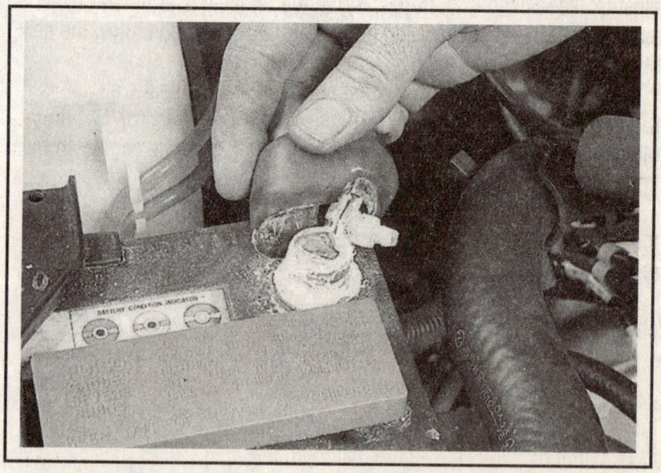

8.6a Battery terminal corrosion usually appears as light, fluffy powder

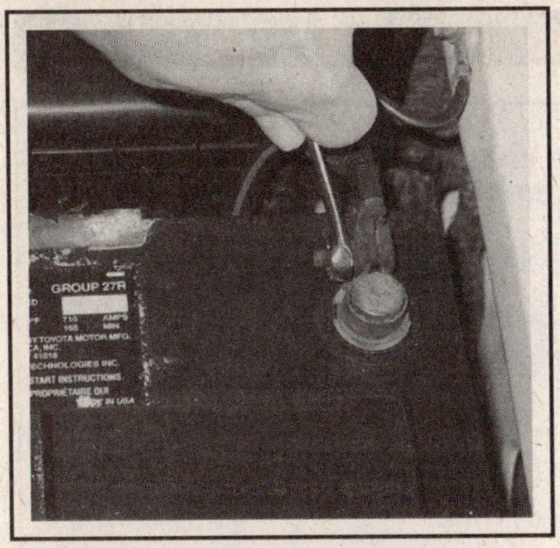

8.6b Removing the cable from a battery post with a wrench - sometimes special battery pliers are required for this procedure if corrosion has caused deterioration of the nut hex (always remove the ground cable first and hook it up last!)

8.7a When cleaning the cable clamps, all corrosion must be removed (the inside of the clamp is tapered to match the taper on the post, so don't remove too much material)

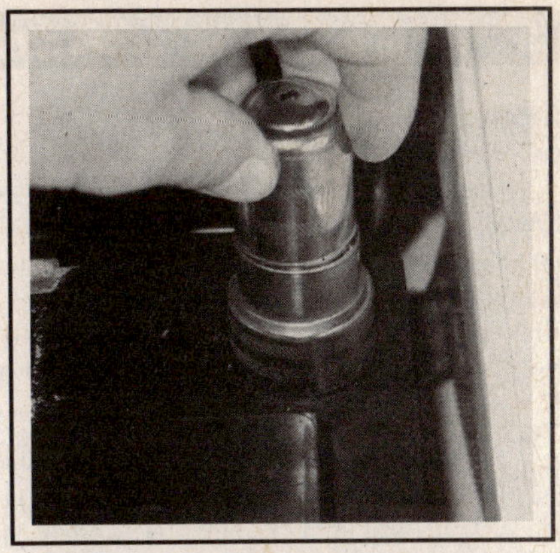

8.7b Regardless of the type of tool used on the battery posts, a clean, shiny surface should be the end result

have been extensively corroded, clean them up with a terminal cleaner (see illustration). Thoroughly wash all cleaned areas with plain water.

8 Make sure that the battery tray is in good condition and the hold-down clamp bolts are tight. If the battery is removed from the tray, make sure no parts remain in the bottom of the tray when the battery is reinstalled. When reinstalling the hold-down clamp bolts, do not over-tighten them.

9 Any metal parts of the vehicle damaged by corrosion should be covered with a zinc-based primer, then painted.

10 Information on removing and installing the battery can be found in Chapter 5. Information on jump starting can be found at the front of this manual.

CHARGING

WARNING:

When batteries are being charged, hydrogen gas, which is very explosive and flammable, is produced. Do not smoke or allow open flames near a battery. Wear eye protection when near the battery during charging. Also, make sure the charger is unplugged before connecting or disconnecting the battery from the charger.

➡Note: The manufacturer recommends the battery be removed from the vehicle for charging because the gas that escapes during this procedure can damage the paint. Fast charging with the battery cables connected can result in damage to the electrical system.

11 Slow-rate charging is the best way to restore a battery that's discharged to the point where it will not start the engine. It's also a good way to maintain the battery charge in a vehicle that's only driven a few miles between starts. Maintaining the battery charge is particularly important in the winter when the battery must work harder to start the engine and electrical accessories that drain the battery are in greater use.

12 It's best to use a one or two-amp battery charger (sometimes called a "trickle" charger). They are the safest and put the least strain on the battery. They are also the least expensive. For a faster charge, you can use a higher amperage charger, but don't use one rated more than 1/10th the amp/hour rating of the battery. Rapid boost charges that claim to restore the power of the battery in one to two hours are hardest on the battery and can damage batteries not in good condition. This type of charging should only be used in emergency situations.

13 The average time necessary to charge a battery should be listed in the instructions that come with the charger. As a general rule, a trickle charger will charge a battery in 12 to 16 hours.

14 Remove all the cell caps (if equipped) and cover the holes with a clean cloth to prevent spattering electrolyte. Disconnect the negative battery cable and hook the battery charger cable clamps up to the battery posts (positive to positive, negative to negative), then plug in the charger. Make sure it is set at 12-volts if it has a selector switch.

15 If you're using a charger with a rate higher than two amps, check the battery regularly during charging to make sure it doesn't overheat. If you're using a trickle charger, you can safely let the battery charge overnight after you've checked it regularly for the first couple of hours.

16 If the battery has removable cell caps, measure the specific gravity with a hydrometer every hour during the last few hours of the charging cycle. Hydrometers are available inexpensively from auto parts stores - follow the instructions that come with the hydrometer. Consider the battery charged when there's no change in the specific gravity reading for two hours and the electrolyte in the cells is gassing (bubbling) freely. The specific gravity reading from each cell should be very close to the others. If not, the battery probably has a bad cell(s).

17 Some batteries with sealed tops have built-in hydrometers on the top that indicate the state of charge by the color displayed in the hydrometer window. Normally, a bright-colored hydrometer indicates a full charge and a dark hydrometer indicates the battery still needs charging.

18 If the battery has a sealed top and no built-in hydrometer, you can hook up a voltmeter across the battery terminals to check the charge. A fully charged battery should read 12.6 volts or higher after the surface charge has been removed.

19 Further information on the battery and jump starting can be found in Chapter 5 and at the front of this manual.

9 Cooling system check (every 7500 miles [12,000 km] or 6 months)

➧ **Refer to illustration 9.4**

1 Many major engine failures can be attributed to a faulty cooling system. If the vehicle is equipped with an automatic transmission, the cooling system also cools the transmission fluid and thus plays an important role in prolonging transmission life.

2 The cooling system should be checked with the engine cold. Do this before the vehicle is driven for the day or after it has been shut off for at least three hours.

3 Remove the radiator cap by turning it to the left until it reaches a stop. If you hear a hissing sound (indicating there is still pressure in the system), wait until this stops. Now press down on the cap with the palm of your hand and continue turning to the left until the cap can be removed. Thoroughly clean the cap, inside and out, with clean water. Also clean the filler neck on the radiator. All traces of corrosion should be removed. The coolant inside the radiator should be relatively transparent. If it is rust colored, the system should be drained and refilled (see Section 25). If the coolant level is not up to the top, add additional antifreeze/coolant mixture (see Section 4).

4 Carefully check the large upper and lower radiator hoses along with the smaller diameter heater hoses which run from the engine to the firewall. Inspect each hose along its entire length, replacing any hose which is cracked, swollen or shows signs of deterioration. Cracks may become more apparent if the hose is squeezed (see illustration). Regardless of condition, it's a good idea to replace hoses with new ones every two years.

5 Make sure all hose connections are tight. A leak in the cooling system will usually show up as white or rust colored deposits on the areas adjoining the leak. If wire-type clamps are used at the ends of the hoses, it may be a good idea to replace them with more secure screw-type clamps.

6 Use compressed air or a soft brush to remove bugs, leaves, etc. from the front of the radiator or air conditioning condenser. Be careful not to damage the delicate cooling fins or cut yourself on them.

7 Every other inspection, or at the first indication of cooling system problems, have the cap and system pressure tested. If you don't have a pressure tester, most gas stations and repair shops will do this for a minimal charge.

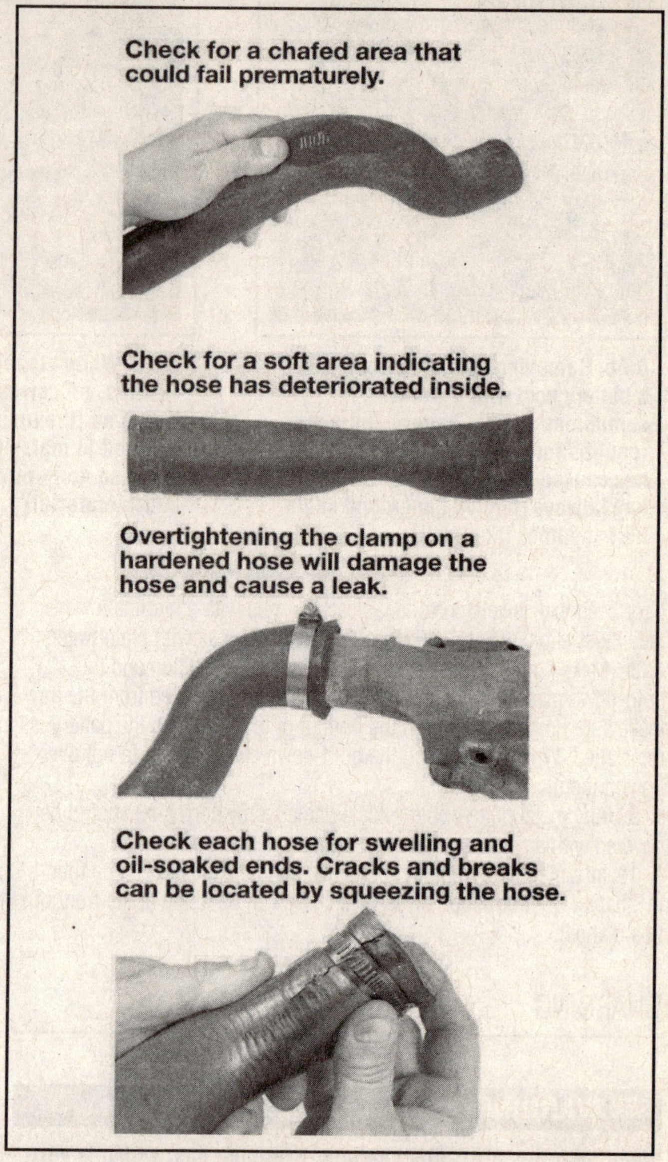

9.4 Hoses, like drivebelts, have a habit of failing at the worst possible time - to prevent the inconvenience of a blown radiator or heater hose, inspect them carefully as shown here

10 Underhood hose check and replacement (every 7500 miles [12,000 km] or 6 months)

GENERAL

⁂ WARNING:

Replacement of air conditioning hoses must be left to a dealer service department or air conditioning shop that has the equipment to depressurize the system safely. Never remove air conditioning components or hoses until the system has been depressurized.

1 High temperatures in the engine compartment can cause the deterioration of the rubber and plastic hoses used for engine, accessory and emission systems operation. Periodic inspection should be made for cracks, loose clamps, material hardening and leaks. Information specific to the cooling system hoses can be found in Section 9.

2 Some, but not all, hoses are secured to the fittings with clamps. Where clamps are used, check to be sure they haven't lost their tension, allowing the hose to leak. If clamps aren't used, make sure the hose has not expanded and/or hardened where it slips over the fitting, allowing it to leak.

VACUUM HOSES

3 It's quite common for vacuum hoses, especially those in the emissions system, to be color coded or identified by colored stripes molded into them. Various systems require hoses with different wall thickness, collapse resistance and temperature resistance. When replacing hoses, be sure the new ones are made of the same material.

4 Often the only effective way to check a hose is to remove it completely from the vehicle. If more than one hose is removed, be sure to label the hoses and fittings to ensure correct installation.

5 When checking vacuum hoses, be sure to include any plastic T-fittings in the check. Inspect the fittings for cracks and the hose where it fits over the fitting for distortion, which could cause leakage.

6 A small piece of vacuum hose (1/4-inch inside diameter) can be used as a stethoscope to detect vacuum leaks. Hold one end of the hose to your ear and probe around vacuum hoses and fittings, listening for the hissing sound characteristic of a vacuum leak.

WARNING:

When probing with the vacuum hose stethoscope, be very careful not to come into contact with moving engine components such as the drivebelt, cooling fan, etc.

FUEL HOSE

WARNING:

Gasoline is extremely flammable, so take extra precautions when you work on any part of the fuel system. Don't smoke or allow open flames or bare light bulbs near the work area, and don't work in a garage where a gas-type appliance (such as a water heater or clothes dryer) is present. Since gasoline is carcinogenic, wear fuel-resistant gloves when there's a possibility of being exposed to fuel, and, if you spill any fuel on your skin, rinse it off immediately with soap and water. Mop up any spills immediately and do not store fuel-soaked rags where they could ignite. The fuel system is under constant pressure, so if any fuel lines are to be disconnected, the fuel pressure in the system must be relieved first (see Chapter 4). When you perform any kind of work on the fuel system, wear safety glasses and have a Class B Type fire extinguisher on hand.

7 Check all rubber fuel lines for deterioration and chafing. Check especially for cracks in areas where the hose bends and just before fittings, such as where a hose attaches to the fuel filter.

8 Only high quality fuel line, designed for high-pressure fuel injection systems, may be used for fuel line replacement. Never, under any circumstances, use standard fuel hose, unreinforced vacuum line, clear plastic tubing or water hose for fuel lines.

9 Spring-type clamps are commonly used on fuel lines. These clamps often lose their tension over a period of time, and can be sprung during removal. Replace all spring-type clamps with screw clamps whenever a hose is replaced.

METAL LINES

10 Sections of metal line are often used in the fuel system. Check carefully to be sure the line has not been bent or crimped and that cracks have not started in the line.

11 If a section of metal fuel line must be replaced, only seamless steel tubing should be used, since copper and aluminum tubing don't have the strength necessary to withstand normal engine vibration.

12 Check the metal brake lines where they enter the master cylinder and brake proportioning unit (if used) for cracks in the lines or loose fittings. Any sign of brake fluid leakage calls for an immediate thorough inspection of the brake system.

11 Wiper blade inspection and replacement (every 7500 miles [12,000 km] or 6 months)

Refer to illustrations 11.3, 11.5 and 11.6

1 The windshield wiper and blade assembly should be inspected periodically for damage, loose components and cracked or worn blade elements.

2 Road film can build up on the wiper blades and affect their efficiency, so they should be washed regularly with a mild detergent solution.

3 The action of the wiping mechanism can loosen bolts, nuts and fasteners, so they should be checked and tightened, as necessary (see illustration), at the same time the wiper blades are checked.

4 If the wiper blade elements are cracked, worn or warped, or no longer clean adequately, they should be replaced with new ones.

5 Lift the arm assembly away from the glass for clearance, press on the release lever, then slide the wiper blade assembly out of the hook in the end of the arm (see illustration).

6 Use needle-nose pliers to compress the blade element, then slide the element out of the frame and discard it (see illustration).

7 Installation is the reverse of removal.

11.3 Pry open the trim cap and check the tightness of the wiper arm retaining nut

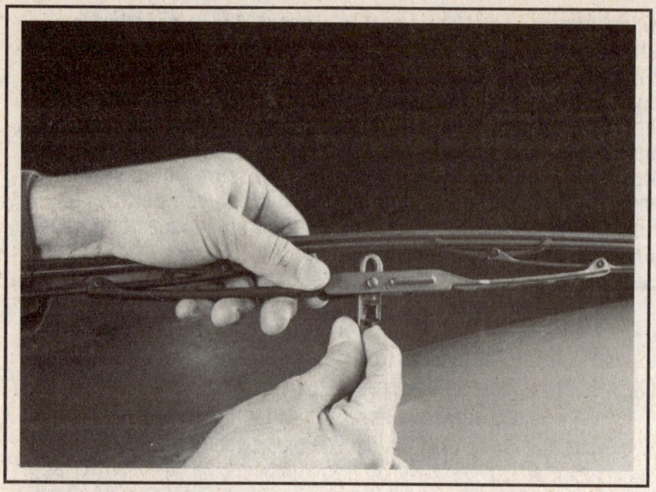

11.5 Press on the release tab, then push the blade assembly down and out of the hook in the arm

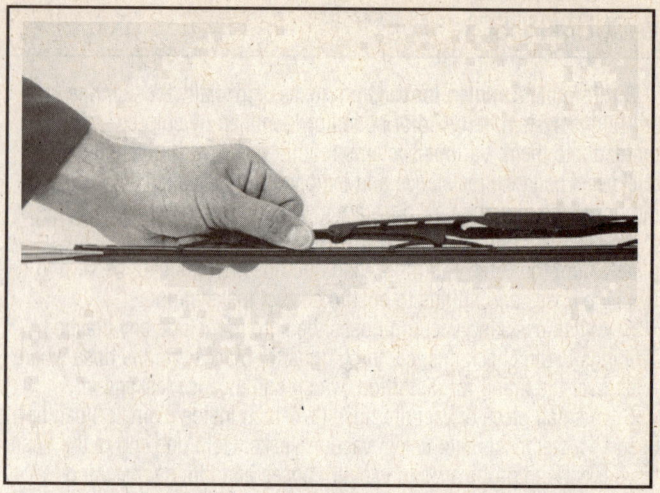

11.6 Use needle-nose pliers to compress the rubber element, then slide the element out - slide the new element in and lock the blade assembly fingers into the notches of the wiper element

12 Tire rotation (every 7500 miles [12,000 km] or 6 months)

Refer to illustrations 12.2a and 12.2b

1 The tires should be rotated at the specified intervals and whenever uneven wear is noticed. Since the vehicle will be raised and the tires removed anyway, check the brakes (see Section 23) at this time.

2 Radial tires must be rotated in a specific pattern (see illustrations).

Note: Models with a spare tire of the same type use the five tire rotation pattern. All other types use the four tire rotation pattern.

3 Refer to the information in *Jacking and towing* at the front of this manual for the proper procedures to follow when raising the vehicle and changing a tire. If the brakes are to be checked, do not apply the parking brake as stated. Make sure the tires are blocked to prevent the vehicle from rolling.

4 Preferably, the entire vehicle should be raised at the same time. This can be done on a hoist or by jacking up each corner, then lowering the vehicle onto jackstands placed under the frame rails. Always use four jackstands and make sure the vehicle is firmly supported.

5 After rotation, check and adjust the tire pressures as necessary and be sure to check the lug nut tightness.

6 For further information on the wheels and tires, refer to Chapter 10.

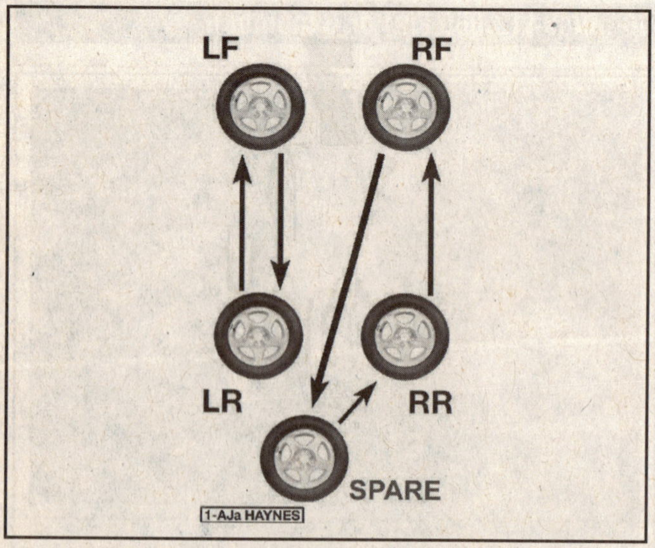

12.2a Tire rotation diagram for radial tires with a spare tire of the same type

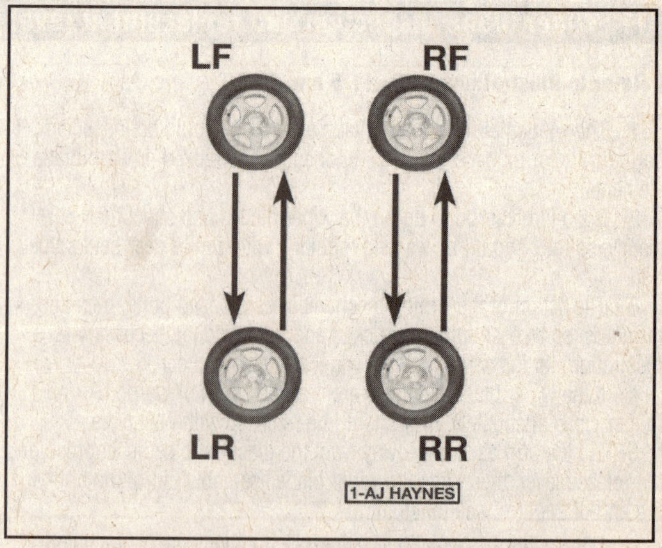

12.2b The recommended rotation pattern for radial tires with a spare tire that is different from the installed tires

13 Suspension and steering check (every 7500 miles or [12,000 km] 6 months)

➡Note: For detailed illustrations of the steering and suspension components, refer to Chapter 10.

WITH THE WHEELS ON THE GROUND

➧ Refer to illustration 13.1

1 With the vehicle stopped and the front wheels pointed straight ahead, rock the steering wheel gently back and forth. If freeplay (see illustration) is excessive, a front wheel bearing, main shaft U-joint, intermediate shaft U-joint or tie rod end is worn or the steering gear is out of adjustment or worn. Refer to Chapter 10 for the appropriate repair procedure.

2 Other symptoms, such as excessive vehicle body movement over rough roads, swaying (leaning) around corners and binding as the steering wheel is turned, may indicate faulty steering and/or suspension components.

3 Check the shock absorbers by pushing down and releasing the vehicle several times at each corner. If the vehicle does not come back to a level position within one or two bounces, the shocks are worn and must be replaced. When bouncing the vehicle up and down, listen for squeaks and noises from the suspension components.

UNDER THE VEHICLE

➧ Refer to illustrations 13.6 and 13.7

4 Raise the vehicle with a floor jack and support it securely on jackstands. See *Jacking and towing* at the front of this book for proper jacking points.

5 Check the tires for irregular wear patterns and proper inflation. See Section 5 in this Chapter for information regarding tire wear.

6 Inspect the universal joint between the steering shaft and the steering rack. Check the steering rack and driveaxle boots for grease leakage. Check the steering linkage for looseness or damage. Check the tie-rod ends for excessive play. Look for loose bolts, broken or disconnected parts and deteriorated rubber bushings on all suspension and steering components (see illustration). While an assistant turns the steering wheel from side to side, check the steering components for free movement, chafing and binding. If the steering components do not seem to be reacting with the movement of the steering wheel, try to determine where the slack is located.

7 Check the wheel bearings. Do this by spinning the front wheels. Listen for any abnormal noises and watch to make sure the wheel spins true (doesn't wobble). Grab the top and bottom of the tire and pull in-and-out on it. Notice any movement which would indicate a loose wheel bearing assembly (see illustration). If the bearings are loose, they are in need of replacement. Refer to Chapter 10 for more information.

8 Check the steering knuckle, moving the knuckle up and down with a prybar to ensure that knuckle bearings have no play. If the bearings are suspect, they should be checked and repacked. Refer to Chapter 10 for more information.

9 Inspect the driveshafts for worn U-joints and for excessive play in the slip yoke and spline area (see Chapter 8).

10 Check the transfer case and differentials for evidence of fluid leakage.

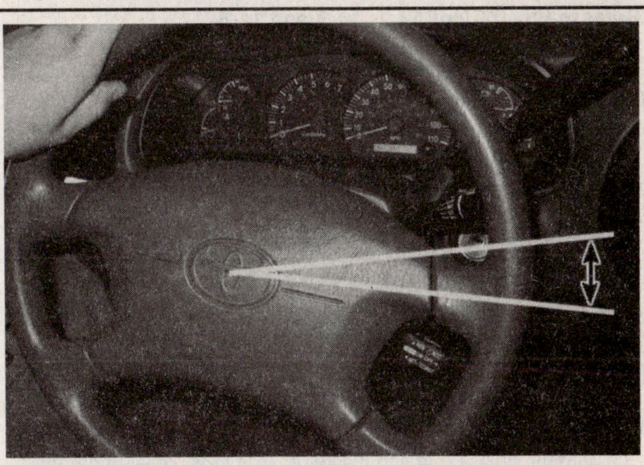

13.1 Steering wheel freeplay is the amount of travel between an initial steering input and the point at which the front wheels begin to turn (indicated by a slight resistance)

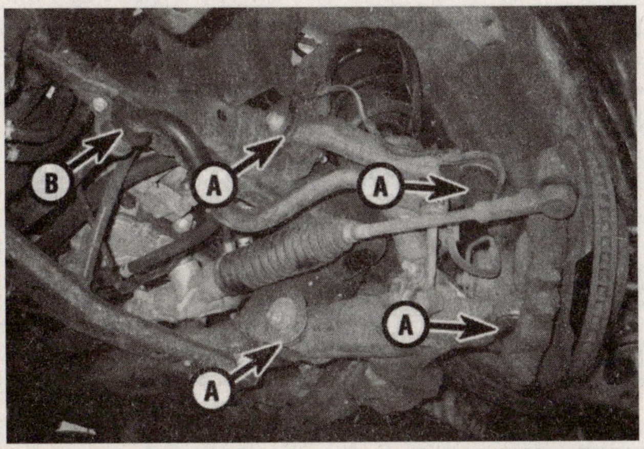

13.6 Inspect the suspension for deteriorated rubber bushings and torn grease seals (A) and check the stabilizer bar bushings for deterioration at the front and the rear of the vehicle (B)

13.7 Grasp the tire as shown and check for endplay at the wheel bearings - if endplay is found, the wheel bearings must be replaced

14 Driveshaft and body lubrication (every 7500 miles [12,000 km] or 6 months)

➧ **Refer to illustrations 14.1, 14.5 and 14.8**

1 Refer to *Recommended lubricants and fluids* in this Chapter's Specifications to obtain the necessary grease, etc. You will also need a grease gun (see illustration).

2 Look under the vehicle for grease fittings on the driveline components (4WD models). They are normally found on the driveshaft universal joints and slip yokes.

3 For easier access under the vehicle, raise it with a jack and place jackstands under the frame. Make sure it is safely supported by the stands. If the wheels are to be removed at this time for tire rotation or brake inspection, loosen the lug nuts slightly while the vehicle is still on the ground.

4 Before beginning, force a little grease out of the nozzle to remove any dirt from the end of the gun. Wipe the nozzle clean with a rag.

5 With the grease gun and plenty of clean rags, crawl under the vehicle and begin lubricating the driveshaft universal joints (see illustration).

6 Wipe the area around the grease fitting free of dirt, then squeeze the trigger on the grease gun to force grease into the component. Continue pumping grease into the fitting until it just oozes out of the bearing cup seals. If it escapes around the grease gun nozzle, the fitting is clogged or the nozzle is not completely seated on the fitting. Resecure the gun nozzle to the fitting and try again. If necessary, replace the fitting with a new one.

7 Wipe the excess grease from the components and the grease fitting. Repeat the procedure for the remaining fittings.

8 Lubricate the driveshaft slip yoke by pumping grease into the fitting until it can be seen coming out of the slip yoke seal (see illustration).

9 While you are under the vehicle, clean and lubricate the parking brake cable along with the cable guides and levers. This can be done by smearing some chassis grease onto the cable and its related parts with your fingers.

10 Lubricate the contact points on the steering knuckle stop and adjustment bolt if equipped.

11 Open the hood and smear a little chassis grease on the hood latch mechanism. Have an assistant pull the hood release lever from inside the vehicle as you lubricate the cable at the latch.

12 Lubricate all the hinges (door, hood, etc.) with engine oil to keep them in proper working order.

13 The key lock cylinders can be lubricated with spray-on graphite or silicone lubricant, which is available at auto parts stores.

14 Lubricate the door weatherstripping with silicone spray. This will reduce chafing and retard wear.

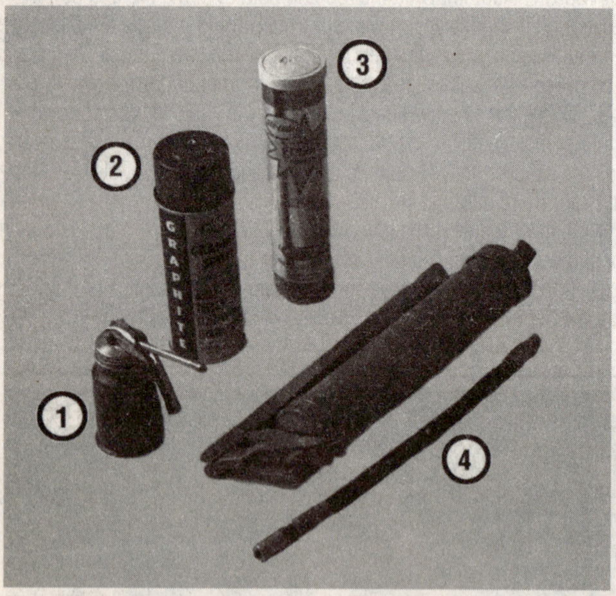

14.1 Materials required for chassis and body lubrication

1 ***Engine oil*** - *Light engine oil in a can like this can be used for door and hood hinges*
2 ***Graphite spray*** - *Used to lubricate lock cylinders*
3 ***Grease*** - *Grease, in a variety of types and weights, is available for use in a grease gun. Check the Specifications for your requirements*
4 ***Grease gun*** - *A common grease gun, shown here with a detachable hose and nozzle, is needed for chassis lubrication. After use, clean it thoroughly*

14.5 Pump grease into the universal joints until it can be seen coming out from the seals

14.8 The slip joint grease fitting is located on the yoke - pump grease into it until it comes out of the slip joint seal

15 Exhaust system check (every 7500 miles [12,000 km] or 6 months)

Refer to illustrations 15.2a and 15.2b

1 With the engine cold (at least three hours after the vehicle has been driven), check the complete exhaust system from the manifold to the end of the tailpipe. Be careful around the catalytic converter (if equipped), which may be hot even after three hours. The inspection should be done with the vehicle on a hoist to permit unrestricted access. If a hoist isn't available, raise the vehicle and support it securely on jackstands.

2 Check the exhaust pipes and connections for signs of leakage and/or corrosion indicating a potential failure. Make sure that all brackets and hangers are in good condition and tight (see illustrations).

3 Inspect the underside of the body for holes, corrosion, open seams, etc. which may allow exhaust gasses to enter the passenger compartment. Seal all body openings with silicone sealant or body putty.

4 Rattles and other noises can often be traced to the exhaust system, especially the hangers, mounts and heat shields. Try to move the pipes, mufflers and catalytic converter. If the components can come in contact with the body or suspension parts, secure the exhaust system with new brackets and hangers.

15.2a Check the exhaust pipes and connections for signs of leakage and corrosion

15.2b Check the exhaust system rubber hangers for cracks and damage

16 Transfer case lubricant level check (4WD models) (every 7500 miles [12,000 km] or 6 months)

Refer to illustration 16.2

1 The transfer case lubricant level is checked by removing the upper plug located in the back of the case.

2 Use a finger to reach inside the housing to determine the lubricant level. The lubricant level should be just at the bottom of the hole (see illustration). If not, add the appropriate lubricant through the opening.

3 Install and tighten the plug and check for leaks after the first few miles of driving.

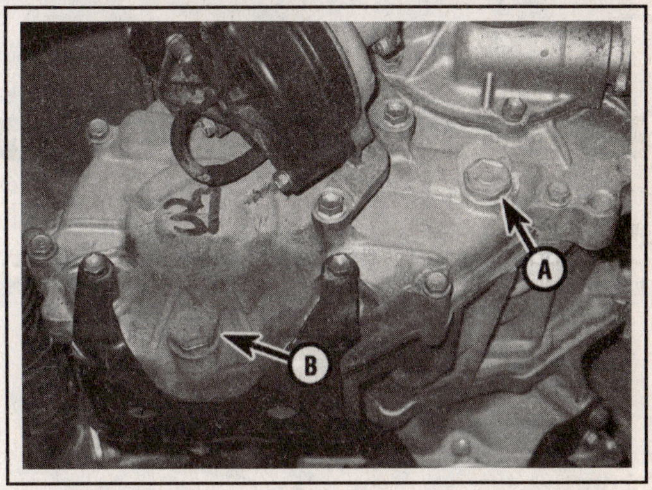

16.2 Typical locations of the transfer case fill plug (A) and drain plug (B) - use your finger as a dipstick to check the lubricant level

17 Differential and manual transmission lubricant level check (every 7500 miles [12,000 km] or 6 months)

DIFFERENTIAL(S)

Refer to illustration 17.2

Note: 4WD models covered by this manual have two differentials; be sure to check the lubricant level in both differentials.

1 The differential has a check/fill plug which must be removed to check the lubricant level. If the vehicle must be raised to gain access to the plug, be sure to support it safely on jackstands - DO NOT crawl under the vehicle when it's supported only by the jack.

2 Remove the oil check/fill plug from the back of the rear differential or the front of the front differential (see illustration). On some models, a tag is located in the area of the plug which gives information regarding lubricant type, particularly on models equipped with a limited-slip differential.

3 Use a finger to reach inside the housing to determine the lubricant level. The oil level should be at the bottom of the plug opening. If it isn't, use a hand pump (available at auto parts stores) to add the specified lubricant until it just starts to run out of the opening.

4 Install the plug and tighten it securely.

MANUAL TRANSMISSION

5 The manual transmission has a filler plug which must be removed to check the lubricant level. If the vehicle is raised to gain access to the plug, be sure to support it safely on jackstands - DO NOT crawl under a vehicle which is supported only by a jack! Be sure the vehicle is level or the check may be inaccurate.

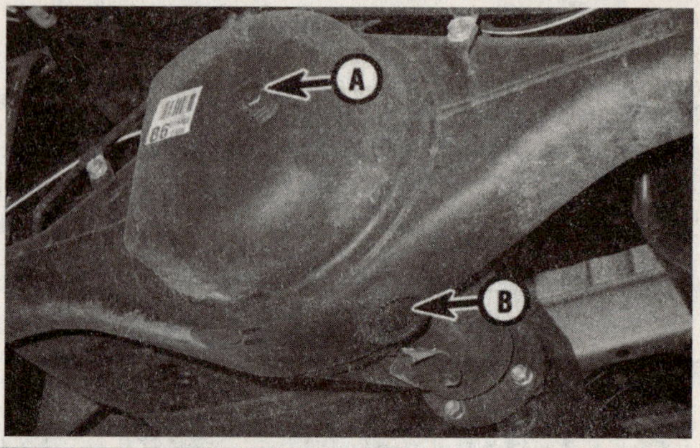

17.2 The differential fill plug (A) and drain plug (B) are located on the axle housing - use your finger as a dipstick to check the lubricant level

6 Unscrew the fill plug from the transmission and use a finger to reach inside the housing to determine the lubricant level. The level should be at or near the bottom of the plug hole.

7 If it isn't, add the recommended lubricant through the plug hole with a pump or squeeze bottle.

8 Install and tighten the plug and check for leaks after the first few miles of driving.

18 Seat belt check (every 7500 miles [12,000 km] or 6 months)

1 Check the seat belts, buckles, latch plates and guide loops for any obvious damage or signs of wear.

2 Make sure the seat belt reminder light comes on when the key is turned on.

3 The seat belts are designed to lock up during a sudden stop or impact, yet allow free movement during normal driving. The retractors should hold the belt against your chest while driving and rewind the belt when the buckle is unlatched.

4 If any of the above checks reveal problems with the seat belt system, replace parts as necessary.

19 Driveaxle boot check (every 15,000 miles [24,000 km] or 12 months)

➧ Refer to illustration 19.2

1 The driveaxle boots are very important because they prevent dirt, water and foreign material from entering and damaging the constant velocity joints (CV).

2 Inspect the boots for tears and cracks as well as loose clamps (see illustration). If there is any evidence of cracks or leaking grease, they must be replaced as described in Chapter 8.

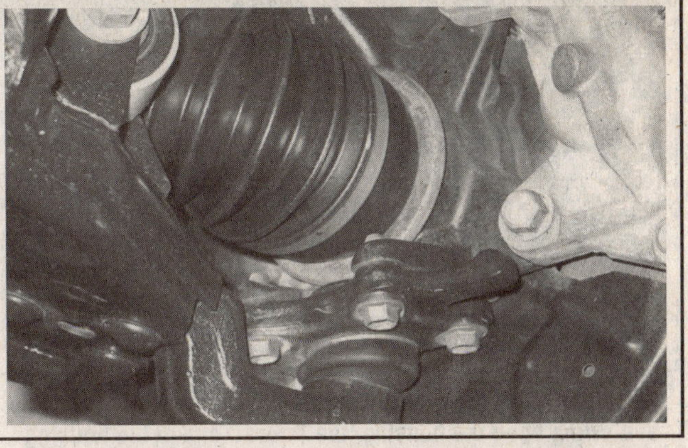

19.2 Check the condition of the driveaxle boot for signs of cracks and grease leaks

20 Air filter check and replacement (every 15,000 miles [24,000 km] or 12 months)

1 At the specified intervals, the air filter should be replaced with a new one. A thorough preventive maintenance schedule would also require the filter to be inspected between filter changes.

FOUR-CYLINDER MODELS

➧ Refer to illustrations 20.3 and 20.4

2 The air filter housing is mounted in the right side of the engine compartment.

3 Unlatch the cover retaining clips (see illustration). Detach all hoses and electrical connectors that would interfere with the removal of the air filter cover from the housing. While the top cover is off, be careful not to drop anything down into the housing.

4 Lift the air filter element out of the housing and wipe out the inside of the housing with a clean rag (see illustration).

5 Inspect the outer surface of the filter element. If it is dirty, replace it. If it is only moderately dusty, it can be reused by blowing it clean from the back to the front surface with compressed air. Because it is a pleated paper type filter, it cannot be washed or oiled. If it cannot be cleaned satisfactorily with compressed air, discard and replace it.

CAUTION:

Never drive the vehicle with the air filter removed. Excessive engine wear could result and backfiring could even cause a fire under the hood.

6 Place the new filter in the air filter housing, making sure it seats properly.

7 The remainder of installation is the reverse of removal.

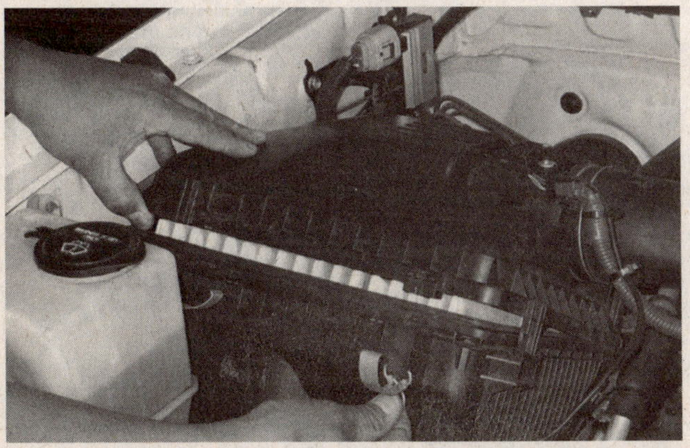

20.3 Unlatch the cover retaining clips (four-cylinder models)

20.4 Lift the air filter element out of the housing (four-cylinder models)

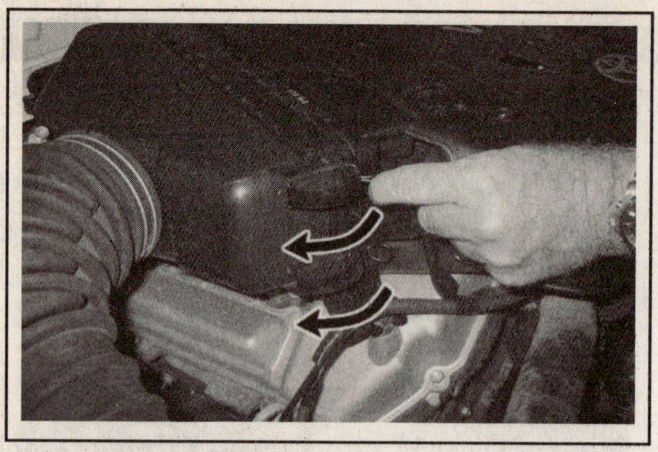

20.9 To access the air filter element, release these two clips and swing open the housing (the firewall end of the housing is hinged)

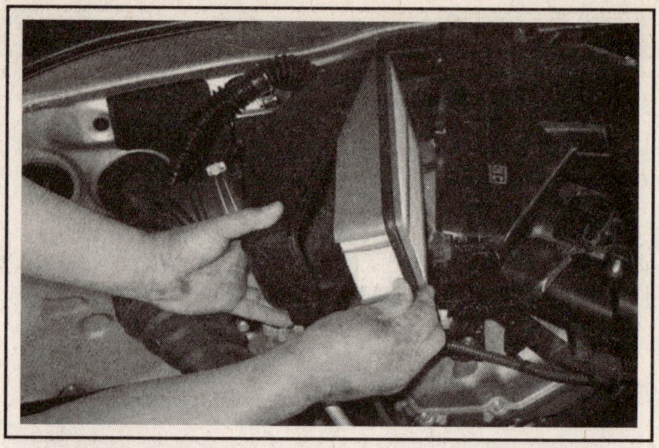

20.10 Note how the old air filter is installed, then pull it out of the air filter housing. The new filter must be installed in exactly the same way

V6 MODELS

Refer to illustrations 20.9 and 20.10

8 The air filter housing is located on the right side of the upper intake manifold.

9 Open the two retaining clips (see illustration) and swing open the air filter housing.

10 Remove the air filter element from the housing (see illustration), then wipe out the inside of the housing with a clean rag.

11 Inspect the outer surface of the filter element. If it's dirty, replace it. If it's only moderately dusty, blow it out from the back to the front surface with low-pressure compressed air. Because the filter element is a pleated paper type filter, it cannot be oiled or washed. If you can't clean the filter element satisfactorily with compressed air, replace it.

***** CAUTION:**

Never drive the vehicle with the air filter removed. Excessive engine wear could result and backfiring could even cause a fire under the hood.

12 Installation is the reverse of removal.

21 Drivebelt check and replacement (every 15,000 miles [24,000 km] or 12 months)

***** WARNING:**

Before checking, adjusting or replacing a drivebelt, make sure the ignition key is not in the ignition lock cylinder.

CHECK

Refer to illustration 21.2

1 The drivebelt is located at the front of the engine and plays an important role in the operation of the vehicle and its components. Due to their function and material makeup, belts are prone to failure after a period of time and should be inspected and adjusted periodically to prevent major damage. All models use a single serpentine belt; no adjustment is necessary because an automatic tensioner is used.

2 With the engine turned off, open the hood and locate the drivebelt(s) at the front of the engine. Use a flashlight to carefully check for a severed core, separation of the adhesive rubber on both sides of the core and for core separation from the belt side. Inspect the ribs for separation from the adhesive rubber and for cracking or separation of the ribs, torn or worn ribs or cracks in the inner ridges of the ribs (see illustration). Also check for fraying and glazing, which gives the belt a

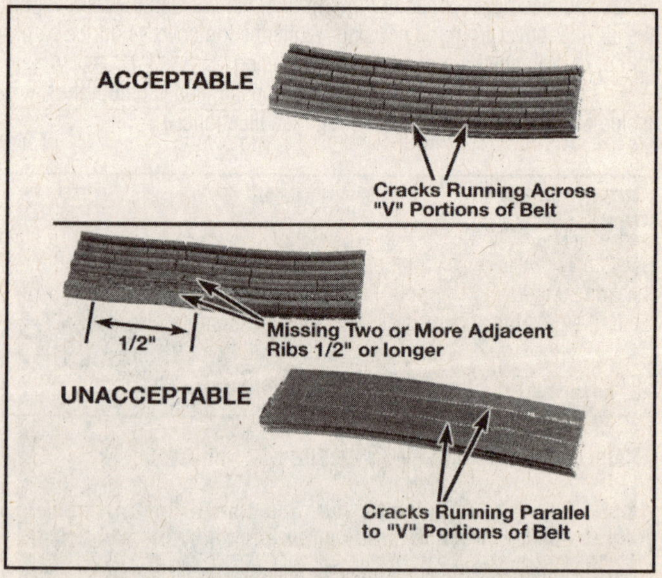

21.2 Small cracks in the underside of the V-ribbed belt are acceptable - lengthwise cracks or missing pieces are cause for replacement

21.3a To release tension on the drivebelt, place a socket on the tensioner's cast hexagonal stud on the arm of the tensioner and rotate it clockwise (four-cylinder models)

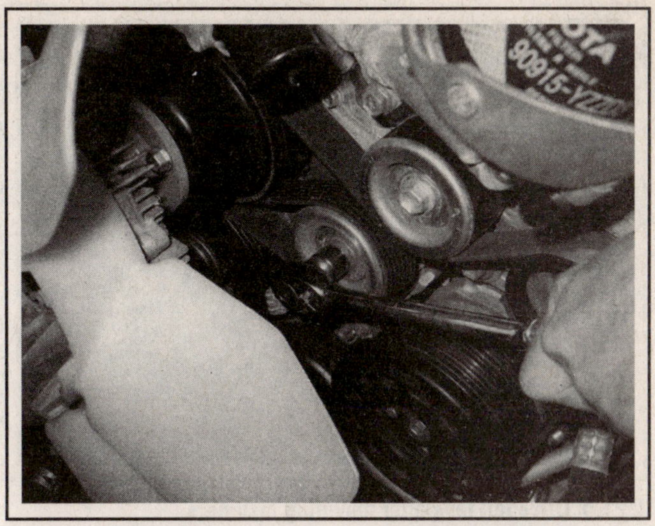

21.3b To release tension on the drivebelt, put a socket on the tensioner pulley bolt and rotate the tensioner counterclockwise (V6 models)

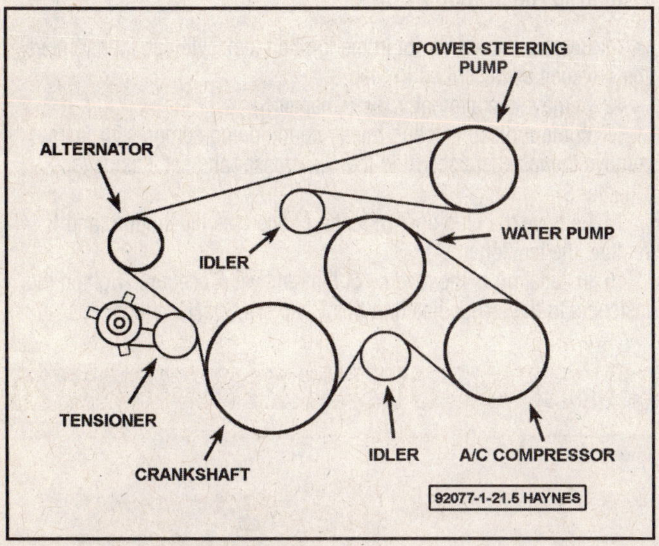

21.5 Drivebelt routing (four-cylinder models)

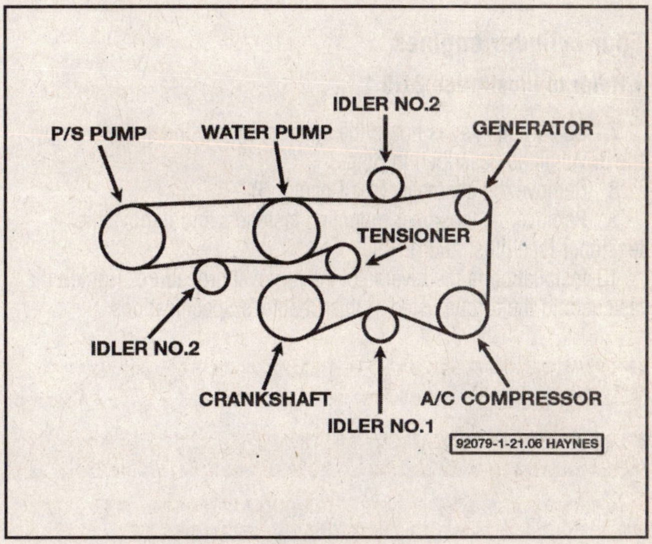

21.6 Drivebelt routing (V6 models)

shiny appearance. Inspect both sides of the belt by twisting the belt to check the underside. Use your fingers to feel the belt where you can't see it. If any of the above conditions are evident, replace the belt(s).

REPLACEMENT

➧ Refer to illustrations 21.3a, 21.3b, 21.5 and 21.6

➡Note: Take the old belt with you when purchasing new ones in order to make a direct comparison for length, width and design.

3 The automatic tensioner must be released to allow drivebelt replacement. Check to make sure the key has been removed from the ignition lock cylinder. If you're working on a four-cylinder engine, place a socket on the tensioner's cast hexagonal stud on the arm of the tensioner and rotate it clockwise to release tension on the belt. If you're working on a V6 engine, place a socket on the bolt in the center of the tensioner pulley and rotate it counterclockwise to release tension on the belt (see illustrations). Remove the belt and slowly release the tensioner.

4 If you are working on a V6 engine, before installing the drivebelt, lock the tensioner as follows: turn the tensioner counterclockwise, align the two holes on the tensioner assembly and insert a 0.24-inch dowel pin or drill bit through the two holes.

5 On four-cylinder engines, install the new belt and rotate the tensioner counterclockwise to allow the belt to slip over it, then release the tensioner slowly until it contacts the drivebelt. Make sure the drivebelt is centered on all of the pulleys (see illustration).

6 On V6 engines, install the drivebelt. Be sure to route it correctly. Remove the dowel pin or drill bit and tension the belt. Make sure that the belt is centered on all of the pulleys (see illustration).

21.9 Drivebelt tensioner mounting fasteners (four-cylinder models)

21.14 Drivebelt tensioner mounting fasteners (V6 models)

TENSIONER REPLACEMENT

Four-cylinder engines

➧ Refer to illustration 21.9

7 Be sure the key is not in the ignition lock cylinder, then remove the drivebelt as described in Step 3.

8 Remove the alternator (see Chapter 5).

9 Remove the tensioner mounting fasteners and remove the tensioner (see illustration).

10 Installation is the reverse of the removal procedure. Tighten the fasteners to the torque listed in this Chapter's Specifications.

V6 engines

➧ Refer to illustration 21.14

11 Be sure the key is not in the ignition lock cylinder, then remove the drivebelt as described in Step 3.

12 Remove the alternator (see Chapter 5).

13 Without disconnecting the air conditioning compressor lines, remove the fasteners securing the compressor and set it aside (see Chapter 3).

14 Remove the tensioner mounting bolts (see illustration) and remove the tensioner.

15 Installation is the reverse of the removal procedure. Tighten the fasteners to the torque listed in this Chapter's Specifications.

22 Fuel system check (every 15,000 miles [24,000 km] or 12 months)

➧ Refer to illustration 22.3

WARNING:

Gasoline is extremely flammable, so take extra precautions when you work on any part of the fuel system. Don't smoke or allow open flames or bare light bulbs near the work area, and don't work in a garage where a gas-type appliance (such as a water heater or clothes dryer) is present. Since gasoline is carcinogenic, wear fuel-resistant gloves when there's a possibility of being exposed to fuel, and, if you spill any fuel on your skin, rinse it off immediately with soap and water. Mop up any spills immediately and do not store fuel-soaked rags where they could ignite. The fuel system is under constant pressure, so, if any fuel lines are to be disconnected, the fuel pressure in the system must be relieved first (see Chapter 4 for more information). When you perform any kind of work on the fuel system, wear safety glasses and have a Class B type fire extinguisher on hand.

1 The fuel system is most easily checked with the vehicle raised on a hoist so the components underneath the vehicle are readily visible and accessible.

2 If the smell of gasoline is noticed while driving or after the vehicle has been in the sun, the system should be thoroughly inspected immediately.

3 Remove the fuel tank cap and check for damage, corrosion and an unbroken sealing imprint on the gasket. Replace the cap with a new one if necessary (see illustration).

4 With the vehicle raised and safely supported, inspect the gas tank and filler neck for punctures, cracks and other damage. The connection

between the filler neck and the tank is particularly critical. Sometimes a rubber filler neck will leak because of loose clamps or deteriorated rubber. These are problems a home mechanic can usually rectify.

WARNING:

Do not, under any circumstances, try to repair a fuel tank (except rubber components). A welding torch or any open flame can easily cause fuel vapors inside the tank to explode.

5 Carefully check all rubber hoses and metal lines leading away from the fuel tank. Check for loose connections, deteriorated hoses, crimped lines and other damage. Follow the lines to the front of the vehicle, carefully inspecting them all the way to the carburetor or fuel injection system. Repair or replace damaged sections as necessary.

6 If a fuel odor is still evident after the inspection, refer to Chapter 6 and check the EVAP system.

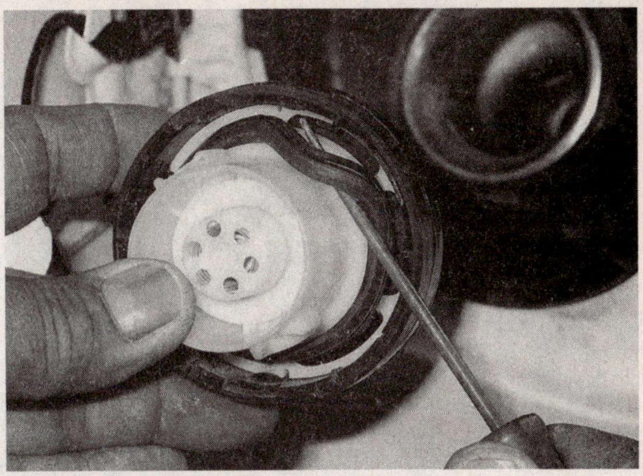

22.3 Use a small screwdriver to carefully pry out the old gasket - take care not to damage the cap

23 Brake check (every 15,000 miles [24,000 km] or 12 months)

WARNING:

The dust created by the brake system is harmful to your health. Never blow it out with compressed air and don't inhale any of it. An approved filtering mask should be worn when working on the brakes. Do not, under any circumstances, use petroleum-based solvents to clean brake parts. Use brake system cleaner only!

➡Note: For detailed photographs of the brake system, refer to Chapter 9.

1 In addition to the specified intervals, the brakes should be inspected every time the wheels are removed or whenever a defect is suspected. Any of the following symptoms could indicate a potential brake system defect: The vehicle pulls to one side when the brake pedal is depressed; the brakes make squealing or dragging noises when applied; brake pedal travel is excessive; the pedal pulsates; brake fluid leaks, usually onto the inside of the tire or wheel.

2 Loosen the wheel lug nuts.

3 Raise the vehicle and place it securely on jackstands.

4 Remove the wheels.

DISC BRAKES

➧ Refer to illustrations 23.5a and 23.5b

5 There are two pads (an outer and an inner) in each caliper. The pads are visible after the wheels are removed (see illustrations).

6 Measure the pad thickness. If the lining material is less than the minimum thickness listed in this Chapter's Specifications, replace the pads.

➡Note: Keep in mind that the lining material is riveted or bonded to a metal backing plate and the metal portion is not included in this measurement.

23.5a With the wheels removed, check the thickness of the inner pad through the inspection hole (front disc shown, rear disc caliper similar)

23.5b The outer pad is checked at the edge of the caliper

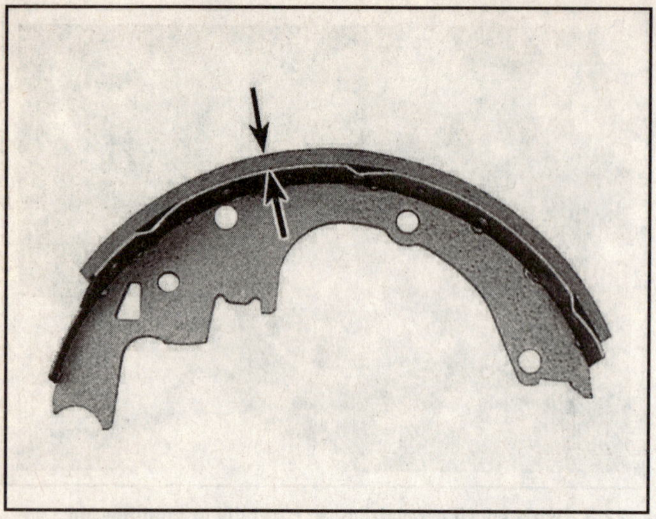

23.12 If the lining is bonded to the brake shoe, measure the lining thickness from the outer surface to the metal shoe, as shown here; if the lining is riveted to the shoe, measure from the lining outer surface to the rivet head

23.14 Carefully peel back the wheel cylinder boots and check for leaking fluid indicating that the cylinder must be replaced

7 If it is difficult to determine the exact thickness of the remaining pad material by the above method, or if you are at all concerned about the condition of the pads, remove the pads for further inspection (see Chapter 9).

8 Once the pads are removed, clean them with brake cleaner and re-measure them.

9 Measure the disc thickness with a micrometer to make sure that it still has service life remaining. If any disc is thinner than the specified minimum thickness, replace it (see Chapter 9). Even if the disc has service life remaining, check its condition. Look for scoring, gouging and burned spots. If these conditions exist, remove the disc and have it resurfaced (see Chapter 9).

10 Before installing the wheels, check all brake lines and hoses for damage, wear, deformation, cracks, corrosion, leakage, bends and twists, particularly in the vicinity of the rubber hoses at the calipers. Check the clamps for tightness and the connections for leakage. Make sure that all hoses and lines are clear of sharp edges, moving parts and the exhaust system. If any of the above conditions are noted, repair, reroute or replace the lines and/or fittings as necessary (see Chapter 9). Be sure to tighten the lug nuts to the torque listed in this Chapter's Specifications after the vehicle has been lowered.

REAR DRUM BRAKES

➧ Refer to illustrations 23.12 and 23.14

11 Refer to Chapter 9 and remove the rear brake drums.

⁂ WARNING:

Brake dust produced by lining wear and deposited on brake components is hazardous to your health. DO NOT blow it out with compressed air and DO NOT inhale it!

12 Note the thickness of the lining material on the rear brake shoes (see illustration) and look for signs of contamination by brake fluid and grease. If the lining material is within 1/16-inch of the recessed rivets or metal shoes, replace the brake shoes with new ones. The shoes should also be replaced if they are cracked, glazed (shiny lining surfaces) or contaminated with brake fluid or grease. See Chapter 9 for the replacement procedure.

13 Check the shoe return and hold-down springs and the adjusting mechanism to make sure they're fitted correctly and in good condition. Deteriorated or distorted springs, if not replaced, could allow the linings to drag and wear prematurely.

14 Check the wheel cylinders for leakage by carefully peeling back the rubber boots (see illustration). If brake fluid is noted behind the boots, the wheel cylinders must be replaced (see Chapter 9).

15 Check the drums for cracks, score marks, deep scratches and hard spots, which will appear as small discolored areas. If imperfections cannot be removed with emery cloth, the drums must be resurfaced by an automotive machine shop (see Chapter 9 for more detailed information).

16 Refer to Chapter 9 and install the brake drums.

17 Install the wheels and snug the wheel nuts finger tight.

18 Remove the jackstands and lower the vehicle.

19 Tighten the wheel nuts to the torque listed in this Chapter's Specifications.

PARKING BRAKE

20 Slowly depress the parking brake pedal and count the number of clicks you hear until maximum travel has been reached. The adjustment should be within the specified number of clicks listed in this Chapter's Specifications. If you hear more or fewer clicks, it's time to adjust the parking brake (see Chapter 9).

21 An alternative method of checking the parking brake is to park the vehicle on a steep hill with the parking brake set and the transmission in Neutral (be sure to stay in the vehicle during this check!). If the parking brake cannot prevent the vehicle from rolling, it is in need of adjustment (see Chapter 9).

24 Spark plug replacement (see Maintenance schedule for service interval)

➧ Refer to illustrations 24.1, 24.4a and 24.4b

1 Spark plug replacement requires a spark plug socket, extension and ratchet. This socket is lined with a rubber grommet to protect the porcelain insulator of the spark plug and to hold the plug while you insert it into the spark plug hole. You will also need a wire-type feeler gauge to check and adjust the spark plug gap and a torque wrench to tighten the new plugs to the specified torque (see illustration).

2 If you are replacing the plugs, purchase the new plugs, adjust them to the proper gap and replace each plug one at a time.

➡Note: When buying new spark plugs, it's essential that you obtain the correct plugs for your specific vehicle. This information can be found in this Chapter's Specifications, on the Vehicle Emissions Control Information (VECI) label located on the underside of the hood or in the owner's manual. If these sources specify different plugs, purchase the spark plug type specified on the VECI label because that information is provided specifically for your engine.

3 Inspect each of the new plugs for defects. If there are any signs of cracks in the porcelain insulator of a plug, don't use it.

4 Check the electrode gaps of the new plugs. Check the gap by inserting the wire gauge of the proper thickness between the electrodes at the tip of the plug (see illustration). The gap between the electrodes should be identical to that listed in this Chapter's Specifications. If the gap is incorrect, use the notched adjuster on the feeler gauge body to bend the curved side electrode slightly (see illustration).

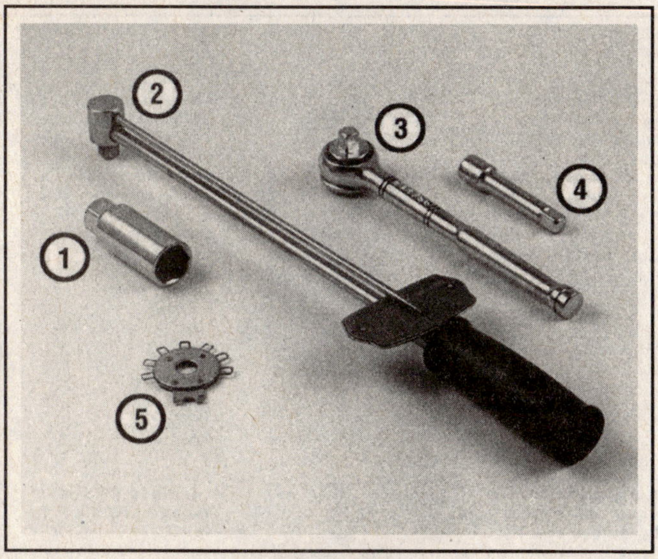

24.1 Tools required for changing spark plugs

1 ***Spark plug socket*** - *This will have special padding inside to protect the spark plug's porcelain insulator*
2 ***Torque wrench*** - *Although not mandatory, using this tool is the best way to ensure the plugs are tightened properly*
3 ***Ratchet*** - *Standard hand tool to fit the spark plug socket*
4 ***Extension*** - *Depending on model and accessories, you may need special extensions and universal joints to reach one or more of the plugs*
5 ***Spark plug gap gauge*** - *This gauge for checking the gap comes in a variety of styles. Make sure the gap for your engine is included*

⁂ CAUTION:

Platinum and iridium spark plugs generally come pre-gapped. If you check the gap, treat them very gently and do not scratch the coating on the electrodes. Also, don't attempt to adjust the gap on used platinum or iridium spark plugs.

5 If the side electrode is not exactly over the center electrode, use the notched adjuster to align them.

24.4a Spark plug manufacturers recommend using a wire type gauge when checking the gap - if the wire does not slide between the electrodes with a slight drag, adjustment is required

24.4b To change the gap, bend the side electrode only, as indicated by the arrows, and be very careful not to crack or chip the porcelain insulator surrounding the center electrode

REMOVAL

▸ Refer to illustration 24.8 and 24.9

6 If compressed air is available, blow any dirt or foreign material away from the coil/spark plug area before proceeding.

7 All models have a separate coil mounted over each spark plug. Remove the ignition coil (see Chapter 5).

8 Remove the spark plug (see illustration).

9 Whether you are replacing the plugs at this time or intend to reuse the old plugs, compare each old spark plug with the chart shown (see illustration) to determine the overall running condition of the engine.

INSTALLATION

▸ Refer to illustrations 24.10a and 24.10b

10 Prior to installation, apply a coat of anti-seize compound to the plug threads. It's often difficult to insert spark plugs into their holes without cross-threading them. To avoid this possibility, fit a short piece of rubber hose over the end of the spark plug (see illustrations). The flexible hose acts as a universal joint to help align the plug with the plug hole. Should the plug begin to cross-thread, the hose will slip on the spark plug, preventing thread damage. Tighten the plug to the torque listed in this Chapter's Specifications.

24.8 Use a socket with a long extension to unscrew the spark plugs (V6 model shown, four-cylinder model similar)

11 Attach the coil to the new spark plug, again using a twisting motion until it is firmly seated on the end of the spark plug. When installing coils, tighten the mounting bolts securely.

12 Follow the above procedure for the remaining spark plugs, replacing them one at a time.

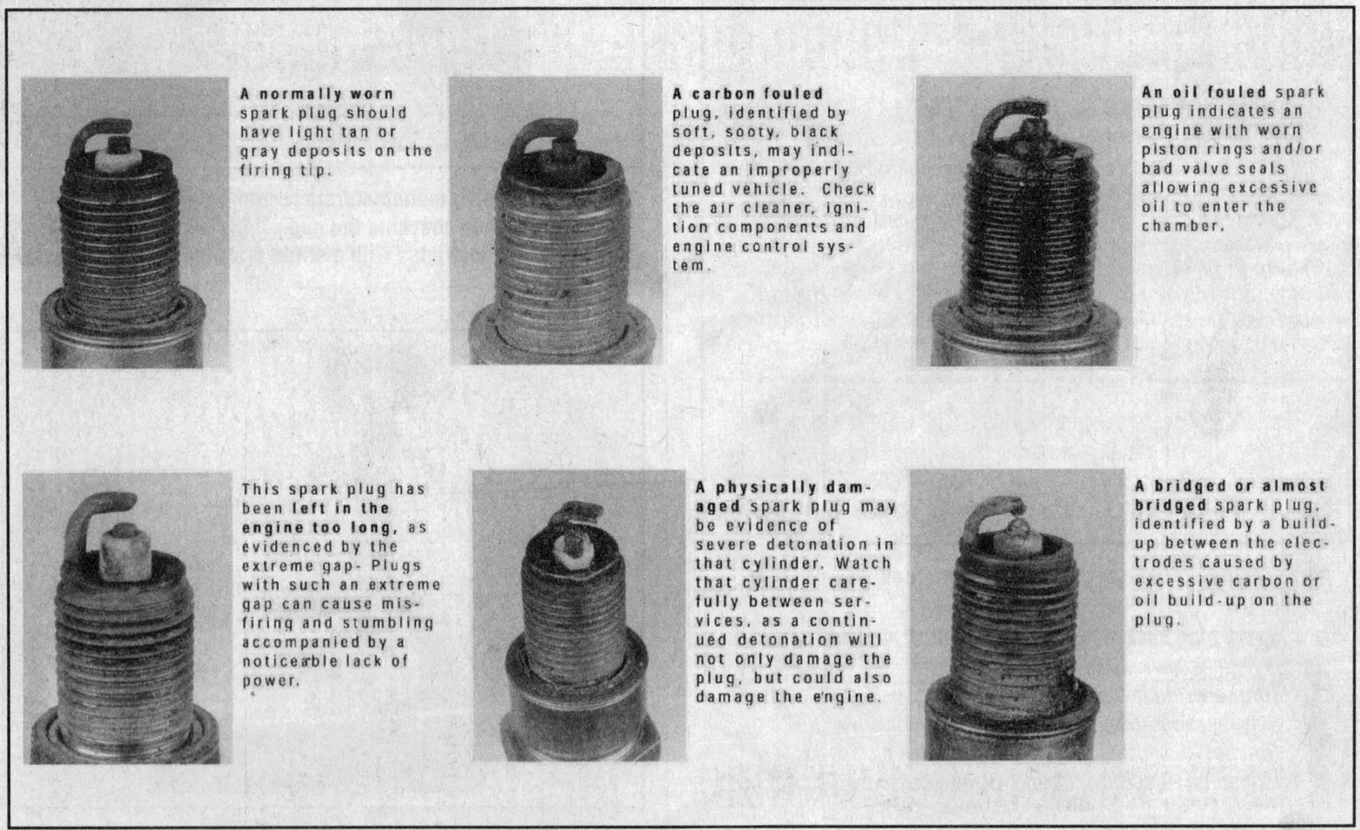

24.9 Inspect the spark plug to determine engine running conditions

24.10a Apply a thin coat of anti-seize compound to the spark plug threads

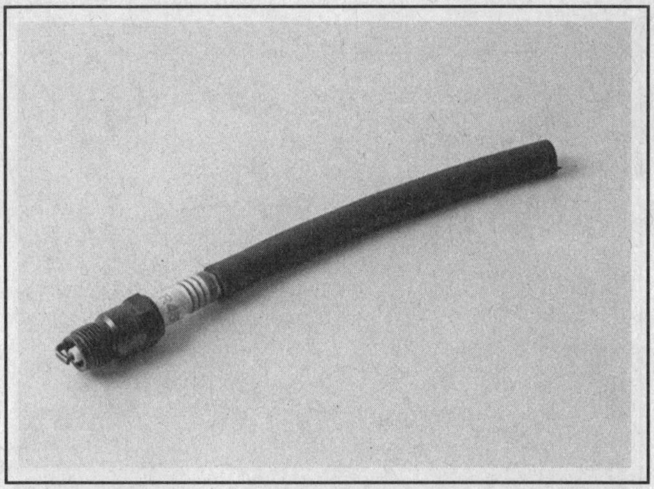

24.10b A length of rubber hose will save time and prevent damaged threads when installing the spark plugs

25 Cooling system servicing (draining, flushing and refilling) (every 30,000 miles [48,000 km] or 24 months)

WARNING:

Do not allow antifreeze to come in contact with your skin or painted surfaces of the vehicle. Rinse off spills immediately with plenty of water. Antifreeze is highly toxic if ingested. Never leave antifreeze lying around in an open container or in puddles on the floor; children and pets are attracted by it's sweet smell and may drink it. Check with local authorities about disposing of used antifreeze. Many communities have collection centers which will see that antifreeze is disposed of safely.

1 Periodically, the cooling system should be drained, flushed and refilled to replenish the antifreeze mixture and prevent formation of rust and corrosion, which can impair the performance of the cooling system and cause engine damage. When the cooling system is serviced, all hoses and the radiator cap should be checked and replaced if necessary.

DRAINING

Refer to illustrations 25.3 and 25.4

2 Apply the parking brake and block the wheels.

WARNING:

If the vehicle has just been driven, wait several hours to allow the engine to coo[illegible]wn before beginning this procedure.

If equipp[illegible] vehicle forward splash shield.

3 Move a large container under the radiator drain to catch the coolant. The radiator drain plug is located on the left (driver's) side lower corner of the radiator (see illustration). Unscrew the drain plug until coolant starts flowing from the drain hole (a pair of pliers may be required to turn it).

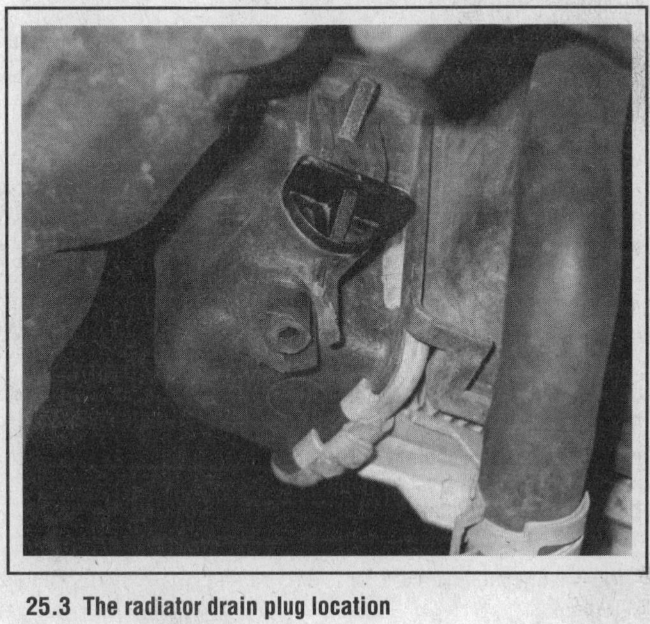

25.3 The radiator drain plug location

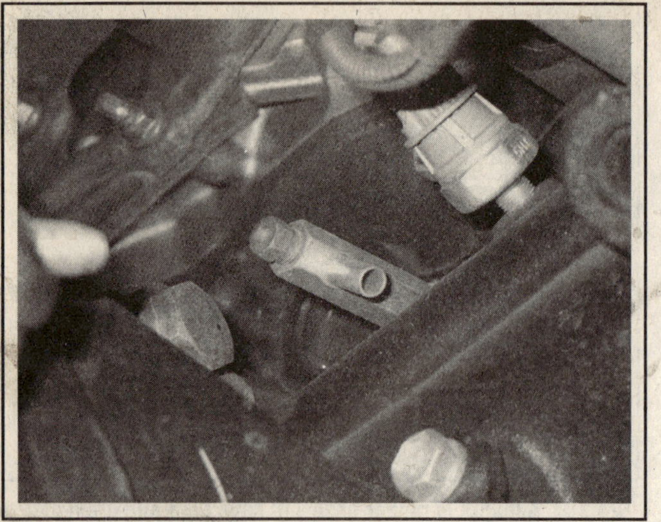

25.4 The engine block drain plug is located on the side of the engine block

4 Remove the radiator cap and allow the radiator to drain, then move the container under the engine. Loosen the engine block drain plug(s) and allow the coolant in the block to drain (see illustration).

5 While the coolant is draining, check the condition of the radiator hoses, heater hoses and clamps (refer to Section 10 if necessary).

6 Replace any damaged clamps or hoses. Close the drain plugs.

FLUSHING

➧ Refer to illustration 25.9

7 Once the system is completely drained, remove the thermostat from the engine (see Chapter 3), then reinstall the thermostat housing without the thermostat. This will allow the system to be thoroughly flushed.

➡Note: On V6 models, the thermostat will have to be separated from the housing. Before removing it, note the position of the jiggle pin.

8 Turn the heating system controls to Hot, so that the heater core will be flushed at the same time as the rest of the cooling system.

9 Disconnect the upper radiator hose from the radiator, then place a garden hose in the upper radiator inlet and flush the system until the water runs clear at the upper radiator hose (see illustration).

10 In severe cases of contamination or clogging of the radiator, remove the radiator (see Chapter 3) and have a radiator repair facility clean and repair it if necessary.

11 Many deposits can be removed by the chemical action of a cleaner available at auto parts stores. Follow the procedure outlined in the manufacturer's instructions.

➡Note: When the coolant is regularly drained and the system refilled with the correct antifreeze/water mixture, there should be no need to use chemical cleaners or descalers.

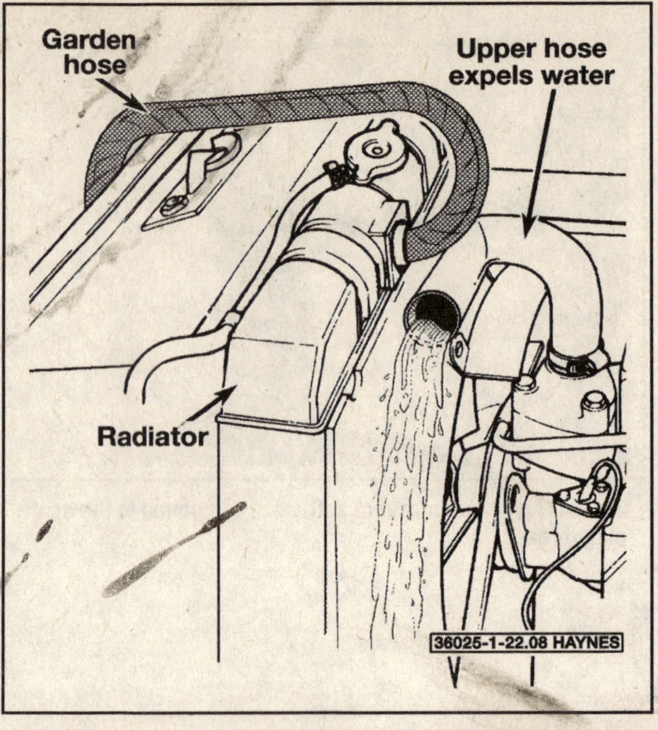

25.9 With the thermostat removed, disconnect the upper radiator hose and flush the radiator and engine block with a garden hose

12 Remove the overflow hose from the coolant recovery reservoir. Drain the reservoir and flush it with clean water, then reconnect the hose.

REFILLING

13 Reconnect the upper radiator hose and reinstall the thermostat. Close the radiator drain.

14 Place the heater temperature control in the maximum heat position, if not already done.

15 Slowly add new coolant (a 50/50 mixture of water and antifreeze) to the radiator until it's full. Add coolant to the reservoir up to the lower mark.

16 Install the radiator cap and run the engine at approximately 2,000 to 2,500 rpm in a well-ventilated area until the thermostat opens (coolant will begin flowing through the radiator and the upper radiator hose will become hot).

17 Turn the engine off and let it cool. Add more coolant mixture to bring the level back up to the lip on the radiator filler neck.

18 Squeeze the upper radiator hose to expel air, then add more coolant mixture if necessary. Reinstall the radiator cap and the under-vehicle splash shield. Add coolant to the reservoir, if necessary.

19 Start the engine, allow it to reach normal operating temperature and check for leaks.

26 Automatic transmission fluid and filter change (every 30,000 miles [48,000 km] or 24 months)

Refer to illustrations 26.5, 26.7 and 26.12

1 At the specified intervals, the transmission fluid should be drained and replaced. Since the fluid will remain hot long after driving, perform this procedure only after the engine has cooled down completely.

2 Before beginning work, purchase the specified transmission fluid (see *Recommended lubricants and fluids* in this Chapter's Specifications) and a new filter.

3 Other tools necessary for this job include a floor jack, jackstands to support the vehicle in a raised position, a drain pan capable of holding at least eight quarts, newspapers and clean rags.

4 Raise the vehicle and support it securely on jackstands.

5 Place the drain pan underneath the transmission pan and remove the drain plug (see illustration). Allow the fluid to completely drain from the transmission, then reinstall the drain plug.

6 Detach the transmission pan rock shield (if equipped) and remove the pan mounting bolts from the outer edges of the pan.

7 Using a rubber mallet, carefully tap on the pan to break the layer of gasket sealant between the pan and the transmission case (see illustration).

➡Note: Prying between the pan and the transmission case with a screwdriver or similar tool may result in damage to the sealing surface on the transmission case.

8 Lower the pan and separate the lower part of the dipstick tube from the upper part. If the union between the two tubes is stuck and won't separate, you'll have to unbolt the bracket at the upper end of the tube from the cylinder head. Once the pan and dipstick tube are free, drain any remaining transmission fluid from the pan.

9 Remove the filter retaining bolts from the valve body and remove the filter.

➡Note: On some models different length bolts are used; be sure to note the length and locations of the bolts as you remove them.

10 Thoroughly inspect the bottom of the pan, the filter and the fluid. Although normally bright red, transmission fluid may turn dark red or brown during normal use. If you find the fluid very dark colored, or if it smells burned, it usually indicates the transmission has been overheated. If you find small pieces of metal or clutch material in the pan or filter, it indicates wear or damage have occurred to the internal parts or clutches. If you have any concerns about the condition of your transmission based on what you find in the fluid, pan and filter, it's a good idea to take your vehicle to your dealer or a transmission shop for further evaluation.

11 Clean the pan with solvent and dry it. Use a gasket scraper to remove any traces of old gasket material remaining on the transmission case.

➡Note: Be very careful not to gouge the delicate aluminum gasket surfaces. Install new gaskets on the filter, then install the filter, tightening the bolts to the torque listed in this Chapter's Specifications.

12 Be sure the gasket surface on the transmission pan is clean, then install a bead of RTV sealant to the pan (see illustration). Reinsert the filler tube onto the dipstick tube and place the pan against the transmission case. Working around the pan, tighten each bolt a little at a time to the torque listed in this Chapter's Specifications.

26.5 The transmission drain plug is located on the bottom of the transmission pan

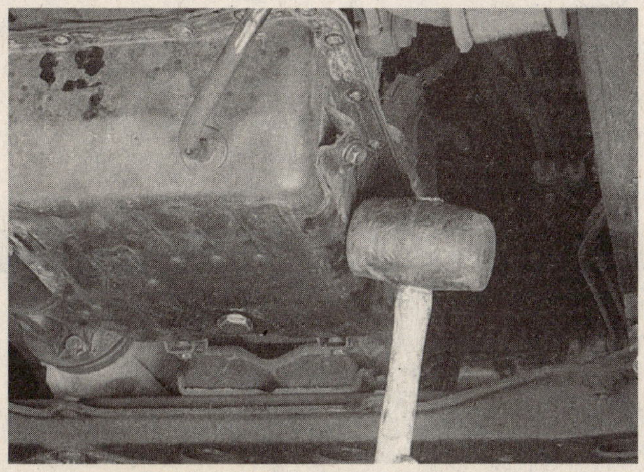

26.7 Using a rubber mallet, carefully tap on the pan to break the gasket seal between the pan and the transmission case

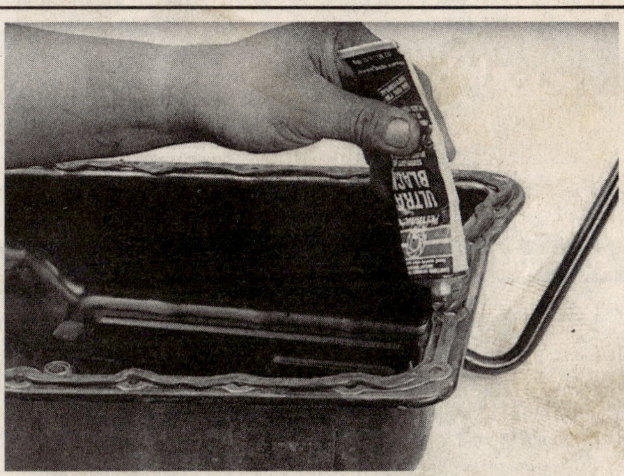

26.12 Apply a bead of RTV sealant all the way around the pan mating surface, inboard of the bolt holes

FOUR-CYLINDER MODELS

13 Lower the vehicle and add approximately 1-1/2 quarts of the specified type of automatic transmission fluid through the filler tube (see Section 6).

14 With the transmission in Park and the parking brake set, start the engine.

15 Move the gear selector through each range and back to Park. Check the fluid level and add fluid, a little at a time, until the level is within the correct range on the dipstick. Between each application of fluid, move the selector lever through each range and back to Park.

V6 MODELS

16 Refer to Section 6 for the refilling procedure.

17 Check under the vehicle for leaks during the first few trips. Check the fluid level again when the transmission is hot (see Section 6).

27 Cabin air filter replacement (every 30,000 miles [48,000 km] or 24 months)

➧ Refer to illustrations 27.2 and 27.3

1 Remove the fastener securing the glove box door stop strut, then push in on the glove box compartment sides at the stops and swing out the compartment to gain access to the filter element access panel

2 Open the filter element access panel by pushing down on the two retaining tabs (see illustration)

3 Pull out and remove the filter element from the blower housing (see illustration).

4 Installation is the reverse of removal.

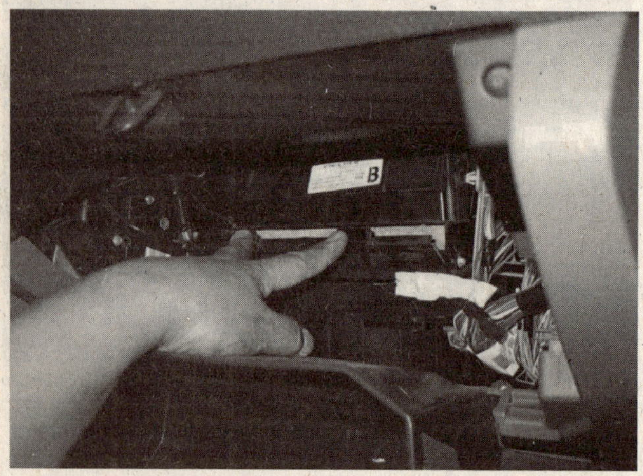

27.2 To open the cabin filter access panel, push down on the two retaining tabs

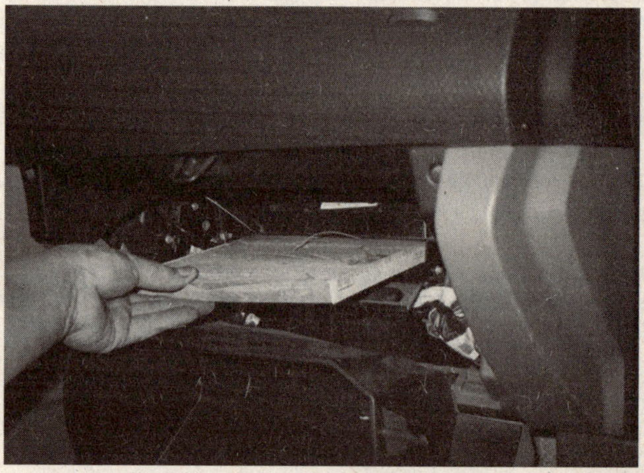

27.3 Pull out and remove the filter element from the blower housing

28 Transfer case lubricant change (4WD models) (every 30,000 miles [48,000 km] or 24 months)

1 Drive the vehicle for at least 15 minutes to warm the lubricant in the case. Perform this warm-up procedure with 4WD engaged, if possible. Use all gears, including Reverse, to ensure the lubricant is sufficiently warm to drain completely.

2 Raise the vehicle and support it securely on jackstands.

3 Remove the filler plug from the case (see illustration 16.2).

4 Remove the drain plug from the lower part of the case and allow the old lubricant to drain completely.

5 After the lubricant has drained completely, reinstall the drain plug and tighten it to the torque listed in this Chapter's Specifications.

6 Fill the case with the specified lubricant until it is level with the lower edge of the filler hole.

7 Install the filler plug and tighten it to the torque listed in this Chapter's Specifications.

8 Drive the vehicle for a short distance and recheck the lubricant level.

29 Differential and manual transmission lubricant change (every 30,000 miles [48,000 km] or 24 months)

DIFFERENTIAL(S)

➡Note: The following procedure applies to the front and rear differentials.

1 Drive the vehicle for several miles to warm up the differential oil, then raise the vehicle and support it securely on jackstands.

2 Move a drain pan, rags, newspapers and the proper tools under the vehicle.

3 Remove the differential check/fill plug (see Section 17). With the drain pan under the differential, use a socket and ratchet to loosen the drain plug. It's the lower of the two plugs (see illustration 17.2).

4 Once loosened, carefully unscrew it with your fingers until you can remove it from the case.

5 Allow all of the oil to drain into the pan, then replace the drain plug and tighten it to the torque listed in this Chapter's Specifications.

6 Feel with your hands along the bottom of the drain pan for any metal bits that may have come out with the oil. If there are any, it's a sign of excessive wear, indicating that the internal components should be carefully inspected in the near future.

7 Using a hand pump, syringe or funnel, fill the differential with the correct amount and grade of oil (see the Specifications) until the level is just at the bottom of the plug hole.

8 Reinstall the plug and tighten it to the torque listed in this Chapter's Specifications.

9 Lower the vehicle. Check for leaks at the drain plug after the first few miles of driving.

MANUAL TRANSMISSION

10 Raise the vehicle and support it securely on jackstands.

11 Move a drain pan, rags, newspapers and wrenches under the transmission.

12 Remove the fill plug from the side of the transmission case. Remove the transmission drain plug and allow the lubricant to drain into the pan.

13 After the lubricant has drained completely, reinstall the plug and tighten it securely.

14 Using a hand pump, syringe or funnel, fill the transmission with the specified lubricant until it is level with the lower edge of the filler hole. Reinstall the fill plug and tighten it securely.

15 Lower the vehicle.

16 Drive the vehicle for a short distance, then check the drain and fill plugs for leakage.

30 Brake fluid change (every 30,000 miles [48,000 km] or 24 months)

⁂ WARNING:

Brake fluid can harm your eyes and damage painted surfaces, so use extreme caution when handling or pouring it. Do not use brake fluid that has been standing open or is more than one year old. Brake fluid absorbs moisture from the air. Excess moisture can cause a dangerous loss of braking effectiveness.

1 At the specified intervals, the brake fluid should be drained and replaced. Since the brake fluid may drip or splash when pouring it, place plenty of rags around the master cylinder to protect any surrounding painted surfaces.

2 Before beginning work, purchase the specified brake fluid (see *Recommended lubricants and fluids* in this Chapter's Specifications).

3 Remove the cap from the master cylinder reservoir.

4 Using a hand suction pump or similar device, withdraw the fluid from the master cylinder reservoir.

5 Add new fluid to the master cylinder until it rises to the base of the filler neck.

6 Bleed the brake system as described in Chapter 9 at all four brakes until new and uncontaminated fluid is expelled from the bleeder screw. Be sure to maintain the fluid level in the master cylinder as you perform the bleeding process. If you allow the master cylinder to run dry, air will enter the system.

7 Refill the master cylinder with fluid and check the operation of the brakes. The pedal should feel solid when depressed, with no sponginess.

⁂ WARNING:

Do not operate the vehicle if you are in doubt about the effectiveness of the brake system.

31 Valve clearance check and adjustment (V6 engine) (every 60,000 miles [96,000 km] or 48 months)

1 Refer to Chapter 2B and position the number 1 piston at TDC on the compression stroke.

2 Disconnect the cable from the negative terminal of the battery.

3 Disconnect the coils and remove any other components that will interfere with valve cover removal.

4 Blow out the recessed area around the spark plug openings with compressed air, if available, to remove any debris that might fall into the cylinders, then remove the spark plugs (see Section 24).

5 Remove the valve covers (see Chapter 2B).

31.6a Check the clearance of each valve with a feeler gauge of the specified thickness - if the clearance is correct, you should feel a slight drag on the gauge as you pull it out

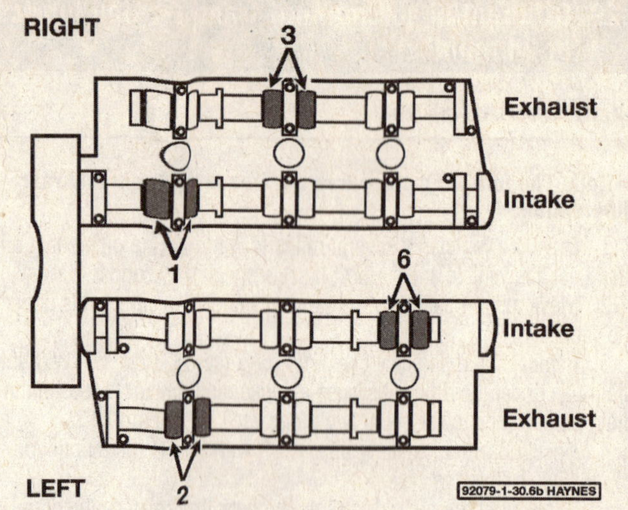

31.6b With the No. 1 piston at TDC on the compression stroke, check the indicated valves

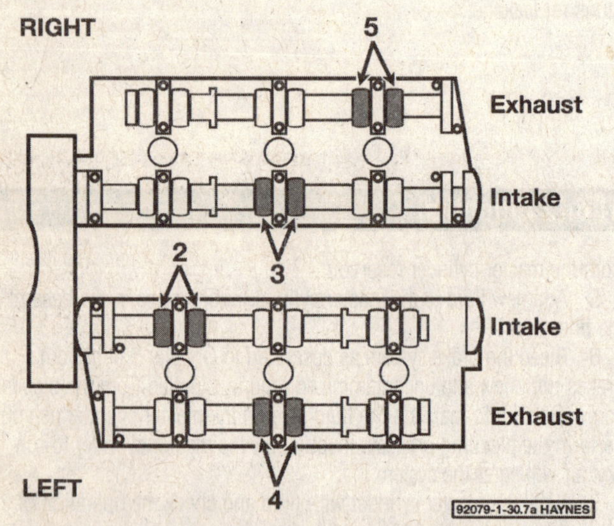

31.7 Rotate the crankshaft 2/3 turn (240 degrees) and check the indicated valves

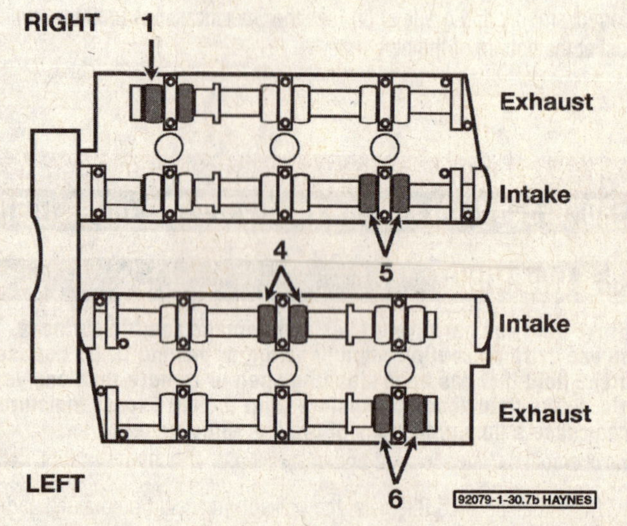

31.8 Rotate the crankshaft 2/3 turn (240 degrees) and check the remaining valves indicated

CHECK

➧ Refer to illustrations 31.6a, 31.6b, 31.7 and 31.8

6 Measure the clearances of the indicated valves with feeler gauges (see illustrations). Record the measurements which are out of specification. They will be used later to determine the required replacement lifters.

7 Turn the crankshaft 2/3 revolution (240 degrees) and check the indicated valves (see illustration).

8 Turn the crankshaft 2/3 revolution again and check the next group of valves (see illustration).

ADJUSTMENT

➧ Refer to illustration 30.10

9 Remove the camshaft(s) for the valve(s) that you intend to adjust (see Chapter 2B).

10 Remove and measure each lifter (whose clearance is not correct) with a micrometer (see illustration). Put each lifter back into its bore in the cylinder head before moving on to the next lifter. Record the measurement for each lifter.

11 To calculate the correct thickness of a replacement lifter that will

put the valve clearance within the specified range, use the following formula:

N = T + (A - V), where:
N = thickness of the new lifter
T = thickness of the old lifter
A = measured valve clearance
V = specified valve clearance (see this Chapter's Specifications)

12 Select a lifter with a thickness as close as possible to the calculated valve clearance. Lifters for this engine are available in 35 sizes in increments of 0.008-inch (0.020 mm), and range in size from 0.1992-inch (5.060 mm) to 0.2260-inch (5.740 mm).

13 Install the camshaft(s) (see Chapter 2B).

14 After the camshafts are reinstalled, check the valve clearances again to verify that they're now within the range of clearance listed in this Chapter's Specifications.

15 Installation of the valve cover(s), spark plugs, ignition coils, air filter housing and/or intake manifold is the reverse of removal.

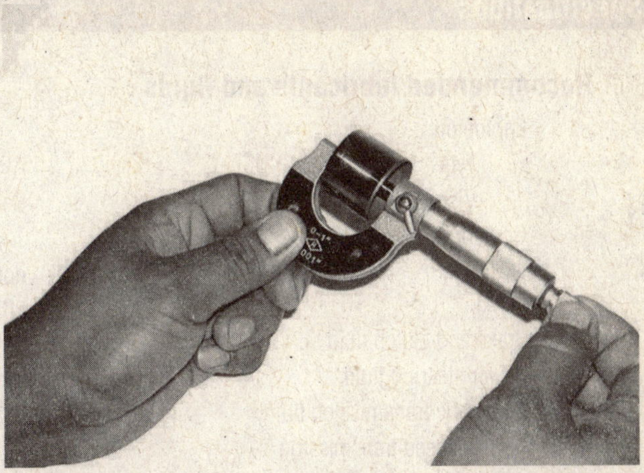

31.10 Measure the thickness of each lifter head with a micrometer

32 Evaporative emissions control system check (every 60,000 miles [96,000 km] or 48 months)

Refer to illustration 32.2

1 The function of the evaporative emissions control system is to draw fuel vapors from the gas tank and fuel system, store them in a charcoal canister and route them to the intake manifold during normal engine operation.

2 The most common symptom of a fault in the evaporative emissions system is a strong fuel odor. If a fuel odor is detected, inspect the charcoal canister, located under the vehicle, directly behind the fuel tank (see illustration). Check the canister and all hoses for damage and deterioration.

3 The evaporative emissions control system is explained in more detail in Chapter 6.

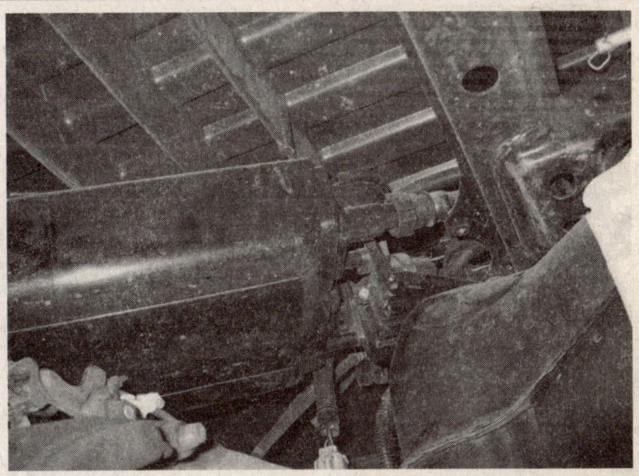

32.2 Check the evaporative emissions control canister and the hose connections for cracks and damage, it's mounted underneath the vehicle, near the fuel tank

Specifications

Recommended lubricants and fluids

Engine oil	
Type	API "certified for gasoline engines"
Viscosity	See accompanying chart
Coolant	Toyota Super Long Life Coolant or equivalent 50/50 mixture of non-silicate antifreeze and water (the genuine Toyota antifreeze is pre-mixed - don't add water to it)
Brake and clutch fluid	DOT 3 brake fluid
Power steering fluid	DEXRON ® II or III automatic transmission fluid
Automatic transmission fluid	
4-speed transmission	TOYOTA Genuine ATF Type T-IV
5-speed transmission	TOYOTA Genuine ATF WS
Manual transmission lubricant	APL GL-4 or APL GI-5 75W-90
Transfer case lubricant	API GL-5 SAE 75W-90 gear oil
Differential lubricant	
Four-cylinder models	
Rear differential	
Above 0 degrees F (-18 degrees C)	API GL-5 SAE 90W gear oil
Below 0 degrees F (-18 degrees C)	API GL-5 SAE 80W-90 gear oil
Front differential (4WD models)	API GL-5 SAE 75W-90 gear oil
V6 models	
Rear differential	
Above 0 degrees F (-18 degrees C)	API GL-5 SAE 90W gear oil
Below 0 degrees F (-18 degrees C)	API GL-5 SAE 80W-90 gear oil
Front differential (4WD models)	
Above 0 degrees F (-18 degrees C)	API GL-5 SAE 90W gear oil
Below 0 degrees F (-18 degrees C)	API GL-5 SAE 80W-90 gear oil
Chassis grease	NLGI No. 2 lithium base chassis grease

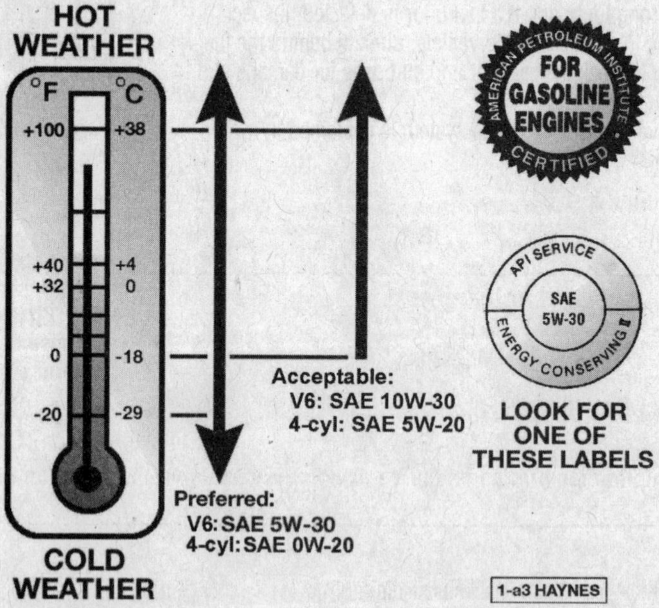

Engine oil viscosity chart - for best fuel economy and cold starting, select the lowest SAE viscosity grade for the expected temperature range

Capacities*

Engine oil (with filter change)	
Four-cylinder engine	6.1 quarts (5.8 liters)
V6 engine	
2WD (except PreRunner)	4.8 quarts (4.5 liters)
4WD (except PreRunner)	5.5 quarts (5.2 liters)
Cooling system	
Four-cylinder engine	9.1 quarts (8.7 liters)
V6 engine	
Manual transmission	10.3 quarts (9.7 liters)
Automatic transmission	10.1 quarts (9.6 liters)
Automatic transmission (drain and refill)	
4-speed transmission	
2008 and earlier models	2.1 quarts (2.0 liters)
2009 and later models	2.9 quarts (2.7 liters)
5-speed transmission	3.2 quarts (3.0 liters)
Manual transmission	
Four-cylinder models	
2WD	2.7 quarts (2.6 liters)
4WD	2.3 quarts (2.2 liters)
V6 models	1.9 quarts (1.8 liters)
Transfer case	1.1 quarts (1.0 liter)
Differential	
2WD models (except PreRunner)	3.5 quarts (3.31 liters)
4WD models and PreRunner	
Front differential	1.6 quarts (1.5 liters)
Rear differential	3.0 quarts (2.8 liters)

All capacities approximate. Add as necessary to bring to appropriate level.

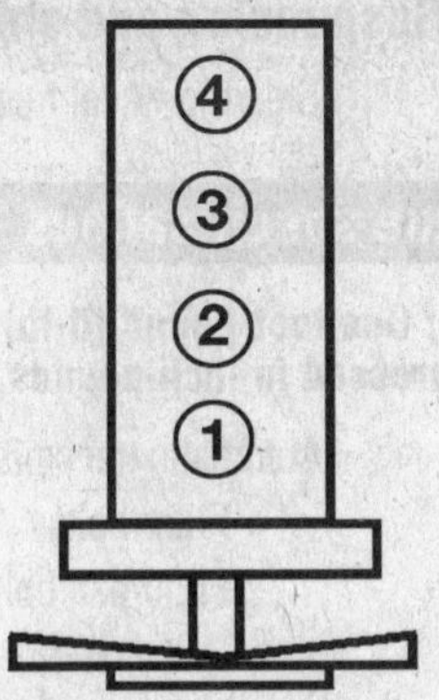

Cylinder location diagram - four-cylinder engine

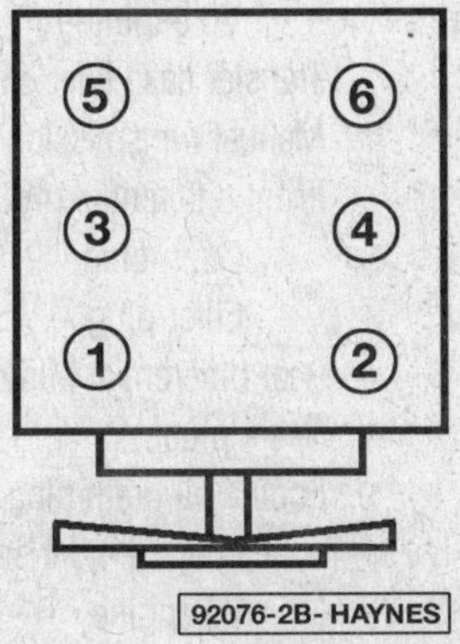

Cylinder location diagram - four-cylinder engine

Spark plug type and gap

Four-cylinder engine	
Type	Denso SK20HR-11 or NGK 1LFR6C11
Gap	0.043 inch (1.1 mm)
V6 engine	
Type	Denso K20HR-U11 or NGK LFR6C11
Gap	0.043 inch (1.1 mm)

Firing order

Four-cylinder engine	1-3-4-2
V6 engine	1-2-3-4-5-6

Valve clearances (engine cold)

Four-cylinder engine	Automatic adjustment
V6 engine	
Intake	0.006 to 0.010 inch (0.15 to 0.25 mm)
Exhaust	0.011 to 0.015 inch (0.29 to 0.39 mm)

➡Note: Use the information printed on the Vehicle Emissions Control Information label, if different than the Specifications listed here.

Brakes

Disc brake pad lining thickness (minimum)	1/16 inch (1.5 mm)
Drum brake shoe lining thickness (minimum)	1/16 inch (1.5 mm)
Parking brake pedal adjustment	7 to 10 clicks

Suspension and steering

Steering wheel freeplay limit	1.2 inches (30.5 mm)

Torque specifications	Ft-lbs (unless otherwise indicated)	Nm

➡**Note: One foot-pound (ft-lb) of torque is equivalent to 12 inch-pounds (in-lbs) of torque. Torque values below approximately 15 ft-lbs are expressed in inch-pounds, since most foot-pound torque wrenches are not accurate at these smaller values.**

Torque specifications	Ft-lbs (unless otherwise indicated)	Nm
Automatic transmission		
Filter bolts	84 in-lbs	9.5
Fluid pan bolts	65 in-lbs	7
Drain plug	15	20
Overflow plug	15	20
Refill plug	29	39
Drivebelt tensioner mounting fasteners		
Four-cylinder engine	32	43
V6 engine	27	37
Transfer case filler plug and drain plug	27	37
Manual transmission filler plug and drain plug	27	37
Front differential (4WD models)		
Drain plug	48	65
Filler plug	29	39
Rear differential fill/drain plug	36	49
Spark plugs	15	20
Engine oil drain plug		
Four-cylinder engine	28	38
V6 engine	30	40
Wheel lug nuts	85	115

2A

FOUR-CYLINDER ENGINE

Section

Reference to other Chapters

CHECK ENGINE light on - See Chapter 6
Cylinder compression check - See Chapter 2C
Drivebelt - check, adjustment and replacement - See Chapter 1
Engine - removal and installation - See Chapter 2C
Engine oil and filter change - See Chapter 1
Engine overhaul - general information - See Chapter 2C
Oil cooler - removal and installation - See Chapter 3
Spark plug replacement - See Chapter 1
Valve clearance check and adjustment - See Chapter 1
Valves - servicing - See Chapter 2C
Water pump - removal and installation - See Chapter 3

1 General information

This Part of Chapter 2 is devoted to in-vehicle repair procedures for the 2.7L DOHC four-cylinder engine. All information concerning engine removal and installation and engine block and cylinder head overhaul can be found in Part C of this Chapter.

The following repair procedures are based on the assumption that the engine is installed in the vehicle. If the engine has been removed from the vehicle and mounted on a stand, many of the steps outlined in this Part of Chapter 2 will not apply.

The Specifications included in this Part of Chapter 2 apply only to the procedures contained in this Part. Part C of Chapter 2 contains the Specifications necessary for cylinder head and engine block rebuilding.

This engine is equipped with overhead camshafts driven by a timing chain. Each cylinder is equipped with two intake valves and two exhaust valves. The left camshaft (driver's side) operates the intake valves, while the right camshaft (passenger's side) operates the exhaust valves. This 2.7L engine is equipped with balance shafts that are driven by another chain and crankshaft sprocket.

2 Repair operations possible with the engine in the vehicle

Many major repair operations can be accomplished without removing the engine from the vehicle.

Clean the engine compartment and the exterior of the engine with some type of degreaser before any work is done. It will make the job easier and help keep dirt out of the internal areas of the engine.

Depending on the components involved, it may be helpful to remove the hood to improve access to the engine as repairs are performed (refer to Chapter 11 if necessary). Cover the fenders to prevent damage to the paint. Special pads are available, but an old bedspread or blanket will also work.

If vacuum, exhaust, oil or coolant leaks develop, indicating a need for gasket or seal replacement, the repairs can generally be made with the engine in the vehicle. The intake and exhaust manifold gaskets, crankshaft oil seals and cylinder head gasket are all accessible with the engine in place. The oil pan can be removed with the engine in place on some models; on others, the engine must be removed.

Exterior engine components, such as the intake and exhaust manifolds, the oil pan, the oil pump, the water pump, the starter motor, the alternator, and the fuel system components can be removed for repair with the engine in place.

Since the cylinder head can be removed without pulling the engine, camshaft and valve component servicing can also be accomplished with the engine in the vehicle. Replacement of the timing chain and sprockets is also possible with the engine in the vehicle. Although the balance shaft retaining bolts are accessible with the timing chains removed, it isn't recommended to remove them with the engine in the vehicle, as it isn't possible to support them sufficiently as they are withdrawn from the cylinder block, which could result in damage.

3 Top Dead Center (TDC) for number one piston - locating

➧ Refer to illustration 3.6

1 Top Dead Center (TDC) is the highest point in the cylinder that each piston reaches as it travels up the cylinder bore. Each piston reaches TDC on the compression stroke and again on the exhaust stroke, but TDC generally refers to piston position on the compression stroke.

2 Positioning the piston(s) at TDC is an essential part of certain procedures, such as camshaft and timing chain/sprocket removal.

3 Before beginning this procedure, be sure to place the transmission in Neutral (manual transmission) or Park (automatic transmission) and apply the parking brake or block the rear wheels. Remove the spark plugs (see Chapter 1). Also, if method b) or c) in Step 4 will be used to rotate the engine, disable the fuel system by removing the circuit opening (C/OPN) relay from the underhood fuse/relay box.

4 In order to bring any piston to TDC, the crankshaft must be turned using one of the methods outlined below. When looking at the front of the engine, normal crankshaft rotation is clockwise.

a) *The preferred method is to turn the crankshaft with a socket and ratchet attached to the bolt threaded into the front of the crankshaft. Turn the bolt in the normal direction of crankshaft rotation (clockwise).*

b) *A remote starter switch, which may save some time, can also be used. Follow the instructions included with the switch. Once the piston is close to TDC, use a socket and ratchet as described in the previous paragraph.*

c) *If an assistant is available to turn the ignition switch to the Start position in short bursts, you can get the piston close to TDC without a remote starter switch. Make sure your assistant is out of the vehicle, away from the ignition switch, then use a socket and ratchet as described in Paragraph (a) to complete the procedure.*

5 Install a compression gauge in the number one spark plug hole (see Chapter 2C). It should be a gauge with a screw-in fitting and a

hose at least six inches long.

6 Rotate the crankshaft using one of the methods described in Step 4 while observing the compression gauge. The moment the gauge begins to register pressure, the compression stroke has begun. Continue to turn the crankshaft until the notch on the crankshaft pulley aligns with the "0" mark on the timing chain cover (see illustration). If you go past the marks, release the gauge pressure and rotate the crankshaft two more revolutions.

7 After the number one piston has been positioned at TDC on the compression stroke, TDC for any of the remaining cylinders can be found by turning the crankshaft 180-degrees at a time and following the firing order. Rotating the engine 180-degrees past TDC for cylinder number one would put cylinder number three at TDC compression, and so on.

3.6 The mark on the crankshaft pulley will align with the "0" mark on the timing cover tab when the engine is at Top Dead Center

4 Valve cover - removal and installation

REMOVAL

➧ Refer to illustration 4.6

1 Disconnect the cable from the negative terminal of the battery (see Chapter 5, Section 1).

2 Remove the air intake duct (see Chapter 4). Detach the PCV hose from the valve cover.

3 Refer to Chapter 5 and remove the ignition coils.

4 Disconnect the wiring from the throttle body, the vacuum switching valve and the camshaft position sensor.

5 Release the engine wiring harness from the clips that secure it to the valve cover.

6 Remove the valve cover mounting fasteners (see illustration), then detach the valve cover and gasket from the cylinder head. If the valve cover is stuck to the cylinder head, bump the end with a wood block and a hammer to jar it loose. If that doesn't work, try to slip a flexible putty knife between the cylinder head and valve cover to break the seal.

⁂ CAUTION:

Don't pry at the valve cover-to-cylinder head joint or damage to the sealing surfaces may occur, leading to oil leaks after the valve cover is reinstalled.

4.6 Valve cover fastener locations

4.7a Make sure the valve cover gasket is seated in its groove

4.7b The spark plug seals must be aligned correctly

INSTALLATION

➧ Refer to illustrations 4.7a and 4.7b

7 The mating surfaces of the cylinder head and valve cover must be clean when the valve cover is installed. The rubber sealing gasket can be reused unless the rubber has hardened or cracked. If necessary, pull out the rubber gasket and clean the mating surfaces. Install a new rubber gasket, pressing it evenly into the groove around the underside of the valve cover (see illustration). If there's residue or oil on the mating surfaces when the valve cover is installed, oil leaks may develop.

➡Note: Make sure the spark plug tube gaskets are in place on the underside of the valve cover before reinstalling it (see illustration).

8 Install the valve cover and bolts.

9 Tighten the bolts to the torque listed in this Chapter's Specifications in three or four equal steps.

10 Reinstall the remaining components, run the engine and check for oil leaks.

5 Intake manifold - removal and installation

⁂ WARNING:

Wait until the engine is completely cool before beginning this procedure.

REMOVAL

➧ Refer to illustrations 5.5, 5.7a, 5.7b, 5.8a and 5.8b

1 Refer to Chapter 4 and relieve the fuel system pressure.

2 Disconnect the cable from the negative terminal of the battery (see Chapter 5, Section 1).

3 Detach the PCV hose connected to the rear of the intake manifold.

4 Remove the intake air connector and the throttle body (see Chapter 4).

➡Note: Be sure to clamp-off the coolant hoses connected to the throttle body, and be prepared for a little spillage when the hoses are disconnected.

5 Disconnect the vacuum hoses and the dipstick tube from the intake manifold (see illustration).

5.5 Disconnect all hoses (A) and the dipstick tube (B) from the intake manifold

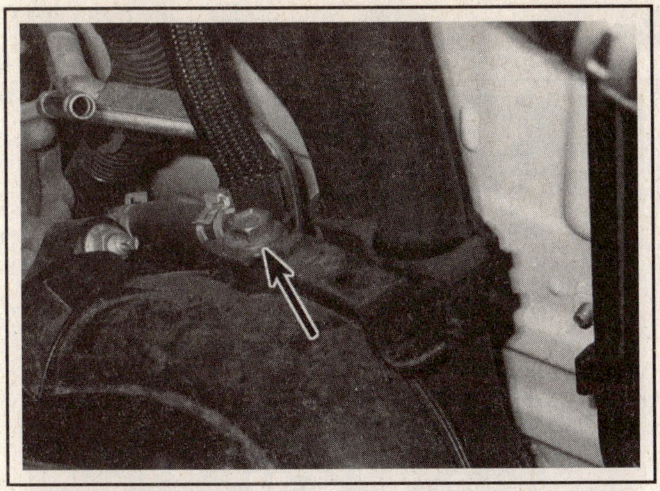

5.7a Unbolt the heater hose bracket from the rear of the manifold

5.7b Detach the coolant hose from the clip at the back of the intake manifold

5.8a Intake manifold upper mounting bolts

5.8b Intake manifold lower mounting bolts

6 Remove the fuel rail and injectors (see Chapter 4).
7 Disconnect any other interfering components, including the wiring harnesses (see illustrations).
8 Remove the bolts from the intake manifold including the two lower ones that secure it to the engine block (see illustrations). Remove the intake manifold from the cylinder head.

INSTALLATION

Refer to illustration 5.10

9 Clean the mating surfaces of the intake manifold and the cylinder head mounting surface.
10 Install a new gasket (see illustration), then position the manifold on the cylinder head and install the bolts and nuts.
11 Tighten the nuts/bolts in three equal steps to the torque listed in this Chapter's Specifications. Work from the center out towards the ends to avoid warping the manifold.
12 Install the remaining parts in the reverse order of removal.
13 Run the engine and check for leaks.
14 Road test the vehicle and check for proper operation.

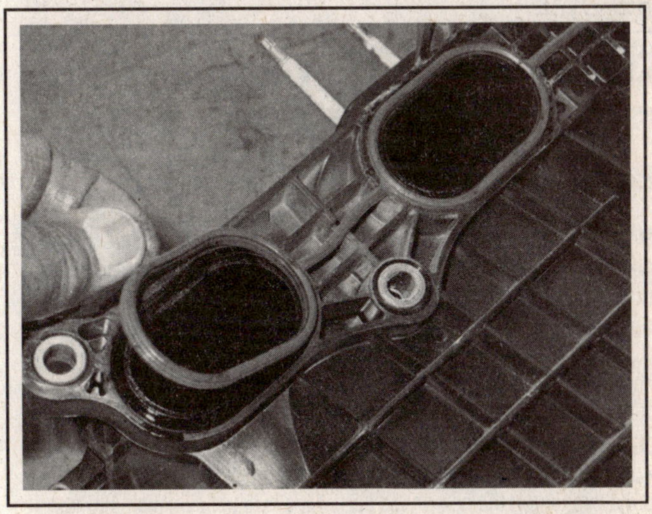

5.10 Carefully place the new intake manifold gasket into the recess

6 Exhaust manifold - removal and installation

WARNING:

The engine must be completely cool before beginning this procedure.

REMOVAL

Refer to illustration 6.6

1 Disconnect the cable from the negative terminal of the battery (see Chapter 5, Section 1).
2 Raise the vehicle and support it securely on jackstands.
3 Disconnect the exhaust pipe from the exhaust manifold. If the nuts are frozen, try soaking them with penetrating oil and letting them soak. Lower the vehicle.
4 Remove the exhaust manifold heat shield.
5 Disconnect the large air hose from the air switching valve.
6 Remove the two nuts from each end of the air tube and remove the tube along with its gaskets (see illustration).
7 Disconnect the vacuum hose and the wiring from the air switching valve, then remove the valve and gasket.
8 Apply penetrating oil to the eight exhaust manifold mounting nuts and allow them to soak for awhile.
9 Remove the nuts and detach the manifold and gasket.

INSTALLATION

10 Use a scraper to remove all traces of old gasket material and carbon deposits from the manifold and the cylinder head mating surface. If the gasket was leaking, have the manifold checked for warpage at an automotive machine shop and resurfaced if necessary.
11 Position a new gasket over the cylinder head studs.
12 Install the manifold and thread the mounting nuts into place.
13 Working from the center out, tighten the nuts to the torque listed in this Chapter's Specifications in three or four equal steps.
14 Reinstall the remaining parts in the reverse order of removal.
15 Run the engine and check for exhaust leaks.

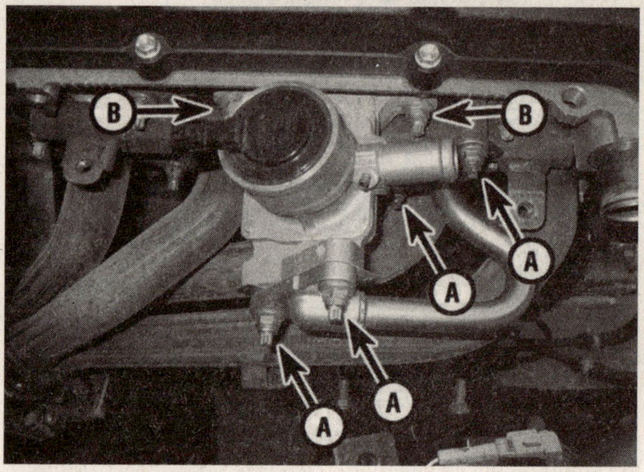

6.6 Remove the four nuts (A) and remove the air tube before removing the air switching valve mounting bolts (B)

7 Camshafts, rocker arms and lash adjusters - removal, inspection and installation

REMOVAL

Refer to illustrations 7.3, 7.4, 7.9 and 7.11

1 Remove the valve cover as described in Section 4.
2 Remove the timing chain guide from the top of the timing chain area, then remove the O-ring under it (see illustration 9.17).
3 Refer to Section 3 and set the engine to TDC on the compression stroke. Check that the timing marks on the camshaft sprockets are aligned with the marks on the camshaft front bearing caps (see illustration). Paint matchmarks on the chain links that coincide with the marks on the sprockets and bearing caps.
4 Remove the plug from the right side of the timing chain cover to access the timing chain tensioner (see illustration).
5 Put a flat-blade screwdriver into the hole and lift the tensioner stopper plate (see illustration 9.18a).
6 Hold the stopper plate up (this will unlock the tensioner), then put a wrench on the hex-shape part of the exhaust camshaft and turn it slightly clockwise; this will depress the plunger of the tensioner.
7 Remove the screwdriver but don't remove the wrench from the camshaft. The stopper plate should now be in the down position and the plunger should be locked.
8 Insert a steel pin of about 0.118-inch diameter into the hole in the stopper plate (see illustration 9.18b). If the pin won't go in, try wiggling the camshaft back and forth until it slides into the hole, then tape it into place so it can't be knocked out.
9 Remove the exhaust camshaft sprocket bolt and lift off the sprocket (see illustration).
10 Loosen the camshaft bearing cap bolts in the reverse of the tightening sequence (see illustration 7.31).
11 Remove the oil tube and its O-ring (see illustration).
12 Remove the front camshaft bearing cap followed by the other caps.
13 Carefully lift out the camshafts, taking care to avoid damaging the aluminum cylinder head.

Note: The intake camshaft sprocket should remain attached to the camshaft. Keep tension on the chain by pulling up on it as the camshafts are removed (have an assistant help you, if necessary).

14 Secure the upper end of the timing chain in position with a wire so that it can't fall into the timing chain cover.

Note: The timing chain cannot be allowed to come off of the crankshaft sprocket.

7.3 With the engine at TDC compression for cylinder no. 1, the marks on the camshaft sprockets should be aligned with the marks on the camshaft front bearing caps. Make a paint mark on the chain links that coincide with the marks

7.4 Remove this plug in the timing chain cover for access to the timing chain tensioner

7.9 Use a large wrench to hold the camshaft while removing the sprocket bolt

7.11 Remove this bolt to remove the oil tube and its O-ring

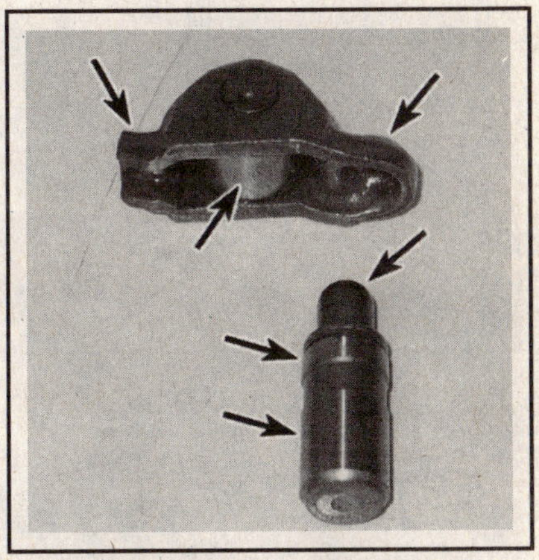

7.18 Check the rocker arms and lash adjusters for wear at the indicated points

15 If necessary, remove all rocker arms and lash adjusters. Make a simple storage box so that all components are labeled and their installed locations are clearly noted. Each rocker arm should be installed in the position in which it was originally used.

➡Note: Store the lash adjusters upright.

INSPECTION

➧ Refer to illustrations 7.18, 7.20, 7.21, 7.22a and 7.22b

16 Position the intake camshaft in a vise, clamping it on the hex portion. Check that the timing gear doesn't rotate.

17 Remove the center intake camshaft sprocket bolt to remove the sprocket.

⁂ CAUTION:

Don't try to remove any of the three outer sprocket bolts.

18 Examine each rocker arm closely for wear and damage (see illustration). Each roller must spin freely and smoothly. If any rocker arm shows signs of wear, replace it.

19 Check each lash adjuster for damage also. The plungers should move, but in a very firm but smooth manner. Replace any lash adjusters that are damaged, locked or have overly loose action.

➡Note: If any plunger offers little resistance when depressing it, try bleeding the air out of it before condemning it as faulty. To do this, place the adjuster upright in a container of clean engine oil, then insert a thin tool, such as a straightened out paper clip or small drill bit into the lash adjuster until it depresses the check ball at the bottom (the tool must be inserted straight down the center). Depress the check ball and actuate the plunger of the adjuster five or six times. Remove the tool and check the operation of the adjuster; if it still offers little resistance, replace it.

20 Visually examine the cam lobes and bearing journals for score marks, pitting, galling and evidence of overheating (blue, discolored areas). Look for flaking away of the hardened surface layer of each lobe.

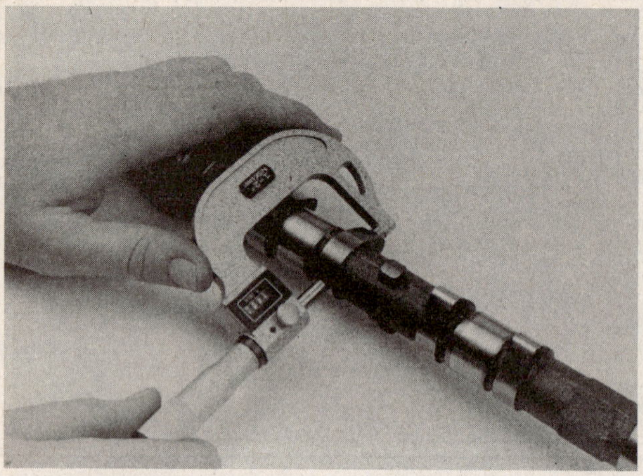

7.20 Measure the lobe heights on each camshaft - if any lobe height is less than the specified allowable minimum, replace that camshaft

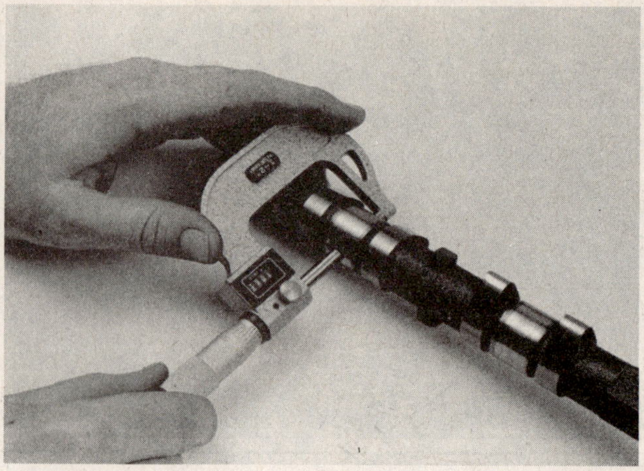

7.21 Measure each journal diameter with a micrometer (if any journal measures less than the specified limit, replace the camshaft)

7.22a The camshaft bearing caps are numbered with an arrow facing the front of the engine (typical cap shown)

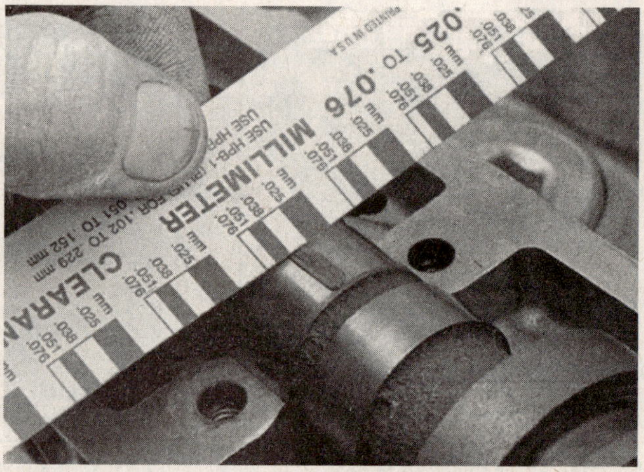

7.22b Compare the width of the crushed Plastigage to the scale on the envelope to determine the oil clearance

Using a micrometer, measure the height of each camshaft lobe (see illustration). Compare your measurements with this Chapter's Specifications. If the height for any one lobe is less than the specified minimum, replace the camshaft.

21 Using a micrometer, measure the diameter of each journal at several points (see illustration). Compare your measurements with this Chapter's Specifications. If the diameter of any one journal is less than specified, replace the camshaft.

22 Check the oil clearance for each camshaft journal as follows:

a) *Clean the bearing caps and the camshaft journals with brake system cleaner.*
b) *Carefully lay the camshaft(s) in place in the cylinder head. Don't install the lifters or intake camshaft sub-gear and don't use any lubrication.*
c) *Lay a strip of Plastigage on each journal.*
d) *Install the bearing caps with the arrows pointing toward the front (timing chain end) of the engine (see illustration).*
e) *Tighten the bolts to the torque listed in this Chapter's Specifications in 1/4-turn increments.*

➡Note: Don't turn the camshaft while the Plastigage is in place.

f) *Remove the bolts and detach the caps.*
g) *Compare the width of the crushed Plastigage (at its widest point) to the scale on the Plastigage envelope (see illustration).*
h) *If the clearance is greater than specified, replace the camshaft and/or cylinder head.*
i) *Scrape off the Plastigage with your fingernail or the edge of a credit card - don't scratch or nick the journals or bearing caps.*

INSTALLATION

➧ Refer to illustration 7.31

23 With the intake camshaft mounted in the vise, align the hole in the intake camshaft sprocket with the knock pin on the camshaft, then install it onto the camshaft. Tighten the bolt to the torque listed in this Chapter's Specifications.

24 Apply camshaft installation lube or clean engine oil to the lobes and bearing journals of both camshafts.

25 Lubricate and install each lash adjuster into the bore from which it was removed.

26 Lubricate the ends and rollers of the rocker arms and install them onto the lash adjusters and the valve stem tips. Make sure that the socket of each rocker arm is seated on its lash adjuster.

27 Install the intake camshaft, with the mark on the sprocket facing up

28 Engage the timing chain with the intake camshaft sprocket, aligning the paint mark on the chain with the mark on the sprocket. Carefully install the intake camshaft/sprocket.

CAUTION:

Keep tension on the timing chain, and be careful not to let it become disengaged from the crankshaft sprocket.

29 Install the exhaust camshaft (without the sprocket), then set the front camshaft bearing cap in place and tighten the bolts finger tight.

30 Install the rest of the camshaft bearing caps. The intake caps are each indicated with an "I," the exhaust caps with an "E." The smaller caps closest to the timing chain are number 2 and those farthest away are number 5. The arrows on all caps point to the timing chain.

31 Tighten the bearing cap bolts a little at a time, and in sequence, to the torque listed in this Chapter's Specifications (see illustration).

32 Place the timing chain onto the exhaust camshaft sprocket, making sure that the chain's paint mark is aligned with the mark on the sprocket.

33 Put the sprocket onto the camshaft, aligning the pin with the hole. Tighten the sprocket bolt to the torque listed in this Chapter's Specifications. Hold the camshaft from turning using a wrench on the camshaft hex area.

34 Verify that all timing marks are correctly aligned, then remove the pin from the tensioner.

➡Note: If the timing chain has been allowed to come off of the crankshaft sprocket, refer to Section 9 for information about correctly aligning the crankshaft and camshaft sprockets.

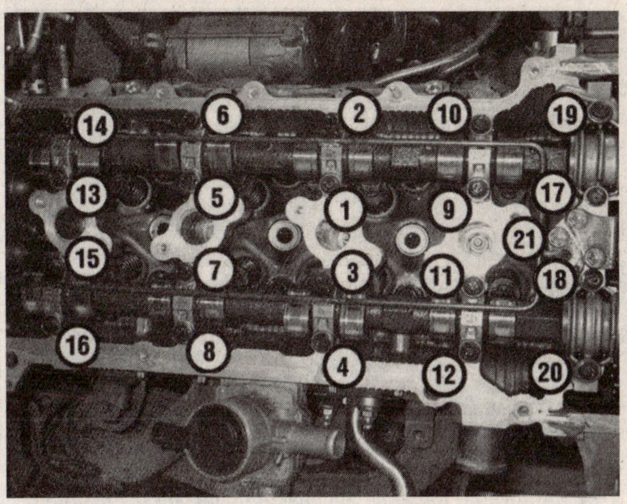

7.31 Camshaft bearing cap bolt tightening sequence

35 Apply a dab of sealant to the tensioner hole plug threads, then install it.

36 Temporarily install the crankshaft pulley bolt and turn the engine in the normal direction of rotation (clockwise) at least two revolutions to verify that the camshafts and chain were installed properly.

CAUTION:

If any resistance is encountered at any point, STOP! There is something wrong - most likely valves are contacting the pistons. You must find the problem before proceeding.

37 The remainder of installation is the reverse of removal. Be sure to replace the O-rings for the timing chain guide and the oil tube with new ones.

8 Cylinder head - removal and installation

REMOVAL

1 Disconnect the cable from the negative terminal of the battery (see Chapter 5, Section 1).

2 Drain the engine coolant (see Chapter 1).

3 Remove the valve cover (see Section 4).

4 Remove the intake manifold (see Section 5).

5 Remove the exhaust manifold (see Section 6).

6 Remove the alternator (see Chapter 5).

7 Refer to Section 9 and remove the timing chain cover.

8 Remove the camshafts, rocker arms and lash adjusters (see Section 7).

9 Disconnect the coolant hoses from the cylinder head. Also disconnect the wiring from the coolant temperature sensor.

10 Label and remove any remaining items, such as coolant fittings, brackets, tubes, cables and hoses.

11 Disconnect and remove the wiring harness brackets.

12 Using a socket bit and a breaker bar, loosen the cylinder head bolts in 1/4-turn increments until they can be removed by hand. Loosen the cylinder head bolts using the reverse of the tightening sequence to avoid warping or cracking the cylinder head (see illustration 8.25).

➡Note: Don't drop any head bolt washers. If they fall into the engine, they are extremely difficult to retrieve.

13 Lift the cylinder head off the engine block. If it's stuck, place a block of wood against it and strike it with a hammer to jar it loose. Don't pry between the gasket mating surfaces.

INSTALLATION

➧ Refer to illustrations 8.15, 8.17 and 8.25

14 The mating surfaces of the cylinder head and block must be perfectly clean when the cylinder head is installed.

8.15 Remove all traces of old gasket material - the cylinder head and block must be perfectly clean to ensure a good gasket seal

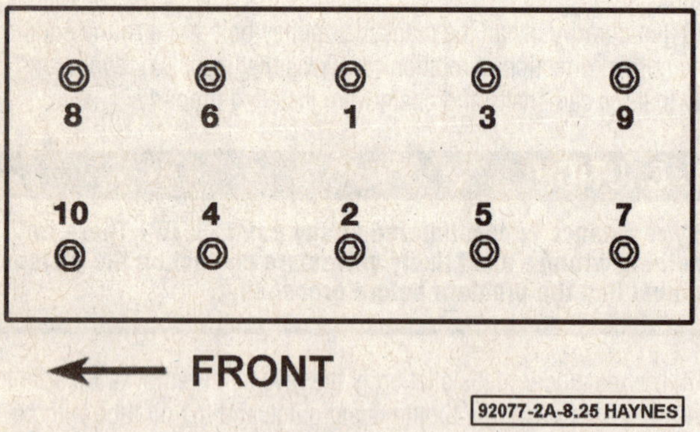

8.25 Cylinder head bolt TIGHTENING sequence

15 Use a gasket scraper to remove all traces of carbon and old gasket material (see illustration), then clean the mating surfaces with brake system cleaner. If there's oil on the mating surfaces when the cylinder head is installed, the gasket may not seal correctly and leaks could develop. When working on the block, stuff the cylinders with clean shop rags to keep out debris. Use a vacuum cleaner to remove material that falls into the cylinders.

16 Check the block and cylinder head mating surfaces for nicks, deep scratches and other damage. If damage is slight, it can be removed with a file; if it's excessive, machining may be the only alternative.

17 Using a precision straightedge and feeler gauges, check the cylinder head mating surface for warpage (see illustration), comparing your measurements with the value listed in this Chapter's Specifications. Also check the surfaces that mate with the intake and exhaust manifolds. If the surface warpage exceeds the limit, take the head to an automotive machine shop for resurfacing.

18 Use a tap of the correct size to chase the threads in the cylinder head bolt holes, then clean the holes with compressed air - make sure that nothing remains in the holes.

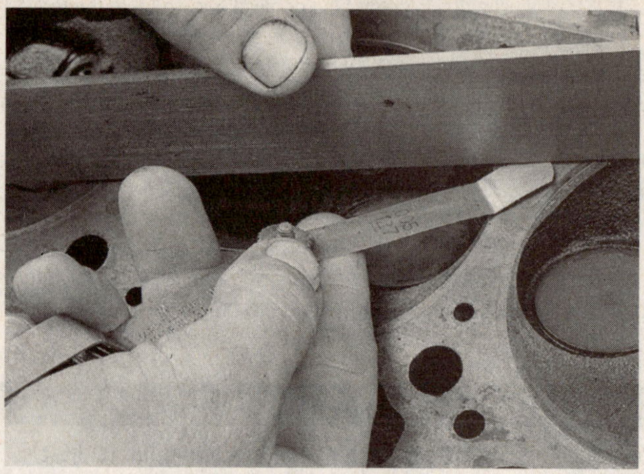

8.17 Use a precision straightedge and feeler gauge to check for warpage

⁂ WARNING:

Wear eye protection when using compressed air!

19 Mount each bolt in a vise and run a die down the threads to remove corrosion and restore the threads. Dirt, corrosion, sealant and damaged threads will affect torque readings.

20 Using a vernier caliper, measure the outside diameter of each cylinder head bolt in several places, comparing your measurements to the minimum allowable diameter listed in this Chapter's Specifications. Replace any bolt with a diameter less than the allowable minimum.

21 Apply a 5/32 to 9/32-inch (4 to 7 mm) dab of RTV sealant to each side of the top of the block and the top of the new head gasket at the front of the engine where the head meets the timing chain cover (25/64-inch/10 mm from the side of the engine block).

➡Note: Install the head gasket within 3 minutes and install the cylinder head within 15 minutes. Wipe off the excess sealant after tightening the head bolts. Don't add coolant or engine oil for at least 4 hours.

22 Position the new gasket over the dowel pins in the block. The number stamped on the gasket should be facing up.

23 Carefully set the cylinder head on the block without disturbing the gasket.

24 Before installing the cylinder head bolts, apply a small amount of clean engine oil to the threads and under the bolt heads.

25 Install the bolts and tighten them finger tight. Following the recommended sequence (see illustration), tighten the bolts in several steps to the Step 1 torque listed in this Chapter's Specifications. Steps 2 and 3 of the tightening sequence require the bolts to be tightened an additional 90-degrees (1/4-turn) each. If you don't have an angle-torque attachment for your torque wrench, simply apply a paint mark at one edge of each cylinder head bolt and tighten the bolt until that mark is 90-degrees from where you started

26 The remaining installation steps are the reverse of removal; refer to the appropriate Sections for component installation.

27 After four hours, refill the cooling system, and change the engine oil and filter (see Chapter 1).

28 Run the engine and check for leaks. Road test the vehicle.

9 Timing chain cover and chains - removal, inspection and installation

WARNING:

The engine must be completely cool before beginning this procedure.

CAUTION:

The timing system is complex, and severe engine damage will occur if you make any mistakes. If you are at all unsure of your abilities, be sure to consult an expert. Double-check all your work and be sure everything is correct before you attempt to start the engine.

REMOVAL

Refer to illustrations 9.14a, 9.14b, 9.17, 9.18a, 9.18b, 9.18c, 9.19a, 9.19b, 9.21, 9.22, 9.23, 9.24a, 9.24b and 9.24c

1 Disconnect the cable from the negative terminal of the battery (see Chapter 5, Section 1).

2 Block the rear wheels and apply the parking brake. Raise the front of the vehicle and support it securely on jackstands.

3 If equipped, remove the protective cover under the front of the engine compartment. Drain the engine coolant (see Chapter 1).

4 Detach the air conditioning compressor without disconnecting the refrigerant lines, then position the compressor aside (see Chapter 3).

CAUTION:

Support the compressor with wire or rope; don't let it hang by the refrigerant lines. Remove the compressor mounting bracket.

5 Lower the vehicle.

6 Detach the power steering pump without disconnecting the hoses. Set it to the side.

7 Remove the alternator (see Chapter 5).

8 Remove the engine cooling fan and fan shroud (see Chapter 3).

9 Remove the spark plugs and drivebelt (see Chapter 1).

10 Refer to Section 4 and remove the valve cover.

11 Remove the spark plugs, then position the number one piston at TDC on the compression stroke (see Section 3).

12 Remove the drivebelt tensioner assembly. Remove the drivebelt idler pulleys and brackets.

13 Remove the camshaft and crankshaft position sensors (see Chapter 6).

14 Loosen the crankshaft pulley bolt, using a pin spanner or a chain wrench to prevent the pulley from turning (see illustration). The crankshaft pulley should slide off the crankshaft. If necessary, use a bolt-type puller to pull it off (see illustration).

15 Remove any other components that interfere with the removal of the timing chain cover.

16 Remove the timing chain cover bolts and nuts. Carefully pry the timing chain cover off with a dull screwdriver, being careful not to nick or gouge the aluminum.

17 Remove the upper timing chain guide along with its O-ring (see illustration).

9.14a A chain wrench can be used to hold the pulley, but only if special precautions are taken; note the sockets used as spacers to raise the chain over the timing scale, and the piece of old drivebelt used to pad the pulley

9.14b If the crankshaft pulley can be removed by hand, use a puller that bolts to the hub of the pulley - not a jaw-type puller. Also, be sure to use the correct adapter between the nose of the crankshaft and the puller screw, so as not to damage the threads in the crankshaft

9.17 Upper timing chain guide bolts

9.18a Rotate the lever upward to unlock the tensioner, then push the plunger in . . .

9.18b . . . then pull the lever down and lock it into place with a drill bit or a similar pin

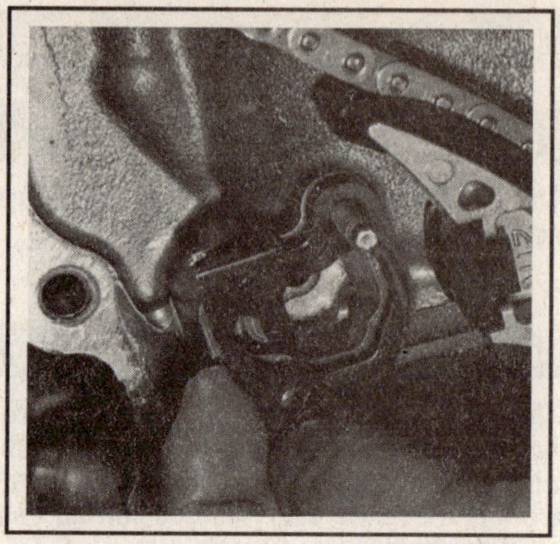

9.18c Remove the tensioner and its gasket

9.19a Right-side (adjusting) chain guide bolt

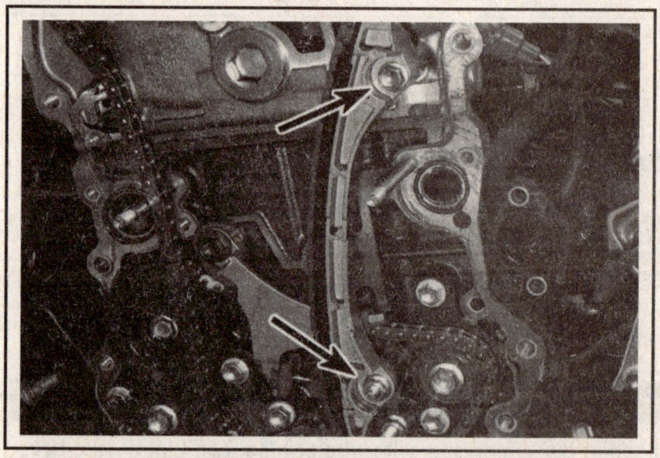

9.19b Left-side (stationary) timing chain guide bolts

9.21 Slide the crankshaft sprocket off

9.22 Balance shaft chain tensioner guide mounting bolt

18 Move the stopper plate on the timing chain tensioner to the up position, then push the plunger in as far as possible (see illustration). Push the plate down to lock the plunger, then insert a steel pin or drill bit (about 0.118 inch diameter) into its hole (see illustration). Remove the timing chain tensioner and the gasket (see illustration).

19 Remove the chain tensioner guide and the fixed guide (see illustrations).

20 Remove the timing chain.

21 Remove the crankshaft sprocket (see illustration).

22 Remove the upper-right side balance shaft chain vibration damper (see illustration).

23 Remove the lower-left side balance shaft chain guide (see illustration).

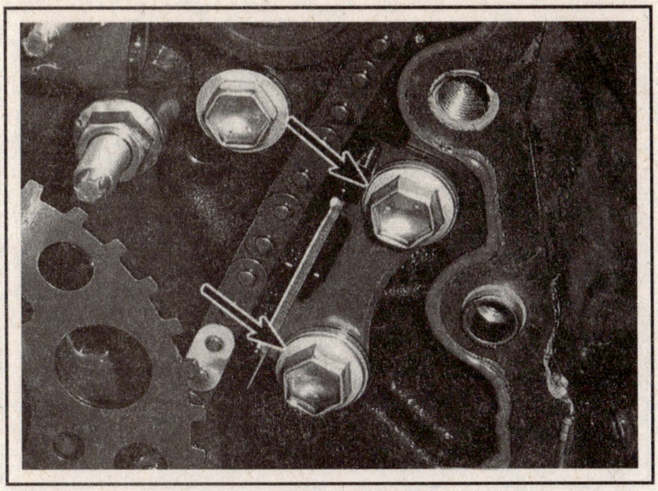

9.23 Left-side balance shaft chain guide

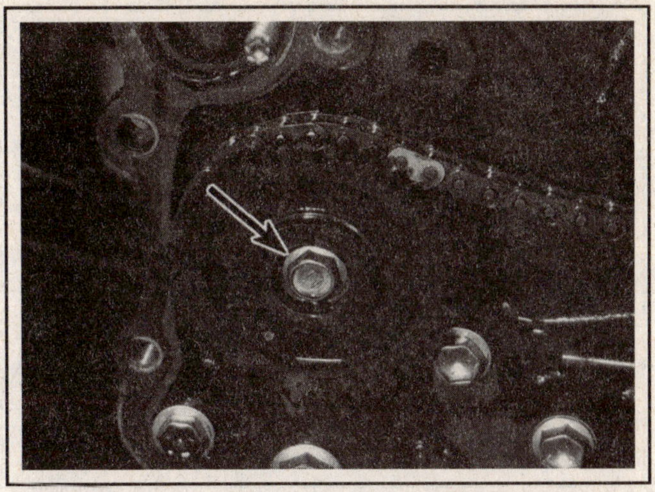

9.24a Remove this bolt from the balance shaft driven gear. . .

9.24b . . . remove this nut and take off the balance shaft chain tensioner . . .

9.24c . . . then lift off the balance shaft chain assembly

24 Remove the bolt from the upper right chain sprocket (see illustration). Remove the tensioner assembly (see illustration), then lift off the balance shaft timing chain (see illustration).

INSPECTION

Refer to illustrations 9.25, 9.27 and 9.28

25 Inspect the individual sprocket teeth and keyways for wear and damage. Check the chains for cracked plates and pitted or worn rollers. Check the chain guides for wear or damage (see illustration). Replace any worn or defective parts with new ones.

26 Check the timing chain tensioners for proper operation:

a) Check that the plunger moves smoothly when the ratchet pawl is raised with your finger.

b) Release the ratchet pawl and make sure the plunger is locked in place by the ratchet pawl and does not move when pushed with your finger.

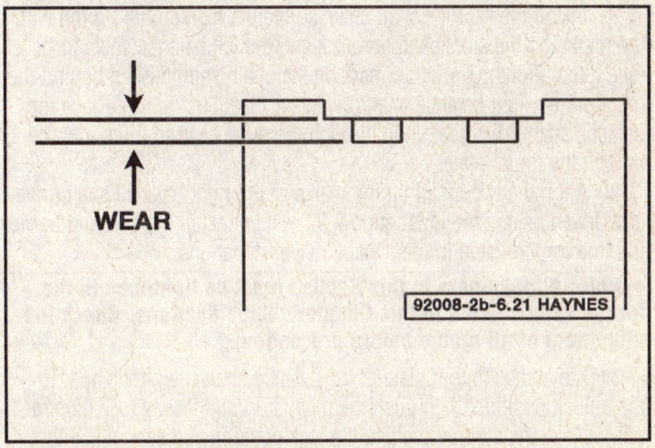

9.25 When inspecting the chain guides, measure wear from the top of the chain contact surface to the bottom of the wear grooves

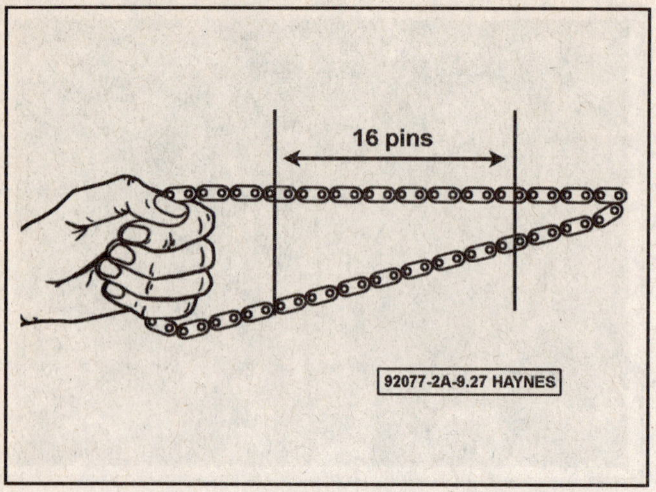

9.27 Timing chain stretch is measured by checking the length of the chain between 16 pins at three or more places

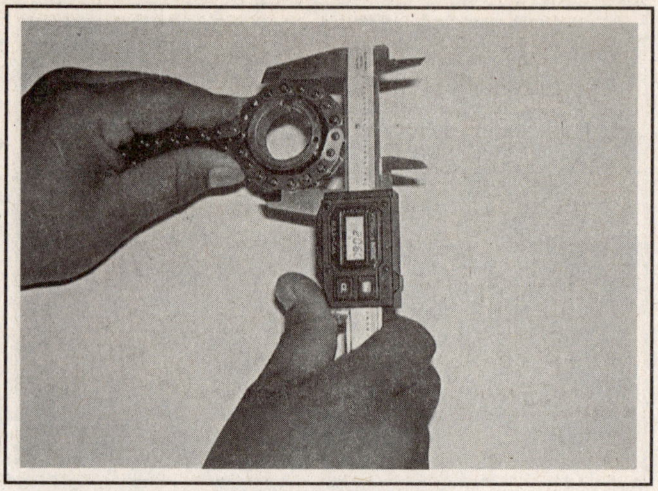

9.28 Wrap the chain around each of the timing sprockets and measure the diameter of the sprockets across the chain rollers - replace the chain and sprockets if the measurement is less than the minimum listed in the Specifications in this Chapter

27 Mount the timing chains in a solid fixture, then pull them tight with a force of about 35 pounds. Measure the distance between any 16 pins using calipers and compare your reading to that listed in this Chapter's Specifications (see illustration). If it's more than specified, replace it.

➡Note: Measure between the pins, not from center of pin to center of pin.

28 Wrap the chains around the sprockets and measure their diameters with calipers (see illustration). Compare your readings to those listed in this Chapter's Specifications. If they're less than the minimums, replace the sprockets.

➡Note: The calipers must be touching the rollers of the chains.

INSTALLATION

➧ Refer to illustrations 9.29a, 9.29b. 9.29c, 9.31a, 9.31b, 9.34a, 9.34b, 9.39a, 9.39b, 9.39c and 9.41

29 Install the balance shaft chain so that its marked links align with the marks on the crankshaft sprocket and the left balance shaft sprocket (see illustrations). The other mark on the chain should align behind the mark on the right balance shaft drive gear (see illustration). Place the balance shaft drive gear shaft through the drive gear so it fits into the hole in the thrust plate.

30 Align the two small timing marks of the right balance shaft drive and driven gears (see illustration 9.29c), then install the bolt and tighten it to the torque listed in this Chapter's Specifications.

➡Note: All fasteners in this Section must be tightened to the torque values listed in this Chapter's Specifications. Check the alignment of all marks before proceeding.

31 Move the stopper plate of the balance shaft chain tensioner to the unlock position, then push the plunger in all the way (see illustration). Move the stopper plate to the lock position and secure it with a 0.118-inch drill bit or steel pin through its hole (see illustration). Install the tensioner and tighten the nut to the torque listed in this Chapter's Specifications.

32 Install the left-side guide shoe and the vibration damper. Remove the pin from the tensioner to allow the plunger to extend. Again check the alignment of all marks.

33 Install the crankshaft sprocket and the left-side chain guide for the camshaft timing chain.

34 Install the camshaft timing chain. The orange marks on the chain links should align with the marks on the camshaft sprockets and the yellow mark should align with the mark on the crankshaft (see illustrations).

35 Secure the chain so it's wrapped tight around the crankshaft sprocket with a plastic wire tie or cord to prevent it from falling off.

36 Install the chain tensioner guide shoe and the tensioner assembly along with a new gasket. Remove the tie from the crankshaft sprocket, hold the chain tight to prevent it from dropping off the crankshaft sprocket, then remove the pin from the tensioner.

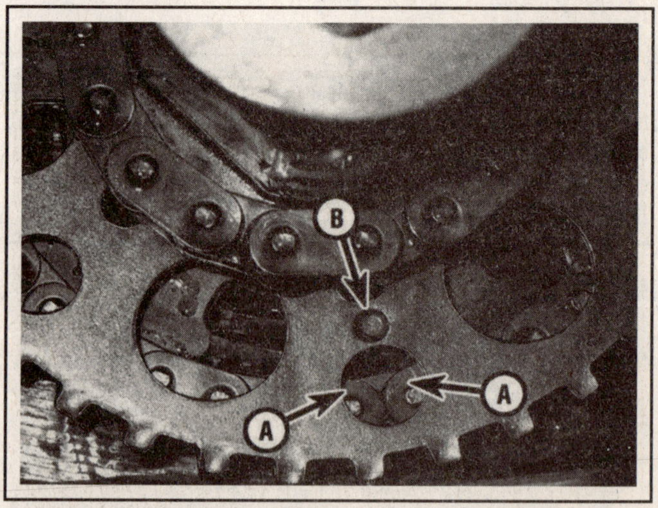

9.29a The two colored links (A) on the balance shaft chain must flank the mark (B) on the crankshaft sensor wheel

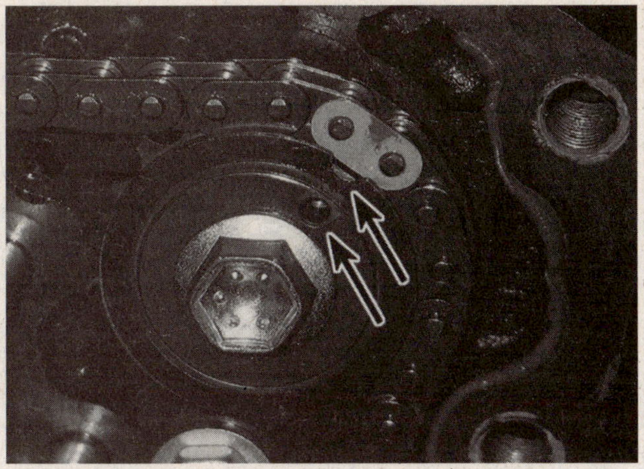

9.29b The marks on the left balance shaft sprocket align with the colored link on the balance shaft chain

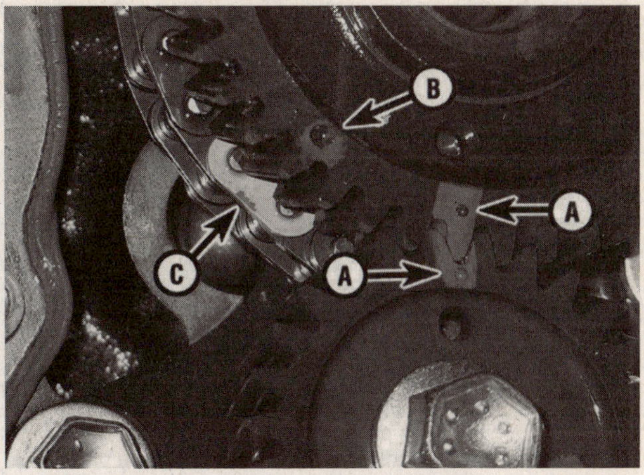

9.29c The marks (A) on the right-side balance shaft drive and driven gears must be aligned; the mark on the drive gear (B) aligns with the colored link of the chain (C)

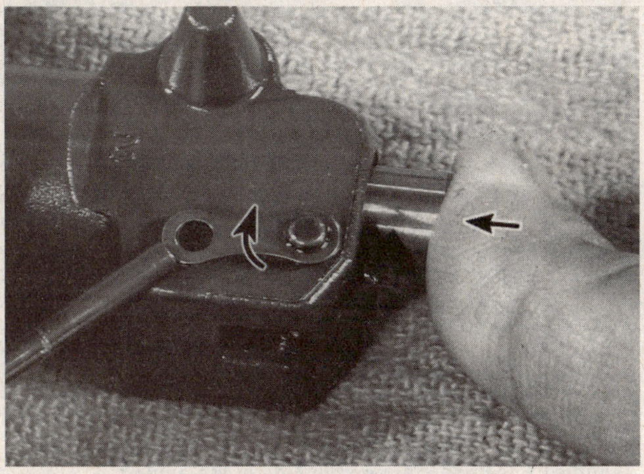

9.31a Rotate the lever, then push in the plunger

9.31b Temporarily lock the tensioner using a drill bit or other small rod

9.34a The orange timing chain links align with the marks on the camshaft sprockets

9.34b The yellow link aligns with the mark on the crankshaft sprocket

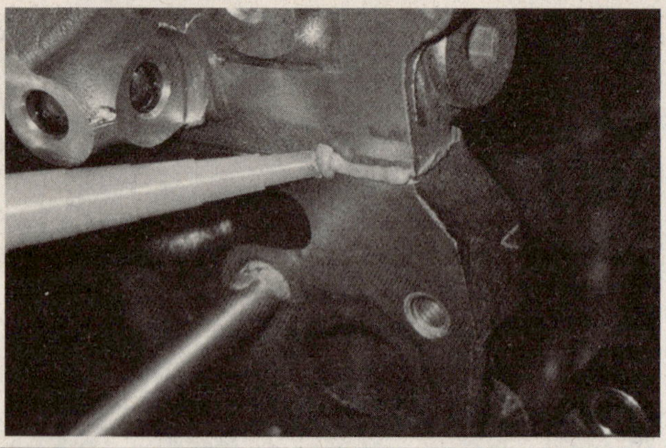

9.39a Apply short beads of sealant at the juncture on each side where the timing chain cover contacts the cylinder head and engine block

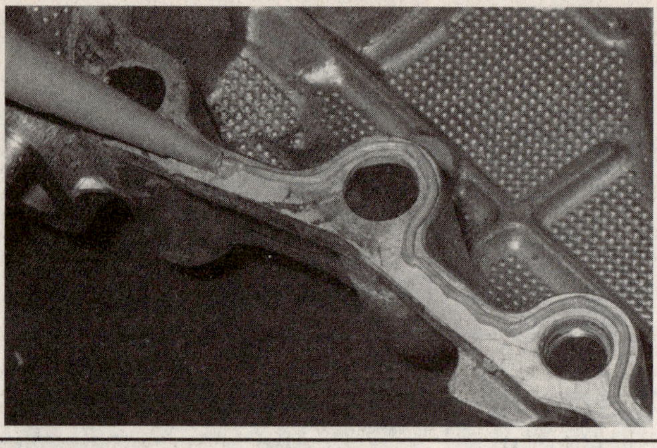

9.39b Apply a continuous bead of RTV sealant, inboard of the bolt holes . . .

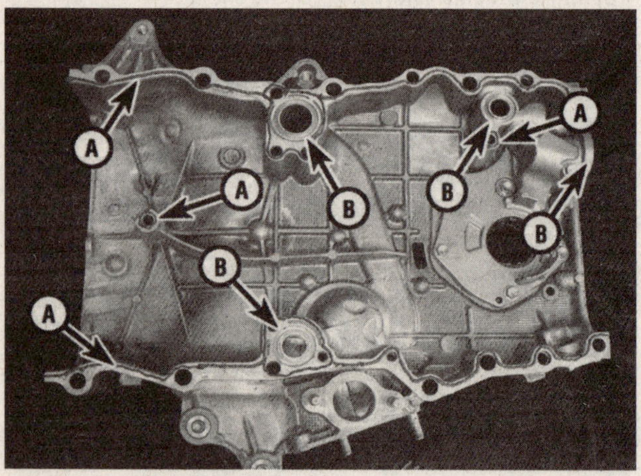

9.39c . . . to the mating surfaces on the perimeter of the cover, as well as around the bolt holes (A); (B) indicates O-ring locations

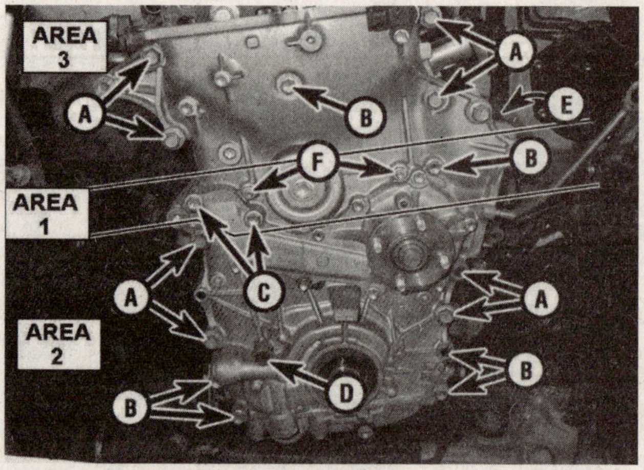

9.41 Timing chain cover fastener tightening areas and bolt lengths - E fasteners are on the side of the engine

A	*2.95 inches (75 mm)*	*D*	*3.74 inches (95 mm)*
B	*2.95 inches (75 mm)*	*E*	*1.37 inches (35 mm)*
C	*3.43 inches (87 mm)*	*F*	*Nuts*

37 Install the upper timing chain guide with a new O-ring.

CAUTION:

Verify that every set of marks is correctly aligned. Severe engine damage can occur if any components are not aligned properly. Temporarily install the crankshaft pulley bolt and rotate the crankshaft in the normal direction of rotation (clockwise) at least two revolutions. If any resistance is encountered at any point, STOP! There is something wrong - most likely, valves are contacting the pistons. You must find the problem before proceeding.

38 Remove all dirt, oil and grease from the timing chain area at the front of the engine and the timing chain cover. Clean the sealing surface of the cover and engine with brake system cleaner and install new O-rings.

39 Apply beads of RTV sealant to the areas shown (see illustrations). Also apply sealant to the bottom of the timing chain cover and the mating lip of the oil pan.

40 Make sure the O-rings are in place and install the timing chain cover onto the engine block, engaging the splined shaft of the oil pump drive rotor with the oil pump drive gear.

Note: There are three O-rings on the rear of the timing cover and one at the bottom of the oil pump.

41 Install the timing chain cover bolts in their proper locations (see illustration).

42 Tighten all of the nuts and bolts (except "A" bolts) in the correct sequence - Area 1 first, followed by Area 3, then Area 2 (see illustration 9.41). When done, tighten all of the "A" bolts in the correct sequence - Area 2 first, followed by Area 3. Tighten the "E" bolts last. These fasteners as well as all others used in this Section must be tightened to the torque listed in this Chapter's Specifications.

43 Reinstall the remaining parts in the reverse order of removal, referring to the appropriate Sections.

44 Change the engine oil and filter, and refill the cooling system (see Chapter 1).

45 Start the engine and check for leaks and proper operation.

10 Crankshaft front oil seal - replacement

Refer to illustrations 10.5 and 10.7

1 Remove the drivebelt (see Chapter 1).
2 Refer to Chapter 3 and remove the fan shroud for clearance.
3 Loosen the crankshaft pulley bolt. It will be necessary to hold the pulley with a strap wrench, a chain wrench or a special tool to keep it from turning (see illustration 9.14a).
4 If the pulley can't be removed by hand, use a special puller.

CAUTION:

Use a puller that has bolts that screw into the pulley - not the type with jaws that grab the outer rim.

Remove the pulley.

5 Note how far the seal is recessed in the bore, then carefully pry it out of the timing chain cover with a screwdriver or seal removal tool (see illustration). Don't scratch the cover bore or damage the crankshaft in the process (if the crankshaft is damaged, the new seal will end up leaking).
6 Clean the bore in the cover and coat the outer edge and the lip of the new seal with engine oil or multi-purpose grease.
7 Using a seal driver (preferably) or a socket with an outside diameter slightly smaller than the outside diameter of the seal, carefully drive the new seal into place (see illustration). Make sure it's installed squarely and driven in to the same depth as the original. Check the seal after installation to make sure the spring didn't pop out of place.
8 Reinstall the crankshaft pulley and drivebelt.
9 Run the engine and check for oil leaks at the front seal.

10.5 Use a seal puller to remove the crankshaft front oil seal from the oil pump housing - do not nick or scratch the crankshaft or the timing chain cover in the process

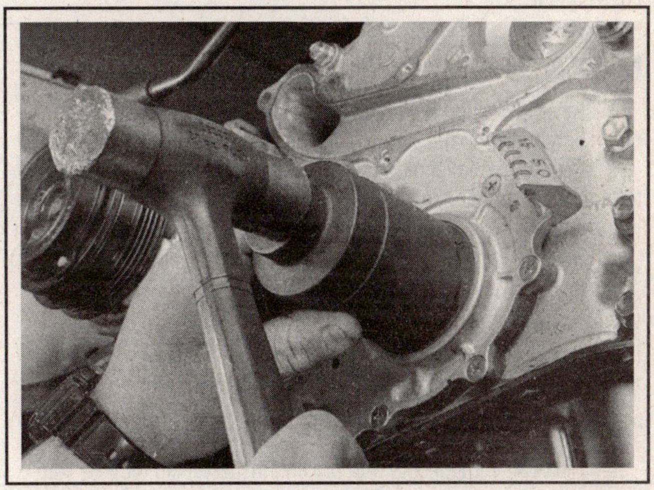

10.7 A large socket with a slightly smaller diameter than the seal bore can be used to carefully and evenly drive the seal into the timing chain cover

11 Oil pan - removal and installation

Note: The oil pan cannot be removed from these vehicles without first removing the engine (see Chapter 2C).

REMOVAL

Lower oil pan

1 Disconnect the cable from the negative terminal of the battery (see Chapter 5, Section 1).
2 Refer to Chapter 2C to remove the engine.
3 Remove the bolts and nuts from the lower oil pan.
4 Strike the oil pan with a soft-face hammer to break it loose. If it's stuck, tap a knife blade into the joint and hammer it around until the seal is broken. Don't pry the oil pan or drive a screwdriver into the joint.
5 The stamped steel lower oil pan can be straightened if the sealing flange is distorted. Place it on a solid, flat surface and use a hammer and punch to flatten any bent areas.

Upper oil pan

6 After removing the lower pan, remove the nuts and bolts securing the oil pump pick-up tube and strainer. Remove the assembly and its gasket.
7 Remove the nuts and bolts retaining the upper oil pan.
8 Strike the oil pan with a soft-face hammer to break it loose. If it's stuck, tap a knife blade into the joint and hammer it around until the seal is broken. Don't pry the oil pan or drive a screwdriver into the joint.

INSTALLATION

9 Use a scraper to remove all traces of old gasket material and sealant from the block and oil pan. Clean the mating surfaces with brake system cleaner.

10 Make sure the threaded bolt holes are clean.

11 Apply a 1/8-inch wide bead of RTV sealant to the oil pan flange. The bead of sealant must travel inboard of the bolt holes.

➡Note: Installation must be completed within 5 minutes of applying the sealant.

12 Install the upper oil pan. Tighten the fasteners, a little at a time and in a criss-cross pattern, to the torque listed in this Chapter's Specifications.

13 Inspect the oil pump pick-up tube assembly for cracks and a blocked strainer. Install the oil pick-up tube with a new gasket. Tighten the fasteners to the torque listed in this Chapter's Specifications.

14 Apply a 1/8-inch bead of RTV sealant to the lower oil pan, inboard of the bolt holes. Install the lower oil pan. Tighten the nuts and bolts, a little at a time and in a criss-cross pattern, to the torque listed in this Chapter's Specifications.

CAUTION:

Don't add oil for at least 4 hours after installation.

15 The remainder of installation is the reverse of removal.

16 Install the engine (see Chapter 2C).

12 Oil pump - removal, inspection and installation

REMOVAL

➧ Refer to illustrations 12.2a, 12.2b and 12.3

1 Refer to Section 9 and remove the timing chain cover (the oil pump is mounted to the inside of the timing cover).

2 Remove the relief plug from the bottom of the pump (see illustration). Lay out the plug, gasket, spring and relief valve in order (see illustration).

3 Remove the bolts and lift off the oil pump cover (see illustration).

4 Lift out the drive and driven rotors.

INSPECTION

➧ Refer to illustrations 12.7a, 12.7b and 12.7c

5 Clean all components with solvent, inspect them for wear and damage.

6 Check the oil pressure relief valve piston sliding surface and valve spring. If either the spring or the valve is damaged, they must be replaced as a set. The valve must slide very freely in its bore.

7 Check the driven rotor-to-body clearance, rotor-to-cover clearance and drive rotor tip clearance with a feeler gauge (see illustrations) and compare the results to this Chapter's Specifications. If any clearance is excessive, replace the rotors as a set. If necessary, replace the oil pump body (timing cover).

INSTALLATION

➧ Refer to illustrations 12.8 and 12.9

8 Lubricate the drive and driven rotors with clean engine oil and place them in the pump body. The small marks on the rotors must face out, away from the timing cover (see illustration).

9 Pack the pump cavities with petroleum jelly - this will help prime the pump when the engine is first started. Install a new O-ring in the groove in the timing chain cover (see illustration) and attach the pump, tightening the screws to the torque listed in this Chapter's Specifications.

10 Lubricate the oil pressure relief valve piston with clean engine oil and reinstall the valve components in the pump body.

11 Install the timing cover onto the engine block, engaging the oil pump drive gear with the oil pump gear on the crankshaft (see Section 9).

12 Reinstall the remaining parts in the reverse order of removal, referring to the appropriate Sections for component installation.

13 Change the engine oil and filter and add coolant (see Chapter 1). Start the engine and check for oil pressure and leaks.

12.2a Remove the relief valve plug slowly to catch the spring

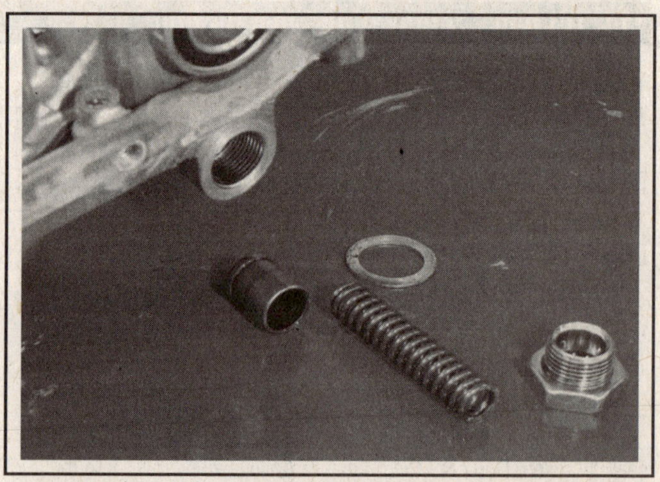

12.2b Relief valve components - the valve must slide freely in the bore

12.3 Oil pump cover bolts

12.7a Measure the driven rotor-to-body clearance with a feeler gauge (typical pump shown)

12.7b Using a straightedge and feeler gauge, measure the rotor-to-cover clearance

12.7c Measure the rotor tip clearance with a feeler gauge

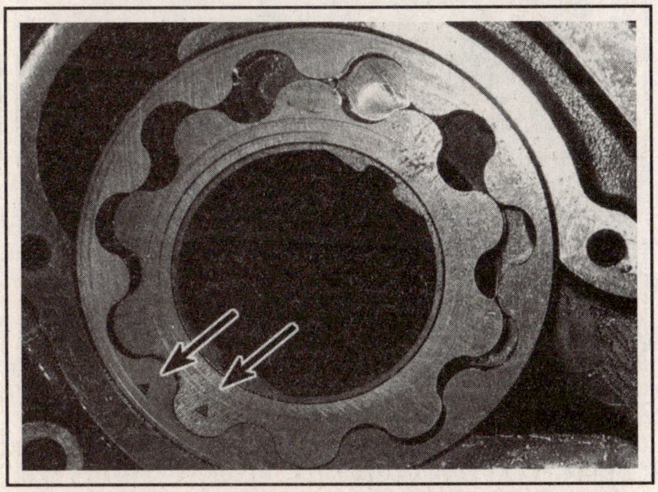

12.8 The marks on the oil pump rotors must face out

12.9 Be sure to replace the O-ring in the bottom of the timing chain cover

13 Flywheel/driveplate - removal and installation

➡Note: Manual transmission-equipped models use flywheels. "Driveplate" is the terminology used for the automatic transmission version of this part.

REMOVAL

1 Disconnect the cable from the negative terminal of the battery (see Chapter 5, Section 1).

2 On manual transmission equipped vehicles, remove the pressure plate and clutch disc (see Chapter 8).

3 Make alignment marks on the flywheel/driveplate and crankshaft to ensure correct alignment during reinstallation.

4 Remove the bolts securing the flywheel/driveplate to the crankshaft. If the crankshaft turns, wedge a screwdriver in the ring gear teeth to hold the flywheel/driveplate.

5 Remove the flywheel/driveplate from the crankshaft. Since it is fairly heavy, be sure to support it while removing the last bolt. Automatic transmission equipped vehicles have spacers on both sides of the driveplate. Keep them with the driveplate.

⁂ WARNING:

The ring-gear teeth may be sharp; wear gloves to protect your hands.

INSTALLATION

6 Clean the flywheel/driveplate to remove grease and oil. Inspect the surface for cracks, rivet grooves, burned areas and score marks. Light scoring can be removed with emery cloth. Check for cracked or broken ring gear teeth. Lay the flywheel on a flat surface and use a straightedge to check for warpage.

7 Clean and inspect the mating surfaces of the flywheel/driveplate and the crankshaft. If the crankshaft rear seal is leaking, replace it before reinstalling the flywheel/driveplate (see Section 14).

8 Position the flywheel/driveplate against the crankshaft. Be sure to align the marks made during removal. Note that some engines have an alignment dowel or staggered bolt holes to ensure correct installation. Before installing the bolts, apply thread-locking compound to the threads.

9 Wedge a screwdriver in the ring gear teeth to keep the flywheel/driveplate from turning and tighten the bolts to the torque listed in this Chapter's Specifications. Follow a criss-cross pattern and work up to the final torque in three or four steps.

10 The remainder of installation is the reverse of the removal procedure.

14 Rear main oil seal - replacement

➧ Refer to illustration 14.6

1 Remove the transmission (see Chapter 7). Remove the rear end plate.

2 The seal can be replaced without removing the oil pan or seal retainer. However, this method is not recommended because the lip of the seal is quite stiff and it's possible to cock the seal in the retainer bore or damage it during installation. If you want to take the chance, pry out the old seal with a screwdriver. Apply multi-purpose grease to the crankshaft seal journal and the lip of the new seal and carefully press the new seal into place. The lip is stiff, so carefully work it onto the seal journal of the crankshaft with a smooth object like the end of an extension as you tap the seal into place. Don't rush it or you may damage the seal.

3 The following method is recommended but requires removing the seal retainer and resealing the rear of the oil pan (see Section 11).

4 After removing the two rearmost oil pan-to-seal retainer bolts, break the seal between the rear of the oil pan and the bottom of the seal retainer with a putty knife. Remove the retainer-to-engine block bolts, detach the seal retainer and remove all the old gasket material. Remove the sealant from the top of the oil pan flange.

➡Note: Cover the open area of the oil pan with clean rags to keep debris out while bracing the pan flange.

5 Position the seal and retainer assembly between two wood blocks on a workbench and drive the old seal out from the back side with a screwdriver (see illustration).

6 Drive the new seal into the retainer with a wood block (see illustration).

7 Lubricate the crankshaft seal journal and the lip of the new seal

14.6 Drive the new seal into the retainer with a wood block or a section of pipe - make sure that you don't cock the seal in the retainer bore

with multi-purpose grease. Position a new gasket on the engine block. Apply a bead of RTV sealant to the exposed portion of oil pan flange and particularly at the pan-to-block mating surface.

8 Slowly and carefully push the seal and retainer onto the crankshaft. The seal lip is stiff, so work it onto the crankshaft with a smooth object such as the end of an extension as you push the retainer against the block.

9 Install and tighten the retainer bolts to the torque listed in this Chapter's Specifications.

10 The remainder of installation is the reverse of removal.

15 Engine mounts - check and replacement

1 Engine mounts seldom require attention, but broken or deteriorated mounts should be replaced immediately or the added strain placed on the driveline components may cause damage or wear.

CHECK

2 During the check, the engine must be raised slightly to remove the weight from the mounts.

3 Raise the vehicle and support it securely on jackstands, then position a jack under the engine oil pan. Place a large wood block between the jack head and the oil pan, then carefully raise the engine just enough to take the weight off the mounts. Do not position the wood block under the drain plug.

⁂ WARNING:

DO NOT place any part of your body under the engine when it's supported only by a jack!

4 Check the mounts to see if the rubber is cracked, hardened or separated from the metal plates. Sometimes the rubber will split down the center.

5 Check for relative movement between the mount plates and the engine or frame (use a large screwdriver or pry bar to attempt to move the mounts). If movement is noted, lower the engine and tighten the mount fasteners.

6 Rubber preservative should be applied to the mounts to slow deterioration.

REPLACEMENT

7 Disconnect the cable from the negative battery terminal (see Chapter 5, Section 1), then raise the vehicle and support it securely on jackstands (if not already done). Support the engine as described in Step 3.

8 To remove an engine mount, remove the fasteners, raise the engine and detach the mount.

9 Installation is the reverse of removal. Use thread locking compound on the mount bolts/nuts and be sure to tighten them securely.

10 See Chapter 7 for transmission mount replacement.

Specifications

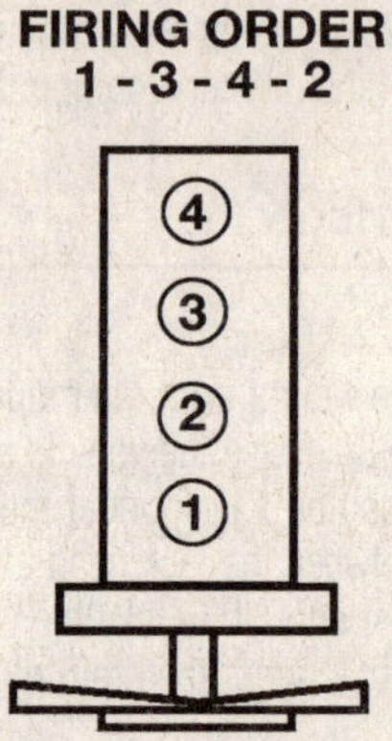

Cylinder location diagram

General

Engine designation	2TR-FE
Displacement	2.7L
Cylinder numbers	
(drivebelt end-to-transmission end)	1-2-3-4
Firing order	1-3-4-2

Camshaft

Journal diameter	
Number 1	1.4153 to 1.4159 inches (35.949 to 35.965 mm)
All others	1.0614 to 1.0620 inches (26.959 to 26.975 mm)
Journal-to-bearing (oil) clearance	
Number 1	
Standard	0.0014 to 0.0029 inch (0.035 to 0.072 mm)
Limit	0.003 inch (0.08 mm)
All others	
Standard	0.0010 to 0.0024 inch (0.025 to 0.062 mm)
Limit	0.003 inch (0.08 mm)
Journal out-of-round limit	0.0012 inch (0.03 mm)
Lobe height	
Standard	1.6872 to 1.6911 inches (42.855 to 42.954 mm)
Limit	1.6872 inches (42.855 mm)
Thrust clearance (endplay)	
Standard	0.004 to 0.009 inch (0.10 to 0.24 mm)
Limit	0.010 inch (0.26 mm)

Cylinder head

Warpage limits	
Cylinder head surface	0.002 inch (0.05 mm)
Intake manifold surface	0.002 inch (0.05 mm)
Exhaust manifold surface	0.002 inch (0.05 mm)
Cylinder head bolt diameter	
Standard	0.4236 to 0.4319 inch (10.76 to 10.97 mm)
Minimum	0.4049 inch (10.40 mm)

Oil pump

Rotor-to-body clearance	
Standard	0.0098 to 0.0128 inch (0.250 to 0.325 mm)
Service limit	0.0167 inch (0.425 mm)
Rotor tip clearance	
Standard	0.0016 to 0.0063 inch (0.040 to 0.160 mm)
Service limit	0.0102 inch (0.26 mm)
Rotor-to-cover clearance	
Standard	0.0010 to 0.0030 inch (0.025 to 0.075 mm)
Service limit	0.0051 inch (0.130 mm)

Timing chains and sprockets

Timing chain stretch limit	
Timing chain	5.807 inches (147.5 mm)
Timing chain number 2	4.866 inches (123.6 mm)
Sprocket wear limit (with chain)	
Camshaft sprocket diameter	4.480 inches (113.8 mm)
Crankshaft sprocket diameter	2.338 inches (59.4 mm)
Crankshaft sprocket number 2 diameter	3.807 inches (96.7 mm)
Balance shaft sprocket diameter	2.988 inches (75.9 mm)

➡Note: One foot-pound (ft-lb) of torque is equivalent to 12 inch-pounds (in-lbs) of torque. Torque values below approximately 15 foot-pounds are expressed in inch-pounds, because most foot-pound torque wrenches are not accurate at these smaller values.

Torque specifications	Ft-lbs (unless otherwise indicated)	Nm
Intake manifold fasteners	18	24
Exhaust manifold nuts	27	37
Exhaust pipe-to-exhaust manifold	32	43
Crankshaft pulley-to-crankshaft bolt	192	260
Flywheel bolts (manual transmission)		
Step 1	20	27
Step 2	Tighten an additional 90-degrees	
Driveplate bolts (automatic transmission)	55	75
Cylinder head bolts		
Step 1	29	39
Step 2	Tighten an additional 90-degrees	
Step 3	Tighten an additional 90-degrees	
Cylinder head-to-timing cover bolts	180 in-lbs	20
Camshaft bearing cap bolts		
Bolt A	108 in-lbs	12
All others	132 in-lbs	15
Camshaft sprocket bolts	58	79
Oil pump bolts	80 in-lbs	9
Oil pick-up tube strainer assembly fasteners	19	26
Oil pan-to-block bolts	19	26
Oil pan-to-lower oil pan bolts	80 in-lbs	9
Oil pump relief valve plug	36	49
Rear main oil seal retainer bolts	180 in-lbs	20
Timing cover bolts and nuts		
Bolt A	34	46
All others	180 in-lbs	20
Timing chain components		
Balance shaft chain		
Drive gear bolt	18	24
Tensioner nut	156 in-lbs	17.5
Guide bolts	156 in-lbs	17.5
Vibration damper bolts	20	27
Timing chain		
Left guide bolts	180 in-lbs	20
Tensioner guide bolts	180 in-lbs	20
Tensioner assembly bolts	84 in-lbs	9.5
Upper guide bolts	84 in-lbs	9.5
Valve cover bolts	80 in-lbs	9

Notes

2B

V6 ENGINE

Reference to other Chapters

CHECK ENGINE light - See Chapter 6
Cylinder compression check - See Chapter 2C
Drivebelt check, adjustment and replacement - See Chapter 1
Engine - removal and installation - See Chapter 2C
Engine oil and filter change - See Chapter 1
Engine overhaul - general information - See Chapter 2C
Spark plug replacement - See Chapter 1
Water pump - removal and installation - See Chapter 3

1 General information

This Part of Chapter 2 is devoted to in-vehicle repair procedures for the 4.0L (1GR-FE) engine. Information concerning engine removal and installation and engine overhaul can be found in Part C of this Chapter.

The following repair procedures are based on the assumption that the engine is installed in the vehicle. If the engine has been removed from the vehicle and mounted on a stand, many of the steps outlined in this Part of Chapter 2 will not apply.

The Specifications included in this Part of Chapter 2 apply only to the procedures contained in this Part. Additional specifications can be found in Chapter 2 Part C.

2 Repair operations possible with the engine in the vehicle

1 Many major repair operations can be accomplished without removing the engine from the vehicle. Clean the engine compartment and the exterior of the engine with some type of degreaser before any work is done. It will make the job easier and help keep dirt out of the internal areas of the engine.

2 Depending on the components involved, it may be helpful to remove the hood to improve access to the engine as repairs are performed (refer to Chapter 11 if necessary). Cover the fenders to prevent damage to the paint. Special pads are available, but an old bedspread or blanket will also work.

3 If vacuum, exhaust, oil or coolant leaks develop, indicating a need for gasket or seal replacement, the repairs can generally be made with the engine in the vehicle. The intake and exhaust manifold gaskets, oil pan gasket, crankshaft oil seals and cylinder head gaskets are all accessible with the engine in place.

4 Exterior engine components, such as the intake and exhaust manifolds, the oil pump, the water pump, the starter motor, the alternator and the fuel system components can be removed for repair with the engine in place.

5 Since the cylinder heads can be removed without pulling the engine, valve component servicing can also be accomplished with the engine in the vehicle. Replacement of the camshafts, timing belt and sprockets is also possible with the engine in the vehicle.

3 Top Dead Center (TDC) for number one piston - locating

➧ Refer to illustration 3.6

1 Top Dead Center (TDC) is the highest point in the cylinder that each piston reaches as it travels up the cylinder bore. Each piston reaches TDC on the compression stroke and again on the exhaust stroke, but TDC generally refers to piston position on the compression stroke.

2 Positioning the piston(s) at TDC is an essential part of certain procedures such as valve adjustment and timing chain/camshaft removal and installation.

3 Before beginning this procedure, remove the Number One spark plug (see Chapter 1, if necessary). Also, be sure to place the transmission in Neutral and apply the parking brake or block the rear wheels. If method b) or c) in Step 4 will be used to rotate the engine, also disable the ignition system by disconnecting the electrical connector from each ignition coil, and disable the fuel pump (see Chapter 4, Section 2).

4 In order to bring any piston to TDC, the crankshaft must be turned using one of the methods outlined below. When looking at the front of the engine, normal crankshaft rotation is clockwise.

- a) *The preferred method is to turn the crankshaft with a socket and ratchet attached to the bolt threaded into the front of the crankshaft. Turn the crankshaft in a clockwise direction only.*
- b) *A remote starter switch, which may save some time, can also be used. Follow the instructions included with the switch. Once the piston is close to TDC, use a socket and ratchet as described in the previous paragraph.*
- c) *If an assistant is available to turn the ignition switch to the Start position in short bursts, you can get the piston close to TDC without a remote starter switch. Make sure your assistant is out of the vehicle, away from the ignition switch, then use a socket and ratchet as described in Paragraph (a) to complete the procedure.*

5 Install a compression pressure gauge in the number one spark plug hole (see Chapter 2C). It should be a gauge with a screw-in fitting and a hose at least six inches long.

6 Rotate the crankshaft using one of the methods described above while observing the compression gauge. When TDC for the compression stroke of number one cylinder is reached, compression pressure will show on the gauge as the marks on the crankshaft pulley are beginning to line up (see illustration). If you go past the marks, release the gauge pressure and rotate the crankshaft around two more revolutions.

7 After the number one piston has been positioned at TDC on the compression stroke, TDC for the next cylinder in the firing order can be located by turning the crankshaft another 120-degrees (refer to the firing order in this Chapter's Specifications).

3.6 Turn the crankshaft until the notch in the pulley aligns with the zero on the timing plate

4 Valve covers - removal and installation

Refer to illustrations 4.7, 4.9, 4.10 and 4.11

1 Disconnect the cable from the negative terminal of the battery (see Chapter 5, Section 1).

2 Remove the engine cover.

3 Remove the air intake duct and the air filter housing (see *Air filter housing - removal and installation* in Chapter 4).

4 If you're going to remove the left valve cover, remove the upper intake manifold (see Section 5).

5 Disconnect the PCV ventilation hose from the valve cover.

6 Remove the ignition coils.

7 Remove the valve cover retaining bolts and nuts and remove the valve cover (see illustration).

8 Remove and discard the valve cover gasket.

9 Installation is the reverse of removal. Inspect the main cover gasket and spark plug tube seals (which are part of the gasket), replacing them if necessary (see illustration).

10 Put new seals on the three center bolts (see illustration).

11 Apply dabs of RTV sealant to the joints where the timing chain cover meets the engine block (see illustration).

12 Evenly tighten the valve cover nuts and bolts to the torque listed in this Chapter's Specifications.

13 Start the engine and check for oil leaks around the edges of the valve cover.

4.7 To detach the valve cover from the cylinder head, remove the ignition coils before removing the valve cover bolts

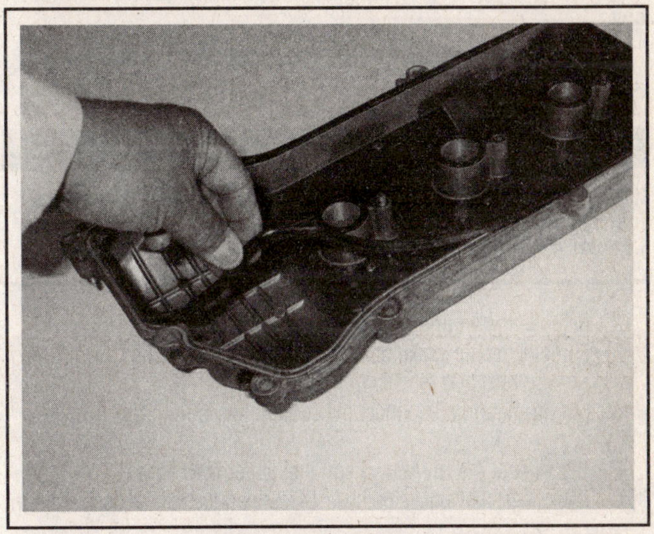

4.9 Make sure that the gasket is fully seated into the valve cover groove, and that the spark plug tube seals are seated as well

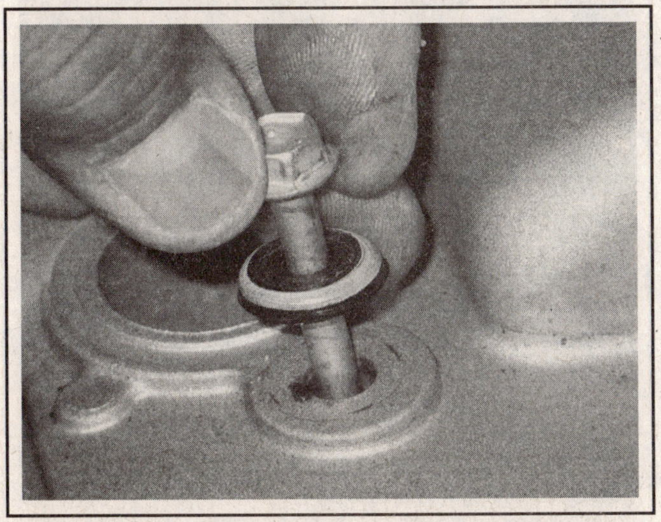

4.10 The center valve cover bolts use sealing washers

4.11 After removing the old sealant, apply a fresh dab of RTV sealant to the joints at the front of the valve cover where the cylinder head meets the timing chain cover

5 Intake manifold - removal and installation

WARNING:

Wait until the engine is completely cool before beginning this procedure.

UPPER INTAKE MANIFOLD

Refer to illustrations 5.11 and 5.12

Note: Toyota refers to the upper intake manifold as the intake air surge tank. If you're buying a gasket for the upper intake manifold at a dealer parts department, use the Toyota terminology.

1 Disconnect the cable from the negative terminal of the battery (see Chapter 5, Section 1).

WARNING:

If you plan on removing the lower intake manifold, relieve the fuel system pressure before disconnecting the battery (see Chapter 4, Section 2).

2 Remove the engine cover.

3 Remove the air intake duct and the air filter housing (see *Air filter housing - removal and installation* in Chapter 4).

4 Clamp-off and disconnect the coolant hoses from the throttle body.

5 Disconnect the (EVAP system) fuel vapor feed hose.

6 Disconnect the ventilation hose. Disconnect the throttle body wiring.

7 Disconnect the two Vacuum Switching Valve (VSV) electrical connectors.

8 Remove the two throttle body bracket bolts and remove the bracket.

9 Remove the baffle plate.

10 Remove both bolts from each intake manifold support bracket (Toyota refers to these two support brackets as surge tank stays) and remove both brackets.

11 Remove the two intake manifold mounting nuts and four mounting bolts and remove the upper intake manifold (see illustration).

12 Check the condition of the upper intake manifold gasket (see illustration). If it isn't cracked, hardened or flattened out, it can be reused.

13 Installation is the reverse of removal. Be sure to tighten the upper intake manifold mounting bolts and nuts to the torque listed in this Chapter's Specifications.

LOWER INTAKE MANIFOLD

Refer to illustration 5.17

Note: Toyota refers to the lower intake manifold as the intake manifold. If you're buying a gasket for the lower intake manifold at a dealer parts department, use the Toyota terminology.

14 Remove the upper intake manifold (see Steps 1 through 11).

15 If you're removing the lower intake manifold to replace it, remove the fuel rail now (see Chapter 4).

16 If you're just removing the lower intake manifold to replace the gaskets, it's not necessary to remove the fuel rail, but you'll have to disconnect the fuel supply and return line connections (see Chapter 4).

5.11 Upper intake manifold fasteners

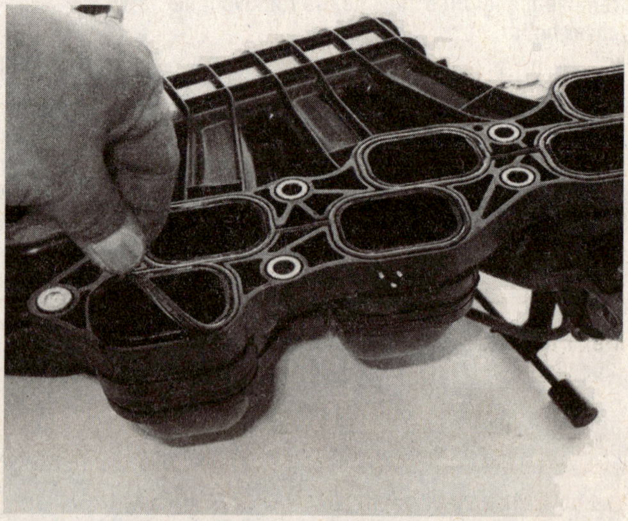

5.12 Check the condition of the upper intake manifold gasket, replacing it if necessary

17 Remove the lower intake manifold bolts and remove the lower intake manifold (see illustration).

18 Remove and discard the old lower intake manifold gaskets. Clean off all traces of old gasket material from the mating surfaces of the manifold and cylinder heads, then wipe the surfaces with brake system cleaner.

19 Installation is the reverse of removal. Be sure to use new gaskets and tighten the lower intake manifold bolts a little at a time, in a criss-cross pattern working from the center bolts outward, to the torque listed in this Chapter's Specifications.

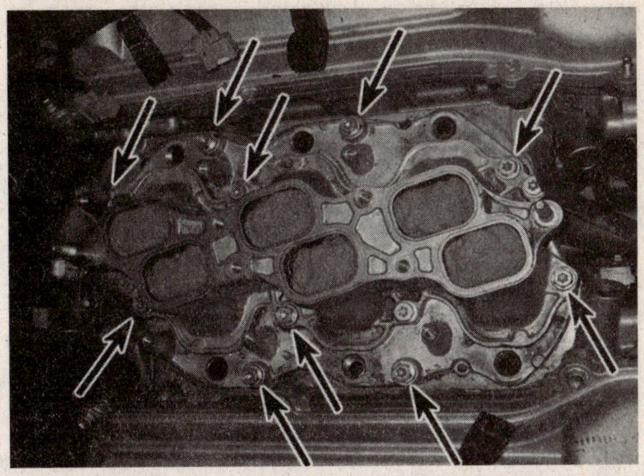

5.17 To detach the lower intake manifold, remove these ten mounting bolts (fuel rail removed for clarity)

6 Exhaust manifolds - removal and installation

WARNING:

The engine must be completely cool before beginning this procedure.

Note: The following procedure applies to either exhaust manifold.

1 Disconnect the cable from the negative terminal of the battery (see Chapter 5, Section 1).

2 Raise the front of the vehicle and place it securely on jackstands.

3 Unbolt and disconnect the front exhaust pipes from the exhaust manifolds.

4 Disconnect the electrical connectors from the upstream oxygen sensors.

5 Remove the three bolts that secure the exhaust manifold support bracket and remove it.

6 Remove the six nuts that secure the exhaust manifold and remove the exhaust manifold.

7 Remove and discard the old exhaust manifold gasket.

8 Installation is the reverse of removal. Be sure to use a new gasket and install it with the oval-shaped protruding tip facing in the correct direction. For the left (driver's side) manifold, the tip must face to the rear; on the right manifold it must face to the front.

9 Tighten the exhaust manifold nuts a little at a time, working from the center nuts outward, to the torque listed in this Chapter's Specifications.

7 Timing chain and sprockets - removal, inspection and installation

REMOVAL

Refer to illustrations 7.17a, 7.17b, 7.17c, 7.20a, 7.20b, 7.20c, 7.21a, 7.21b, 7.21c, 7.22, 7.23, 7.24, 7.28 and 7.29

WARNING:

Wait until the engine is completely cool before beginning this procedure.

CAUTION:

The timing system is complex, and severe engine damage will occur if you make any mistakes. Do not attempt this procedure unless you are highly experienced with this type of repair. If you are at all unsure of your abilities, be sure to consult an expert. Double-check all your work and be sure everything is correct before you attempt to start the engine.

Note 1: There is an access plate for the timing chain tensioner in the right side of the timing chain cover.

Note 2: The manufacturer recommends that the front axle assembly be removed from 4WD models in order to remove the oil pans, and that the entire engine be removed from 2WD models for this reason. We have found that it is possible to leave the oil pans in place for this procedure.

1 Disconnect the cable from the negative terminal of the battery (see Chapter 5, Section 1).

2 Drain the engine oil and coolant (see Chapter 1).

3 Remove the battery (see Chapter 5).

4 Remove the engine cover.

7.17a Hold the crankshaft pulley with a pin spanner while removing the bolt

7.17b A chain wrench can also be used to hold the pulley, but only if special precautions are taken; note the sockets used as spacers to raise the chain over the timing scale, and the piece of old drivebelt used to pad the pulley

7.17c If the crankshaft pulley can't be removed by hand, use a puller that bolts to the hub of the pulley - not a jaw-type puller. Also, be sure to use the correct adapter between the nose of the crankshaft and the puller screw, so as not to damage the threads in the crankshaft

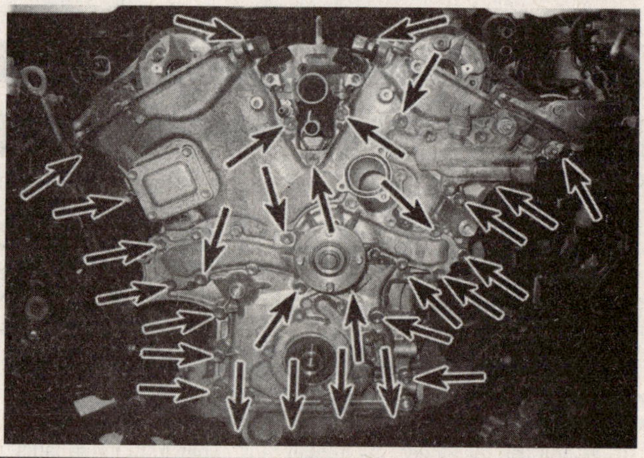

7.20a Remove the timing chain cover fasteners - make sure you keep track of the locations of the long and short bolts

5 Remove the air filter housing (see Chapter 1).

6 Remove the cooling fan and radiator (see Chapter 3).

7 Remove the drivebelt (see Chapter 1).

8 Remove the upper intake manifold (see Section 5).

9 Remove the spark plugs (see Chapter 1) and the valve covers (see Section 4).

10 Remove the Variable Valve Timing (VVT) sensor (see *Variable Valve Timing (VVT) system - description and component replacement* in Chapter 6).

11 Remove the dipstick, then remove the dipstick tube retaining bolt and the dipstick tube. Remove and discard the old dipstick tube O-ring.

12 Disconnect the electrical connector from the power steering pressure switch, then detach the power steering pump (see Chapter 10) and set it aside. Don't disconnect the power steering fluid hoses from the pump!

13 Remove the alternator (see Chapter 5).

14 Detach the air conditioning compressor (see Chapter 3). Do NOT disconnect the air conditioning refrigerant hoses!

15 Unbolt the drivebelt tensioner (see *Drivebelt check and replacement* in Chapter 1).

16 Remove the idler pulley bolts and remove the two idler pulleys.

17 Remove the crankshaft pulley (see illustrations).

18 Remove the four timing cover bolts from the front of the upper oil pan.

19 Remove the water inlet (see illustration 8.16 in Chapter 3). Remove the O-ring and gasket from the water inlet and discard them.

20 Remove the timing chain cover mounting fasteners and remove the timing chain cover (see illustration). There are only four spots where you can pry between the timing chain cover and the engine (see illustrations). Do NOT pry the timing chain cover loose at any other spot or you will damage the sealing surface of the cover. After removing the timing chain·cover, carefully pry out the old crankshaft seal with a

7.20b The timing chain cover can be pried loose at the lower corners . . .

7.20c . . . and at the upper corners; prying anywhere else may damage the cover

7.21a To bring the No. 1 cylinder to TDC on the compression stroke, rotate the crankshaft until the crankshaft key (A) is aligned with the timing line on the cylinder block (B)

7.21b The cast dot on the left cylinder head should line up with the center of the marks on the sprocket assembly . . .

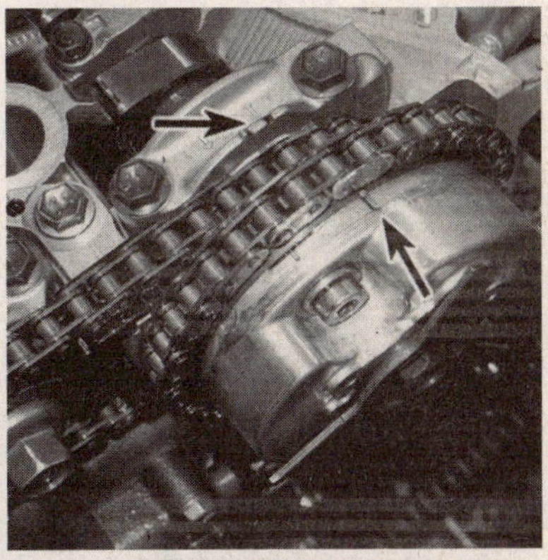

7.21c . . . and the cast dots on the right cylinder head should line up with the mark on the sprocket assembly

screwdriver (see Section 8). Make sure that you don't scratch the seal bore. If you want to inspect or replace any oil pump parts, refer to Section 12.

➡**Note: Keep track of the locations of all of the bolts. They are of two different lengths and can't be interchanged.**

21 Bring the piston in the No. 1 cylinder to TDC on its compression stroke. Install the crankshaft pulley bolt, then rotate the crankshaft until the crankshaft set key is aligned with the timing line on the cylinder block (see illustration). Verify that the timing marks on the camshaft timing sprockets are aligned with their corresponding marks on top of the front camshaft bearing caps (see illustrations). If the marks are not aligned, rotate the crankshaft another 360-degrees and recheck the marks.

22 Turn the stopper plate on the No. 1 tensioner clockwise and push in the tensioner plunger (see illustration). To lock the plunger in this position, turn the stopper plate counterclockwise and insert a drill or punch (0.138-inch diameter) through the holes in the stopper plate and the tensioner body. Remove the two tensioner mounting bolts and remove the No. 1 tensioner.

7.22 To lock the tensioner in the retracted position, rotate the stopper plate clockwise and push the plunger in, then rotate the stopper plate counterclockwise and insert a pin through the hole in the stopper plate and the tensioner body

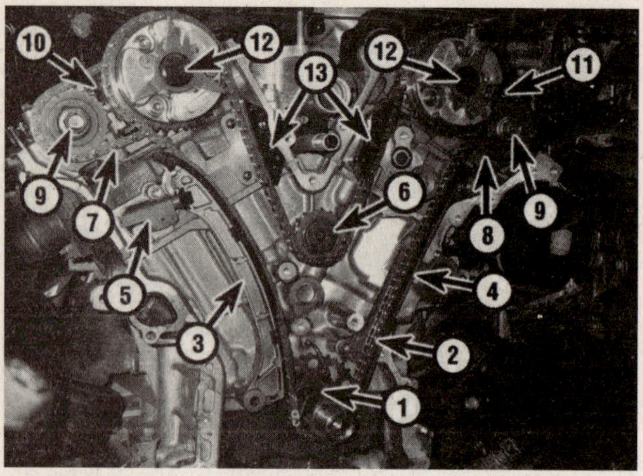

7.23 Details of the timing chains and related components

1	*Crankshaft sprocket*	9	*Exhaust camshaft sprocket bolt*
2	*No. 1 timing chain*	10	*No. 2 timing chain tensioner*
3	*Chain tensioner slipper*	11	*No. 3 timing chain tensioner*
4	*Chain guide*	12	*Intake camshaft sprocket bolt*
5	*No. 1 timing chain tensioner*	13	*Chain vibration dampers*
6	*Idler sprocket and shaft*		
7	*No. 2 timing chain*		
8	*No. 3 timing chain*		

7.24 Use a 10 mm hex bit to remove the idler shaft and sprocket

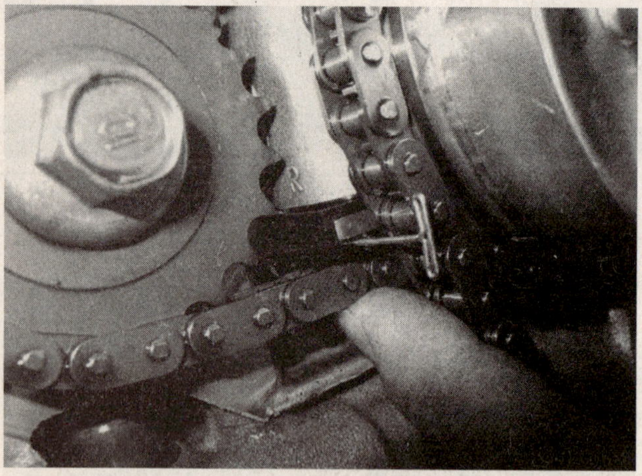

7.28 Insert a pin through the small tensioners to lock them in the retracted position

7.29 Hold the camshafts with a wrench to remove the sprocket bolts; never try to disassemble the intake camshaft sprocket assembly - only remove or install it as an assembly

23 Remove the chain tensioner slipper (see illustration).

24 Using a 10 mm hex bit, unscrew the idler sprocket shaft and remove the idler shaft, sprocket and collar (see illustration). Note which side of the sprocket faces out.

25 Remove the two chain vibration dampers.

26 Remove the No. 1 timing chain.

CAUTION:

While the No. 1 timing chain is removed, DO NOT ROTATE THE CRANKSHAFT!

27 Remove the crankshaft timing chain sprocket.

28 Compress chain tensioner No. 2 and insert a drill or punch (0.039-inch diameter) into the hole (see illustration).

➡Note: Timing chain No. 2 and the No. 2 chain tensioner are on the right (passenger's side) cylinder head. Timing chain No. 3 and the No. 3 tensioner are on the left (driver's side) cylinder head.

29 Hold the hex on the exhaust camshaft with a wrench and unscrew the two bolts that secure the camshaft timing sprockets to the camshafts (see illustration). Remove the sprockets and chain as an assembly. Keep the components in a resealable plastic bag to ensure that none of these components is mixed with the other timing chain set.

CAUTION:

Don't attempt to disassemble the adjustable intake sprocket assembly. If disassembled, it will have to be replaced.

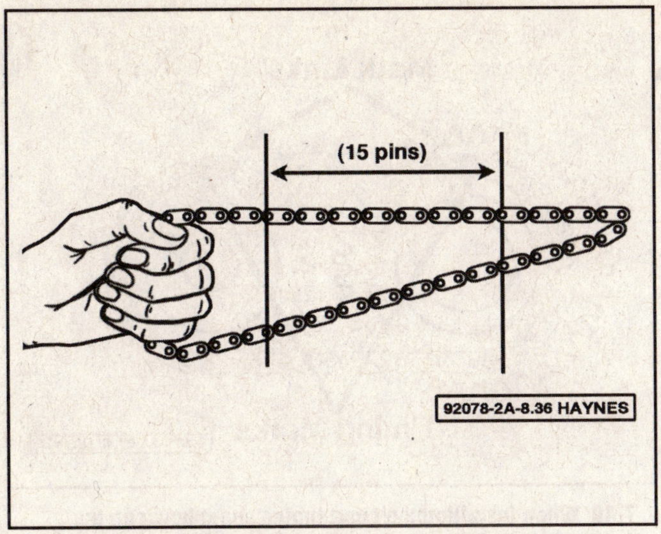

7.33 Measure timing chain stretch by measuring the distance between 15 pins at three or more places around the length of the chain

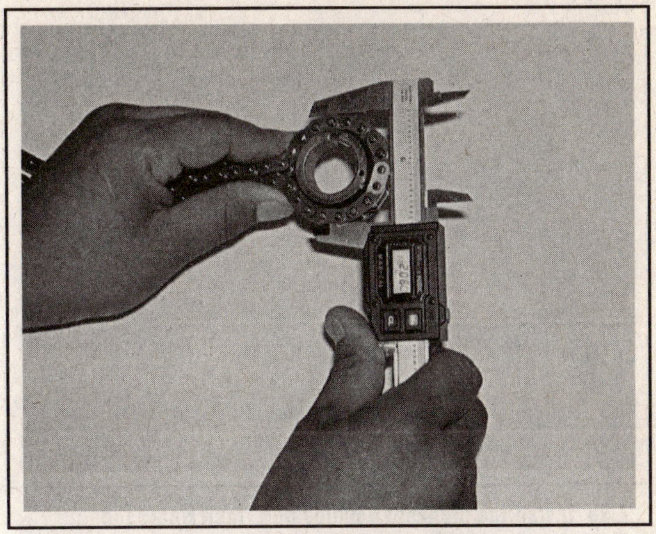

7.34 Wrap the chain around each of the timing chain sprockets and measure the diameter of the sprockets across the chain rollers. If the measurement is less than the minimum sprocket diameter, replace the chain and the timing sprockets

30 Remove the chain tensioner No. 2 mounting bolt and remove chain tensioner No. 2. Store the tensioner in the plastic bag with the other No. 2 timing chain components.

31 To remove the No. 3 timing chain and tensioner, repeat Steps 28 through 30. Again, store the components in a resealable plastic bag.

CAUTION:

While the timing chains are removed, DO NOT ROTATE THE CRANKSHAFT!

INSPECTION

Refer to illustrations 7.33, 7.34 and 7.36

32 Inspect all parts of the timing chain assembly for wear and damage. Inspect the three timing chains for loose pins, cracks, and worn rollers and side plates. Inspect the sprockets for hook-shaped, chipped and/or broken teeth.

33 Inspect timing chain No. 1 for stretching. To measure timing chain stretch, measure the distance between 15 pins at three or more places around the length of the chain (see illustration). Compare your measurements with the distance listed in this Chapter's Specifications.

34 Measure the diameter of each timing chain sprocket and idler sprocket with the appropriate timing chain installed on the sprocket (see illustration). The sprocket diameter, with the chain in place, should not exceed the dimensions listed in this Chapter's Specifications.

35 Measure the idler sprocket oil clearance as follows. First, measure the diameter of the idler sprocket collar with a micrometer and record your measurement. Then measure the inside diameter of the idler sprocket and record that measurement as well. Subtract the idler sprocket collar diameter from the inside diameter of the idler sprocket and compare the result with the clearance listed in this Chapter's Specifications. If the clearance is excessive, replace the idler sprocket and/or collar, as necessary.

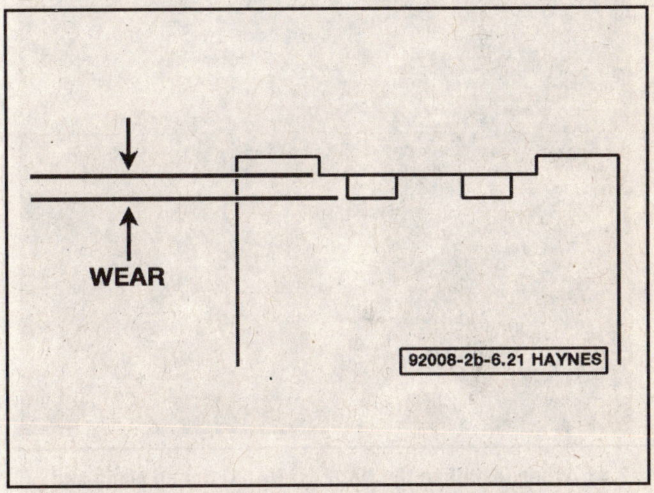

7.36 When inspecting the chain tensioners, the chain tensioner slipper and the chain vibration dampers, measure timing chain wear from the top of the chain contact surface to the bottom of the wear grooves

36 Some scoring and wear of the timing chain tensioners and vibration dampers is normal, but excessive wear will increase chain noise, accelerate chain and sprocket wear and could damage the engine if a chain jumps timing. Inspect chain tensioners No. 2 and 3, the timing chain tensioner slipper and the timing chain vibration dampers for excessive wear (see illustration). If the measured chain wear for any of these components exceeds the depth listed in this Chapter's Specifications, replace the component.

37 Check the chain tensioners for correct operation. On the No. 1 tensioner, raise the ratchet pawl and verify that the plunger moves smoothly in and out of the tensioner, then release the ratchet pawl and verify that it prevents the plunger from sliding back into the tensioner. Also verify that the plungers on the No. 2 and No. 3 tensioners move in and out smoothly.

7.38 Install the crankshaft pulley bolt, then rotate the crankshaft in a counterclockwise direction until the crankshaft set key is at the 270-degree position (on the left and aligned with an imaginary horizontal line), as you're looking at the front of the engine

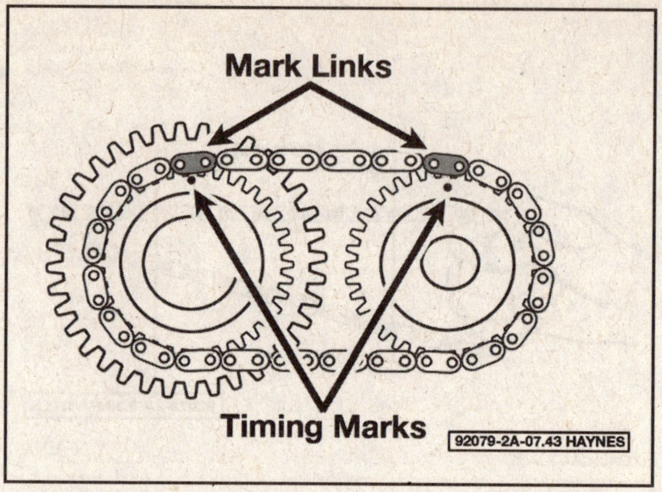

7.40 When installing the short timing chain between the camshaft sprockets, make sure that the yellow mark links on the chain are aligned with the timing marks (dots) on the sprockets

7.41 When installing the No. 2 (or No. 3) timing chain and sprocket, be sure to align the yellow links on the chain with the timing marks on the bearing caps

INSTALLATION

➧ Refer to illustrations 7.38, 7.40, 7.41, 7.49a, 7.49b, 7.53 and 7.54

38 Install the crankshaft pulley bolt, then rotate the crankshaft in a counterclockwise direction until the crankshaft set key is at the 270-degree (9 o'clock) position (and aligned with an imaginary horizontal line), as you're looking at the front of the engine (see illustration).

39 Push in the tensioner plunger on chain tensioner No. 2 and insert a drill bit or punch (0.039 inch diameter) into the hole of each tensioner to lock the plunger in the retracted position (see illustration 7.28). Install the tensioners and tighten the mounting bolts to the torque listed in this Chapter's Specifications.

40 Install the No. 2 timing chain on the camshaft sprockets. Make sure that the yellow mark links on the chain are aligned with the timing marks (dots) on the camshaft sprockets (see illustration).

41 Align the yellow links on the No. 2 timing chain with the timing marks on the bearing caps (see illustration) and install timing chain No. 2 and the camshaft sprockets as an assembly. Install the two bolts that secure the timing sprockets to the camshafts. Immobilize the hex on the exhaust camshaft with an adjustable wrench and tighten these two bolts to the torque listed in this Chapter's Specifications. Remove the drill bit or punch that was used to lock the tensioner plunger in its retracted position and verify that the plunger tensions the chain.

42 Install the No. 3 timing chain (repeat Steps 42 through 44).

43 Install the chain guide and tighten the bolts to the torque listed in this Chapter's Specifications.

44 Install the crankshaft timing sprocket on the crankshaft. Be sure to align the timing sprocket keyway with the key on the crankshaft, and make sure that the sprocket end of the crank timing sprocket faces in, toward the engine.

45 Install the chain tensioner slipper.

46 Turn the stopper plate on the No. 1 tensioner clockwise and push in the tensioner plunger. To lock the plunger in this position, turn the stopper plate counterclockwise and insert a drill bit or punch (0.138-inch diameter) through the holes in the stopper plate and the tensioner. Install the tensioner and tighten the two tensioner mounting bolts to the torque listed in this Chapter's Specifications.

47 Verify that the timing marks on the camshaft timing sprockets are aligned with their corresponding marks on top of the front camshaft bearing caps (see illustrations 7.21b and 7.21c).

48 Using the crankshaft pulley bolt, rotate the crankshaft clockwise until the crankshaft set key is aligned with the timing line on the cylinder block (see illustration 7.21a).

49 Install the long timing chain (No. 1) on the camshaft timing sprockets and on the crankshaft timing sprocket. Make sure that the yellow mark link is aligned with the timing mark (dot) on the crankshaft timing sprocket (see illustration) and that the orange mark links are aligned with the timing marks on the intake camshaft sprockets (see illustration).

50 Apply a light coat of engine oil to the bearing surface of the idler sprocket collar. Install the idler sprocket collar, sprocket and shaft. Make sure that the teeth on the idle sprocket are facing forward. Tighten the idler sprocket shaft to the torque listed in this Chapter's Specifications.

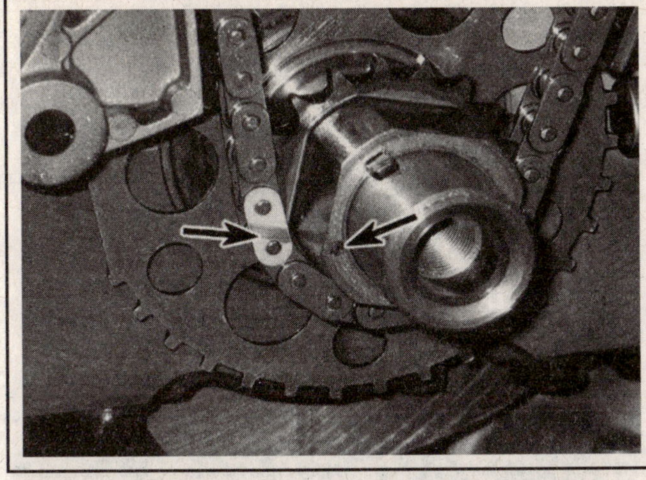

7.49a When installing timing chain No. 1 (the long timing chain) on the camshaft timing sprockets and the crankshaft timing sprocket, make sure that the yellow mark link is aligned with the timing mark (dot) on the crankshaft timing sprocket . . .

7.49b . . . and the orange mark links are aligned with mark(s) on the intake camshaft timing sprockets

7.53 Be sure to install a new O-ring in the front of the left cylinder head

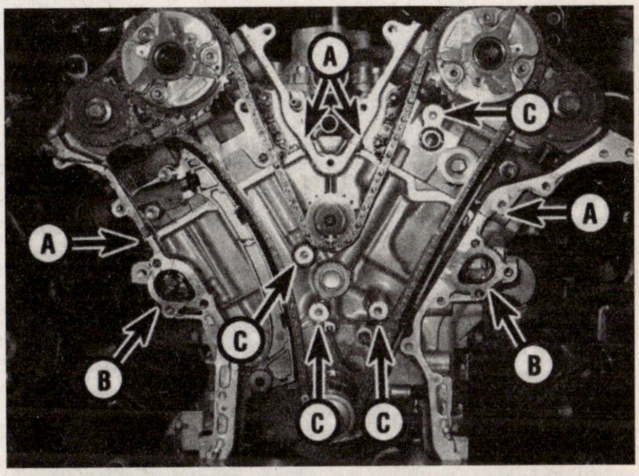

7.54 Apply a continuous bead of gray RTV sealant along the edges of the timing chain cover and across the top of the oil pan, then apply dabs at the cylinder head-to-block joints (A), around the water pump passages (B) and at the bosses (C) - but don't get any on the O-ring seal below the topmost boss

Remove the drill bit or punch that you inserted into chain tensioner No. 1 (in Step 46) and verify that it tensions timing chain No. 1.

⁂ CAUTION:

Carefully rotate the crankshaft by hand through at least two full revolutions (use a socket and breaker bar on the crankshaft pulley center bolt). If you feel any resistance, STOP! There is something wrong - most likely valves are contacting the pistons. You must find the problem before proceeding. Check your work to make sure all timing marks line-up properly, and see if any updated repair information is available.

51 Remove all old RTV sealant from the gasket mating surfaces of the timing chain cover and from the front of the cylinder heads and engine block.

52 Install a new crankshaft oil seal in the timing chain cover (see Section 8).

53 Install a new O-ring on the left cylinder head (see illustration).

54 Apply gray RTV sealant on the timing chain cover and to the indicated locations of the mating surfaces on the engine (see illustration). These beads should be about 1/8 to 3/16-inch (3 to 4.5 mm) wide.

⁂ CAUTION:

Once you have installed the sealant on the engine and timing chain cover you have three minutes to install the cover. If you take longer than that, the sealant might not set up properly, so you'll have to remove the sealant and re-apply it.

Be sure to get sealant into the corners where the oil pan meets the engine block and avoid getting any on the O-rings.

55 Rotate the flats on the oil drive rotor about 15-degrees to the right of vertical to align it with the square part of the crankshaft timing sprocket and slide the timing chain cover into place.

56 Install the timing chain cover mounting bolts and nuts and tighten all fasteners gradually and evenly, in a criss-cross fashion until they're all snug.

CAUTION:

Do not put long bolts in short holes or vice versa.

When all of the fasteners are snug, tighten them to the torque listed in this Chapter's Specifications.

57 Install and tighten the four oil pan bolts that go into the timing chain cover.

58 The remainder of installation is the reverse of the removal procedure.

59 Refill the engine with oil and coolant (see Chapter 1).

60 Reconnect the cable to the negative battery terminal.

61 Start the engine check for leaks.

8 Crankshaft front oil seal - replacement

Refer to illustrations 8.2 and 8.4

1 Remove the crankshaft pulley (see Section 7).

2 Carefully pry the seal out of the timing chain cover with a screwdriver or seal removal tool (see illustration). If you use a screwdriver, wrap tape around the tip - don't scratch the housing bore or damage the crankshaft (if the crankshaft is damaged, the new seal will end up leaking).

3 Clean the bore in the timing chain cover and coat the lip and the outer edge of the new seal with engine oil or multi-purpose grease.

4 Using a seal driver or a socket with an outside diameter slightly smaller than the outside diameter of the seal, carefully drive the new seal into place with a hammer (see illustration). Make sure it's installed squarely and driven in flush with the surface of the timing chain cover. Check the seal after installation to make sure the spring didn't pop out of place.

5 Reinstall the crankshaft pulley, tightening the bolt to the torque listed in this Chapter's Specifications.

6 Run the engine and check for oil leaks at the front seal.

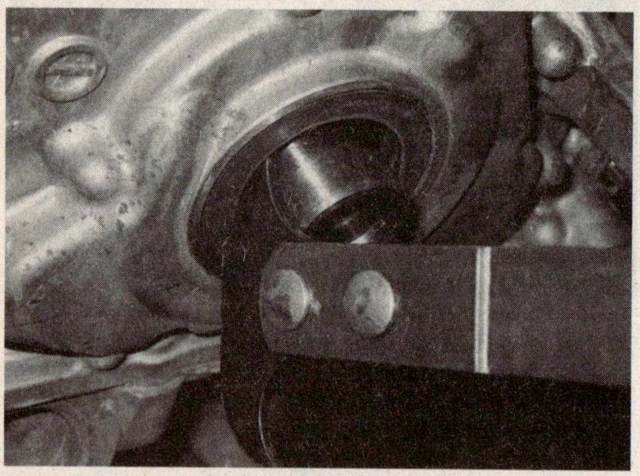

8.2 Pry the seal out of the timing chain cover, being careful not to scratch the crankshaft

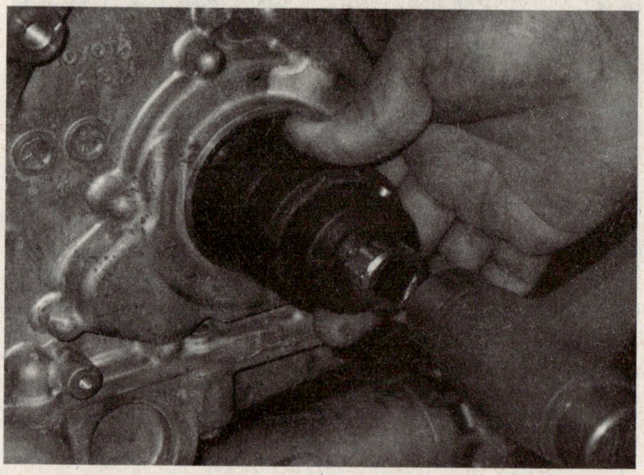

8.4 Lubricate the seal lip and drive the new crankshaft seal into place with a large socket or piece of pipe and a hammer

9 Camshafts and lifters - removal, inspection and installation

REMOVAL

Refer to illustration 9.4

Note: The following procedure is not for beginners. Please read the entire procedure carefully before deciding whether this is a job that you want to tackle at home.

1 Disconnect the cable from the negative terminal of the battery (see Chapter 5, Section 1).

2 Drain the engine oil and coolant (see Chapter 1).

3 Remove the timing chains (see Section 7).

4 To remove the camshafts from the right (passenger's side) cylinder head, rotate the camshafts counterclockwise so that the nose of the intake cam lobe for the No. 1 cylinder is facing toward 7 o'clock, and

the nose of the exhaust cam lobe is facing 12 o'clock, or straight up (see illustration). Use an open-end wrench on the integral hex cast into each camshaft to rotate the cams.

➡Note: This step is NOT necessary for the camshafts on the left cylinder head.

5 Gradually and evenly loosen all 16 camshaft bearing cap bolts in the opposite order of the tightening sequence (see illustrations 9.15 and 9.16).

6 Remove all eight bearing caps and remove the intake and exhaust camshafts.

7 Store the bearing caps in the correct order. One way to do this is to put them in a box and label the cap numbers with a utility marker pen.

8 Remove the lifters with a magnet and put them in the same box with the cam bearing caps. Again, be sure to label each lifter with a marker. You can also write the lifter number directly on top of the lifter.

9 If you're removing the camshafts from the other cylinder head, repeat Steps 5 through 8. Don't forget that Step 4 applies only to camshaft removal for the right cylinder.

⁂ CAUTION:

While the timing chains and camshafts are removed, DO NOT ROTATE THE CRANKSHAFT!

INSPECTION

➧ Refer to illustrations 9.10, 9.11, 9.12, 9.13a and 9.13b

10 Inspect each lifter for scuffing and score marks (see illustration).

11 Visually examine the cam lobes and bearing journals for score marks, pitting, galling and evidence of overheating (blue, discolored areas). Look for flaking away of the hardened surface layer of each lobe. Using a micrometer, measure the height of each camshaft lobe (see illustration). Compare your measurements with this Chapter's Specifications. If the height for any one lobe is less than the specified minimum, replace the camshaft.

12 Using a micrometer, measure the diameter of each journal at several points (see illustration). Compare your measurements with this Chapter's Specifications. If the diameter of any one journal is less than specified, replace the camshaft.

13 Check the oil clearance for each camshaft journal as follows:

a) *Clean the bearing caps and the camshaft journals with brake system cleaner.*

b) *Carefully lay the camshaft(s) in place in the cylinder head. Don't install the lifters and don't use any lubrication.*

c) *Lay a strip of Plastigage on each journal.*

9.4 Correct camshaft position for removal of the right cylinder head camshafts - the camshafts on the left cylinder head can be removed in any position

9.10 Inspect each lifter for wear and scuffing

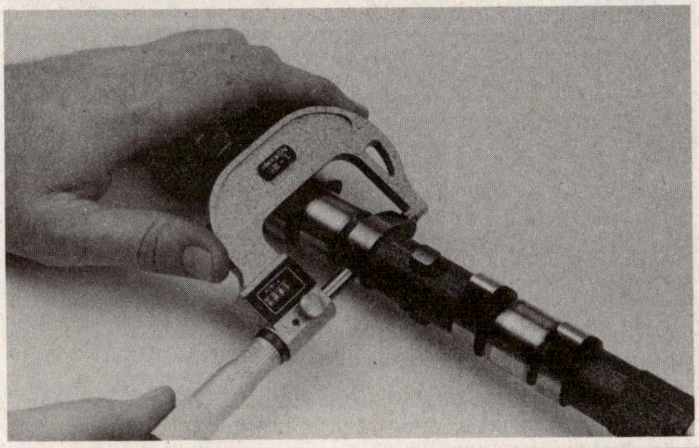

9.11 Measure the lobe heights on each camshaft - if any lobe height is less than the specified allowable minimum, replace that camshaft

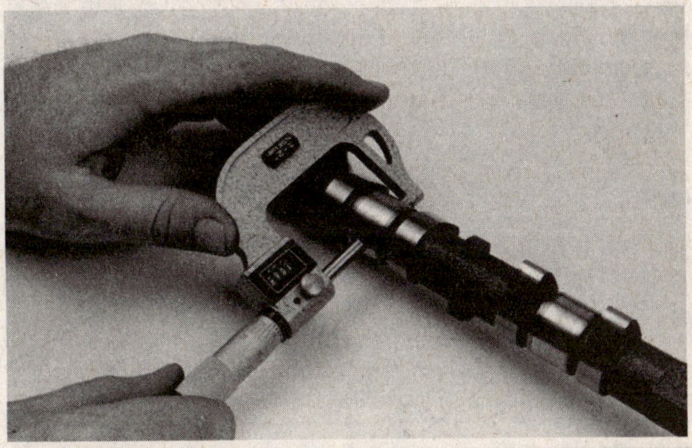

9.12 Measure each journal diameter with a micrometer - if any journal measures less than the specified limit, replace the camshaft

d) Install the bearing caps with the arrows pointing toward the front (timing chain end) of the engine (see illustration).

e) Tighten the bolts to the torque listed in this Chapter's Specifications in 1/4-turn increments.

➡**Note: Don't turn the camshaft while the Plastigage is in place.**

f) Remove the bolts and detach the caps.

g) Compare the width of the crushed Plastigage (at its widest point) to the scale on the Plastigage envelope (see illustration).

h) If the clearance is greater than specified, replace the camshaft and/or cylinder head.

i) Scrape off the Plastigage with your fingernail or the edge of a credit card - don't scratch or nick the journals or bearing caps.

INSTALLATION

➧ Refer to illustrations 9.15 and 9.16

14 Lightly lubricate the lifter bores and the lifters with clean engine oil, then install the lifters in the same bores from which they were removed. Verify that each lifter rotates smoothly in its bore.

15 Right side: Lightly lubricate the camshaft journals with clean engine oil. Install the camshafts on the right cylinder head so that the nose of the intake cam lobe for the No. 1 cylinder is facing toward 7 o'clock, and the nose of the exhaust cam lobe for No. 1 is facing 12 o'clock, or straight up (see illustration 9.4). Apply a light coat of engine oil to the upper bearings, then install the eight bearing caps in their correct locations. Apply a light coat of engine oil to the threads of the bearing cap bolts and install the bolts. Gradually and evenly tighten the bearing cap bolts, in the indicated sequence (see illustration), to the torque listed in this Chapter's Specifications. Rotate the camshafts 90-degrees clockwise until the knock pins on the front end of the camshafts are at a 90-degree position in relation to the cylinder head mating surface.

16 Left side: Lightly lubricate the camshaft journals with clean engine oil. Install the camshafts on the left cylinder head so that the nose of the intake cam lobe for the No. 2 cylinder is facing toward 7 o'clock, and the nose of the exhaust cam lobe for No. 2 is facing 12 o'clock, or straight up (see illustration). Apply a light coat of engine oil to the upper bearings, then install the eight bearing caps in their correct locations. Apply a light coat of engine oil to the threads of the bearing cap bolts and install the bolts. Then gradually and evenly tighten the bearing cap bolts, in the indicated sequence, to the torque listed in this Chapter's Specifications.

9.13a The camshaft bearing caps are numbered with an arrow facing the front of the engine

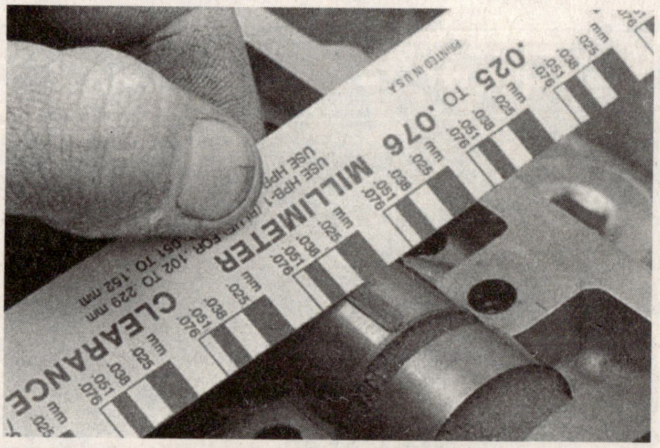

9.13b Compare the width of the crushed Plastigage to the scale on the envelope to determine the oil clearance

9.15 Camshaft bearing cap bolt tightening sequence (right cylinder head)

9.16 Tighten the camshaft bearing caps on the left cylinder head in this sequence; the front lobes of the exhaust camshaft must be facing upward, and the front lobes of the intake camshaft must be in the 7 o'clock position

17 The remainder of installation is the reverse of removal. Install the timing chains, the timing chain cover and all of the components attached to the timing cover (see Section 7).

CAUTION:

Carefully rotate the crankshaft by hand through at least two full revolutions (use a socket and breaker bar on the crankshaft pulley center bolt). If you feel any resistance, STOP! There is something wrong - most likely valves are contacting the pistons. You must find the problem before proceeding. Check your work to make sure all timing marks line-up properly, and see if any updated repair information is available.

18 Refill the engine with oil and coolant (see Chapter 1), reconnect the cable to the negative battery terminal, start the engine and check for leaks.

Note: It may take a few minutes for lifter clatter to disappear.

10 Cylinder heads - removal and installation

WARNING:

The engine must be completely cool before starting this procedure.

Note: This procedure applies to either cylinder head.

REMOVAL

Refer to illustration 10.13

1 Position the engine at TDC compression for cylinder no. 1 (see Section 3). Disconnect the cable from the negative terminal of the battery (see Chapter 5, Section 1).
2 Drain the engine oil and coolant (see Chapter 1).
3 Remove the engine cover.
4 Remove the air filter housing (see Chapter 4).
5 Remove the upper intake manifold (see Section 5).
6 Remove the valve cover(s) (see Section 4).
7 Disconnect the fuel supply and return lines from the fuel rail (see Chapter 4). Remove the lower intake manifold and the fuel rail as a single assembly (see Section 5).
8 Locate the coolant passage on the back of the engine. It is bolted to the upper rear part of the block and is attached to both cylinder heads by a pair of nuts at each end. Disconnect the Engine Coolant Temperature (ECT) sensor (see *Engine Coolant Temperature sensor - replacement* in Chapter 6). Disconnect the heater hose from the coolant passage. Remove the two bolts and four nuts and remove the coolant passage. Remove and discard the old gaskets. Remove the old O-ring from the coolant outlet pipe and discard it.
9 Disconnect the front exhaust pipe from the exhaust manifold and remove the exhaust manifold (see Section 6).
10 Remove the timing chain (see Section 7).
11 Remove the camshaft timing oil control valve, the oil control valve filter and the VVT-i sensor (see *Variable Valve Timing-intelligent [VVT-I] - description and component replacement* in Chapter 6).
12 Remove the camshafts (see Section 9).
13 If you're removing the left cylinder head, remove the two front bolts in the indicated sequence (see illustration).
14 Remove the cylinder head bolts gradually and evenly, in the order opposite that of the tightening sequence (see illustrations 10.24a and 10.24b), then remove the cylinder head from the engine block.
15 Remove and discard the old cylinder head gasket.

INSTALLATION

Refer to illustrations 10.22, 10.24a and 10.24b

16 The mating surfaces of the cylinder heads and the block must be perfectly clean as the heads are installed.
17 Use a gasket scraper to remove all traces of carbon and old gasket material, then clean the mating surfaces with brake system cleaner. If there's oil on the mating surfaces when the head is installed, the gasket may not seal correctly and leaks could develop.
18 When working on the block, stuff the cylinders with clean shop rags to keep out debris. Use a vacuum cleaner to remove material that falls into the cylinders.

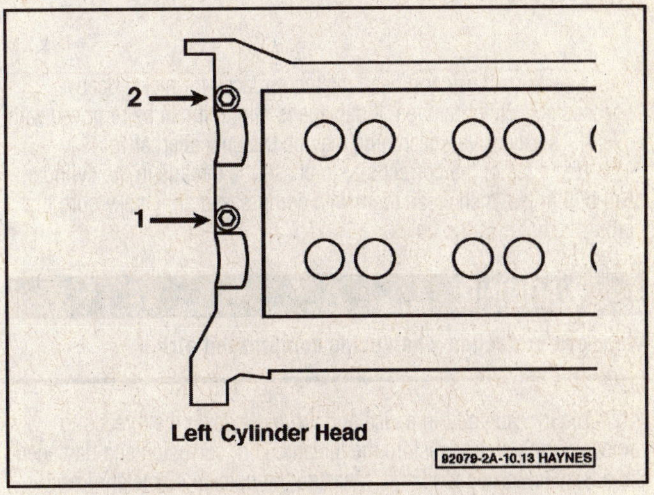

10.13 If you're removing the left cylinder head, remove these two bolts first, in this order. When installing the left head, install these two bolts, in reverse order, AFTER you have installed the rest of the cylinder head bolts

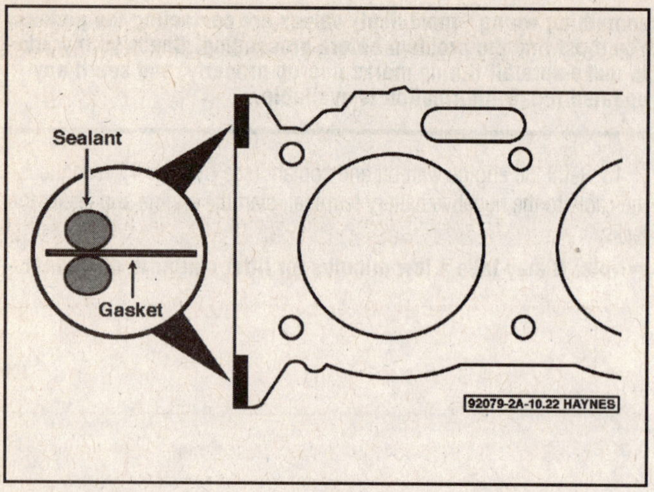

10.22 Apply small beads of sealant to the top and bottom of the head gasket at these two points

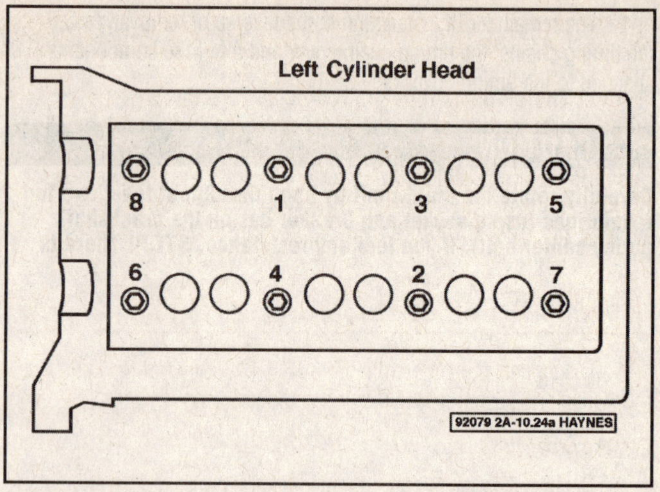

10.24a Cylinder head bolt tightening sequence - left cylinder head

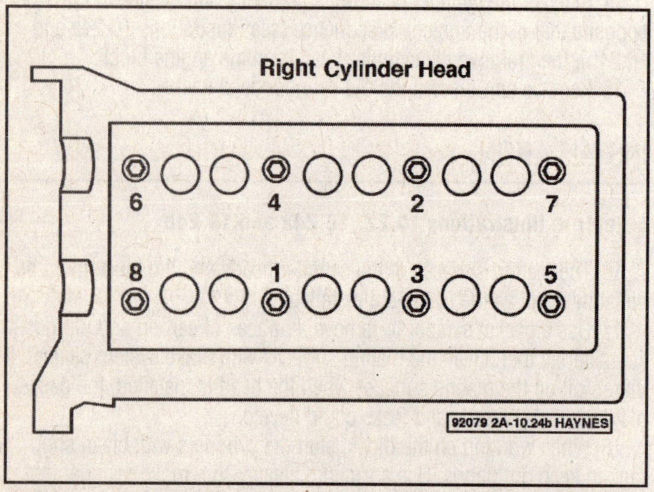

10.24b Cylinder head bolt tightening sequence - right cylinder head

19 Check the block and head mating surfaces for nicks, deep scratches and other damage. If damage is slight, it can be removed with a file; if it's excessive, machining may be the only alternative.

20 Use a tap of the correct size to chase the threads in the cylinder head bolt holes, then clean them with compressed air - make sure that nothing remains in the holes.

⁂ WARNING:

Wear eye protection when using compressed air!

21 Mount each bolt in a vise and run a die down the threads to remove corrosion and restore the threads. Dirt, corrosion and damaged threads affect torque readings. Measure the outside diameter of each cylinder head bolt in several places, comparing your measurements to the minimum allowable diameter listed in this Chapter's Specifications. Replace any bolt with a diameter less than the allowable minimum.

22 Apply two small beads of RTV sealant to the two indicated areas of the new cylinder head gasket (see illustration). Apply the sealant to the top and the bottom of these two spots.

⁂ CAUTION:

Once you have applied the RTV sealant to the cylinder head gasket, you must install the cylinder head within three minutes, and you must tighten and torque the cylinder head bolts within 15 minutes. If you don't, you must remove the RTV sealant and reapply it.

23 Position the cylinder head gasket on the engine block so that the lot number stamp is on the center upper edge of the gasket (next to the intake manifold), facing up. Carefully place the cylinder head on the head gasket.

24 Apply a light coat of oil to the cylinder head bolts, then install and tighten them (don't forget the washers!), gradually and evenly, in the proper sequence (see illustrations), to the initial torque listed in this Chapter's Specifications.

25 After tightening all eight bolts to the initial torque, put a paint mark on the front edge of each bolt (the edge facing toward the front of the engine), then retighten each bolt, in the same sequence, another 180-degrees.

26 If you're installing the left cylinder head, install the two front head bolts and tighten them to the torque listed in this Chapter's Specifications, in the opposite order of the sequence indicated in illustration 10.13.

27 If you removed both cylinder heads, install the other cylinder head now (see Steps 17 through 25).

28 Install the intake and exhaust camshafts (see Section 9).

29 Install the camshaft timing oil control valve, the oil control valve filter and the VVT-i sensor (see *Variable Valve Timing-intelligent (VVT-i) - description and component replacement* in Chapter 6).

30 Install the timing chains, the timing chain cover and all components attached to the cover (see Section 7).

31 The remainder of installation is the reverse of the removal procedure.

32 Refill the engine with oil and coolant (see Chapter 1).

33 Reconnect the cable to the negative battery terminal, start the engine and check for leaks.

11 Oil pan - removal and installation

➡Note: The oil pan cannot be removed from these vehicles without first getting adequate clearance. On 2WD models, the engine must be removed entirely. On some models it may be possible to simply raise the engine enough to make room for oil pan removal. On some 4WD models, it may be possible to remove the front differential to obtain sufficient clearance.

REMOVAL

1 Disconnect the cable from the negative terminal of the battery (see Chapter 5, Section 1). Drain the engine oil (see Chapter 1).

2WD models

2 Refer to Chapter 2C and remove the engine.

➡Note: It's possible to lift the engine above the crossmember enough to get clearance for removing the oil pan without having to remove the engine from the engine bay.

4WD models

3 Refer to Chapter 8 and remove the front differential/axle assembly for clearance below the oil pan.

All models

4 Remove the bolts and nuts that secure oil pan No. 2 (the smaller stamped steel pan) to the larger cast aluminum oil pan.

5 Oil pan No. 2 will probably be stuck to the oil pan with RTV sealant. Try tapping it loose with a rubber-tipped mallet. If you're unable to knock it loose, carefully cut the sealant with a putty knife and a hammer. Make sure that you don't damage the mating surfaces of the two pans.

6 Remove the two oil pump pickup tube/strainer mounting nuts and remove the pickup/strainer assembly.

7 Remove the four bolts that attach the flywheel housing cover and remove the cover.

8 Remove the 17 bolts and 2 nuts that secure the oil pan to the engine block. Remove the 4 stud bolts.

9 Carefully pry the oil pan loose from the engine block.

⁂ CAUTION:

Only pry in the small cutout areas along the side of the pan.

INSTALLATION

10 Use a scraper to remove all traces of old sealant from the block and oil pan. Clean the mating surfaces with brake system cleaner.

11 Make sure the threaded holes in the block are clean.

12 Check the flange of the steel oil pan for distortion around the bolt holes. If necessary, place it on a wood block and use a hammer to flatten and restore the gasket surface.

13 Inspect the strainer for cracks or blockage. Clean it with solvent and install it using a new gasket. Tighten the fasteners to the torque listed in this Chapter's Specifications.

14 Apply a 1/8 inch bead of RTV sealant to the upper oil pan flange.

15 Position the pan onto the block and install the fasteners. Working from the center out, tighten the fasteners to the torque listed in this Chapter's Specifications in several steps.

16 After you have installed the aluminum portion of the oil pan, apply a bead of RTV sealant to the flange of the No. 2 oil pan, carefully position it on the upper oil pan and install the bolts. Working from the center out, tighten them to the torque listed in this Chapter's Specifications in several steps.

17 The remainder of installation is the reverse of removal. Add oil and install a new filter (see Chapter 1). When you're done, run the engine and check for leaks.

12 Oil pump - removal, inspection and installation

REMOVAL

➧ Refer to illustrations 12.2 and 12.3

1 Remove the timing chain cover (see Section 7).

2 Remove the oil pipe mounting bolts (see illustration) and remove the oil pipe. Remove and discard the two old oil pipe O-rings.

3 Remove the oil pump cover bolts and remove the oil pump cover (see illustration). Remove the oil pump drive rotor and driven rotor.

4 Remove the oil pressure relief plug, valve spring and valve from

12.2 Oil pump tube bolts

12.3 Oil pump cover bolts

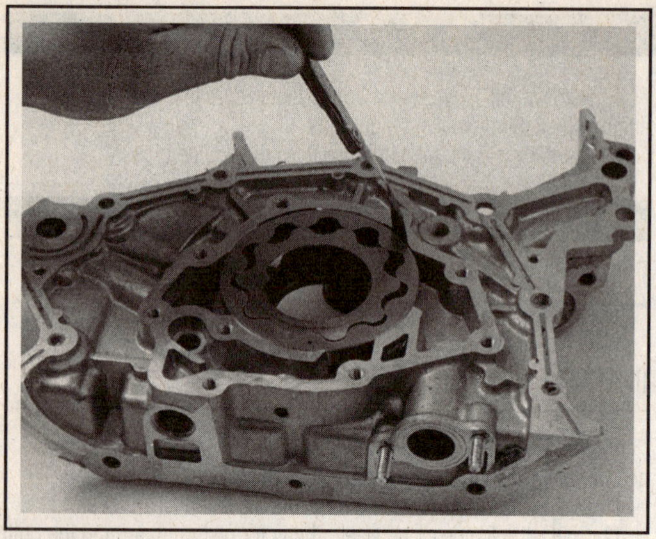

12.7a Measure the driven rotor-to-body clearance with a feeler gauge

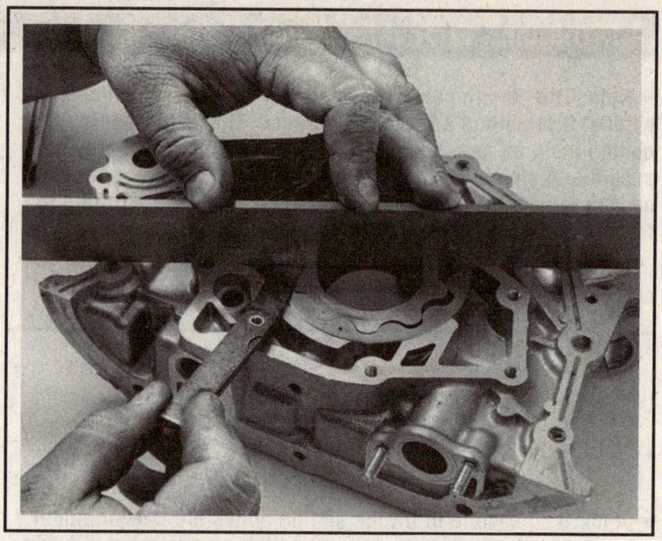

12.7b Measure the rotor side clearance with a precision straightedge and feeler gauge

the oil pump cover.

INSPECTION

➧ **Refer to illustrations 12.7a, 12.7b and 12.7c**

5 Clean all components with solvent, then inspect them for wear and damage. Check that the oiled relief valve falls easily through its bore without sticking.

6 Check the oil pressure relief valve sliding surface and valve spring. If either the spring or the valve is damaged, they must be replaced as a set.

7 Check the clearance of the following components with a feeler gauge and compare the measurements to this Chapter's Specifications (see illustrations):

a) Driven rotor-to-oil pump body
b) Rotor side clearance
c) Rotor tip clearance

INSTALLATION

8 Pry the old crankshaft seal out of the timing chain cover with a screwdriver.

9 Apply multi-purpose grease or engine oil to the outer edge of the new crank seal and carefully drive it into place with a deep socket and a hammer. Apply multi-purpose grease or engine oil to the seal lip.

10 Apply a coat of petroleum jelly to the pump drive and driven rotors, then place the two rotors into position in the timing chain cover. Make sure that the pump marks (dimples) are facing out, toward the pump cover and away from the timing chain cover.

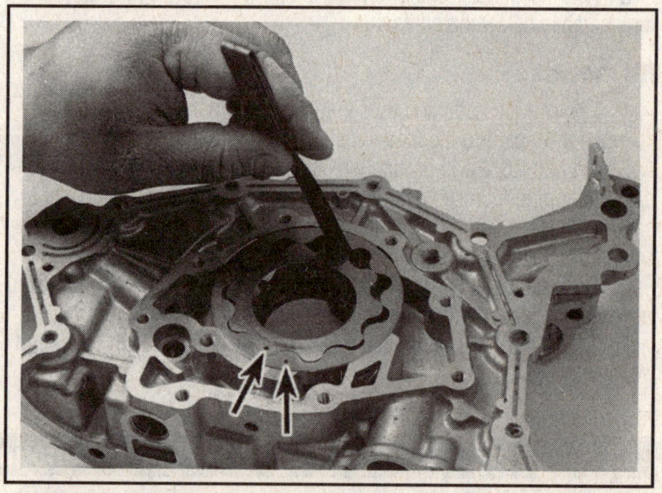

12.7c Measure the rotor tip clearance with a feeler gauge - note the rotor marks are facing out (when the pump body cover is installed, the marks will be against the cover)

11 Pack the pump cavity with petroleum jelly (this will help to prime the pump) and install the cover. Tighten the cover bolts to the torque listed in this Chapter's Specifications.

12 Lubricate the oil pressure relief valve with clean engine oil and insert the valve, then the spring, into the pump cover. Screw in the plug and tighten it to the torque listed in this Chapter's Specifications.

13 Install the timing chain cover (see Section 7).

14 The remainder of installation is the reverse of removal. Add oil and install a new filter (see Chapter 1). When you're done, run the engine and check for leaks.

13 Flywheel/driveplate - removal and installation

Note: Driveplates (also known as flexplates) are used with automatic transmissions; flywheels are used with manual transmissions.

REMOVAL

1 Disconnect the cable from the negative terminal of the battery (see Chapter 5, Section 1). Raise the vehicle and support it securely on jackstands, then refer to Chapter 7 and remove the transmission.

2 Remove the clutch from vehicles equipped with manual transmissions.

3 Make alignment marks on the flywheel/driveplate and crankshaft to ensure correct alignment during reinstallation.

4 Remove the bolts securing the flywheel/driveplate to the crankshaft. If the crankshaft turns, wedge a screwdriver in the ring gear teeth to hold the driveplate.

5 Remove the flywheel/driveplate from the crankshaft. Be sure to support it while removing the last bolt. Automatic transmission equipped vehicles have spacers on both sides of the driveplate. Keep them with the driveplate.

INSTALLATION

6 Clean the flywheel/driveplate to remove grease and oil. Inspect the surface for cracks. Check for cracked or broken ring gear teeth.

7 Clean and inspect the mating surfaces of the flywheel/driveplate and the crankshaft. If the crankshaft rear seal is leaking, replace it before reinstalling the flywheel/driveplate (see Section 14).

8 Position the flywheel/driveplate against the crankshaft. Be sure to align the marks made during removal. Note that some engines have an alignment dowel or staggered bolt holes to ensure correct installation. Before installing the bolts, apply thread-locking compound to the threads.

9 Wedge a screwdriver in the ring gear teeth to keep it from turning and tighten the bolts to the torque listed in this Chapter's Specifications. Follow a criss-cross pattern and work up to the final torque in three or four steps.

10 The remainder of installation is the reverse of the removal procedure.

14 Rear main oil seal - replacement

Refer to illustration 14.6

1 Remove the transmission (see Chapter 7). Refer to Section 13 and remove the flywheel/driveplate.

2 The seal can be replaced without removing the oil pan or seal retainer. However, this method is not recommended because the lip of the seal is quite stiff and it's possible to cock the seal in the retainer bore or damage it during installation. If you want to take the chance, pry out the old seal with a screwdriver. Apply multi-purpose grease to the crankshaft seal journal and the lip of the new seal and carefully press the new seal into place. The lip is stiff, so carefully work it onto the seal journal of the crankshaft with a smooth object like the end of an extension as you tap the seal into place. Don't rush it, or you may damage the seal.

3 The following method is recommended but requires removing the seal retainer and resealing the rear of the oil pan.

4 After removing the two rearmost oil pan-to-seal retainer bolts, break the seal between the rear of the oil pan and the bottom of the seal retainer with a putty knife. Remove the retainer-to-engine block bolts, detach the seal retainer and remove all the old gasket material. Remove the sealant from the top of the oil pan flange.

Note: Cover the open area of the oil pan with clean rags to keep debris out while bracing the pan flange.

5 Position the seal and retainer assembly between two wood blocks on a workbench and drive the old seal out from the back side with a screwdriver.

6 Drive the new seal into the retainer with a wood block (see illustration) or a section of pipe slightly smaller in diameter than the outside diameter of the seal.

7 Lubricate the crankshaft seal journal and the lip of the new seal with multi-purpose grease. Note that the engine doesn't have a gasket between the seal retainer and the engine block. Instead, apply a 2 to 3 mm wide bead of RTV sealant to the retainer flange before attaching the retainer to the block.

8 Slowly and carefully push the seal and retainer onto the crankshaft. The seal lip is stiff, so work it onto the crankshaft with a smooth object such as the end of an extension as you push the retainer against the block.

9 Install and tighten the retainer bolts to the torque listed in this Chapter's Specifications.

10 The remainder of installation is the reverse of removal.

14.6 Drive the new seal into the retainer with a wood block or a section of pipe - make sure that you don't cock the seal in the retainer bore

15 Engine mounts - check and replacement

1 Engine mounts seldom require attention, but broken or deteriorated mounts should be replaced immediately or the added strain placed on the driveline components may cause damage or wear.

CHECK

2 During the check, the engine must be raised slightly to remove the weight from the mounts.

3 Raise the vehicle and support it securely on jackstands, then position a jack under the engine oil pan. Place a large wood block between the jack head and the oil pan, then carefully raise the engine just enough to take the weight off the mounts. Do not position the wood block under the drain plug.

WARNING:

DO NOT place any part of your body under the engine when it's supported only by a jack!

4 Check the mounts to see if the rubber is cracked, hardened or separated from the metal plates. Sometimes the rubber will split down the center.

5 Check for relative movement between the mount plates and the engine or frame (use a large screwdriver or prybar to attempt to move the mounts).

6 If movement is noted, lower the engine and tighten the mount fasteners.

REPLACEMENT

Refer to illustrations 15.8a and 15.8b

7 Disconnect the cable from the negative terminal of the battery (see Chapter 5, Section 1), then raise the vehicle and support it securely on jackstands (if not already done). Support the engine as described in Step 3.

8 To remove an engine mount, remove the fasteners, raise the engine and detach the mount (see illustrations). The engine can be raised with an engine hoist, or with a floor jack and wood block placed under the oil pan.

Note: Even if only one mount is being replaced, remove the mount-to-engine bracket nut from the other mount (this will allow the engine to be raised far enough for mount removal).

9 Installation is the reverse of removal. Use non-hardening thread locking compound on the mount bolts/nuts and be sure to tighten them securely.

10 See Chapter 7 for transmission mount replacement.

15.8a Engine mount-to-engine bracket nut

15.8b Engine mount-to-frame bracket bolts (A) and nuts (B, on the other side of the bracket)

Specifications

General

Engine designation	1GR-FE
Displacement	4.0 liters
Cylinder numbers (timing chain end-to-transmission end)	
Right (passenger) side	1-3-5
Left (driver) side	2-4-6
Firing order	1-2-3-4-5-6

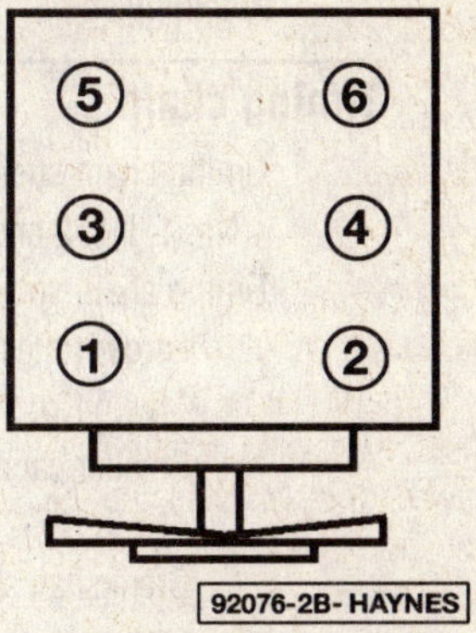

Cylinder locations

Warpage limits

Cylinder head	0.0039 inch (1 mm)
Intake manifold	
Intake plenum side	0.031 inch (0.8 mm)
Cylinder head side	0.008 inch (0.2 mm)
Exhaust manifolds	0.028 inch (0.7 mm)

Camshaft and related components

Valve clearance (engine cold)	See Chapter 1
Bearing journal diameter	
No. 1 journal	1.4162 to 1.4167 inches (35.97 to 35.98 mm)
Other journals	0.9039 to 0.9045 inch (22.95 to 22.97 mm)
Bearing oil clearance	
Standard	
Right side No. 1 (intake)	0.0003 to 0.0015 inch (0.008 to 0.038 mm)
Right side No. 1 (exhaust)	0.0016 to 0.0031 inch (0.040 to 0.079 mm)
Left side No. 1 (intake)	0.0016 to 0.0031 inch (0.040 to 0.079 mm)
Left side No. 1 (exhaust)	0.0016 to 0.0031 inch (0.040 to 0.079 mm)
Others	0.0010 to 0.0024 inch (0.025 to 0.062 mm)
Service limit	
Right side No. 1 (intake)	0.0028 inch (0.07 mm)
Others	0.0039 inch (0.10 mm)
Lobe height	
Intake	
Standard	1.7389 to 1.7428 inches (44.168 to 44.268 mm)
Service limit	1.7330 inches (44.018 mm)
Exhaust	
Standard	1.7551 to 1.7591 inches (44.580 to 44.680 mm)
Service limit	1.7492 inches (44.430 mm)
Thrust clearance (endplay)	
Standard	0.016 to 0.035 inch (0.04 to 0.09 mm)
Service limit	0.0043 inch (0.11 mm)
Runout limit (total indicator reading)	0.0024 inch (0.06 mm)
Lifters	
Lifter outside diameter	1.2191 to 1.2195 inches (30.966 to 30.976 mm)
Lifter bore diameter	1.2208 to 1.2215 inches (31.009 to 31.025 mm)
Lifter-to-bore (oil) clearance	
Standard	0.0013 to 0.0023 inch (0.033 to 0.059 mm)
Service limit	0.0031 inch (0.08 mm)

Cylinder head bolt diameter	
Standard	0.4272 to 0.4331 inch (10.85 to 11.0 mm)
Minimum	0.421 inch (10.7 mm)

Timing chain	
Timing chain stretch limit (15 pins)	
(No. 1, No. 2 and No. 3 chains)	5.780 inches (146.8 mm)
Timing chain sprocket wear limits	
Larger (intake) camshaft sprocket	
(with No. 1 chain installed)	4.547 inches (115.5 mm)
Smaller camshaft sprockets	
(with No. 2 or No. 3 chain installed)	2.878 inches (73.1 mm)
Crankshaft sprocket (with No. 1 chain installed)	2.402 inches (61.0 mm)
Idler sprocket wear limits	
With No. 1 chain installed	2.402 inches (61.0 mm)
Idler sprocket collar diameter	0.9050 to 0.9055 inch (22.987 to 23.0 mm)
Idler sprocket inside diameter	0.9063 to 0.9067 inch (23.02 to 23.03 mm)
Oil clearance	
Standard	0.0008 to 0.0017 inch (0.020 to 0.043 mm)
Maximum	0.0037 inch (0.093 mm)
Chain tensioner No. 2 wear limit	0.039 inch (1.0 mm)
Chain tensioner slipper wear limit	0.039 inch (1.0 mm)
Vibration damper No. 1 and No. 2 wear limit	0.039 inch (1.0 mm)

Oil pump	
Driven rotor-to-pump body clearance	
Standard	0.0098 to 0.0128 inch (0.250 to 0.325 mm)
Service limit	0.0128 inch (0.325 mm)
Rotor tip clearance	
Standard	0.0024 to 0.0063 inch (0.06 to 0.16 mm)
Service limit	0.0063 inch (0.16 mm)
Rotor side clearance	
Standard	0.0012 to 0.0035 inch (0.03 to 0.09 mm)
Service limit	0.0035 inch (0.09 mm)

Note: One foot-pound (ft-lb) of torque is equivalent to 12 inch-pounds (in-lbs) of torque. Torque values below approximately 15 ft-lbs are expressed in inch-pounds, since most foot-pound torque wrenches are not accurate at these smaller values

Torque specifications	Ft-lbs (unless otherwise indicated)	Nm
Intake manifold assembly		
Upper intake manifold bolts/nuts	21	28
Lower intake manifold bolts	19	26
Valve cover bolts/nuts		
Center bolts (with sealing washers)	80 in-lbs	9
Perimeter bolts/nuts	84 in-lbs	9.5
Exhaust manifold nuts	16	22
Crankshaft pulley bolt	185	251
Timing chain cover bolts/nuts	80 in-lbs	9
Drivebelt idler pulley bolts		
Idler pulley No. 1 bolt (lower left of engine)	40	54
Idler pulley No. 2 bolt	29	39
Drivebelt tensioner mounting bolts	See Chapter 1	

Torque specifications (continued)	Ft-lbs (unless otherwise indicated)	Nm
➡Note: One foot-pound (ft-lb) of torque is equivalent to 12 inch-pounds (in-lbs) of torque. Torque values below approximately 15 ft-lbs are expressed in inch-pounds, since most foot-pound torque wrenches are not accurate at these smaller values		
Camshaft timing sprocket bolt	74	100
Camshaft bearing cap bolts		
10 mm bolts	80 in-lbs	9
12 mm bolts	18	24
Cylinder head bolts (in sequence - see illustrations 10.24a and 10.24b)		
Initial torque	27	37
Final torque	Tighten an additional 180-degrees	
Left cylinder head (two front 14 mm bolts)	22	30
Oil filter assembly		
Oil filter bracket mounting bolts	168 in-lbs	19
Oil cooler-to-oil filter bracket bolt	50	68
Oil pan bolts/nuts		
Oil pan No. 1 (to engine block and front cover)		
10 mm head bolts	80 in-lbs	9
12 mm head bolts	16	22
Oil pan No. 2 (to oil pan No. 1)		
Bolts	80 in-lbs	9
Nuts	84 in-lbs	9.5
Oil pan assembly-to-transmission	27	37
Oil pick-up tube nuts	80 in-lbs	9
Oil pump assembly		
Oil pump cover bolts	80 in-lbs	9
Oil pipe flange bolts	80 in-lbs	9
Oil pump relief valve plug	36	49
Driveplate/flywheel-to-crankshaft bolts	61	83
Engine rear oil seal retainer		
Bolts	84 in-lbs	9.5
Nuts	80 in-lbs	9
Timing chain tensioner bolts		
Chain tensioner No. 1	84 in-lbs	9.5
Chain tensioner No. 2	168 in-lbs	19
Chain tensioner No. 3	168 in-lbs	19
Timing chain guide bolts	168 in-lbs	19
Timing chain idler sprocket shaft	44	60

Notes

2C

GENERAL ENGINE OVERHAUL PROCEDURES

Section

Reference to other Chapters

CHECK ENGINE light on - See Chapter 6

1 General information - engine overhaul

Refer to illustrations 1.1, 1.2, 1.3, 1.4, 1.5 and 1.6

Included in this portion of Chapter 2 are general information and diagnostic testing procedures for determining the overall mechanical condition of your engine.

The information ranges from advice concerning preparation for an overhaul and the purchase of replacement parts and/or components to detailed, step-by-step procedures covering removal and installation.

The following Sections have been written to help you determine whether your engine needs to be overhauled and how to remove and install it once you've determined it needs to be rebuilt. For information concerning in-vehicle engine repair, see Chapter 2A or 2B.

The Specifications included in this Part are general in nature and include only those necessary for testing the oil pressure, checking the engine compression, and bottom-end torque specifications. Refer to Chapter 2A or 2B for additional engine Specifications.

It's not always easy to determine when, or if, an engine should be completely overhauled, because a number of factors must be considered.

High mileage is not necessarily an indication that an overhaul is needed, while low mileage doesn't preclude the need for an overhaul. Frequency of servicing is probably the most important consideration. An engine that's had regular and frequent oil and filter changes, as well as other required maintenance, will most likely give many thousands of miles of reliable service. Conversely, a neglected engine may require an overhaul very early in its service life.

Excessive oil consumption is an indication that piston rings, valve seals and/or valve guides are in need of attention. Make sure that oil leaks aren't responsible before deciding that the rings and/or guides are bad. Perform a cylinder compression check to determine the extent of the work required (see Section 3). Also check the vacuum readings under various conditions (see Section 4).

Check the oil pressure with a gauge installed in place of the oil pressure sending unit and compare it to this Chapter's Specifications (see Section 2). If it's extremely low, the bearings and/or oil pump are probably worn out.

Loss of power, rough running, knocking or metallic engine noises, excessive valve train noise and high fuel consumption rates may also point to the need for an overhaul, especially if they're all present at the same time. If a complete tune-up doesn't remedy the situation, major mechanical work is the only solution.

An engine overhaul involves restoring the internal parts to the specifications of a new engine. During an overhaul, the piston rings are replaced and the cylinder walls are reconditioned (rebored and/or honed) (see illustrations 1.1 and 1.2). If a rebore is done by an automotive machine shop, new oversize pistons will also be installed. The main

1.1 An engine block being bored - an engine rebuilder will use special machinery to recondition the cylinder bores

1.2 If the cylinders are bored, the machine shop will normally hone the engine on a machine like this

1.3 A crankshaft having a main bearing journal ground

1.4 A machinist checks for a bent connecting rod, using specialized equipment

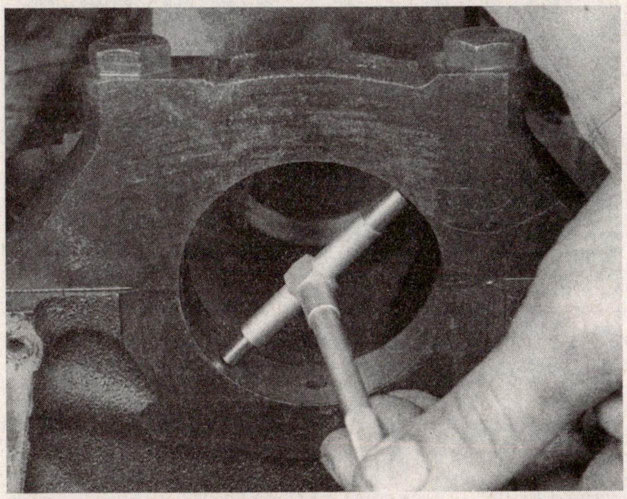

1.5 A bore gauge being used to check the main bearing bore

1.6 Uneven piston wear like this indicates a bent connecting rod

bearings and connecting rod bearings are generally replaced with new ones and, if necessary, the crankshaft may be reground to restore the journals (see illustration 1.3). Generally, the valves are serviced as well, since they're usually in less-than-perfect condition at this point. While the engine is being overhauled, other components, such as the starter and alternator, can be rebuilt as well. The end result should be a like-new engine that will give many trouble-free miles.

➡Note: Critical cooling system components such as the hoses, drivebelts, thermostat and water pump should be replaced with new parts when an engine is overhauled. The radiator should be checked carefully to ensure that it isn't clogged or leaking (see Chapter 3). If you purchase a rebuilt engine or short block, some rebuilders will not warranty their engines unless the radiator has been professionally flushed. Also, we don't recommend overhauling the oil pump - always install a new one when an engine is rebuilt.

Overhauling the internal components on today's engines is a difficult and time-consuming task that requires a significant amount of specialty tools and is best left to a professional engine rebuilder (see illustrations 1.4, 1.5 and 1.6). A competent engine rebuilder will handle the inspection of your old parts and offer advice concerning the reconditioning or replacement of the original engine. Never purchase parts or have machine work done on other components until the block has been thoroughly inspected by a professional machine shop. As a general rule, time is the primary cost of an overhaul, especially since the vehicle may be tied up for a minimum of two weeks or more. Be aware that some engine builders only have the capability to rebuild the engine you bring them while other rebuilders have a large inventory of rebuilt exchange engines in stock. Also be aware that many machine shops could take as much as two weeks time to completely rebuild your engine depending on shop workload. Sometimes it makes more sense to simply exchange your engine for another engine that's already rebuilt to save time.

2 Oil pressure check

➧ Refer to illustrations 2.2a and 2.2b

1 Low engine oil pressure can be a sign of an engine in need of rebuilding. A low oil pressure indicator (often called an "idiot light") is not a test of the oiling system. Such indicators only come on when the oil pressure is dangerously low. Even a factory oil pressure gauge in the instrument panel is only a relative indication, although much better for driver information than a warning light. A better test is with a mechanical (not electrical) oil pressure gauge.

2 Locate the oil pressure indicator sending unit on the engine block.

a) On four-cylinder models, the oil pressure sending unit is on the right (passenger) side of the engine block (see illustration).

b) On V6 models, the oil pressure sending unit is located next to the oil filter (see illustration).

3 Unscrew and remove the oil pressure sending unit, then screw in the hose for your oil pressure gauge. If necessary, install an adapter fitting. Use Teflon tape or thread sealant on the threads of the adapter and/or the fitting on the end of your gauge's hose.

4 Connect an accurate tachometer to the engine, according to the tachometer manufacturer's instructions.

5 Check the oil pressure with the engine running (normal operating temperature) at the specified engine speed, and compare it to this Chapter's Specifications. If it's extremely low, the bearings and/or oil pump are probably worn out.

2.2a The four-cylinder engine's oil pressure sending unit is on the right side of the engine block

2.2b V6 engine oil pressure sending unit

3 Cylinder compression check

➧ Refer to illustration 3.6

1 A compression check will tell you what mechanical condition the upper end of your engine (pistons, rings, valves, head gaskets) is in. Specifically, it can tell you if the compression is down due to leakage caused by worn piston rings, defective valves and seats or a blown head gasket.

➡Note: The engine must be at normal operating temperature and the battery must be fully charged for this check.

2 Begin by cleaning the area around the spark plugs before you remove them (compressed air should be used, if available). The idea is to prevent dirt from getting into the cylinders as the compression check is being done.

3 Remove all of the spark plugs from the engine (see Chapter 1).

4 Block the throttle wide open.

5 Disable the fuel pump circuit by removing the circuit opening relay from the fuse/relay center in the engine compartment (see Chapter 4, Section 2).

6 Install a compression gauge in the spark plug hole (see illustration).

7 Crank the engine over at least seven compression strokes and watch the gauge. The compression should build up quickly in a healthy engine. Low compression on the first stroke, followed by gradually increasing pressure on successive strokes, indicates worn piston rings. A low compression reading on the first stroke, which doesn't build up during successive strokes, indicates leaking valves or a blown head gasket (a cracked head could also be the cause). Deposits on the undersides of the valve heads can also cause low compression. Record the highest gauge reading obtained.

8 Repeat the procedure for the remaining cylinders and compare the results to this Chapter's Specifications.

9 Add some engine oil (about three squirts from a plunger-type oil can) to each cylinder, through the spark plug hole, and repeat the test.

10 If the compression increases after the oil is added, the piston rings are definitely worn. If the compression doesn't increase significantly, the leakage is occurring at the valves or head gasket. Leakage past the valves may be caused by burned valve seats and/or faces or warped, cracked or bent valves.

11 If two adjacent cylinders have equally low compression, there's a strong possibility that the head gasket between them is blown. The appearance of coolant in the combustion chambers or the crankcase would verify this condition.

12 If one cylinder is slightly lower than the others, and the engine has a slightly rough idle, a worn lobe on the camshaft could be the cause.

13 If the compression is unusually high, the combustion chambers are probably coated with carbon deposits. If that's the case, the cylinder head(s) should be removed and decarbonized.

14 If compression is way down or varies greatly between cylinders, it would be a good idea to have a leak-down test performed by an automotive repair shop. This test will pinpoint exactly where the leakage is occurring and how severe it is.

3.6 Use a compression gauge with a threaded fitting for the spark plug hole, not the type that requires hand pressure to maintain the seal

4 Vacuum gauge diagnostic checks

➧ **Refer to illustrations 4.4 and 4.6**

1 A vacuum gauge provides inexpensive but valuable information about what is going on in the engine. You can check for worn rings or cylinder walls, leaking head or intake manifold gaskets, restricted exhaust, stuck or burned valves, weak valve springs, improper ignition or valve timing and ignition problems.

2 Unfortunately, vacuum gauge readings are easy to misinterpret, so they should be used in conjunction with other tests to confirm the diagnosis.

3 Both the absolute readings and the rate of needle movement are important for accurate interpretation. Most gauges measure vacuum in inches of mercury (in-Hg). The following references to vacuum assume the diagnosis is being performed at sea level. As elevation increases (or atmospheric pressure decreases), the reading will decrease. For every 1,000-foot increase in elevation above approximately 2000 feet, the gauge readings will decrease about one inch of mercury.

4 Connect the vacuum gauge directly to the intake manifold vacuum, not to ported (throttle body) vacuum (see illustration). Be sure no hoses are left disconnected during the test or false readings will result.

5 Before you begin the test, allow the engine to warm up completely. Block the wheels and set the parking brake. With the transmission in Park, start the engine and allow it to run at normal idle speed.

4.4 A simple vacuum gauge can be handy in diagnosing engine condition and performance. Connect it to an intake manifold vacuum source, not a ported (throttle body) source

⁂ WARNING:

Keep your hands and the vacuum gauge clear of the fans and drivebelts.

6 Read the vacuum gauge; an average, healthy engine should normally produce about 17 to 22 in-Hg with a fairly steady needle (see illustration). Refer to the following vacuum gauge readings and what they indicate about the engine's condition:

7 A low steady reading usually indicates a leaking gasket between the intake manifold and cylinder head(s) or throttle body, a leaky vacuum hose, late ignition timing or incorrect camshaft timing. Check ignition timing with a timing light and eliminate all other possible causes, utilizing the tests provided in this Chapter before you remove the timing belt cover to check the timing marks.

8 If the reading is three to eight inches below normal and it fluctuates at that low reading, suspect an intake manifold gasket leak at an intake port or a faulty fuel injector.

9 If the needle has regular drops of about two-to-four inches at a steady rate, the valves are probably leaking. Perform a compression check or leak-down test to confirm this.

10 An irregular drop or down-flick of the needle can be caused by a sticking valve or an ignition misfire. Perform a compression check or leak-down test and read the spark plugs.

11 A rapid vibration of about four in-Hg variation at idle combined with exhaust smoke indicates worn valve guides. Perform a leak-down test to confirm this. If the rapid vibration occurs with an increase in engine speed, check for a leaking intake manifold gasket or head gasket, weak valve springs, burned valves or ignition misfire.

12 A slight fluctuation, say one inch up and down, may mean ignition problems. Check all the usual tune-up items and, if necessary, run the engine on an ignition analyzer.

13 If there is a large fluctuation, perform a compression or leak-down test to look for a weak or dead cylinder or a blown head gasket.

14 If the needle moves slowly through a wide range, check for a clogged PCV system, incorrect idle fuel mixture, throttle body or intake manifold gasket leaks.

15 Check for a slow return after revving the engine by quickly snapping the throttle open until the engine reaches about 2,500 rpm and let it shut. Normally the reading should drop to near zero, rise above normal idle reading (about 5 in-Hg over) and then return to the previous idle reading. If the vacuum returns slowly and doesn't peak when the throttle is snapped shut, the rings may be worn. If there is a long delay, look for a restricted exhaust system (often the muffler or catalytic converter). An easy way to check this is to temporarily disconnect the exhaust ahead of the suspected part and redo the test.

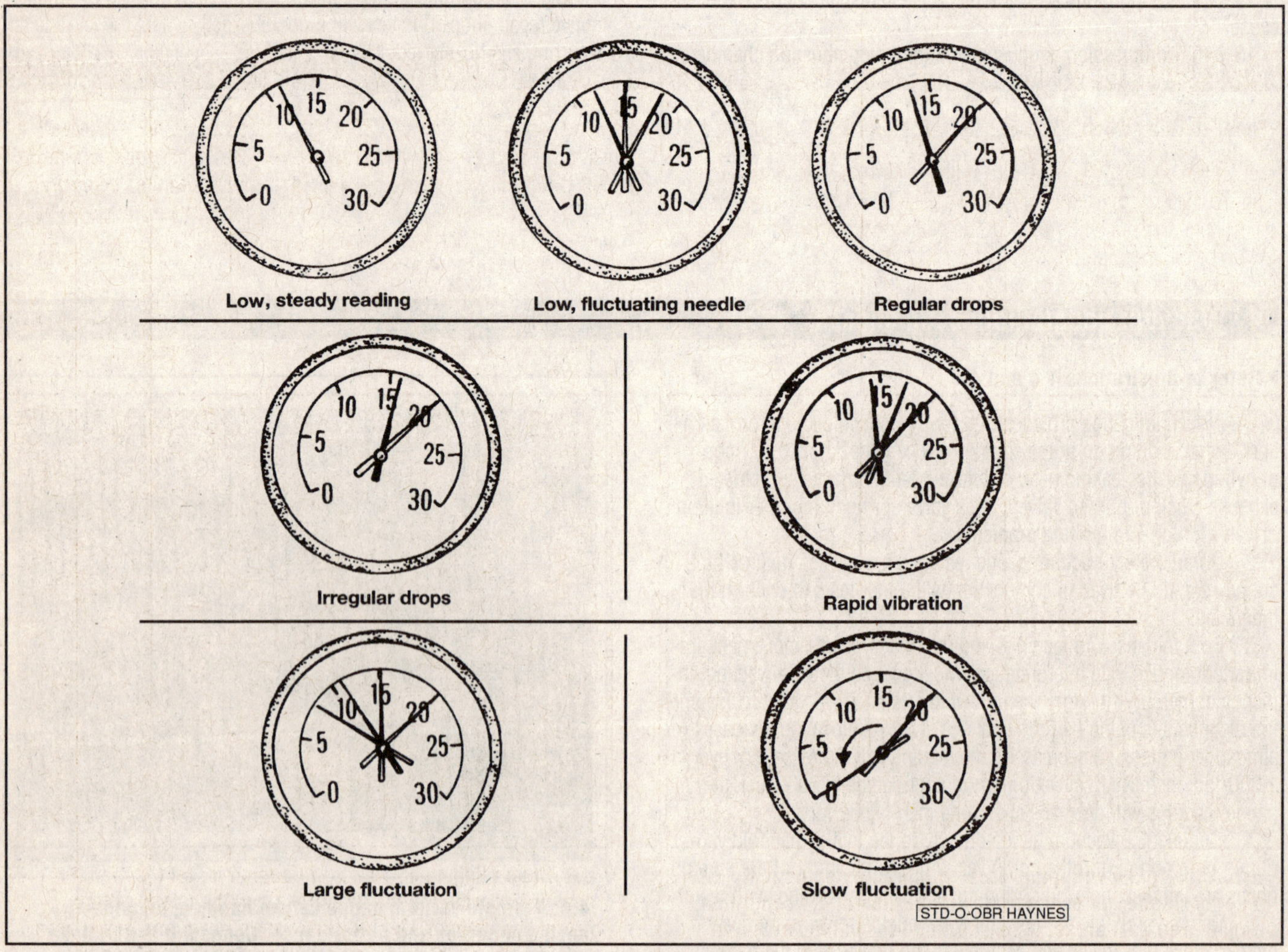

4.6 Typical vacuum gauge readings

5 Engine rebuilding alternatives

The do-it-yourselfer is faced with a number of options when purchasing a rebuilt engine. The major considerations are cost, warranty, parts availability and the time required for the rebuilder to complete the project. The decision to replace the engine block, piston/connecting rod assemblies and crankshaft depends on the final inspection results of your engine. Only then can you make a cost effective decision whether to have your engine overhauled or simply purchase an exchange engine for your vehicle.

Some of the rebuilding alternatives include:

Individual parts - If the inspection procedures reveal that the engine block and most engine components are in reusable condition, purchasing individual parts and having a rebuilder rebuild your engine may be the most economical alternative. The block, crankshaft and piston/connecting rod assemblies should all be inspected carefully by a machine shop first.

Short block - A short block consists of an engine block with a crankshaft and piston/connecting rod assemblies already installed. All new bearings are incorporated and all clearances will be correct. The existing camshafts, valve train components, cylinder head and external parts can be bolted to the short block with little or no machine shop work necessary.

Long block - A long block consists of a short block plus an oil pump, oil pan, cylinder head, valve cover, camshaft and valve train components, timing sprockets and belt or gears and timing cover. All components are installed with new bearings, seals and gaskets incorporated throughout. The installation of manifolds and external parts is all that's necessary.

Low mileage used engines - Some companies now offer low mileage used engines that are a very cost effective way to get your vehicle up and running again. These engines often come from vehicles that have been in totaled in accidents or come from other countries that have a higher vehicle turn over rate. A low mileage used engine also usually has a similar warranty like the newly remanufactured engines.

Give careful thought to which alternative is best for you and discuss the situation with local automotive machine shops, auto parts dealers and experienced rebuilders before ordering or purchasing replacement parts.

6 Engine removal - methods and precautions

Refer to illustrations 6.1, 6.2, 6.3, 6.4 and 6.5

If you've decided that an engine must be removed for overhaul or major repair work, several preliminary steps should be taken. Read all removal and installation procedures carefully prior to committing this job. Some engines are removed by lowering them to the floor, then raising the vehicle sufficiently to slide it out; this will require a vehicle hoist.

Locating a suitable place to work is extremely important. Adequate work space, along with storage space for the vehicle, will be needed. If a shop or garage isn't available, at the very least a flat, level, clean work surface made of concrete or asphalt is required. Cleaning the engine compartment and engine before beginning the removal procedure will help keep tools clean and organized (see illustrations 6.1 and 6.2).

An engine hoist or A-frame will also be necessary. Make sure the equipment is rated in excess of the combined weight of the engine and transmission. Safety is of primary importance, considering the potential hazards involved in lifting the engine out of the vehicle.

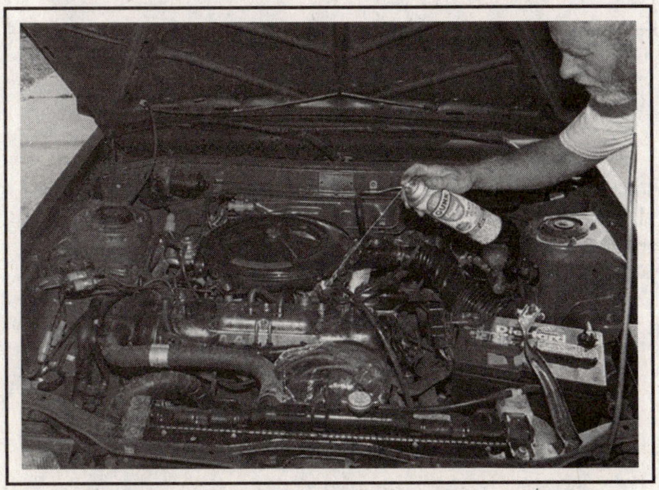

6.1 After tightly wrapping water-vulnerable components, use a spray cleaner on everything, with particular concentration on the greasiest areas, usually around the valve cover and lower edges of the block. If one section dries out, apply more cleaner

6.2 Depending on how dirty the engine is, let the cleaner soak in according to the directions and then hose off the grime and cleaner. Get the rinse water down into every area you can get at; then dry important components with a hair dryer or paper towels

6.3 Get an engine hoist that's strong enough to easily lift your engine in and out of the engine compartment; an adapter, like the one shown here (arrow), can be used to change the angle of the engine as it's being removed or installed

If you're a novice at engine removal, get at least one helper. One person cannot easily do all the things you need to do to lift a big heavy engine out of the engine compartment. Also helpful is to seek advice and assistance from someone who's experienced in engine removal.

Plan the operation ahead of time. Arrange for or obtain all of the tools and equipment you'll need prior to beginning the job (see illustrations 6.3, 6.4 and 6.5). some of the equipment necessary to perform engine removal and installation safely and with relative ease are (in addition to an engine hoist) a heavy duty floor jack, complete sets of wrenches and sockets as described in the front of this manual, wooden blocks, plenty of rags and cleaning solvent for mopping up spilled oil, coolant and gasoline. If the hoist must be rented, make sure that you arrange for it in advance and have everything disconnected and/or removed before bringing the hoist home. This will save you money and time.

6.4 Get an engine stand sturdy enough to firmly support the engine while you're working on it. Stay away from three-wheeled models; they have a tendency to tip over more easily, so get a four-wheeled unit

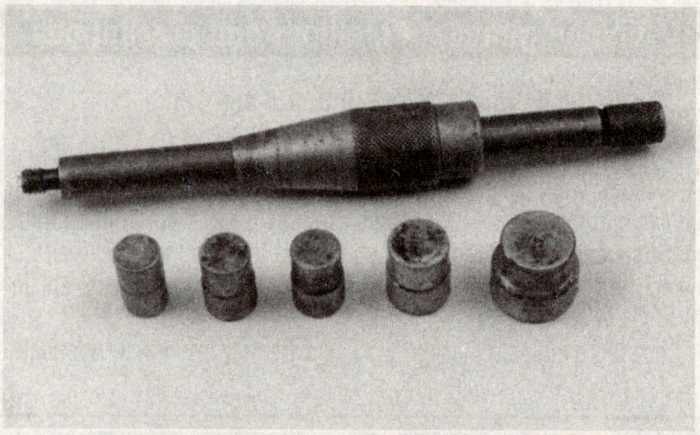

6.5 A clutch alignment tool is necessary if you plan to install a rebuilt engine mated to a manual transmission

Plan for the vehicle to be out of use for quite a while. A machine shop can do the work that is beyond the scope of the home mechanic. Machine shops often have busy schedules, so before removing the engine, consult the shop for an estimate of how long it will take to rebuild or repair the components that may need work.

7 Engine - removal and installation

✲✲ WARNING 1:

Wait until the engine is completely cool before beginning this procedure.

✲✲ WARNING 2:

DO NOT use a cheap engine hoist designed for lifting small engines. Obtain a heavy-duty hoist designed for lifting heavy engines. And always be extremely careful when removing and installing the engine. Serious injury can result from careless actions.

✲✲ WARNING 3:

The models covered by this manual are equipped with Supplemental Restraint Systems (SRS), more commonly known as airbags. Always disable the airbag system before working in the vicinity of any airbag system component to avoid the possibility of accidental deployment of the airbag, which could cause personal injury (see Chapter 12).

✲✲ WARNING 4:

Gasoline is extremely flammable, so take extra precautions when you work on any part of the fuel system. Don't smoke or allow open flames or bare light bulbs near the work area, and don't work in a garage where a gas-type appliance (such as a water heater or a clothes dryer) is present. Since gasoline is carcinogenic, wear fuel-resistant gloves when there's a possibility of being exposed to fuel, and, if you spill any fuel on your skin, rinse it off immediately with soap and water. Mop up any spills immediately and do not store fuel-soaked rags where they could ignite. The fuel system is under constant pressure, so, if any fuel lines are to be disconnected, the fuel pressure in the system must be relieved first (see Chapter 4 for more information). When you perform any kind of work on the fuel system, wear safety glasses and have a Class B type fire extinguisher on hand.

REMOVAL

➧ Refer to illustrations 7.4, 7.8, 7.20a, 7.20b, 7.20c and 7.23

1 Relieve the fuel system pressure (see Chapter 4).

2 Disconnect the battery cables and remove the battery (see Chapter 5). Remove the engine cover from V6 models.

3 Remove the hood (see Chapter 11) and cover the fenders and cowl. Special pads are available to protect the fenders, but an old bedspread or blanket will also work.

4 Raise the vehicle and support it securely on jackstands (see illustration).

➡Note: On 4WD models, and on models with large tires, this step may not be necessary, because some models already have sufficient ground clearance to allow disconnection of the exhaust system, the engine mounts, etc. from underneath the vehicle. Raising these vehicles any higher might even make engine removal more difficult because it might position the vehicle too high to lift the engine out of the engine compartment with a hoist.

5 Drain the engine oil and remove the oil filter (see Chapter 1).

6 Drain the cooling system (see Chapter 1).

7 Remove the air cleaner assembly and the air intake duct (see Chapter 4).

8 Label all vacuum lines, emissions system hoses, wiring harness electrical connectors and ground straps to ensure correct reinstallation, then disconnect them. Pieces of masking tape with numbers or letters written on them work well (see illustration). So does colored electrical tape. If there's any possibility of confusion, make a sketch of the engine compartment and clearly label the lines, hoses and wires. You can also use an inexpensive disposable or digital camera to take photos of connectors, grounds, harness routing, etc.

9 Remove the glove box door and the lower right instrument panel cover for access to the PCM and the 4WD ECU if so equipped (see Chapter 11). Disconnect the wiring harnesses from the PCM and pull them through the firewall into the engine compartment.

10 Disconnect the wiring from the front differential on 4WD models. Remove the nut, then disconnect the three wiring harnesses from the

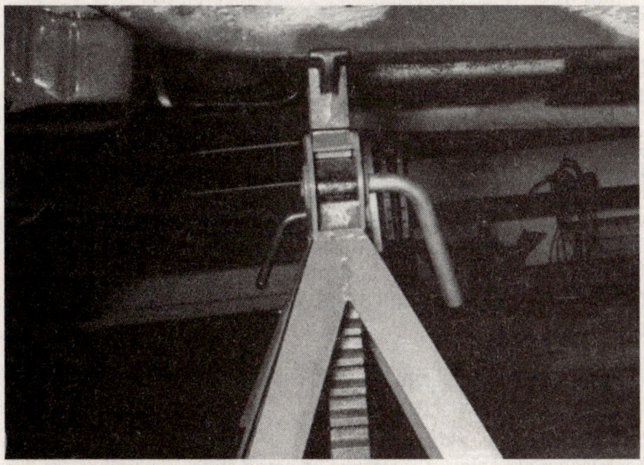

7.4 Place sturdy jackstands under the frame of the vehicle and set them both at uniform height

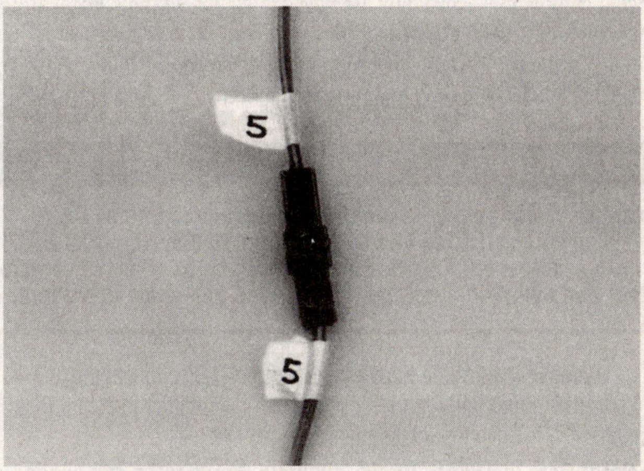

7.8 Label both ends of each wire and hose before disconnecting it - do the same for vacuum hoses

7.20a Engine lifting hanger on the right side of the engine (typical)

7.20c Secure the engine hangers to the lifting device with heavy chain (typical)

engine compartment relay box. Remove the engine wiring harness from the relay box.

11 Disconnect the fuel lines running from the engine to the chassis (see Chapter 4). Plug or cap all open fittings/lines.

✲✲ WARNING:

Gasoline is extremely flammable, so extra precautions must be taken when working on any part of the fuel system. DO NOT smoke or allow open flames or bare light bulbs near the vehicle. Also, don't work in a garage if a gas-type appliance is present.

12 Remove the throttle body (see Chapter 4). Disconnect the fuel lines from the fuel rail, then remove the fuel rail and the injectors (see Chapter 4). Remove the intake manifold (see Chapter 2A or 2B).

➡Note: Only the upper intake manifold needs to be removed from V6 engines. Remove all fuel and/or emission control components that might be damaged during engine removal (see Chapters 4 and 6).

7.20b Engine lifting hanger on the left side of the engine (typical)

13 Clearly label and disconnect all coolant and heater hoses. Remove the cooling fan, shroud and radiator (see Chapter 3).

14 Remove the drivebelt (see Chapter 1), then remove the alternator if you're working on a V6 engine (see Chapter 5).

15 Unbolt the air conditioning compressor (see Chapter 3). Don't disconnect the hoses. Secure it with wire to make sure that it won't be damaged by the engine when the engine is lifted out.

16 Unbolt the power steering pump (see Chapter 10) and secure it with wire so that it won't interfere with engine removal. Don't disconnect the hoses.

17 On V6 models, remove the part of the exhaust system that's routed underneath the engine, between the exhaust manifolds and the downstream catalytic converter (see Chapter 4). It's not absolutely necessary to remove the exhaust manifolds in order to remove the engine, but removing the manifolds will shave a little weight off the engine.

18 On four-cylinder models, disconnect the exhaust pipe from the exhaust manifold. Use wire to support the pipe while it's disconnected.

19 Refer to Chapter 7 and remove the transmission. Support the engine while the transmission is being removed and after it is out.

20 Locate the lifting brackets on the engine (see illustrations).

➡Note: If the engine is not equipped with lifting brackets, they can be obtained through a Toyota dealer parts department. Roll a heavy-duty hoist into position and attach it to the lifting brackets with a couple pieces of heavy-duty chain (see illustration). Take up the slack in the sling or chain, but don't lift the engine.

✲✲ WARNING:

DO NOT use a cheap hoist designed to lift small engines. Obtain a heavy-duty hoist designed for lifting heavy engines. And DO NOT place any part of your body under the engine when it's supported only by a hoist or other lifting device.

21 Remove the engine mount fasteners (see *Engine mounts - check and replacement* in Chapter 2A).

22 Recheck to be sure nothing is still connecting the engine to the transmission or vehicle. Disconnect anything still remaining. Raise the engine slightly and inspect it thoroughly once more to make sure that nothing is still attached, then slowly lift the engine out of the engine compartment. Check carefully to make sure nothing is hanging up as

you go.

23 Remove the driveplate or flywheel (see Chapter 2A or 2B) and mount the engine on an engine stand (see illustration).

24 Inspect the engine and transmission mounts (see "Engine mounts - check and replacement" in Chapter 2A or 2B). If they're worn or damaged, replace them.

INSTALLATION

25 Install the driveplate or flywheel (see Chapter 2A or 2B).

26 Carefully lower the engine into the engine compartment, then reattach it to the engine mounts.

27 Install the transmission (see Chapter 7A or 7B).

28 Reinstall the remaining components in the reverse order of removal.

29 Add coolant, oil and transmission fluid as needed (see Chapter 1).

30 Run the engine and check for leaks and proper operation of all accessories, then install the hood and test drive the vehicle.

7.23 Use long high-strength bolts (arrows) to hold the engine block on the engine stand - make sure they are tight before resting all the weight on the stand

8 Engine overhaul - disassembly sequence

1 It's much easier to disassemble the engine if it's mounted on a portable engine stand. A stand can often be rented quite cheaply from an equipment rental yard. Before the engine is mounted on a stand, the driveplate should be removed from the engine.

2 If a stand isn't available, it's possible to remove the external engine components with it blocked up on the floor. Be extra careful not to tip or drop the engine when working without a stand.

3 If you're going to obtain a rebuilt engine, all external components must come off first, to be transferred to the replacement engine. These components include:

Driveplate
Ignition system components
Emissions-related components
Engine mounts and mount brackets
Fuel injection components
Intake/exhaust manifolds
Oil filter
Thermostat and housing assembly
Water pump

➡**Note: When removing the external components from the engine, pay close attention to details that may be helpful or important during installation. Note the installed position of gaskets, seals, spacers, pins, brackets, washers, bolts and other small items.**

4 If you're going to obtain a short block (assembled engine block, crankshaft, pistons and connecting rods), then you should remove the timing belt, cylinder head, oil pan, oil pump pick-up tube, oil pump and water pump from your engine so that you can turn in your old short block to the rebuilder as a core. See *Engine rebuilding alternatives* for additional information regarding the different possibilities to be considered.

9 Pistons and connecting rods - removal and installation

REMOVAL

Refer to illustrations 9.1, 9.3 and 9.4

➡**Note: Prior to removing the piston/connecting rod assemblies, remove the cylinder head and oil pan (see Chapter 2A or 2B).**

1 Use your fingernail to feel if a ridge has formed at the upper limit of ring travel (about 1/4-inch down from the top of each cylinder). If carbon deposits or cylinder wear have produced ridges, they must be completely removed with a special tool (see illustration). Follow the manufacturer's instructions provided with the tool. Failure to remove the ridges before attempting to remove the piston/connecting rod assemblies may result in piston breakage.

2 After the cylinder ridges have been removed, turn the engine so the crankshaft is facing up.

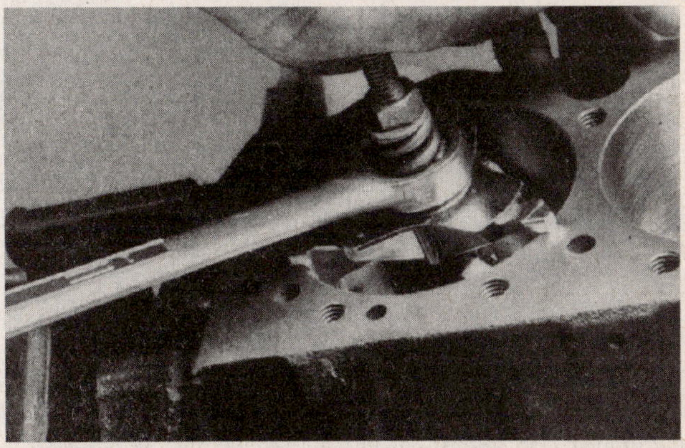

9.1 Before you try to remove the pistons from engines with very worn cylinders, use a ridge reamer to remove the raised material (ridge) from the top of the cylinders

9.3 Checking the connecting rod endplay (side clearance)

9.4 If the connecting rods and caps are not marked, use a center punch or numbered impression stamps to mark the caps to the rods by cylinder number (for example, this would be the No. 4 connecting rod)

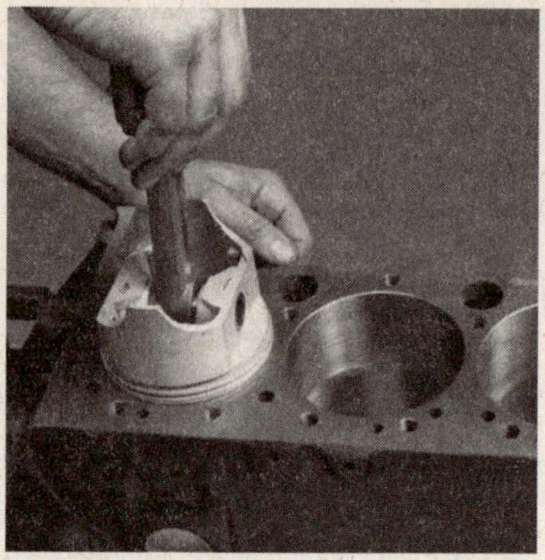

9.13 Install the piston ring into the cylinder, then push it down into position using a piston, so the ring will be square in the cylinder

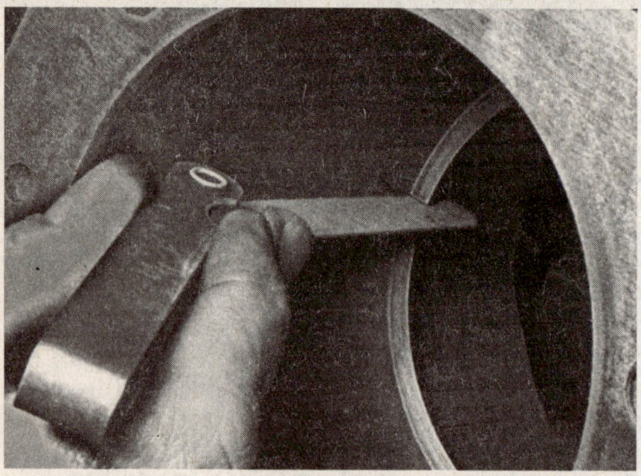

9.14 With the ring square in the cylinder, measure the ring end gap with a feeler gauge

3 Before the connecting rods are removed, check the connecting rod endplay with feeler gauges. Slide them between the first connecting rod and the crankshaft throw until the play is removed (see illustration). Repeat this procedure for each connecting rod. The endplay is equal to the thickness of the feeler gauge(s). Check with an automotive machine shop for the endplay service limit. If the play exceeds the service limit, new connecting rods will be required. If new rods (or a new crankshaft) are installed, the endplay may fall under the minimum allowable clearance. If it does, the rods will have to be machined to restore it. If necessary, consult an automotive machine shop for advice.

4 Check the connecting rods and caps for identification marks (see illustration). If they aren't plainly marked, use a small center-punch to make the appropriate number of indentations on each rod and cap (1, 2, 3, etc., depending on the cylinder they're associated with). Rod caps on four-cylinder engines have a mark that faces forward.

5 Loosen each of the connecting rod cap bolts 1/2-turn at a time until they can be removed by hand. Remove the number one connecting rod cap and bearing insert. Don't drop the bearing insert out of the cap.

6 Remove the bearing insert and push the connecting rod/piston assembly out through the top of the engine. Use a wooden or plastic hammer handle to push on the upper bearing surface in the connecting rod. If resistance is felt, double-check to make sure that all of the ridge was removed from the cylinder.

7 Repeat the procedure for the remaining cylinders.

8 Using a vernier caliper or a micrometer, measure the diameter of each connecting rod bolt in the necked-down area above the threads. Compare your measurements with the values listed in this Chapter's Specifications. If the diameter of any bolt falls under the minimum, replace it.

9 Reassemble the connecting rod caps and bearing inserts in their respective connecting rods and install the cap bolts finger tight. Leaving the old bearing inserts in place until reassembly will help prevent the connecting rod bearing surfaces from being accidentally nicked or gouged.

10 The pistons and connecting rods are now ready for inspection and overhaul at an automotive machine shop.

PISTON RING INSTALLATION

➧ Refer to illustrations 9.13, 9.14, 9.15, 9.19a, 9.19b and 9.22

11 Before installing the new piston rings, the ring end gaps must be checked. It's assumed that the piston ring side clearance has been checked and verified correct.

12 Lay out the piston/connecting rod assemblies and the new ring sets so the ring sets will be matched with the same piston and cylinder during the end gap measurement and engine assembly.

13 Insert the top (number one) ring into the first cylinder and square it up with the cylinder walls by pushing it in with the top of the piston (see illustration). The ring should be near the bottom of the cylinder, at the lower limit of ring travel.

14 To measure the end gap, slip feeler gauges between the ends of the ring until a gauge equal to the gap width is found (see illustration). The feeler gauge should slide between the ring ends with a slight amount of drag. Check with an automotive machine shop for the correct end gap for your engine. If the gap is larger or smaller than specified, double-check to make sure you have the correct rings before proceeding.

9.15 If the ring end gap is too small, clamp a file in a vise and file the piston ring ends - file from the outside of the ring inward only

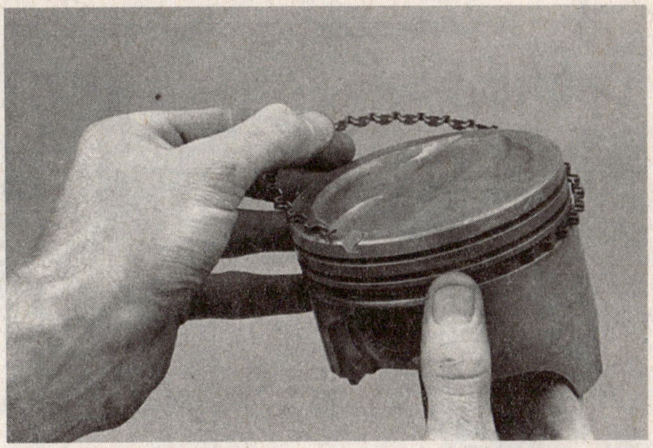

9.19a Installing the spacer/expander in the oil ring groove

9.19b DO NOT use a piston ring installation tool when installing the oil control side rails

9.22 Use a piston ring installation tool to install the number 2 and the number 1 (top) rings - be sure the directional mark on the piston ring(s) is facing toward the top of the piston

15 If the gap is too small, it must be enlarged or the ring ends may come in contact with each other during engine operation, which can cause serious damage to the engine. The end gap can be increased by filing the ring ends very carefully with a fine file. Mount the file in a vise equipped with soft jaws, slip the ring over the file with the ends contacting the file face and slowly move the ring to remove material from the ends. When performing this operation, file only by pushing the ring from the outside end of the file towards the vise (see illustration). Be sure to remove all raised material.

16 Excess end gap isn't critical unless it's greater than approximately 0.040-inch. Again, double-check to make sure you have the correct ring type and that you are referencing the correct section and category of specifications.

17 Repeat the procedure for each ring that will be installed in the first cylinder and for each ring in the remaining cylinders. Remember to keep rings, pistons and cylinders matched up.

18 Once the ring end gaps have been checked/corrected, the rings can be installed on the pistons.

19 The oil control ring (lowest one on the piston) is usually installed first. It's composed of three separate components. Slip the spacer/expander into the groove (see illustration). If an anti-rotation tang is used, make sure it's inserted into the drilled hole in the ring groove. Next, install the upper side rail in the same manner (see illustration). Don't use a piston ring installation tool on the oil ring side rails, as they may be damaged. Instead, place one end of the side rail into the groove between the spacer/expander and the ring land, hold it firmly in place and slide a finger around the piston while pushing the rail into the groove. Finally, install the lower side rail.

20 After the three oil ring components have been installed, check to make sure that both the upper and lower side rails can be rotated smoothly inside the ring grooves.

21 The number two (middle) ring is installed next. It's usually stamped with a mark that must face up, toward the top of the piston. Do not mix up the top and middle rings, as they have different cross-sections.

➡Note: Always follow the instructions printed on the ring package or box - different manufacturers may require different approaches.

22 Use a piston ring installation tool and make sure the identification mark is facing the top of the piston, then slip the ring into the middle groove on the piston (see illustration). Don't expand the ring any more than necessary to slide it over the piston.

ENGINE BEARING ANALYSIS

Debris

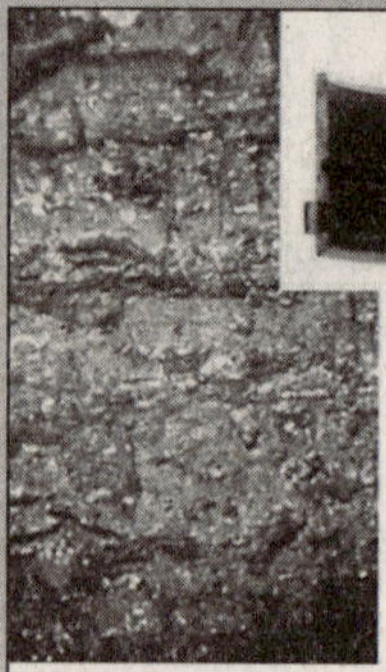
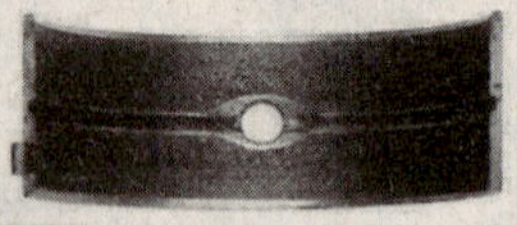

Babbitt bearing embedded with debris from machinings

Microscopic detail of debris

Microscopic detail of gouges

Overplated copper alloy bearing gouged by cast iron debris

Aluminum bearing embedded with glass beads

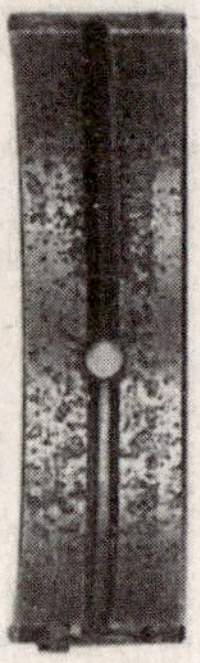

Microscopic detail of glass beads

Damaged lining caused by dirt left on the bearing back

Misassembly

Result of a lower half assembled as an upper - blocking the oil flow

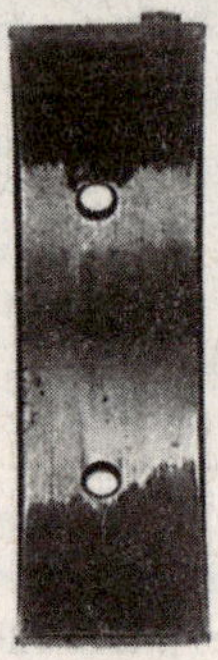

Excessive oil clearance is indicated by a short contact arc

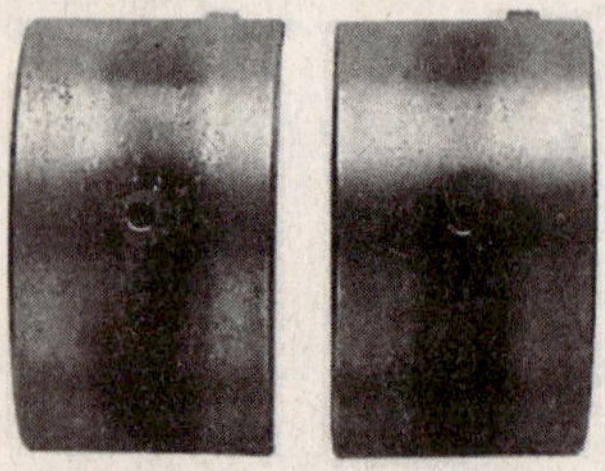

Polished and oil-stained backs are a result of a poor fit in the housing bore

Result of a wrong, reversed, or shifted cap

Overloading

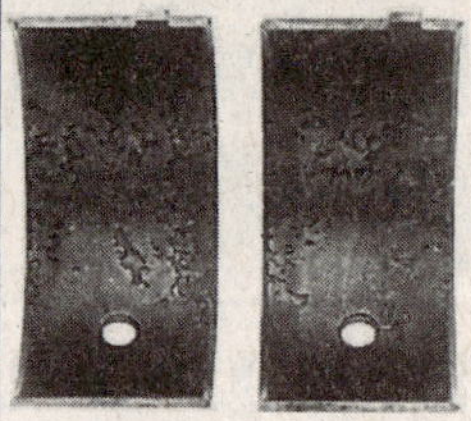

Damage from excessive idling which resulted in an oil film unable to support the load imposed

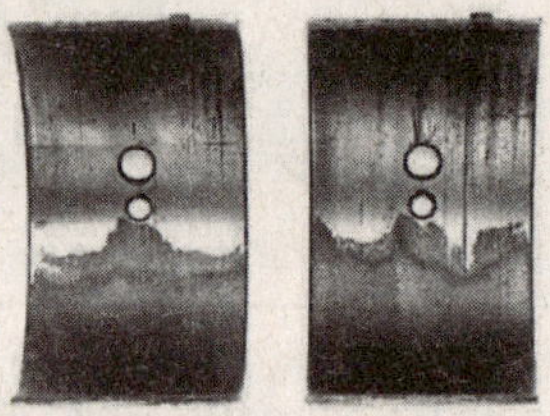

Damaged upper connecting rod bearings caused by engine lugging; the lower main bearings (not shown) were similarly affected

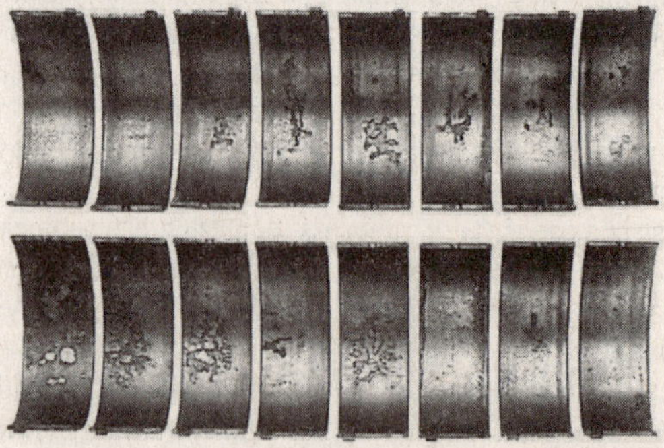

The damage shown in these upper and lower connecting rod bearings was caused by engine operation at a higher-than-rated speed under load

Misalignment

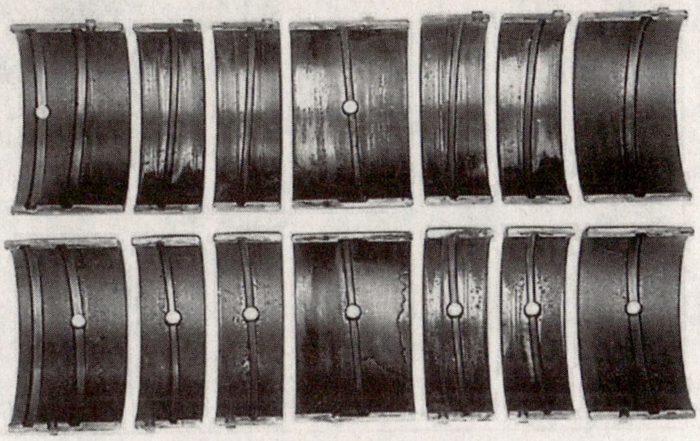

A warped crankshaft caused this pattern of severe wear in the center, diminishing toward the ends

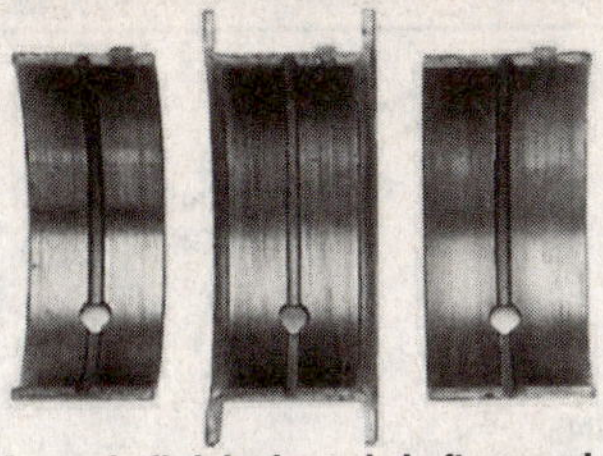

A poorly finished crankshaft caused the equally spaced scoring shown

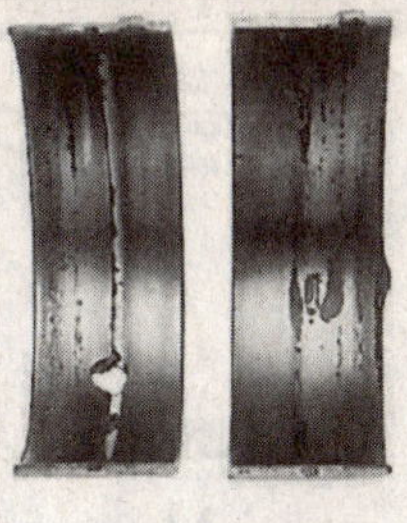

A tapered housing bore caused the damage along one edge of this pair

A bent connecting rod led to the damage in the "V" pattern

Corrosion

Microscopic detail of corrosion

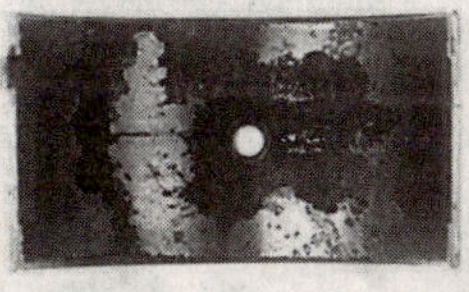

Corrosion is an acid attack on the bearing lining generally caused by inadequate maintenance, extremely hot or cold operation, or inferior oils or fuels

Lubrication

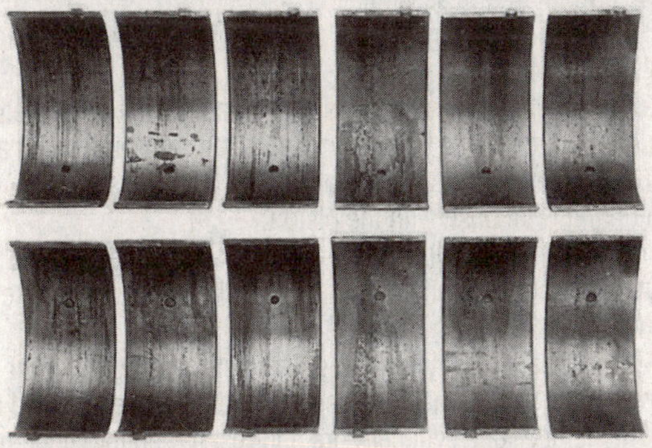

Result of dry start: The bearings on the left, farthest from the oil pump, show more damage

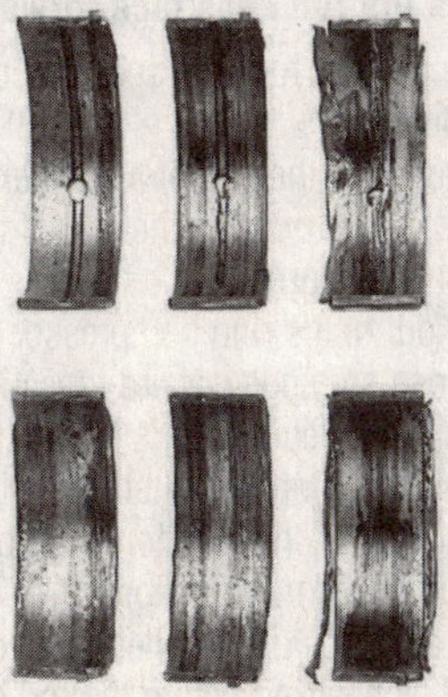

Result of a low oil supply or oil starvation

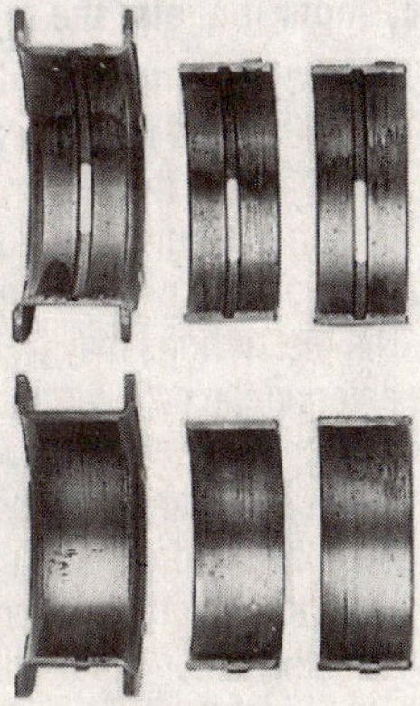

Severe wear as a result of inadequate oil clearance

Microscopic detail of cavitation

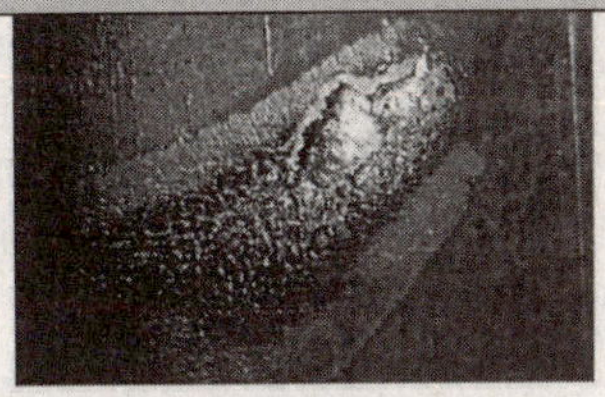

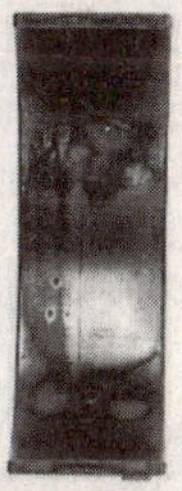

Example of cavitation - a surface erosion caused by pressure changes in the oil film

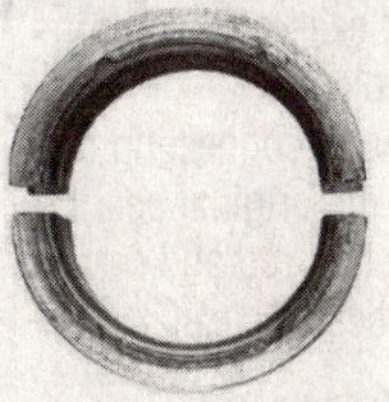

Damage from excessive thrust or insufficient axial clearance

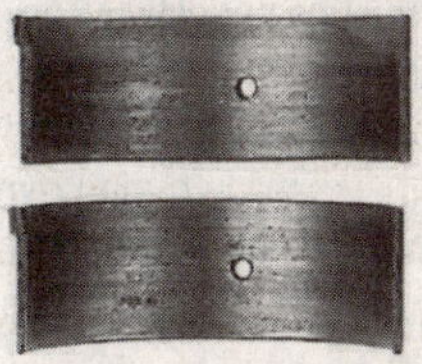

Bearing affected by oil dilution caused by excessive blow-by or a rich mixture

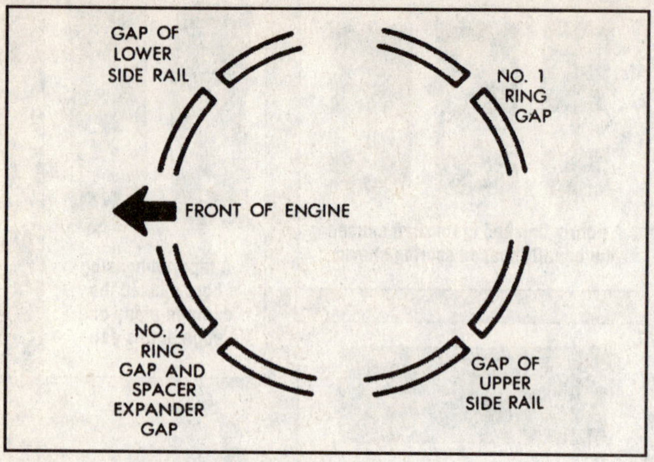

9.29 Position the piston ring end gaps as shown

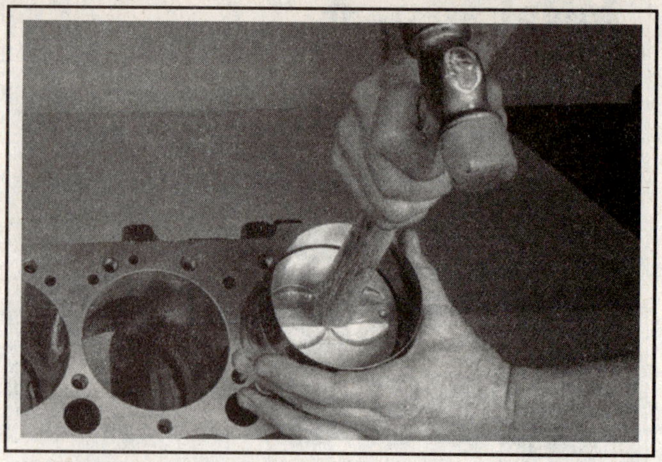

9.34 Use a plastic or wooden hammer handle to push the piston into the cylinder

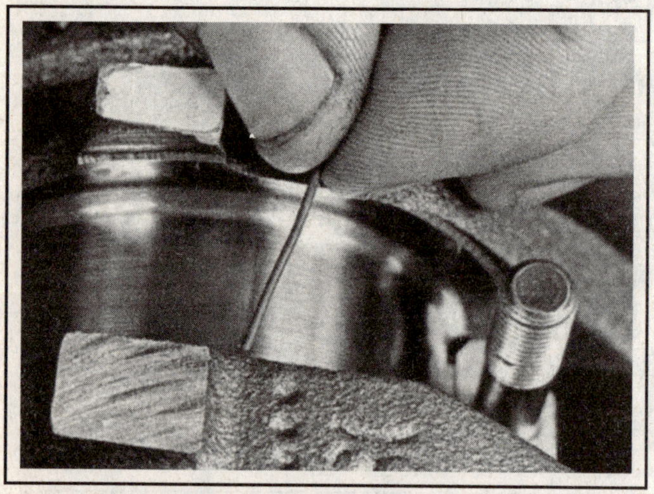

9.37 Place Plastigage on each connecting rod bearing journal parallel to the crankshaft centerline

23 Install the number one (top) ring in the same manner. Make sure the mark is facing up. Be careful not to confuse the number one and number two rings.

24 Repeat the procedure for the remaining pistons and rings.

INSTALLATION

25 Before installing the piston/connecting rod assemblies, the cylinder walls must be perfectly clean, the top edge of each cylinder bore must be chamfered, and the crankshaft must be in place.

26 Remove the cap from the end of the number one connecting rod (refer to the marks made during removal). Remove the original bearing inserts and wipe the bearing surfaces of the connecting rod and cap with a clean, lint-free cloth. They must be kept spotlessly clean.

Connecting rod bearing oil clearance check

➧ **Refer to illustrations 9.29, 9.34, 9.37, 9.38 and 9.41**

27 Clean the back side of the new upper bearing insert, then lay it in place in the connecting rod. Make sure the tab on the bearing fits into the recess in the rod. Don't hammer the bearing insert into place and be very careful not to nick or gouge the bearing face. Don't lubricate the bearing at this time.

28 Clean the back side of the other bearing insert and install it in the rod cap. Again, make sure the tab on the bearing fits into the recess in the cap, and don't apply any lubricant. It's critically important that the mating surfaces of the bearing and connecting rod are perfectly clean and oil free when they're assembled.

29 Position the piston ring gaps at 90-degree intervals around the piston as shown (see illustration).

30 Lubricate the piston and rings with clean engine oil and attach a piston ring compressor to the piston. Leave the skirt protruding about 1/4-inch to guide the piston into the cylinder. The rings must be compressed until they're flush with the piston.

31 Rotate the crankshaft until the number one connecting rod journal is at BDC (bottom dead center) and apply a liberal coat of engine oil to the cylinder walls.

32 With the mark (cavity) on top of the piston facing the front (timing chain end) of the engine, gently insert the piston/connecting rod assembly into the number one cylinder bore and rest the bottom edge of the ring compressor on the engine block.

➡**Note: The connecting rod also has a mark on it that must face the correct direction. On V6 models, the marks on the connecting rods face the front (timing chain end) of the engine while on four-cylinder models, the marks on the rod caps face forward.**

33 Tap the top edge of the ring compressor to make sure it's contacting the block around its entire circumference.

34 Gently tap on the top of the piston with the end of a wooden or plastic hammer handle (see illustration) while guiding the end of the connecting rod into place on the crankshaft journal.

35 The piston rings may try to pop out of the ring compressor just before entering the cylinder bore, so keep some downward pressure on the ring compressor. Work slowly, and if any resistance is felt as the piston enters the cylinder, stop immediately. Find out what's hanging up and fix it before proceeding. Do not, for any reason, force the piston into the cylinder - you might break a ring and/or the piston.

36 Once the piston/connecting rod assembly is installed, the connecting rod bearing oil clearance must be checked before the rod cap is permanently installed.

37 Cut a piece of the appropriate size Plastigage slightly shorter than the width of the connecting rod bearing and lay it in place on the number one connecting rod journal, parallel with the journal axis (see illustration).

38 Clean the connecting rod cap bearing face and install the rod cap. Make sure the mating mark on the cap is on the same side as the mark on the connecting rod (see illustration).

39 Install the rod bolts and tighten them to the torque listed in this Chapter's Specifications.

➡Note: Use a thin-wall socket to avoid erroneous torque readings that can result if the socket is wedged between the rod cap and the bolt. If the socket tends to wedge itself between the fastener and the cap, lift up on it slightly until it no longer contacts the cap. DO NOT rotate the crankshaft at any time during this operation.

40 Remove the fasteners and detach the rod cap, being very careful not to disturb the Plastigage.

41 Compare the width of the crushed Plastigage to the scale printed on the Plastigage envelope to obtain the oil clearance (see illustration). The connecting rod oil clearance is usually about 0.002 inch. Consult an automotive machine shop for the clearance specified for the rod bearings on your engine.

42 If the clearance is not as specified, the bearing inserts may be the wrong size (which means different ones will be required). Before deciding that different inserts are needed, make sure that no dirt or oil was between the bearing inserts and the connecting rod or cap when the clearance was measured. Also, recheck the journal diameter. If the Plastigage was wider at one end than the other, the journal may be tapered. If the clearance still exceeds the limit specified, the bearing will have to be replaced with an undersize bearing.

⁂ CAUTION:

When installing a new crankshaft, always use a standard size bearing.

Final installation

⁂ CAUTION:

Recheck the diameter of the connecting rod bolts at this time, making sure that none of them have stretched so much as to cause the necked-down area of the bolt to fall under the minimum allowable diameter listed in this Chapter's Specifications.

43 Carefully scrape all traces of the Plastigage material off the rod journal and/or bearing face. Be very careful not to scratch the bearing - use your fingernail or the edge of a plastic card.

44 Make sure the bearing faces are perfectly clean, then apply a uniform layer of clean moly-base grease or engine assembly lube to both of them. You'll have to push the piston into the cylinder to expose the face of the bearing insert in the connecting rod.

45 Slide the connecting rod back into place on the journal, install the rod cap, install the bolts and tighten them to the torque listed in this Chapter's Specifications.

46 Repeat the entire procedure for the remaining pistons/connecting rods.

47 The important points to remember are:

a) Keep the back sides of the bearing inserts and the insides of the connecting rods and caps perfectly clean when assembling them.
b) Make sure you have the correct piston/rod assembly for each cylinder.

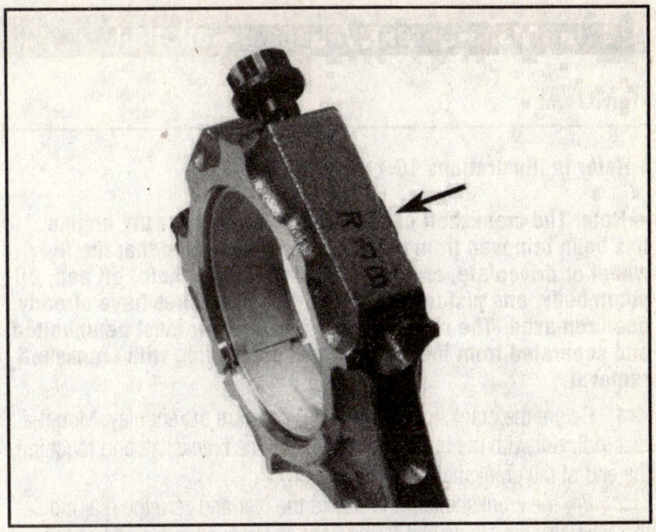

9.38 Install the connecting rod cap, making sure the cap and rod identification numbers match

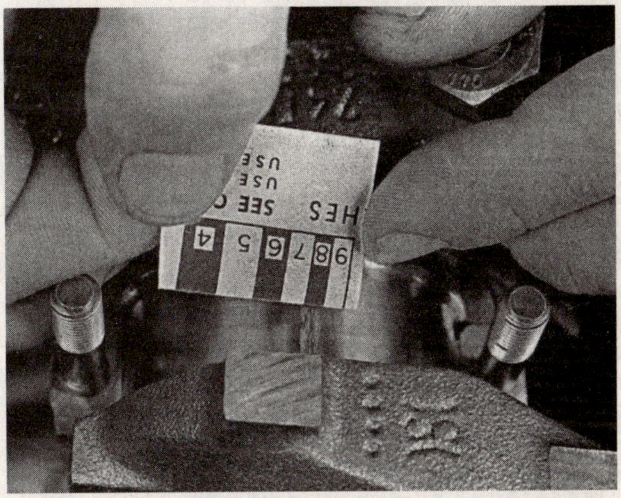

9.41 Use the scale on the Plastigage package to determine the bearing oil clearance - be sure to measure the widest part of the Plastigage and use the correct scale; it comes with both standard and metric scales

c) The mark on the piston must face the front of the engine.
d) Lubricate the cylinder walls lightly with clean oil.
e) Lubricate the bearing faces when installing the rod caps after the oil clearance has been checked.

48 After all the piston/connecting rod assemblies have been correctly installed, rotate the crankshaft a number of times by hand to check for any obvious binding.

49 As a final step, check the connecting rod endplay again. If it was correct before disassembly and the original crankshaft and rods were reinstalled, it should still be correct. If new rods or a new crankshaft were installed, the endplay may be inadequate. If so, the rods will have to be removed and taken to an automotive machine shop for resizing.

10 Crankshaft - removal and installation

REMOVAL

➧ Refer to illustrations 10.1 and 10.3

➡Note: The crankshaft can be removed only after the engine has been removed from the vehicle. It's assumed that the flywheel or driveplate, crankshaft pulley, timing chain, oil pan, oil pump body, and piston/connecting rod assemblies have already been removed. The rear main oil seal retainer must be unbolted and separated from the block before proceeding with crankshaft removal.

1 Before the crankshaft is removed, measure the endplay. Mount a dial indicator with the indicator in line with the crankshaft and touching the end of the crankshaft (see illustration).

2 Pry the crankshaft all the way to the rear and zero the dial indicator. Next, pry the crankshaft to the front as far as possible and check the reading on the dial indicator. The distance traveled is the endplay. A typical crankshaft endplay will fall between 0.003 to 0.010-inch. If it's greater than that, check the crankshaft thrust surfaces for wear after it's removed. If no wear is evident, new main bearings should correct the endplay.

3 If a dial indicator isn't available, feeler gauges can be used. Gently pry the crankshaft all the way to the front of the engine. Slip feeler gauges between the crankshaft and the front face of the thrust bearing or washer to determine the clearance (see illustration).

4 Loosen the main bearing cap bolts 1/4-turn at a time each, until they can be removed by hand.

➡Note: If you're working on a V6 engine, first remove the main bearing cap side bolts in the order opposite that of the tightening sequence (see illustration 10.19b).

Gently tap the main bearing cap(s) with a soft-face hammer. Pull the main bearing cap(s) straight up and off the cylinder block. Try not to drop the bearing inserts if they come out with the assembly.

5 Using a vernier caliper or a micrometer, measure the diameter of each main bearing cap bolt (in several places within the threaded area). Compare your measurements with the values listed in this Chapter's Specifications. If the diameter of any bolt falls under the minimum, replace it.

6 Carefully lift the crankshaft out of the engine. It may be a good idea to have an assistant available, since the crankshaft is quite heavy and awkward to handle. With the bearing inserts in place inside the engine block and main bearing caps, reinstall the main bearing cap assembly onto the engine block and tighten the bolts finger tight. Make sure you install the main bearing cap(s) with the arrow facing the front end of the engine.

INSTALLATION

7 Crankshaft installation is the first step in engine reassembly. It's assumed at this point that the engine block and crankshaft have been cleaned, inspected and repaired or reconditioned.

8 Position the engine block with the bottom facing up.

9 Remove the mounting bolts and lift off the main bearing caps.

10 If they're still in place, remove the original bearing inserts from the block and from the main bearing cap. Wipe the bearing surfaces of the block and main bearing caps with a clean, lint-free cloth. They must be kept spotlessly clean. This is critical for determining the correct bearing oil clearance.

Main bearing oil clearance check

➧ Refer to illustrations 10.17, 10.19a, 10.19b and 10.21

11 Without mixing them up, clean the back sides of the new upper main bearing inserts (with grooves and oil holes) and lay one in each main bearing saddle in the block. Each upper bearing has an oil groove and oil hole in it.

CAUTION:

The oil holes in the block must line up with the oil holes in the upper bearing inserts.

The separate thrust washers on four-cylinder engines are on the center bearing. The thrust washer on V6 models is located on the number 2 crankshaft journal. Install the thrust washers with the grooved side facing out. Install the thrust washers so that one set is located in the block and the other set is with the main bearing cap. Clean the back sides of the lower main bearing inserts (without grooves) and lay them in the corresponding main bearing cap. Make sure the tab on the bearing insert fits into the recess in the block or main bearing cap.

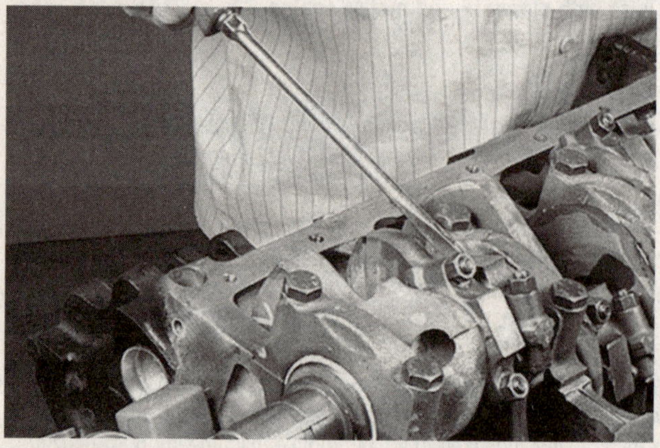

10.1 Checking crankshaft endplay with a dial indicator

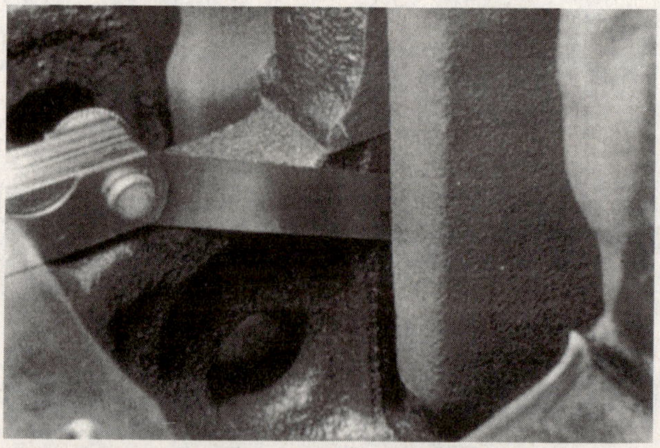

10.3 Checking crankshaft endplay with feeler gauges at the thrust bearing journal

⁂ CAUTION:

Do not hammer the bearing insert into place and don't nick or gouge the bearing faces. DO NOT apply any lubrication at this time.

12 Clean the faces of the bearing inserts in the block and the crankshaft main bearing journals with a clean, lint-free cloth.

13 Check or clean the oil holes in the crankshaft, as any dirt here can go only one way - straight through the new bearings.

14 Once you're certain the crankshaft is clean, carefully lay it in position in the cylinder block.

15 Before the crankshaft can be permanently installed, the main bearing oil clearance must be checked.

16 Cut several strips of the appropriate size of Plastigage (they must be slightly shorter than the width of the main bearing journal).

17 Place one piece on each crankshaft main bearing journal, parallel with the journal axis (see illustration).

18 Clean the faces of the bearing inserts in the main bearing caps. Hold the bearing inserts in place and install the caps onto the crankshaft and cylinder block. DO NOT disturb the Plastigage. Make sure you install each main bearing cap with the arrow facing the front of the engine. The caps on all engines are numbered - number 1 is the closest to the timing chain end of the engine.

19 Apply clean engine oil to all bolt threads prior to installation, then install all bolts finger-tight. Tighten the main bearing cap bolts (in the sequence shown, on V6 models; see illustrations) progressing in two steps, to the torque listed in this Chapter's Specifications. DO NOT rotate the crankshaft at any time during this operation.

➡Note: If you're working on a V6 engine, tighten the main bearing cap bolts first, then the main bearing cap side bolts.

20 Remove the bolts a little at a time (and in the reverse order of the tightening sequence on V6 engines) and carefully lift each main bearing cap straight up and off the block. Do not disturb the Plastigage or rotate the crankshaft. If a main bearing cap is difficult to remove, tap it gently from side-to-side with a soft-face hammer to loosen it.

21 Compare the width of the crushed Plastigage on each journal to the scale printed on the Plastigage envelope to determine the main bearing oil clearance (see illustration). A typical main bearing oil clearance should fall between 0.0015 to 0.0023-inch. Check with an automotive machine shop for the clearance specified for your engine.

10.17 Place the Plastigage (arrow) onto the crankshaft bearing journal as shown

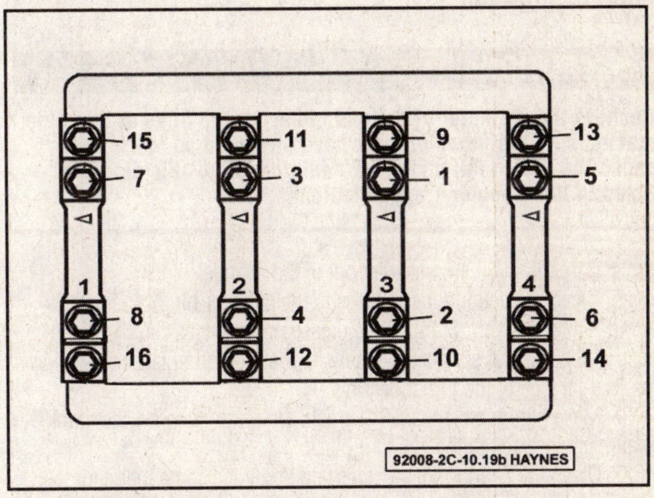

10.19a Main bearing cap bolt tightening sequence (V6 engine)

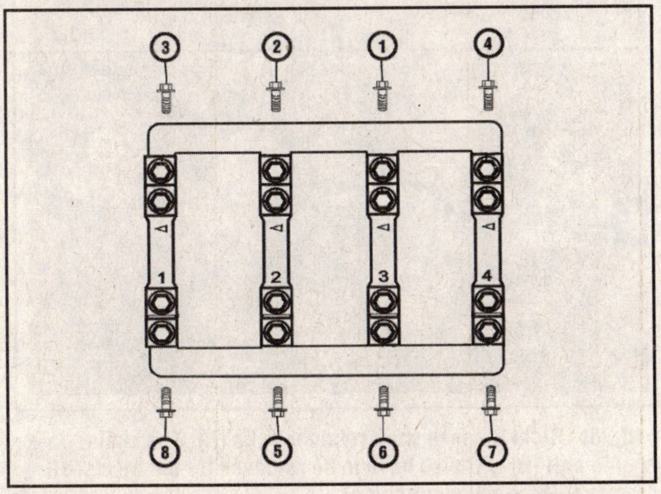

10.19b Main bearing cap side bolt tightening sequence (V6 engine)

10.21 Use the scale on the Plastigage package to determine the bearing oil clearance - be sure to measure the widest part of the Plastigage and use the correct scale; it comes with both standard and metric scales

22 If the clearance is not as specified, the bearing inserts might be the wrong size (which means different ones will be required). Before deciding if different inserts are needed, make sure that no dirt or oil was between the bearing inserts and the cap assembly or block when the clearance was measured. If the Plastigage was wider at one end than the other, the crankshaft journal may be tapered. If the clearance still exceeds the limit specified, the bearing insert(s) will have to be replaced with an undersize bearing insert(s).

CAUTION:

When installing a new crankshaft, always install a standard bearing insert set.

23 Carefully scrape all traces of the Plastigage material off the main bearing journals and/or the bearing insert faces. Be sure to remove all residue from the oil holes. Use your fingernail or the edge of a plastic card - don't nick or scratch the bearing faces.

Final installation

CAUTION:

Recheck the diameter of the main bearing cap bolts at this time, making sure that none of them have stretched so much as to cause the bolt to fall under the minimum allowable diameter listed in this Chapter's Specifications.

24 Carefully lift the crankshaft out of the cylinder block.
25 Clean the bearing insert faces in the cylinder block, then apply a thin, uniform layer of moly-base grease or engine assembly lube to each of the bearing surfaces. Be sure to coat the thrust faces as well as the journal face of the thrust bearing.
26 Make sure the crankshaft journals are clean, then lay the crankshaft back in place in the cylinder block.
27 Clean the bearing insert faces and apply the same lubricant to them.
28 Hold the bearing inserts in place and install the main bearing caps on the crankshaft and cylinder block. Tap the bearing caps into place with a brass punch or a soft-face hammer.
29 Apply clean engine oil to the bolt threads, wipe off any excess oil, then install the bolts finger-tight.
30 Push the crankshaft forward using a screwdriver or prybar to seat the thrust bearing. Once the crankshaft is pushed fully forward to seat the thrust bearing, leave the screwdriver in position so that force stays on the crankshaft until after all main bearing cap bolts have been tightened.
31 On four-cylinder models, tighten the main bearing cap bolts in two steps to the torque and angle listed in this Chapter's Specifications. On V6 models, tighten the main bearing cap bolts in two steps in the indicated sequence (see illustration 10.19b) and to the torque and angle listed in this Chapter's Specifications.
32 On V6 engines, install and tighten the main bearing cap side bolts in the correct sequence (see illustration 10.19b).
33 Recheck crankshaft endplay with a feeler gauge or a dial indicator. The endplay should be correct if the crankshaft thrust faces aren't worn or damaged and if new bearings have been installed.
34 Rotate the crankshaft a number of times by hand to check for any obvious binding. It should rotate with a running torque of 50 in-lbs or less. If the running torque is too high, correct the problem at this time.
35 Install the new rear main oil seal (see Chapter 2A or 2B).

Balance shaft installation (four-cylinder engines)

Refer to illustrations 10.38a and 10.38b

36 Have the balance shafts and the engine block inspected by a qualified machine shop if you suspect damage or excessive wear. If there is too much bearing clearance, the block and balance shafts should be replaced.
37 Install the balance shaft gear or sprocket, then secure the shafts in a vise with padded jaws. Tighten the gear or sprocket bolt to the torque listed in this Chapter's Specifications.
38 Carefully install each balance shaft assembly into the block and tighten the retainer's bolts to the torque listed in this Chapter's Specifications (see illustrations).
39 Refer to Chapter 2A for information on installing the balance shaft timing chain.

10.38a Left balance shaft retainer bolts (four-cylinder engine)

10.38b Right balance shaft retainer bolts (A). The chain guide bolt (B) will also have to be removed for balance shaft removal (four-cylinder engine)

11 Engine overhaul - reassembly sequence

1 Before beginning engine reassembly, make sure you have all the necessary new parts, gaskets and seals as well as the following items on hand:

Common hand tools
A 1/2-inch drive torque wrench
New engine oil
Gasket sealant
Thread locking compound

2 If you obtained a short block it will be necessary to install the cylinder head(s), the timing chain and cover, the oil pump, pick-up tube, oil pans, the water pump and the valve cover(s) (see Chapter 2A or 2B). In order to save time and avoid problems, the external components should be installed in the following general order:

Water pump
Intake and exhaust manifolds
Fuel injection components
Emission control components
Spark plugs
Ignition coils
Oil filter
Engine mounts and mount brackets
Driveplate (automatic transmission)
Flywheel (manual transmission)

12 Initial start-up and break-in after overhaul

WARNING:

Have a fire extinguisher handy when starting the engine for the first time.

1 Once the engine has been installed in the vehicle, double-check the engine oil and coolant levels.

2 With the spark plugs out of the engine and the ignition system and fuel pump disabled (see Chapter 4, Section 2), crank the engine until oil pressure registers on the gauge or the light goes out.

3 Install the spark plugs, hook up the ignition coils and restore the ignition system and fuel pump functions.

4 Start the engine. It may take a few moments for the fuel system to build up pressure, but the engine should start without a great deal of effort.

5 After the engine starts, it should be allowed to warm up to normal operating temperature. While the engine is warming up, make a thorough check for fuel, oil and coolant leaks.

6 Shut the engine off and recheck the engine oil and coolant levels.

7 Drive the vehicle to an area with minimum traffic, accelerate from 30 to 50 mph, then allow the vehicle to slow to 30 mph with the throttle closed. Repeat the procedure 10 or 12 times. This will load the piston rings and cause them to seat properly against the cylinder walls. Check again for oil and coolant leaks.

8 Drive the vehicle gently for the first 500 miles (no sustained high speeds) and keep a constant check on the oil level. It is not unusual for an engine to use oil during the break-in period.

9 At approximately 500 to 600 miles, change the oil and filter.

10 For the next few hundred miles, drive the vehicle normally. Do not pamper it or abuse it.

11 After 2000 miles, change the oil and filter again and consider the engine broken in.

GLOSSARY

B

Backlash - The amount of play between two parts. Usually refers to how much one gear can be moved back and forth without moving the gear with which it's meshed.

Bearing Caps - The caps held in place by nuts or bolts which, in turn, hold the bearing surface. This space is for lubricating oil to enter.

Bearing clearance - The amount of space left between shaft and bearing surface. This space is for lubricating oil to enter.

Bearing crush - The additional height which is purposely manufactured into each bearing half to ensure complete contact of the bearing back with the housing bore when the engine is assembled.

Bearing knock - The noise created by movement of a part in a loose or worn bearing.

Blueprinting - Dismantling an engine and reassembling it to EXACT specifications.

Bore - An engine cylinder, or any cylindrical hole; also used to describe the process of enlarging or accurately refinishing a hole with a cutting tool, as to bore an engine cylinder. The bore size is the diameter of the hole.

Boring - Renewing the cylinders by cutting them out to a specified size. A boring bar is used to make the cut.

Bottom end - A term which refers collectively to the engine block, crankshaft, main bearings and the big ends of the connecting rods.

Break-in - The period of operation between installation of new or rebuilt parts and time in which parts are worn to the correct fit. Driving at reduced and varying speed for a specified mileage to permit parts to wear to the correct fit.

Bushing - A one-piece sleeve placed in a bore to serve as a bearing surface for shaft, piston pin, etc. Usually replaceable.

C

Camshaft - The shaft in the engine, on which a series of lobes are located for operating the valve mechanisms. The camshaft is driven by gears or sprockets and a timing chain. Usually referred to simply as the cam.

Carbon - Hard, or soft, black deposits found in combustion chamber, on plugs, under rings, on and under valve heads.

Cast iron - An alloy of iron and more than two percent carbon, used for engine blocks and heads because it's relatively inexpensive and easy to mold into complex shapes.

Chamfer - To bevel across (or a bevel on) the sharp edge of an object.

Chase - To repair damaged threads with a tap or die.

Combustion chamber - The space between the piston and the cylinder head, with the piston at top dead center, in which air-fuel mixture is burned.

Compression ratio - The relationship between cylinder volume (clearance volume) when the piston is at top dead center and cylinder volume when the piston is at bottom dead center.

Connecting rod - The rod that connects the crank on the crankshaft with the piston. Sometimes called a con rod.

Connecting rod cap - The part of the connecting rod assembly that attaches the rod to the crankpin.

Core plug - Soft metal plug used to plug the casting holes for the coolant passages in the block.

Crankcase - The lower part of the engine in which the crankshaft rotates; includes the lower section of the cylinder block and the oil pan.

Crank kit - A reground or reconditioned crankshaft and new main and connecting rod bearings.

Crankpin - The part of a crankshaft to which a connecting rod is attached.

Crankshaft - The main rotating member, or shaft, running the length of the crankcase, with offset throws to which the connecting rods are attached; changes the reciprocating motion of the pistons into rotating motion.

Cylinder sleeve - A replaceable sleeve, or liner, pressed into the cylinder block to form the cylinder bore.

D

Deburring - Removing the burrs (rough edges or areas) from a bearing.

Deglazer - A tool, rotated by an electric motor, used to remove glaze from cylinder walls so a new set of rings will seat.

E

Endplay - The amount of lengthwise movement between two parts. As applied to a crankshaft, the distance that the crankshaft can move forward and back in the cylinder block.

F

Face - A machinist's term that refers to removing metal from the end of a shaft or the face of a larger part, such as a flywheel.

Fatigue - A breakdown of material through a large number of loading and unloading cycles. The first signs are cracks followed shortly by breaks.

Feeler gauge - A thin strip of hardened steel, ground to an exact thickness, used to check clearances between parts.

Free height - The unloaded length or height of a spring.

Freeplay - The looseness in a linkage, or an assembly of parts, between the initial application of force and actual movement. Usually perceived as slop or slight delay.

Freeze plug - See Core plug.

G

Gallery - A large passage in the block that forms a reservoir for engine oil pressure.

Glaze - The very smooth, glassy finish that develops on cylinder walls while an engine is in service.

H

Heli-Coil - A rethreading device used when threads are worn or damaged. The device is installed in a retapped hole to reduce the thread size to the original size.

I

Installed height - The spring's measured length or height, as installed on the cylinder head. Installed height is measured from the spring seat to the underside of the spring retainer.

J

Journal - The surface of a rotating shaft which turns in a bearing.

K

Keeper - The split lock that holds the valve spring retainer in position on the valve stem.

Key - A small piece of metal inserted into matching grooves machined into two parts fitted together - such as a gear pressed onto a shaft - which prevents slippage between the two parts.

Knock - The heavy metallic engine sound, produced in the combustion chamber as a result of abnormal combustion - usually detonation. Knock is usually caused by a loose or worn bearing. Also referred to as detonation, pinging and spark knock. Connecting rod or main bearing knocks are created by too much oil clearance or insufficient lubrication.

L

Lands - The portions of metal between the piston ring grooves.

Lapping the valves - Grinding a valve face and its seat together with lapping compound.

Lash - The amount of free motion in a gear train, between gears, or in a mechanical assembly, that occurs before movement can begin. Usually refers to the lash in a valve train.

Lifter - The part that rides against the cam to transfer motion to the rest of the valve train.

M

Machining - The process of using a machine to remove metal from a metal part.

Main bearings - The plain, or babbitt, bearings that support the crankshaft.

Main bearing caps - The cast iron caps, bolted to the bottom of the block, that support the main bearings.

O

O.D. - Outside diameter.

Oil gallery - A pipe or drilled passageway in the engine used to carry engine oil from one area to another.

Oil ring - The lower ring, or rings, of a piston; designed to prevent excessive amounts of oil from working up the cylinder walls and into the combustion chamber. Also called an oil-control ring.

Oil seal - A seal which keeps oil from leaking out of a compartment. Usually refers to a dynamic seal around a rotating shaft or other moving part.

O-ring - A type of sealing ring made of a special rubberlike material; in use, the O-ring is compressed into a groove to provide the sealing action.

Overhaul - To completely disassemble a unit, clean and inspect all parts, reassemble it with the original or new parts and make all adjustments necessary for proper operation.

P

Pilot bearing - A small bearing installed in the center of the flywheel (or the rear end of the crankshaft) to support the front end of the input shaft of the transmission.

Pip mark - A little dot or indentation which indicates the top side of a compression ring.

Piston - The cylindrical part, attached to the connecting rod, that moves up and down in the cylinder as the crankshaft rotates. When the fuel charge is fired, the piston transfers the force of the explosion to the connecting rod, then to the crankshaft.

Piston pin (or wrist pin) - The cylindrical and usually hollow steel pin that passes through the piston. The piston pin fastens the piston to the upper end of the connecting rod.

Piston ring - The split ring fitted to the groove in a piston. The ring contacts the sides of the ring groove and also rubs against the cylinder wall, thus sealing space between piston and wall. There are two types of rings: Compression rings seal the compression pressure in the combustion chamber; oil rings scrape excessive oil off the cylinder wall.

Piston ring groove - The slots or grooves cut in piston heads to hold piston rings in position.

Piston skirt - The portion of the piston below the rings and the piston pin hole.

Plastigage - A thin strip of plastic thread, available in different sizes, used for measuring clearances. For example, a strip of plastigage is laid across a bearing journal and mashed as parts are assembled. Then parts are disassembled and the width of the strip is measured to determine clearance between journal and bearing. Commonly used to measure crankshaft main-bearing and connecting rod bearing clearances.

Press-fit - A tight fit between two parts that requires pressure to force the parts together. Also referred to as drive, or force, fit.

Prussian blue - A blue pigment; in solution, useful in determining the area of contact between two surfaces. Prussian blue is commonly used to determine the width and location of the contact area between the valve face and the valve seat.

R

Race (bearing) - The inner or outer ring that provides a contact surface for balls or rollers in bearing.

Ream - To size, enlarge or smooth a hole by using a round cutting tool with fluted edges.

Ring job - The process of reconditioning the cylinders and installing new rings.

Runout - Wobble. The amount a shaft rotates out-of-true.

S

Saddle - The upper main bearing seat.

Scored - Scratched or grooved, as a cylinder wall may be scored by abrasive particles moved up and down by the piston rings.

Scuffing - A type of wear in which there's a transfer of material between parts moving against each other; shows up as pits or grooves in the mating surfaces.

Seat - The surface upon which another part rests or seats. For example, the valve seat is the matched surface upon which the valve face rests. Also used to refer to wearing into a good fit; for example, piston rings seat after a few miles of driving.

Short block - An engine block complete with crankshaft and piston and, usually, camshaft assemblies.

Static balance - The balance of an object while it's stationary.

Step - The wear on the lower portion of a ring land caused by excessive side and back-clearance. The height of the step indicates the ring's extra side clearance and the length of the step projecting from the back wall of the groove represents the ring's back clearance.

Stroke - The distance the piston moves when traveling from top dead center to bottom dead center, or from bottom dead center to top dead center.

Stud - A metal rod with threads on both ends.

T

Tang - A lip on the end of a plain bearing used to align the bearing during assembly.

Tap - To cut threads in a hole. Also refers to the fluted tool used to cut threads.

Taper - A gradual reduction in the width of a shaft or hole; in an engine cylinder, taper usually takes the form of uneven wear, more pronounced at the top than at the bottom.

Throws - The offset portions of the crankshaft to which the connecting rods are affixed.

Thrust bearing - The main bearing that has thrust faces to prevent excessive endplay, or forward and backward movement of the crankshaft.

Thrust washer - A bronze or hardened steel washer placed between two moving parts. The washer prevents longitudinal movement and provides a bearing surface for thrust surfaces of parts.

Tolerance - The amount of variation permitted from an exact size of measurement. Actual amount from smallest acceptable dimension to largest acceptable dimension.

U

Umbrella - An oil deflector placed near the valve tip to throw oil from the valve stem area.

Undercut - A machined groove below the normal surface.

Undersize bearings - Smaller diameter bearings used with re-ground crankshaft journals.

V

Valve grinding - Refacing a valve in a valve-refacing machine.

Valve train - The valve-operating mechanism of an engine; includes all components from the camshaft to the valve.

Vibration damper - A cylindrical weight attached to the front of the crankshaft to minimize torsional vibration (the twist-untwist actions of the crankshaft caused by the cylinder firing impulses). Also called a harmonic balancer.

W

Water jacket - The spaces around the cylinders, between the inner and outer shells of the cylinder block or head, through which coolant circulates.

Web - A supporting structure across a cavity.

Woodruff key - A key with a radiused backside (viewed from the side).

Specifications

General

Displacement	
Four-cylinder engine	164 cubic inches (2.7 liters)
V6 engine	241 cubic inches (4.0 liters)
Bore and stroke	
Four-cylinder engine	3.74 x 3.74 inches (95.0 x 95.0 mm)
V6 engine	3.70 x 3.74 inches (94.0 x 95.0 mm)
Cylinder compression pressure	
Four-cylinder engine	
Standard	178 psi (1,230 kPa)
Minimum	128 psi (880 kPa)
Difference between cylinders (maximum)	10 psi (68 kPa)
V6 engine	
Standard	189 psi (1300 kPa)
Minimum	145 psi (1000 kPa)
Difference between cylinders (maximum)	15 psi (100 kPa)
Oil pressure	
Four-cylinder engine	
At curb idle	4.3 psi minimum
At 3000 rpm	23 to 75 psi
V6 engine	
At curb idle	4.2 psi minimum
At 3000 rpm	43 to 85 psi
Connecting rod bolt diameter (in the necked-down portion) (all engines)	
Standard	0.283 to 0.287 inch (7.2 to 7.3 mm)
Minimum	0.276 inch (7.0 mm)
Main bearing cap bolt diameter	
Four-cylinder engine	
Standard	0.4236 to 0.4319 inch (10.76 to 10.97 mm)
Minimum	0.4197 inch (10.66 mm)
V6 engine	
Standard	0.393 to 0.402 inch (7.2 to 7.3 mm)
Minimum	Any less than 0.393 inch (7.2 mm)

Torque specifications	Ft-lbs (unless otherwise indicated)	Nm

➡Note: One foot-pound (ft-lb) of torque is equivalent to 12 inch-pounds (in-lbs) of torque. Torque values below approximately 15 foot-pounds are expressed in inch-pounds, because most foot-pound torque wrenches are not accurate at these smaller values.

Torque specifications	Ft-lbs (unless otherwise indicated)	Nm
Balance shafts (four-cylinder models)		
Gear mounting bolts	26	35
Mounting plate bolts	156 in-lbs	17.5
Connecting rod bearing cap bolts (all models)		
Step 1	18	25
Step 2	Tighten an additional 90 degrees	
Main bearing cap bolts		
Four-cylinder models		
Step 1	29	39
Step 2	Tighten an additional 90 degrees	
V6 models		
Vertical bolts (see illustration 10.19a for tightening sequence)		
Step 1	45	61
Step 2	Tighten an additional 90 degrees	
Side bolts (see illustration 10.19b for tightening sequence)	18	25

Notes

3

COOLING, HEATING AND AIR CONDITIONING SYSTEMS

Section

Reference to other Chapters

Coolant level check - See Chapter 1
Cooling system check - See Chapter 1
Cooling system servicing (draining, flushing and refilling) - See Chapter 1
Drivebelt check, adjustment and replacement - See Chapter 1
Underhood hose check and replacement - See Chapter 1

1 General information

ENGINE COOLING SYSTEM

➧ Refer to illustrations 1.1 and 1.2

All vehicles covered by this manual employ a pressurized engine cooling system with thermostatically controlled coolant circulation (see illustration). An impeller type water pump mounted on the front of the block pumps coolant through the engine. The coolant flows around each cylinder and toward the rear of the engine. Cast-in coolant passages direct coolant around the intake and exhaust ports, near the spark plug areas and in proximity to the exhaust valve guides. The water pump on all models is driven by the drivebelt.

A wax-pellet type thermostat (see illustration) is located in the thermostat housing on front of the engine. During warm up, the closed thermostat prevents coolant from circulating through the radiator. When the engine reaches normal operating temperature, the thermostat opens and allows hot coolant to travel through the radiator, where it is cooled before returning to the engine.

The cooling system is sealed by a pressure-type radiator cap, which raises the boiling point of the coolant. If the system pressure exceeds the spring pressure of the cap pressure-relief valve, the excess pressure in the system forces the spring-loaded valve inside the cap off its seat and allows the coolant to escape through the overflow tube into a coolant overflow reservoir. When the system cools, the excess coolant is automatically drawn from the reservoir back into the radiator.

The coolant reservoir serves as both the point at which fresh coolant is added to the cooling system to maintain the proper fluid level and as a holding tank for overheated coolant.

This type of cooling system is known as a closed design because coolant that escapes past the pressure cap is saved and reused.

TRANSMISSION COOLING SYSTEM

Automatic transmission models are equipped with a transmission cooler, located inside the radiator, which cools the transmission fluid. The transmission is connected to the cooler by a pair of hoses: one delivers hot transmission fluid to the radiator and the other brings the cooled fluid back to the transmission. Some models are equipped with an auxiliary transmission oil cooler mounted in front of the radiator and condenser.

1.1 Typical cooling system component locations (four-cylinder shown, V6 models similar)

1 *Radiator cap*
2 *Radiator*
3 *Air conditioning Low Side Port*
4 *Air conditioning pressure cycling switch (below radiator support)*
5 *Condenser*
6 *Coolant reservoir*
7 *Thermostat (below power steering fluid reservoir)*

ENGINE OIL COOLING SYSTEM (V6 MODELS)

Besides the engine and transmission cooling systems, on some models engine heat is also dissipated through an external oil cooler. The oil cooling system consists of a housing mounted between the oil filter and the filter adapter. Hoses connected to the oil cooler circulate coolant through the oil cooler housing.

HEATING SYSTEM

The heating system consists of a blower fan and heater core located within the heater box under the dashboard, the inlet and outlet hoses connecting the heater core to the engine cooling system and the heater/air conditioning control head on the dashboard. Hot engine coolant is circulated through the heater core. When the heater mode is activated, a flap opens to expose the heater box to the passenger compartment. A fan switch on the control head activates the blower motor, which forces air through the core, heating the air.

AIR CONDITIONING SYSTEM

The air conditioning system consists of a condenser mounted in front of the radiator, an evaporator mounted adjacent to the heater core, a compressor mounted on the engine, a filter-drier which is mounted directly on the side of the condenser and the plumbing connecting all of the above components together as one system.

A blower fan forces the warmer air of the passenger compartment through the evaporator core (sort of a radiator-in-reverse), transferring the heat from the air to the refrigerant. The liquid refrigerant boils off into low pressure vapor, taking the heat with it when it leaves the evaporator. The compressor keeps refrigerant circulating through the system, pumping the warmed coolant through the condenser where it is cooled then circulated back to the evaporator.

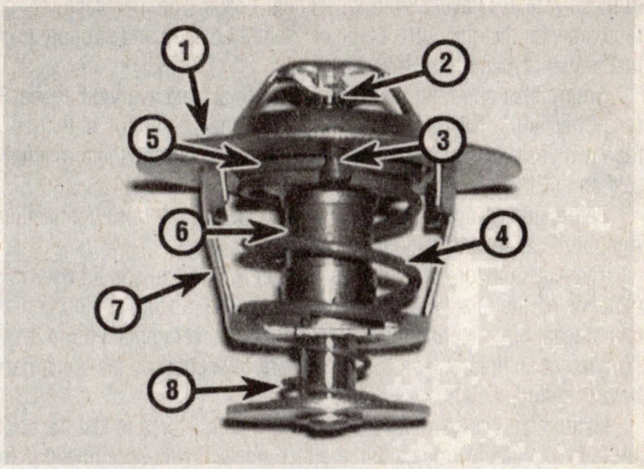

1.2 Typical thermostat details

1 Flange
2 Piston
3 Jiggle valve
4 Main coil spring
5 Valve seat
6 Valve
7 Frame
8 Secondary coil spring

2 Antifreeze - general information

Refer to illustration 2.5

WARNING:

Do not allow antifreeze to come in contact with your skin or painted surfaces of the vehicle. Rinse off spills immediately with plenty of water. Antifreeze is highly toxic if ingested. Never leave antifreeze lying around in an open container or in puddles on the floor; children and pets are attracted by it's sweet smell and may drink it. Check with local authorities about disposing of used antifreeze. Many communities have collection centers which will see that antifreeze is disposed of safely. Never dump used antifreeze on the ground or pour it into drains.

CAUTION:

Do not mix coolants of different colors. Doing so might damage the cooling system and/or the engine. Read the warning label in the engine compartment for additional information.

➡Note: Non-toxic antifreeze is now manufactured and available at local auto parts stores, but even this type must be disposed of properly.

The cooling system should be filled with a water/ethylene glycol based antifreeze solution, which will prevent freezing down to at least -20-degrees F (even lower in cold climates). It also provides protection against corrosion and increases the coolant boiling point. The engines in these vehicles have aluminum cylinder heads. The manufacturer recommends that the correct type of coolant be used and strongly urges that coolant types not be mixed.

Drain, flush and refill the cooling system at least every other year (see Chapter 1). The use of antifreeze solutions for periods of longer than two years is likely to cause damage and encourage the formation of rust and scale in the system.

Before adding antifreeze to the system, inspect all hose connections. Antifreeze can leak through very minute openings.

The exact mixture of antifreeze to water, which you should use, depends on the relative weather conditions. The mixture should contain at least 50-percent antifreeze, but should never contain more than 70-percent antifreeze. Consult the mixture ratio chart on the container before adding coolant.

Hydrometers are available at most auto parts stores to test the coolant (see illustration). Use antifreeze that meets Toyota specifications for engines with aluminum components.

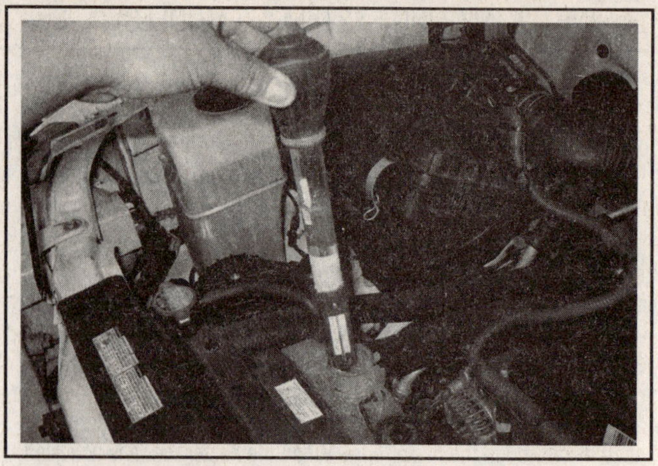

2.5 An inexpensive hydrometer can be used to test the condition of your coolant

3 Thermostat - check and replacement

WARNING:

Do not allow antifreeze to come in contact with your skin or painted surfaces of the vehicle. Rinse off spills immediately with plenty of water. Antifreeze is highly toxic if ingested. Never leave antifreeze lying around in an open container or in puddles on the floor; children and pets are attracted by its sweet smell and may drink it. Check with local authorities on disposing of used antifreeze. Many communities have collection centers, which will see that antifreeze is disposed of safely. Never dump used antifreeze on the ground or into drains.

CHECK

1 Before proceeding, check the coolant level and the drivebelt tension (see Chapter 1) then check the operation of the temperature gauge (see Section 9).

2 If the engine takes a long time to warm up, the thermostat is probably stuck open. Replace the thermostat with a new one.

3 If the engine runs hot, check the temperature of the lower radiator hose. If the hose isn't hot but the engine is, the thermostat is probably stuck in the closed position. Replace the thermostat with a new one.

CAUTION:

Do not drive the vehicle without a thermostat. The computer may stay in open loop and the fuel economy and emissions will suffer.

4 If the lower radiator hose is hot, then the coolant is circulating and the thermostat is open. Refer to the *Troubleshooting* Section at the front of this manual for the cause of overheating.

5 If the engine has overheated, it might have leaking cylinder head gaskets, scuffed pistons and/or warped or cracked cylinder heads.

REPLACEMENT

Refer to illustrations 3.9a, 3.9b, 3.10, 3.11 and 3.12

WARNING:

Wait until the engine is completely cool before beginning this procedure.

6 Disconnect the cable from the negative terminal of the battery (see Chapter 5, Section 1).

7 Drain the cooling system (see Chapter 1).

8 If you're working on a V6 engine, remove the engine cover.

9 Detach the thermostat housing from the engine (see illustrations). Be prepared for some coolant to spill as the gasket seal is broken. The radiator hose can be left attached to the housing, unless the housing itself is to be replaced.

10 Remove the thermostat, noting the direction in which it was installed in the housing, and thoroughly clean the sealing surfaces.

Note: On V6 models, the thermostat is part of the housing (see illustration).

11 If you're working on a four-cylinder engine, install a new gasket onto the thermostat (see illustration). Make sure it fits evenly all the way around. On V6 models, install a new O-ring in the housing groove.

12 On four-cylinder models, install the thermostat, positioning the jiggle pin, if equipped, at the highest point (see illustration).

13 Install the thermostat housing. On four-cylinder models, be sure to install a new thermostat housing gasket.

14 Tighten the housing fasteners to the torque listed in this Chapter's Specifications and reinstall the remaining components in the reverse order of removal.

15 Refill the cooling system (see Chapter 1), run the engine and check for leaks and proper operation.

3.9a Thermostat housing fastener location on a four-cylinder engine

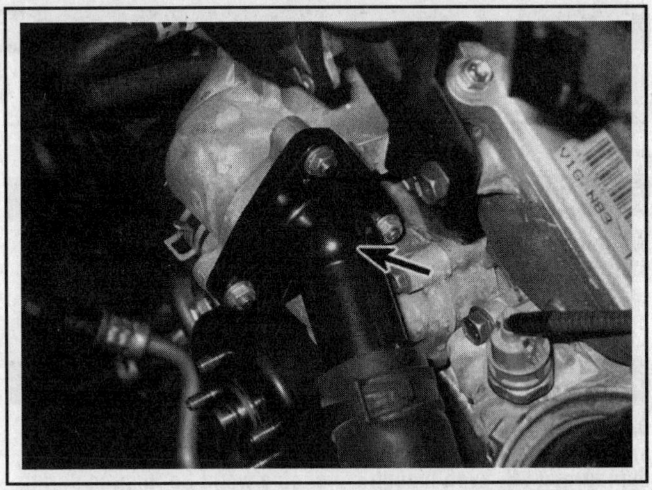

3.9b Thermostat housing location on the V6 engine

3.10 On V6 models, the thermostat is part of the housing (and the housing is sealed by an O-ring)

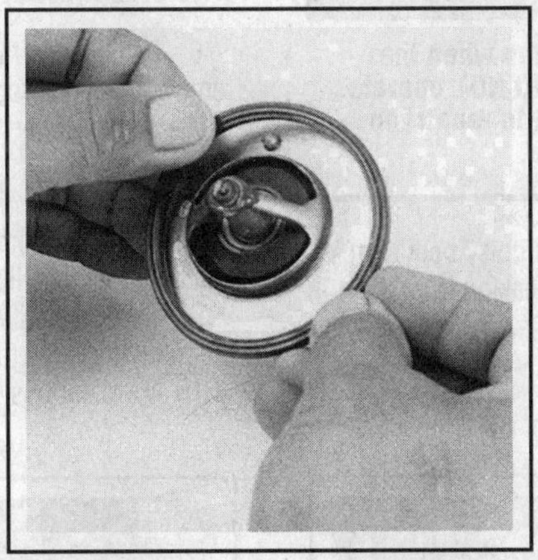

3.11 The thermostat gasket, which is actually a grooved sealing ring, fits around the edge of the thermostat (four-cylinder models)

3.12 Thermostat installed in the timing chain cover on a four-cylinder model - note location of the jiggle pin

4 Coolant reservoir - removal and installation

**** WARNING 1:**

Wait until the engine is completely cool before beginning this procedure.

**** WARNING 2:**

Do not allow antifreeze to come in contact with your skin or painted surfaces of the vehicle. Rinse off spills immediately with plenty of water. Antifreeze is highly toxic if ingested. Never leave antifreeze lying around in an open container or in puddles on the floor; children and pets are attracted by its sweet smell and may drink it. Check with local authorities on disposing of used antifreeze. Many communities have collection centers, which will see that antifreeze is disposed of safely. Never dump used antifreeze on the ground or into drains.

1 Remove the hose clamp, then disconnect the overflow hose from the reservoir.

2 Remove the engine cooling fan shroud (see Section 5).

3 Pour the coolant into a container. Wash out and inspect the reservoir for cracks and chafing, replacing it as necessary.

4 Installation is the reverse of removal.

5 Cooling fan and clutch - check, removal and installation

CHECK

⁂ WARNING 1:

While checking the fan, make sure that the engine is NOT started. If it is, you could be severely injured. As a safeguard, remove the key from the ignition switch.

⁂ WARNING 2:

Before the fan clutch operation can be checked in Step 5, the engine must be warmed up to its normal operating temperature, then turned off. Even though the engine won't be running during this check, it's HOT! Make sure that you don't touch the engine itself during this check, or you could be burned.

⁂ WARNING 3:

Keep hands, tools and clothing away from the fan when the engine is running. To avoid injury or damage DO NOT operate the engine with a damaged fan. Do not attempt to repair fan blades - replace a damaged fan with a new one.

1 Symptoms of fan clutch failure are continuous noisy operation, looseness, vibration and/or silicone fluid or grease leaking from the clutch.

Cold engine checks

2 Rock the fan back and forth by hand to check for excessive bearing play.

3 With the engine cold, turn the blades by hand. The fan should turn freely.

4 Visually inspect for substantial fluid leakage from the fan clutch assembly, or grease leakage from the cooling fan bearing. If any of these conditions exist, replace the fan clutch.

Hot engine check

5 Start the engine and allow it to warm up to its normal operating temperature. When the engine is fully warmed up, turn off the ignition switch. Turn the fan by hand. Some resistance should be felt. If the fan turns easily, replace the fan clutch.

REMOVAL

➧ Refer to illustrations 5.6, 5.8a, 5.8b and 5.9

6 Remove the radiator support shield (see illustration 6.3), then remove the fan shroud upper mounting bolts (see illustration).

7 Separate any hoses and harness wires from the lower section of the fan shroud.

8 Remove the fan clutch mounting nuts (see illustration) and separate the fan from the pulley. Lift the fan and shroud from the engine compartment (see illustration).

9 To separate the fan blades from the viscous clutch hub, remove the mounting nuts (see illustration).

10 Installation is the reverse of removal.

5.6 Location of the fan shroud upper mounting fasteners

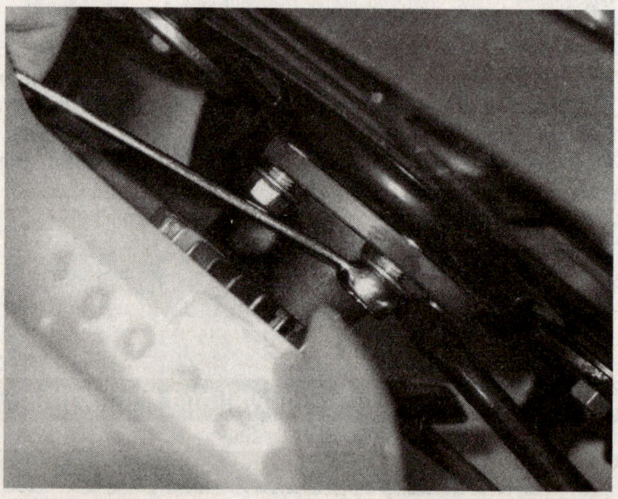

5.8a Use a large screwdriver wedged between a nut and the fan hub to prevent it from turning, then unscrew the nuts

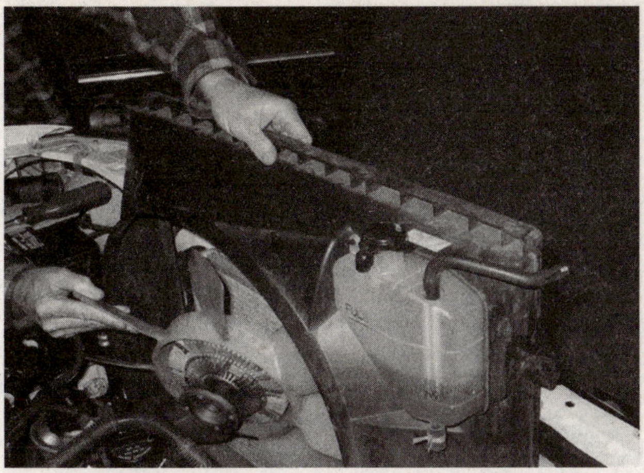

5.8b Remove the fan shroud and fan at the same time. Be careful not to nick the radiator

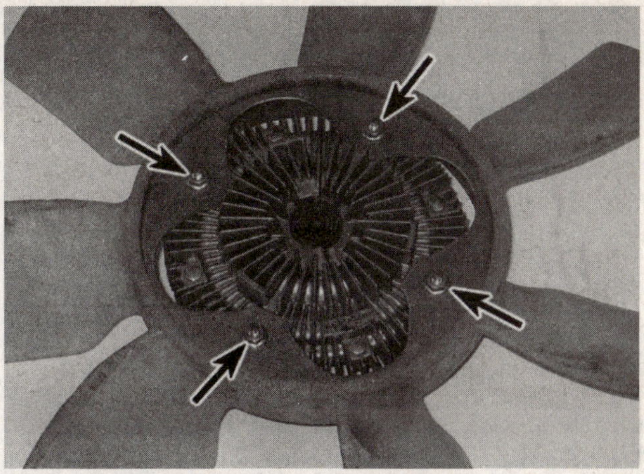

5.9 Remove these four nuts to separate the fan blades from the clutch

6 Radiator - removal and installation

Refer to illustrations 6.3, 6.5a, 6.5b, 6.6 and 6.7

WARNING 1:

Wait until the engine is completely cool before beginning this procedure.

WARNING 2:

Do not allow antifreeze to come in contact with your skin or painted surfaces of the vehicle. Rinse off spills immediately with plenty of water. Antifreeze is highly toxic if ingested. Never leave antifreeze lying around in an open container or in puddles on the floor; children and pets are attracted by its sweet smell and may drink it. Check with local authorities on disposing of used antifreeze. Many communities have collection centers, which will see that antifreeze is disposed of safely. Never dump used antifreeze on the ground or into drains.

1 Disconnect the cable from the negative terminal of the battery (see Chapter 5, Section 1).
2 Drain the engine coolant (see Chapter 1).
3 Remove the radiator support shield (see illustration).

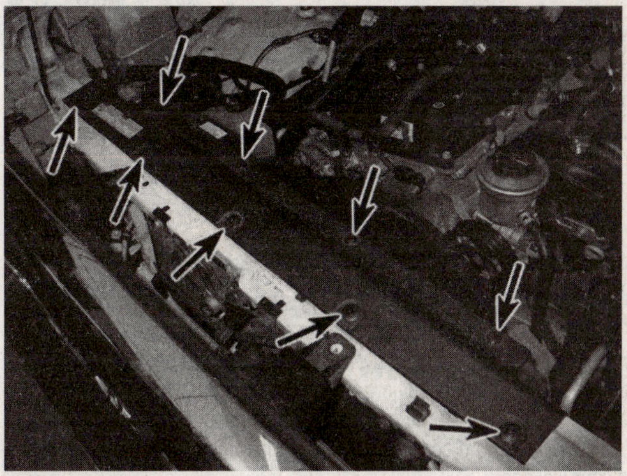

6.3 Remove the fasteners securing the radiator support shield

6.5a At the right side of the radiator, detach the upper radiator hose and the coolant reservoir hose

6.5b At the left side of the radiator, detach the lower radiator hose and, on automatic transmission models, the fluid cooler hoses

6.6 Radiator mounting bolts

6.7 Lift the radiator from the engine compartment carefully to avoid damaging the aluminum cooling fins

4 Remove the cooling fan and fan shroud (see Section 5).

5 Detach the coolant reservoir hose and the upper and lower radiator hoses from the radiator (see illustrations). If the hoses stick, twist them with a pair of large pliers to break the bond, but be careful not to damage the fittings on the radiator. On automatic transmission models, also detach the fluid cooler hoses from the radiator and plug them to prevent leakage.

6 Remove the radiator mounting bolts (see illustration).

7 Lift out the radiator (see illustration). Be aware of dripping fluids and the sharp fins.

8 With the radiator removed, it can be inspected for leaks, damage and internal blockage. If in need of repairs, have a radiator shop perform the work, as special techniques are required.

9 Bugs and dirt can be cleaned from the radiator with compressed air and a soft brush. Don't bend the cooling fins as this is done.

WARNING:

Wear eye protection when using compressed air.

10 Installation is the reverse of the removal procedure. Tighten the radiator mounting bolts to the torque listed in this Chapter's Specifications.

11 After installation, fill the cooling system (see Chapter 1).

12 Start the engine and check for leaks. Allow the engine to reach normal operating temperature, indicated by both radiator hoses becoming hot. Recheck the coolant level and add more if required.

13 Check and add transmission fluid as needed (see Chapter 1).

7 Water pump - check

Refer to illustrations 7.2a and 7.2b

1 A failure in the water pump can cause serious engine damage due to overheating.

2 Water pumps on all models are equipped with a weep hole and a vent hole (see illustrations). If coolant leaks from any of these holes, replace the water pump. In most cases it will be necessary to use a flashlight and a mirror to find the hole(s) by looking behind the water pump pulley.

3 If the water pump shaft bearings fail, there may be a howling sound at the front of the engine while it is running. Bearing wear can be felt if the water pump pulley is rocked up and down. Do not mistake drivebelt slippage, which causes a squealing sound, for water pump failure. Spray automotive drivebelt dressing on the belts to eliminate the belt as a possible cause of the noise.

7.2a Four-cylinder water pump air hole (A) and weep hole (B) (fan and pulley removed for clarity)

7.2b V6 water pump air hole (A) and weep hole (B) (fan and pulley removed for clarity)

8 Water pump - removal and installation

⁂ WARNING 1:

Wait until the engine is completely cool before beginning this procedure.

⁂ WARNING 2:

Do not allow antifreeze to come in contact with your skin or painted surfaces of the vehicle. Rinse off spills immediately with plenty of water. Antifreeze is highly toxic if ingested. Never leave antifreeze lying around in an open container or in puddles on the floor; children and pets are attracted by its sweet smell and may drink it. Check with local authorities on disposing of used antifreeze. Many communities have collection centers, which will see that antifreeze is disposed of safely. Never dump used antifreeze on the ground or into drains.

1 Disconnect the cable from the negative terminal of the battery (see Chapter 5, Section 1).

2 If you're working on a PreRunner or a 4WD model, remove the forward under-vehicle splash shield.

3 Drain the cooling system (see Chapter 1).

4 If you're working on a V6 model, remove the engine cover.

5 Remove the drivebelt (see Chapter 1).

6 Remove the engine cooling fan (see Section 5).

FOUR-CYLINDER MODELS

Refer to illustration 8.9

7 Remove the alternator (see Chapter 5).

8 Remove the drivebelt tensioner (see Chapter 1).

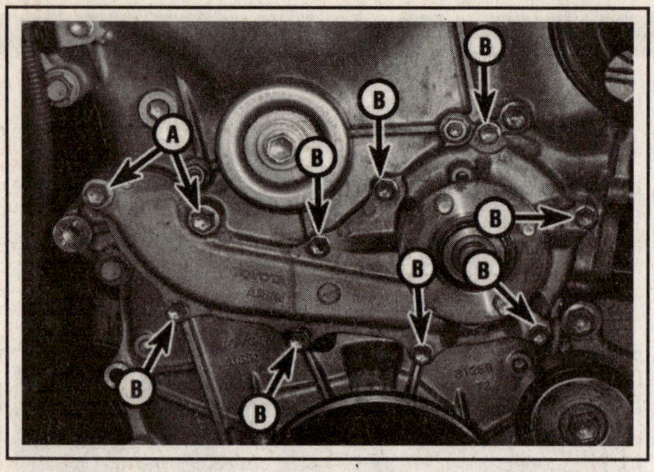

8.9 Water pump mounting bolts on the four-cylinder engine

A Long bolts *B Short bolts*

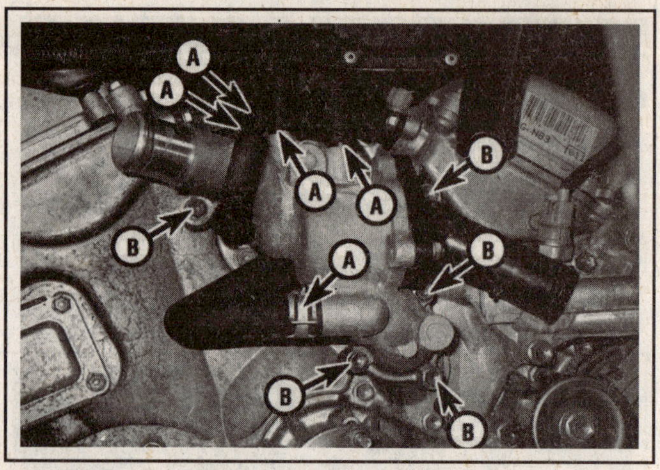

8.16 Water inlet assembly details on the V6 engine

A Bypass hose clamp locations
B Water inlet assembly mounting bolts

8.17 Remove the number 1 and number 2 idler pulleys

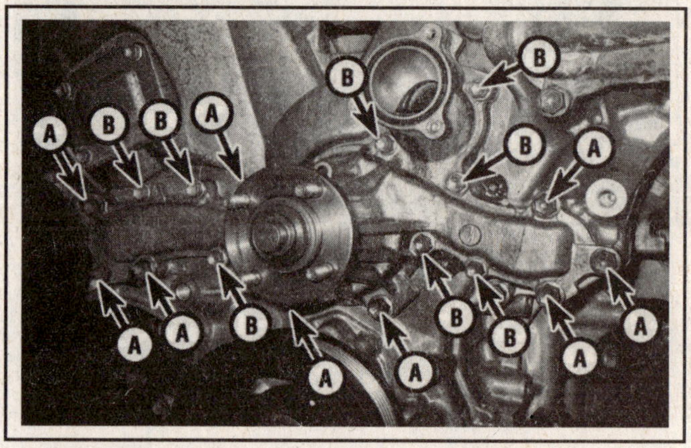

8.21a Water pump mounting bolts on V6 engines

A Long bolts *B Short bolts*

9 Remove the water pump mounting bolts (see illustration).

➡**Note: Be sure to mark the location of the large bolts and small bolts for proper reassembly.**

10 Thoroughly clean the gasket mating surfaces.

11 Install a new gasket and water pump. Tighten the water pump bolts to the torque listed in this Chapter's Specifications.

12 The remainder of installation is the reverse of removal.

13 Refill the cooling system (see Chapter 1), run the engine and check for leaks.

V6 MODELS

➧ **Refer to illustrations 8.16, 8.17, 8.21a, 8.21b and 8.24**

14 Remove the air filter assembly (see Chapter 4).

15 Remove the oil cooler hoses, if equipped, from the timing chain cover and the coolant bypass hoses from the water inlet assembly.

16 Remove the water inlet assembly (see illustration).

17 Remove the number 1 and number 2 idler pulleys (see illustration).

18 Remove the alternator (see Chapter 5).

19 Unbolt the air conditioning compressor and set it aside (see Section 15). Support it with wire or rope.

⁂ WARNING:

Don't disconnect the refrigerant lines.

20 Remove the drivebelt tensioner (see Chapter 1).

21 Remove the water pump mounting bolts (see illustrations).

➡**Note: Be sure to mark the location of the large bolts and small bolts for proper reassembly.**

22 Thoroughly clean the gasket mating surfaces.

23 Install a new gasket and water pump. Tighten the water pump bolts to the torque listed in this Chapter's Specifications.

24 Install a new O-ring on the water pump (see illustration).

25 Install the water inlet assembly. Be sure to install new O-rings, then tighten the water inlet assembly mounting fasteners to the torque listed in this Chapter's Specifications.

26 The remainder of installation is the reverse of removal.

27 Refill the cooling system (see Chapter 1), run the engine and check for leaks.

8.21b Be sure to install a new gasket and clean the surface of the timing chain cover thoroughly

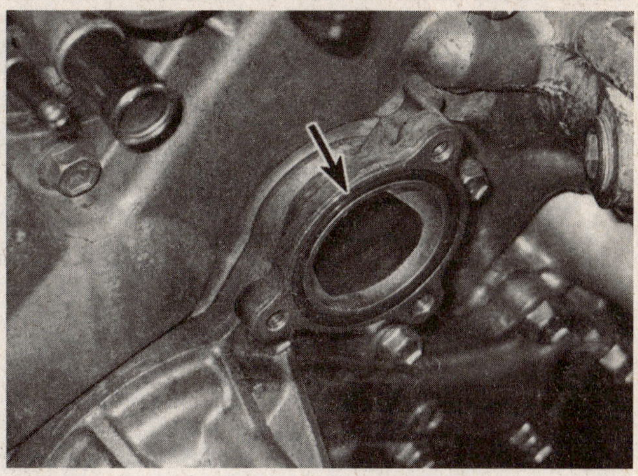

8.24 Install a new O-ring on the water pump

9 Coolant temperature indicator system - check

WARNING:

Wait until the engine is completely cool before beginning this procedure.

1 The coolant temperature indicator system consists of a temperature gauge on the dash and a sensor mounted on the engine. On all models, an Engine Coolant Temperature (ECT) sensor (see Chapter 6) - which is an information sensor for the Powertrain Control Module (PCM) - provides a signal to the PCM which controls and actuates the temperature gauge.

2 If an overheating indication has occurred, first check the coolant level in the system (see Chapter 1) and that the coolant mixture is correct (see Section 2). Also, refer to the *Troubleshooting* Section at the beginning of this book before assuming that the temperature indicator is faulty.

3 Start the engine and warm it up for 10 minutes. If the temperature gauge has not moved from the C position, check the wiring harness connections going to the instrument cluster.

4 If there is a problem with the ECT sensor, it is very likely that the CHECK ENGINE light will be illuminated and the sensor or circuit will need repair (see Chapter 6).

10 Blower motor and resistor - removal and installation

WARNING:

The models covered by this manual are equipped with Supplemental Restraint systems (SRS), more commonly known as airbags. Always disarm the airbag system before working in the vicinity of any airbag system components to avoid the possibility of accidental deployment of the airbag, which could cause personal injury (see Chapter 12).

BLOWER MOTOR

Refer to illustrations 10.2 and 10.4

1 The blower unit is located under the dash, behind the glovebox.

2 Disconnect the blower motor electrical connector, then remove the three blower mounting screws (see illustration).

10.2 Blower motor mounting screws

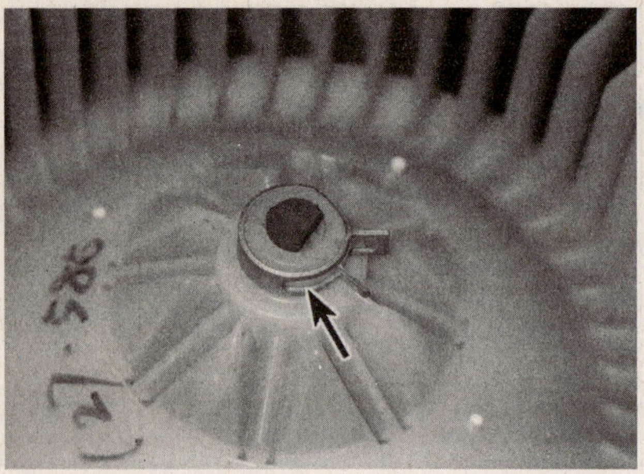

10.4 Remove the clamp from the shaft and separate the fan from the blower motor

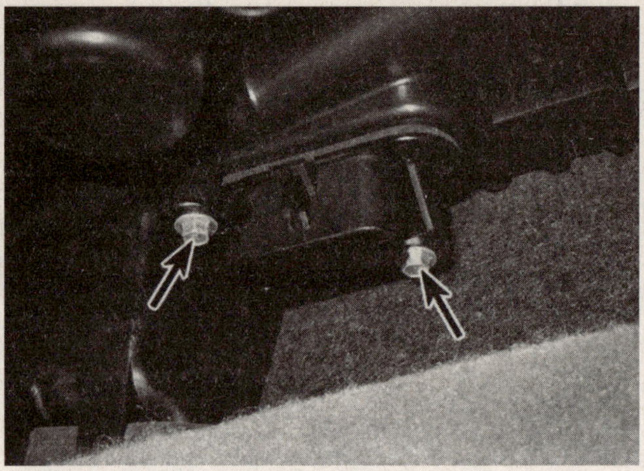

10.7 Blower motor resistor mounting screws

3 Lower the blower assembly from the housing.

4 Remove the spring clamp from the fan (see illustration) and separate the fan from the blower motor.

5 Installation is the reverse of removal.

RESISTOR

➧ Refer to illustration 10.7

6 The resistor is located on the blower duct under the dash, directly below the glovebox.

7 Disconnect the resistor electrical connector and remove the mounting screws (see illustration).

8 Installation is the reverse of removal.

11 Heater core - removal and installation

➧ Refer to illustrations 11.4, 11.5, 11.8, 11.9, 11.10, 11.12a, 11.12b, 11.13a, 11.13b, 11.16, 11.18, 11.19a, 11.19b, 11.19c, 11.19d, 11.19e, 11.20, 11.21a, 11.21b and 11.22

⁂ WARNING 1:

Wait until the engine is completely cool before beginning this procedure.

⁂ WARNING 2:

The models covered by this manual are equipped with Supplemental Restraint systems (SRS), more commonly known as airbags. Always disarm the airbag system before working in the vicinity of any airbag system components to avoid the possibility of accidental deployment of the airbag, which could cause personal injury (see Chapter 12).

⁂ WARNING 3:

Do not allow antifreeze to come in contact with your skin or painted surfaces of the vehicle. Rinse off spills immediately with plenty of water. Antifreeze is highly toxic if ingested. Never leave antifreeze lying around in an open container or in puddles on the floor; children and pets are attracted by it's sweet smell and may drink it. Check with local authorities about disposing of used antifreeze. Many communities have collection centers which will see that antifreeze is disposed of safely. Never dump used antifreeze on the ground or pour it into drains.

⁂ WARNING 4:

The air conditioning system is under high pressure. Do not loosen any hose fittings or remove any components until the system has been discharged. Air conditioning refrigerant should be properly discharged into an EPA-approved recovery/recycling unit by a dealer service department or an automotive air conditioning repair facility. Always wear eye protection when disconnecting air conditioning system fittings.

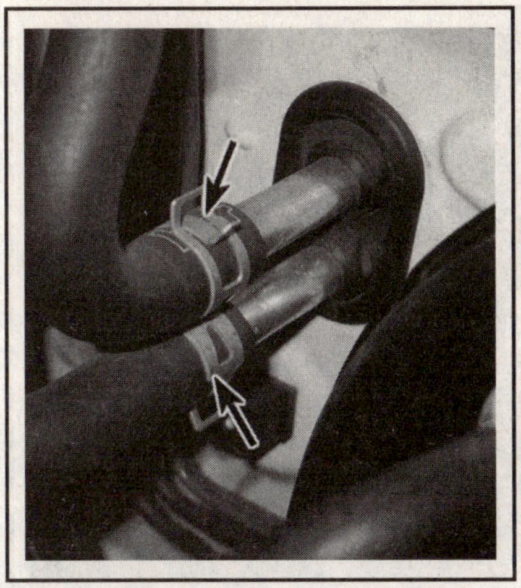

11.4 Squeeze the spring clamps and slide them back on the heater hoses, then disconnect the hoses from the heater core pipes

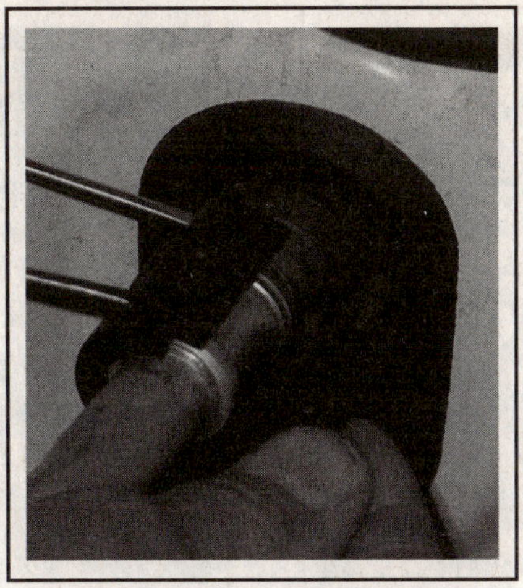

11.5 Insert two narrow screwdrivers into the holes to release the locking tabs and open the clamp

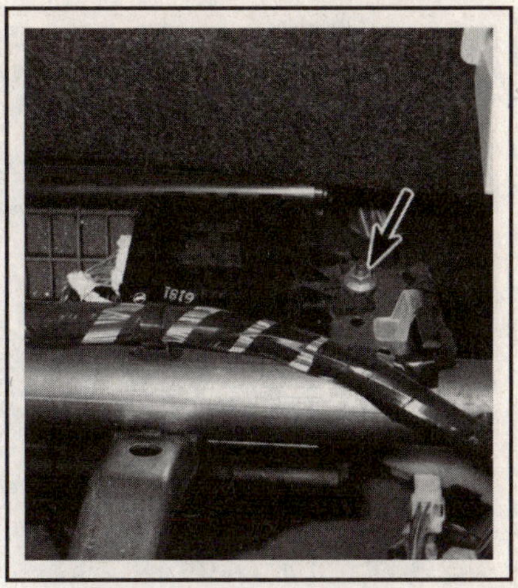

11.8 Disconnect the harness connector, remove the mounting bolts and separate the air conditioning amplifier module from the HVAC housing

11.9 Left side heater to register duct mounting nut

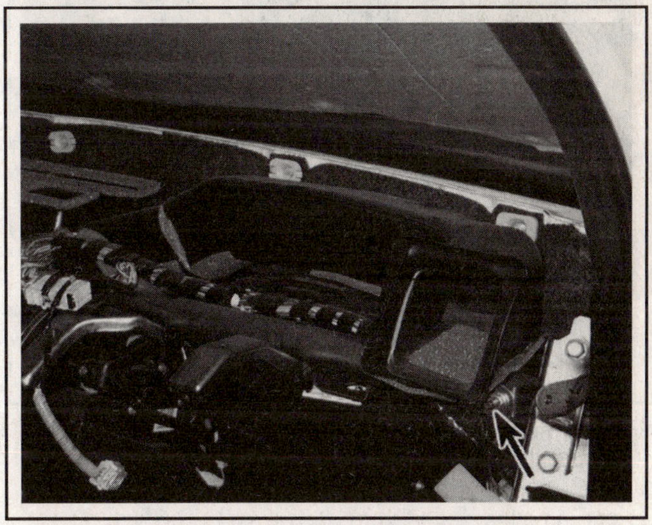

11.10 Right side heater-to-register duct mounting nut

➡**Note: The manufacturer recommends removal of the entire instrument panel to remove the heater core. This involves disconnecting numerous electrical connectors and there is the potential for breakage of delicate plastic tabs on various components. This is a difficult job for the average home mechanic.**

1 Take the vehicle to a dealer service department or automotive air conditioning shop and have the air conditioning system discharged and the refrigerant recovered.

2 Disconnect the cable from the negative terminal of the battery (see Chapter 5, Section 1).

3 Wait until the engine is completely cool, then drain the cooling system (see Chapter 1).

4 Working in the engine compartment, disconnect the heater hoses at the firewall (see illustration). Push the rubber grommet around the hoses toward the inside of the vehicle, releasing it from the sheetmetal. Plug the heater core pipes to prevent leakage when it is removed.

5 Disconnect the refrigerant lines from the evaporator tubes at the firewall (see illustration).

6 Remove the instrument panel (see Chapter 11).

7 Remove the Powertrain Control Module (PCM) (see Chapter 6) and the mounting bracket.

8 Remove the air conditioning amplifier module (see illustration).

9 Working on the left side of the interior, remove the left side heater to register duct (see illustration).

10 Working on the right side of the interior, remove the right side heater to register duct (see illustration).

11 On extra-cab models, remove the left side floor duct and the right side floor duct.

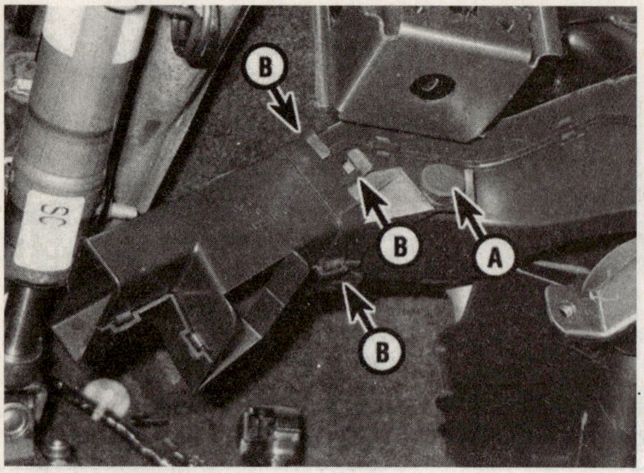

11.12a Left side lower air duct details

A Plastic button B Locating clips and tabs

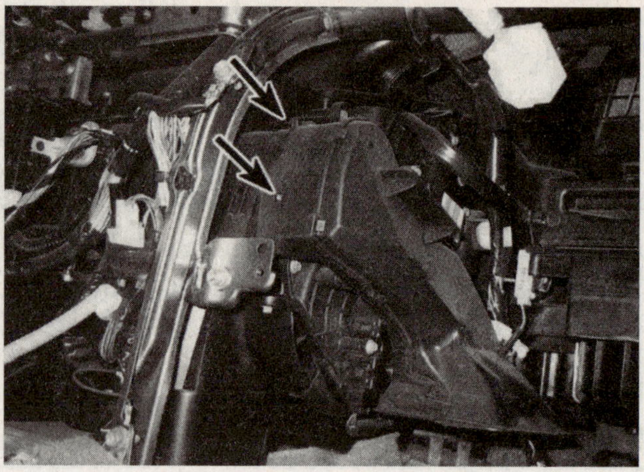

11.12b Location of the clips and tabs on the right side air duct - note here that the rear clip is hidden

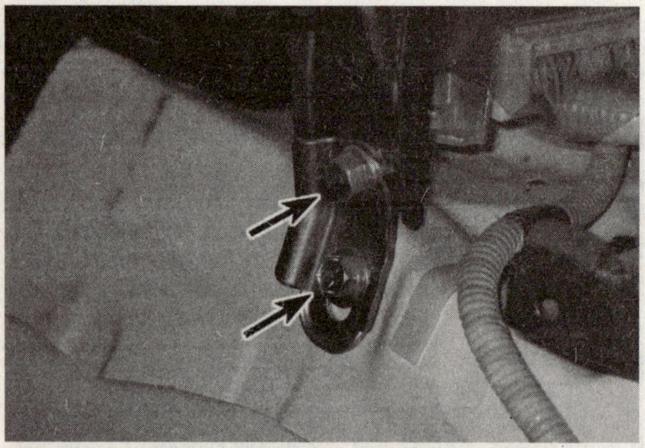

11.13a Remove the left-side reinforcement brace mounting fasteners . . .

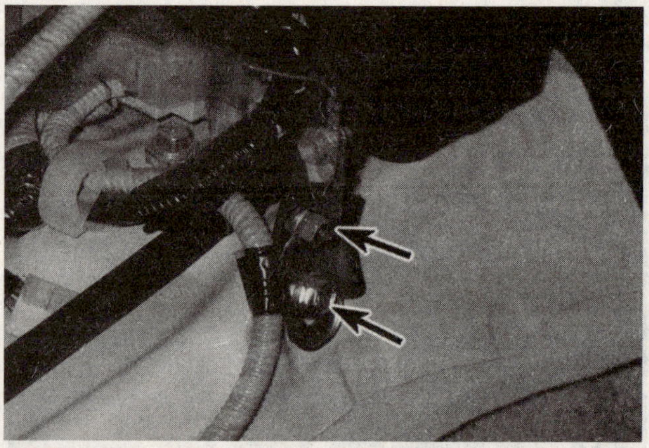

11.13b . . . and the right side reinforcement brace mounting fasteners

11.16 Remove the mounting fasteners and lower the fuse and relay box

12 Remove the left side air duct assembly (see illustration) and the right side air duct assembly (see illustration).

13 Remove the reinforcement brace mounting brackets at the center console (see illustrations).

14 Detach the steering column from the reinforcement brace (see Chapter 10).

15 Remove the parking brake control lever (see Chapter 9).

16 Remove the fuse and relay box from the left side of the reinforcement brace (see illustration).

17 Remove the transponder key module, if equipped, from the console bracket.

18 Release the wiring harness clamps and the ground strap bolts, disconnect the electrical connectors and separate the harness from the reinforcement brace (see illustration).

19 Remove the reinforcement brace mounting fasteners (see illustration) and heater/air conditioning unit mounting fasteners (see illustrations).

➡Note: The exterior mounting bolt is accessible in the engine compartment at the firewall. Remove the windshield wipers (see Chapter 12) and cowl (see Chapter 11) and locate the rubber plug that covers the bolt (see illustration).

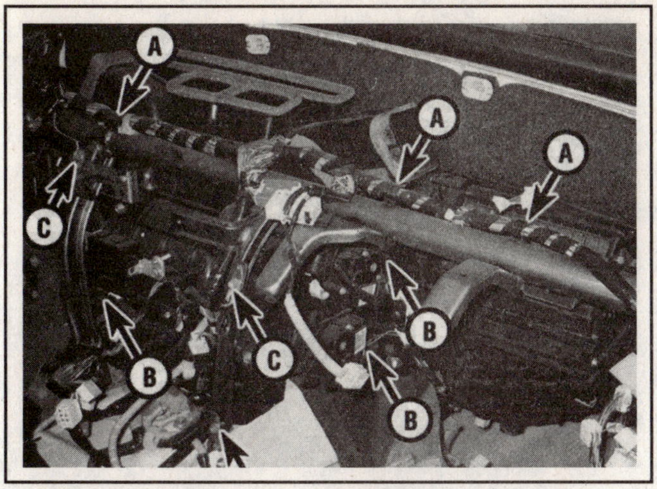

11.18 Remove all the harness mounting clamps (A), the ground wires (C) and the electrical connectors (B) and separate the harness from the reinforcement brace

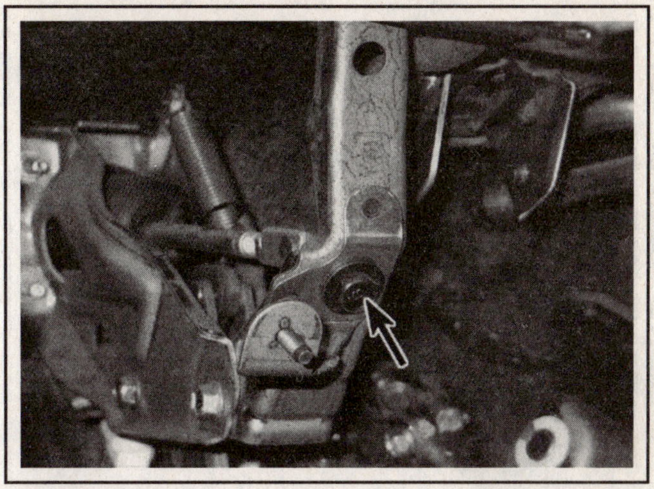

11.19a Location of the left-side reinforcement brace mounting bolt

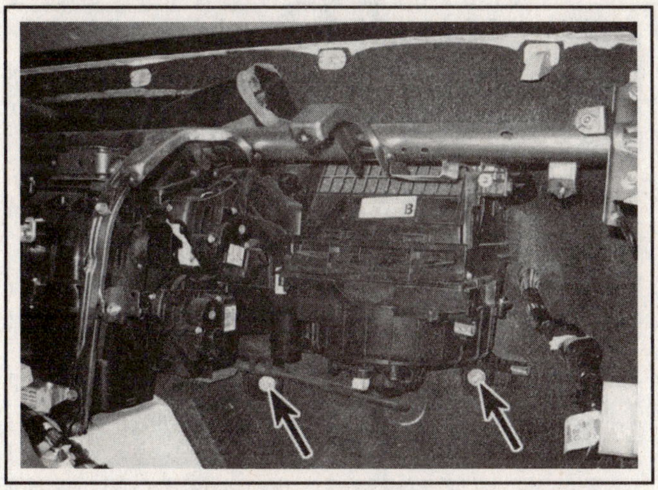

11.19b Location of the right-side heater/air conditioning assembly mounting nuts

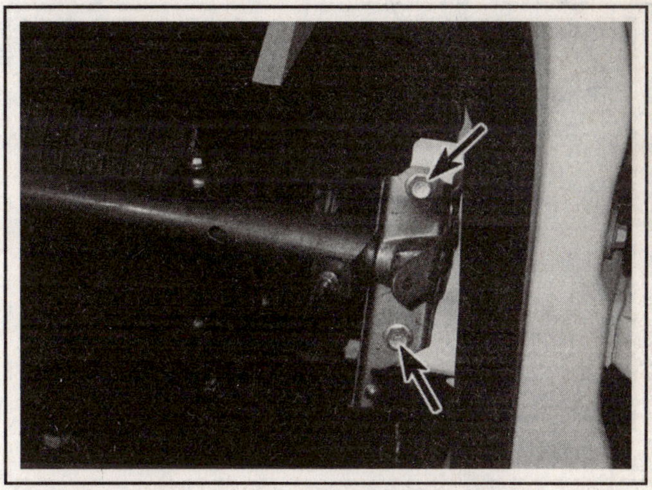

11.19c Location of the right-side main reinforcement brace mounting bolts

11.19d Open the left door to access the left-side main reinforcement brace mounting bolts - two of the bolts will have rubber plugs that must be removed from the door jamb

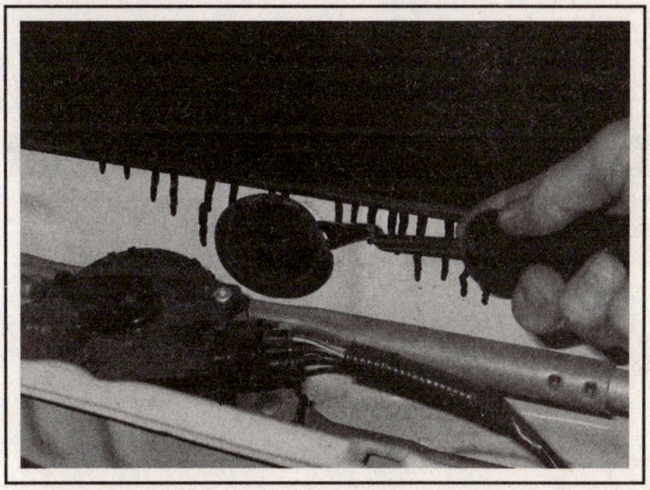

11.19e After the wiper arms and motor are removed, remove the cowl and rubber plug to access the last reinforcement brace mounting bolt

11.20 Lift the reinforcement brace and the air conditioning/heater assembly from the passenger compartment

11.21a Working at the bottom of the air conditioning/heater assembly, remove the cover mounting screws . . .

11.21b . . . and the single screw on the side from the actuator

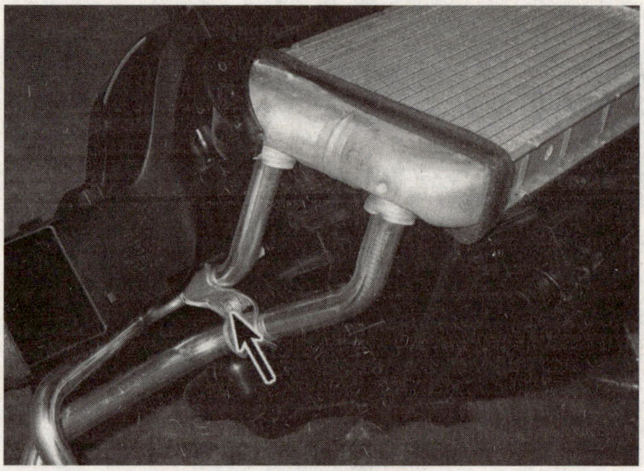

11.22 Remove the screw from the heater core pipe clamp

20 Lift the reinforcement brace and the air conditioning/heater unit from the passenger compartment (see illustration).

⁂ CAUTION:

Work slowly and carefully to avoid breaking any plastic components during removal.

21 Working on the underside of the heater core housing, remove the bolts from the lower cover (see illustrations).

22 Remove the bolt from the heater core inlet and outlet pipe clamp (see illustration).

23 Slide the heater core from the air conditioning/heater unit.

24 Installation is the reverse order of removal.

25 Refill the cooling system (see Chapter 1), reconnect the battery and run the engine. Check for leaks and proper system operation.

12 Heater and air conditioning control assembly - removal and installation

Refer to illustrations 12.2 and 12.3

WARNING:

The models covered by this manual are equipped with Supplemental Restraint systems (SRS), more commonly known as airbags. Always disarm the airbag system before working in the vicinity of any airbag system components to avoid the possibility of accidental deployment of the airbag, which could cause personal injury (see Chapter 12).

1 Disconnect the cable from the negative terminal of the battery (see Chapter 5, Section 1).
2 Carefully pry the control assembly away from the center dash (see illustration).
3 Disconnect the control assembly electrical connectors (see illustration).
4 Installation is the reverse of removal.

12.2 Carefully pry the corners of the control assembly and release the plastic tabs from the instrument panel

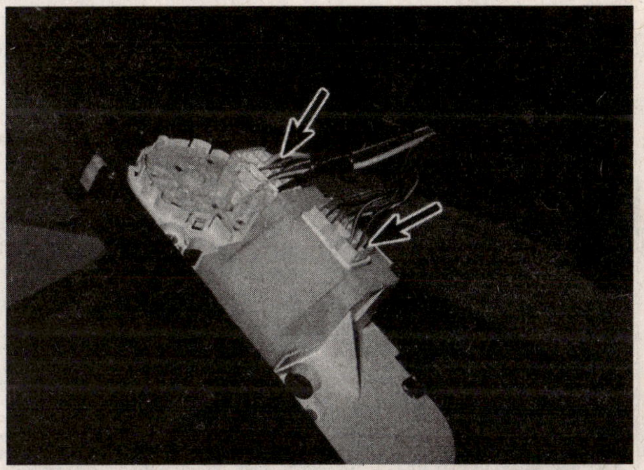

12.3 Disconnect the electrical connectors from the control assembly

13 Air conditioning and heating system - check and maintenance

AIR CONDITIONING SYSTEM

WARNING:

The air conditioning system is under high pressure. Do not loosen any hose fittings or remove any components until the system has been discharged. Air conditioning refrigerant must be properly discharged into an EPA-approved recovery/recycling unit by a dealer service department or an automotive air conditioning repair facility. Always wear eye protection when disconnecting air conditioning system fittings.

1 The following maintenance checks should be performed on a regular basis to ensure that the air conditioner continues to operate at peak efficiency.

a) *Inspect the condition of the compressor drivebelt. If it is worn or deteriorated, replace it (see Chapter 1).*
b) *Check the drivebelt tension and, if necessary, adjust it (see Chapter 1).*
c) *Inspect the system hoses. Look for cracks, bubbles, hardening and deterioration. Inspect the hoses and all fittings for oil bubbles or seepage. If there is any evidence of wear, damage or leakage, replace the hose(s).*
d) *Inspect the condenser fins for leaves, bugs and any other foreign material that may have embedded itself in the fins. Use a fin comb or compressed air to remove debris from the condenser.*
e) *Make sure the system has the correct refrigerant charge.*

2 It's a good idea to operate the system for about ten minutes at least once a month. This is particularly important during the winter months because long term non-use can cause hardening, and subsequent failure, of the seals.

3 Because of the complexity of the air conditioning system and the special equipment necessary to service it, in-depth troubleshooting and repairs are beyond the scope of this manual. However, simple component replacement procedures are provided in this Chapter.

4 The most common cause of poor cooling is simply a low system refrigerant charge. If a noticeable drop in system cooling ability occurs, one of the following quick checks will help you determine whether the refrigerant level is low.

CHECK

Refer to illustration 13.8

5 Warm the engine up to normal operating temperature.

6 Place the air conditioning temperature selector at the coldest setting and put the blower at the highest setting. Open the doors (to make sure the air conditioning system doesn't cycle off as soon as it cools the passenger compartment).

7 After the system reaches operating temperature, feel the two pipes connected to the evaporator at the firewall.

13.8 Check the temperature of the output air in the center register with a thermometer - it should be approximately 35 to 40-degrees F below the ambient air temperature

13.9 A basic charging kit is available at most auto parts stores - it must say R-134a and so must the cans of refrigerant you buy

8 The pipe (thinner tubing) leading from the condenser outlet to the evaporator should be cold, and the evaporator outlet line (the thicker tubing that leads back to the compressor) should be slightly colder (3 to 10 degrees F). If the evaporator outlet is considerably warmer than the inlet, the system needs a charge. Insert a thermometer in the center air distribution duct (see illustration) while operating the air conditioning system - the temperature of the output air should be 35 to 40 degrees F below the ambient air temperature (down to approximately 40 degrees F). If the ambient (outside) air temperature is very high, say 110 degrees F, the duct air temperature may be as high as 60 degrees F, but generally the air conditioning is 35 to 40 degrees F cooler than the ambient air. If the air isn't as cold as it used to be, the system probably needs a charge. Further inspection or testing of the system is beyond the scope of the home mechanic and should be left to a professional.

ADDING REFRIGERANT

Refer to illustrations 13.9 and 13.12

Note: All models covered by this manual use R-134a refrigerant. When recharging or replacing air conditioning components, use only refrigerant, refrigerant oil and seals compatible with this system. The seals and compressor oil used with older, conventional R-12 refrigerant are not compatible with the components in this system.

9 Buy an automotive charging kit at an auto parts store. A charging kit includes a 12-ounce can of R-134a refrigerant, a tap valve and a short section of hose that can be attached between the tap valve and the system low side service valve (see illustration).

10 Connect the charging kit by following the manufacturer's instructions.

11 Back off the valve handle on the charging kit and screw the kit onto the refrigerant can, making sure first that the O-ring or rubber seal inside the threaded portion of the kit is in place.

WARNING:

Wear protective eyewear when dealing with pressurized refrigerant cans.

12 Remove the dust cap from the low-side charging port and attach the quick-connect fitting on the kit hose (see illustration).

WARNING:

DO NOT hook the charging kit hose to the system high side! The fittings on the charging kit are designed to fit only on the low side of the system.

13 Warm the engine to normal operating temperature and turn on the air conditioning. Keep the charging kit hose away from the fan and other moving parts. In some cases, if the refrigerant charge is low enough, the air conditioning system pressure switch may prevent the compressor from operating.

Note: The charging process requires that the compressor be running. If the clutch cycles off, you can switch the A/C controls to High and leave the vehicle's doors open to keep the clutch on the compressor working.

14 Turn the valve handle on the kit until the stem pierces the can,

then back the handle out to release the refrigerant. You should be able to hear the rush of gas. Add refrigerant to the low side of the system until both the outlet and the evaporator inlet pipe feel about the same temperature. Allow stabilization time between each addition.

CAUTION:

Never add more than one can of refrigerant to the system. If more refrigerant than that is required, the system should be evacuated and leak tested.

15 The can may tend to frost up, slowing the procedure. Wet a shop towel with hot water and wrap it around the bottom of the can to keep it from frosting.

16 Put your thermometer back in the center register and check that the output air is getting colder.

17 When the can is empty, turn the valve handle to the closed position and release the connection from the low-side port. Reinstall the dust cap.

18 Remove the charging kit from the can and store the kit for future use with the piercing valve in the UP position, to prevent inadvertently piercing the can on the next use.

HEATING SYSTEMS

Refer to illustration 13.23

19 If the air coming out of the heater vents isn't hot, the problem could stem from any of the following causes:

a) *The thermostat is stuck open, preventing the engine coolant from warming up enough to carry heat to the heater core. Replace the thermostat (see Section 3).*
b) *A heater hose is blocked, preventing the flow of coolant through the heater core. Feel both heater hoses at the firewall. They should be hot. If one of them is cold, there is an obstruction in one of the hoses or in the heater core, or the heater control valve is shut. Detach the hoses and back flush the heater core with a water hose. If the heater core is clear but circulation is impeded, remove the two hoses and flush them out with a garden hose.*
c) *If flushing fails to remove the blockage from the heater core, the core must be replaced (see Section 11).*

20 If the blower motor speed does not correspond to the setting selected on the blower switch, the problem could be a bad fuse, circuit, blower relay, speed switch or blower resistor.

21 If there isn't any air coming out of the vents:

a) *Turn the ignition ON and activate the fan control. Place your ear at the heating/air conditioning register (vent) and listen. Most motors are audible. Can you hear the motor running?*
b) *If you can't (and have already verified that the blower switch and the blower motor resistor are good), the blower motor itself is probably bad (see Section 10).*

22 If the carpet under the heater core is damp, or if antifreeze vapor or steam is coming through the vents, the heater core is leaking. Remove it (see Section 11) and install a new unit (most radiator shops will not repair a leaking heater core).

23 Inspect the drain hose from the heater/evaporator, which exits the body under the floor (see illustration). If there is a humid mist coming from the system ducts, this hose may be plugged with leaves or road debris.

ELIMINATING AIR CONDITIONING ODORS

Refer to illustration 13.27

24 Unpleasant odors that often develop in air conditioning systems are caused by the growth of a fungus, usually on the surface of the evaporator core. The warm, humid environment there is a perfect breeding ground for mildew to develop.

25 The evaporator core on most vehicles is difficult to access, and factory dealerships have a lengthy, expensive process for eliminating the fungus by opening up the evaporator case and using a powerful disinfectant and rinse on the core until the fungus is gone. You can service your own system at home, but it takes something much stronger than basic household germ-killers or deodorizers.

26 Aerosol disinfectants for automotive air-conditioning systems are available in most auto parts stores, but remember when shopping for them that the most effective treatments are also the most expensive. The basic procedure for using these sprays is to start by running the system

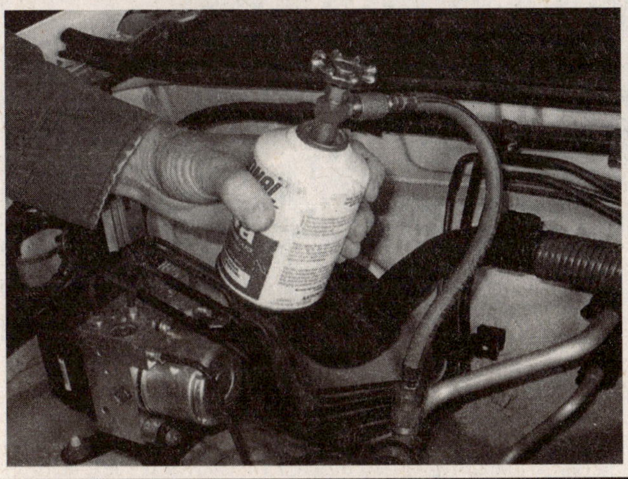

13.12 Add R-134a refrigerant to the low-side port only - the procedure will go faster if you wrap the can with a warm, wet towel to prevent icing

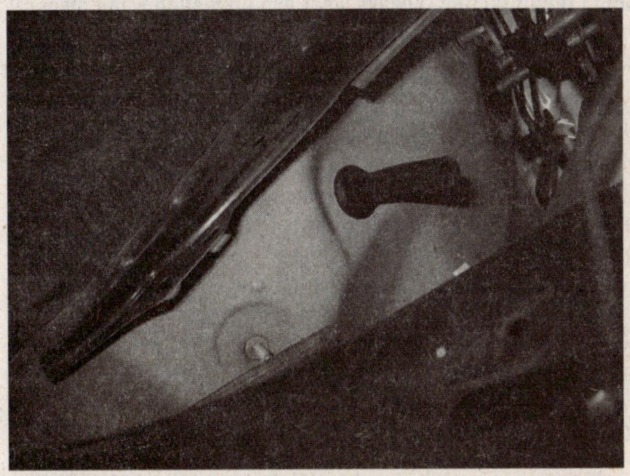

13.23 Location of the evaporator drain at the right side of the firewall

in the RECIRC mode for ten minutes with the blower on its highest speed. Use the highest heat mode to dry out the system and keep the compressor from engaging by disconnecting the wiring connector at the compressor (see Section 15).

27 The disinfectant can usually comes with a long spray hose. Remove the cabin air filter (see Chapter 1), point the nozzle inside the hole and spray according to the manufacturer's recommendations (see illustration). Follow the manufacturer's recommendations for the length of spray and waiting time between applications.

28 Once the evaporator has been cleaned, the best way to prevent the mildew from coming back again is to make sure your evaporator housing drain tube is clear (see illustration 13.23) and to run the defrost cycle briefly to dry the evaporator out after a long drive with the air conditioning on.

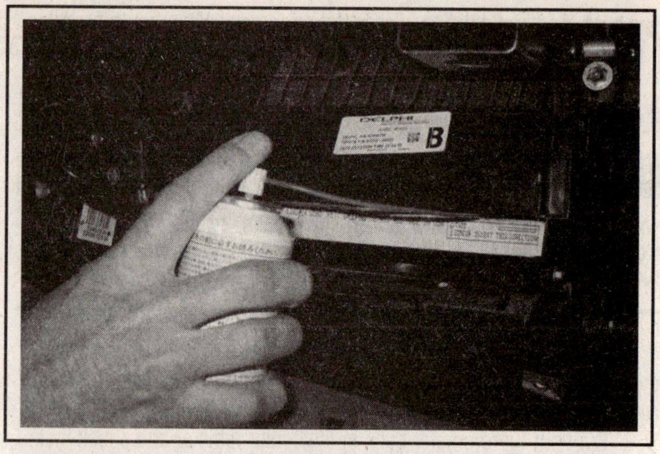

13.27 Remove the cabin air filter and insert the disinfectant spray nozzle - be sure to support the spray nozzle so it does not get tangled in the blower fan

14 Air conditioning receiver/drier - removal and installation

➧ Refer to illustration 14.3

✲✲ WARNING:

The air conditioning system is under high pressure. Do not loosen any hose fittings or remove any components until the system has been discharged. Air conditioning refrigerant should be properly discharged into an EPA-approved recovery/recycling unit by a dealer service department or an automotive air conditioning repair facility. Always wear eye protection when disconnecting air conditioning system fittings.

1 Take the vehicle to a dealer service department or automotive air conditioning shop and have the air conditioning system discharged and the refrigerant recovered.

2 On some models the receiver/drier plug is accessible from under the front of the vehicle, just behind the bumper on the right side, and can be removed without removing or repositioning the condenser. On other models, the condenser will have to be removed, or unbolted from its mounts and repositioned, for access to the plug.

3 Using an Allen wrench, detach the end plug (see illustration) and remove the drier from the condenser with a pair of needle-nose pliers.

4 Use new O-rings when installing the new drier and tighten the end plug securely. Be sure to lubricate the O-rings with R-134a compatible refrigerant oil.

5 Installation is the reverse of removal.

6 Have the system evacuated, charged and leak tested by the shop that discharged it.

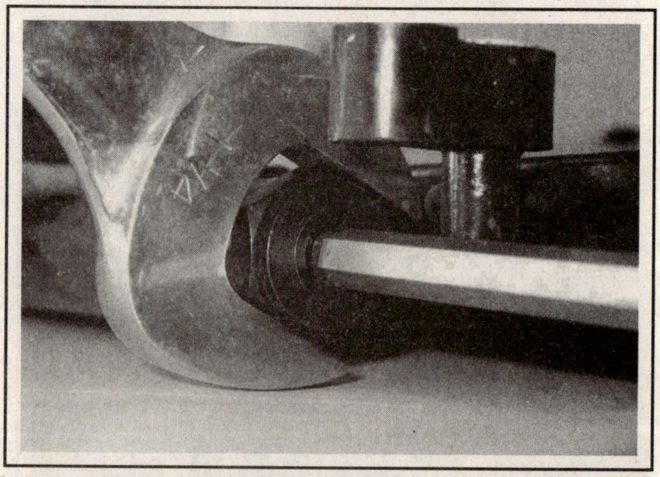

14.3 After the system has been discharged, remove the Allen plug and pull the drier from the tube on the condenser - be sure to hold the fitting with a wrench to prevent twisting the metal housing

15 Air conditioning compressor - removal and installation

⁂ WARNING:

The air conditioning system is under high pressure. Do not loosen any hose fittings or remove any components until the system has been discharged. Air conditioning refrigerant should be properly discharged into an EPA-approved recovery/recycling unit by a dealer service department or an automotive air conditioning repair facility. Always wear eye protection when disconnecting air conditioning system fittings.

➡Note: The receiver-drier should be replaced whenever the compressor is replaced.

1 Take the vehicle to a dealer service department or automotive air conditioning shop and have the air conditioning system discharged and the refrigerant recovered.

2 Remove the drivebelt (see Chapter 1).

3 Remove the engine cooling fan (see Section 5).

4 Loosen the left front wheel lug nuts. Raise the front of the vehicle and support it securely on jackstands.

5 Remove the engine splash shield.

6 Remove the left front wheel and the inner fender splash shield (see Chapter 11).

FOUR CYLINDER MODELS

➧ Refer to illustration 15.7

7 Detach the electrical connector and disconnect the refrigerant lines, then unbolt the compressor and guide it out from under the vehicle (see illustration).

8 If a new or rebuilt compressor is being installed, follow the directions supplied with the compressor to properly adjust the oil level before installing it.

V6 MODELS

➧ Refer to illustration 15.12

9 Remove the engine cover (see Chapter 2B).

10 Remove the alternator (see Chapter 5).

11 Detach the electrical connector and disconnect the refrigerant lines.

12 Unbolt the compressor (see illustration) and guide it out from under the vehicle.

13 If a new or rebuilt compressor is being installed, follow the directions supplied with the compressor to properly adjust the oil level before installing it.

ALL MODELS

14 Installation is the reverse of removal. Replace any O-rings with new ones specifically made for the type of refrigerant in your system and lubricate them with refrigerant oil, also designed specifically for your system (R-134a).

15 Have the system evacuated, recharged and leak tested by the shop that discharged it.

15.7 Air conditioning compressor mounting details - four-cylinder engine

A Refrigerant line fasteners
B Compressor mounting bolts (lower bolts not visible here)
C Electrical connector (already unplugged here)

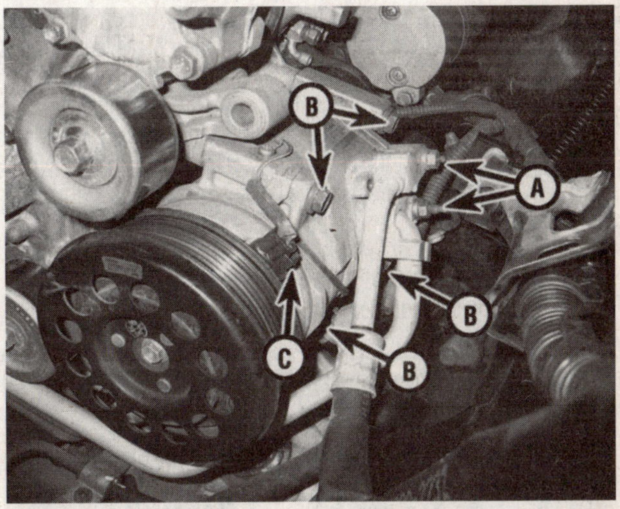

15.12 Air conditioning compressor mounting details - V6 engine

A Refrigerant line fasteners
B Compressor mounting bolts (lower bolts not visible here)
C Electrical connector

16 Air conditioning condenser - removal and installation

➧ **Refer to illustrations 16.6a, 16.6b and 16.8**

WARNING:

The air conditioning system is under high pressure. Do not loosen any hose fittings or remove any components until the system has been discharged. Air conditioning refrigerant should be properly discharged into an EPA-approved recovery/recycling unit by a dealer service department or an automotive air conditioning repair facility. Always wear eye protection when disconnecting air conditioning system fittings.

➡**Note 1: Read through this procedure carefully, especially Step 2, before beginning this procedure.**

➡**Note 2: The receiver-drier should be replaced whenever the condenser is replaced due to a leak. If you're replacing the condenser and the new one isn't equipped with a receiver drier, be sure to install a new one (don't transfer the old one to the new condenser).**

1 Take the vehicle to a dealer service department or automotive air conditioning shop and have the air conditioning system discharged and the refrigerant recovered.

2 The factory-recommended procedure entails draining the cooling system, removing the fan and shroud, removing the radiator, and removing the condenser out towards the rear of the vehicle. We have found that it is possible, if you work very carefully, to just remove the fan and shroud, unbolt the radiator, remove the condenser mounting brackets and detach the refrigerant lines, then remove the condenser out towards the rear, between the radiator and the radiator support. Some aftermarket condensers have non-removable mounting brackets; this type of condenser can be removed towards the front of the vehicle, after removing the grille, hood latch, center support brace, and horns, and repositioning the radiator. Other types of aftermarket condensers can be removed toward the rear, similar to an original equipment condenser; the determining factor is the size of the mounting brackets and the size (top to bottom) of the condenser. So, depending on the type of condenser being removed and installed, some of the steps in this procedure can be bypassed.

3 Drain the engine coolant (see Chapter 1).

4 Remove the radiator grille (see Chapter 11).

5 Remove the radiator (see Section 6).

16.6b Location of the lower refrigerant line mounting nut on the condenser

6 Disconnect the upper and lower refrigerant lines on the front of the condenser (see illustrations). Cap the open fittings immediately to keep moisture and contamination out of the system.

7 If you're removing (or are going to install) an aftermarket condenser, remove the horns (see Chapter 12), the hood latch (see Chapter 11) and the center support brace.

8 Remove the condenser mounting bolts and brackets (see illustration).

9 Pull the condenser up to detach its lower mounts, then maneuver it out. Again, the type of condenser will determine in which direction it will be removed.

10 Install the condenser, brackets and bolts, making sure the rubber cushions fit on the mounting points properly.

11 Reconnect the refrigerant lines, using new O-rings on the refrigerant line fittings. Make sure the O-rings are specifically made for the type of refrigerant in your system and lubricate them with refrigerant oil, also designed specifically for your system (R-134a).

12 Reinstall the remaining parts in the reverse order of removal.

13 If the cooling system was drained, refill it (see Chapter 1).

14 Have the system evacuated, charged and leak tested by the shop that discharged it.

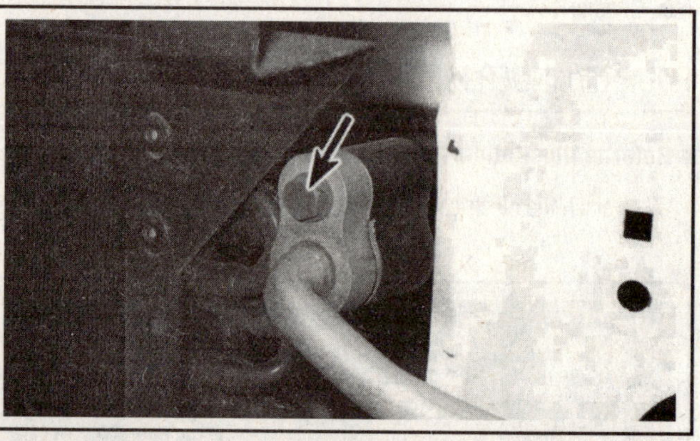

16.6a Location of the upper refrigerant line mounting nut on the condenser

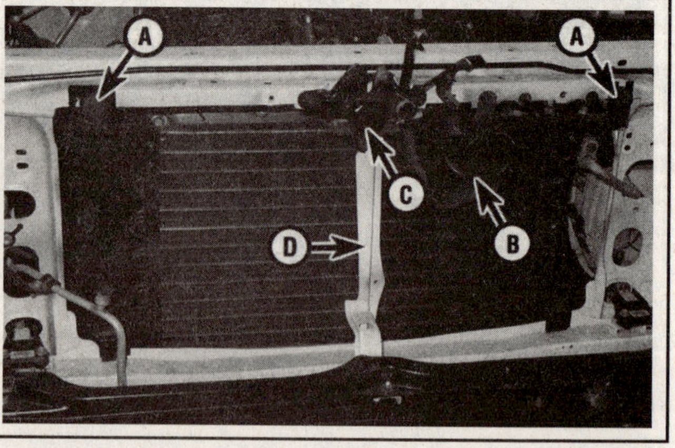

16.8 Condenser mounting details (items B, C and D must be removed if the condenser is being removed toward the front)

A Condenser bracket mounting fasteners
B Horn
C Hood latch
D Center support brace

17 Air conditioning pressure cycling switch - removal and installation

➧ Refer to illustration 17.2

⁂ WARNING:

The air conditioning system is under high pressure. Do not loosen any hose fittings or remove any components until the system has been discharged. Air conditioning refrigerant should be properly discharged into an EPA-approved recovery/recycling unit by a dealer service department or an automotive air conditioning repair facility. Always wear eye protection when disconnecting air conditioning system fittings.

1 Take the vehicle to a dealer service department or automotive air conditioning shop and have the air conditioning system discharged and the refrigerant recovered.

2 Unplug the electrical connector from the pressure cycling switch (see illustration).

3 Unscrew the pressure cycling switch. Be sure to hold the fitting with another wrench to prevent deforming the refrigerant line.

4 Lubricate the switch O-ring with clean refrigerant oil of the correct type.

5 Screw the new switch onto the threads, then tighten it securely.

6 Reconnect the electrical connector.

7 Have the system evacuated, charged and leak tested by the shop that discharged it.

17.2 The pressure cycling switch is located on the discharge (high-pressure) refrigerant line, just inboard of the right-side headlight

18 Oil cooler (V6 models) - removal and installation

⁂ WARNING:

Wait until the engine is completely cool before beginning this procedure.

1 Drain the engine oil and remove the oil filter (see Chapter 1).

2 Drain the engine coolant (see Chapter 1).

3 Remove the engine cover.

4 Disconnect the oil cooler hoses from the oil cooler. Be prepared for spillage, and plug the lines to prevent leakage.

5 Remove the union bolt and separate the oil cooler from the oil filter adapter.

6 Installation is the reverse of removal.

7 Be sure to install a new O-ring and plate washer.

8 Tighten the union bolt to the torque listed in this Chapter's Specifications.

9 Install a new oil filter. Refill the engine with oil and the cooling system with the proper type and concentration of antifreeze (see Chapter 1) and check for coolant leakage and proper gauge function.

Specifications

General

Radiator cap pressure rating	13.5 to 17.8 psi (93 to 122 kPa)
Thermostat rating	176 to 183-degrees F (80 to 84-degrees C)
Refrigerant type	R-134a
Refrigerant capacity	20 to 22 ounces (567 to 624 grams)

Torque specifications	Ft-lbs (unless otherwise indicated)	Nm

➡Note: One foot-pound (ft-lb) of torque is equivalent to 12 inch-pounds (in-lbs) of torque. Torque values below approximately 15 foot-pounds are expressed in inch-pounds, because most foot-pound torque wrenches are not accurate at these smaller values.

Torque specifications	Ft-lbs (unless otherwise indicated)	Nm
Idler pulley bolts (V6 engine)	29	39
Oil cooler union bolt (V6 engine)	50	68
Radiator mounting bolts	156 in-lbs	17.5
Thermostat housing mounting fasteners		
Four-cylinder engine	180 in-lbs	20
V6 engine	80 in-lbs	9
Water inlet assembly mounting bolts (V6 engine)	80 in-lbs	9
Water pump mounting bolts		
Four-cylinder engine		
Longer bolts	180 in-lbs	20
Shorter bolts	80 in-lbs	9
V6 engine		
Longer bolts	17	23
Shorter bolts	80 in-lbs	9

4 FUEL AND EXHAUST SYSTEMS

Section

Reference to other Chapters

Air filter replacement - See Chapter 1
CHECK ENGINE light - See Chapter 6
Exhaust system check - See Chapter 1
Fuel system check - See Chapter 1
Underhood hose check and replacement - See Chapter 1

1 General information

All models are equipped with an electronic Sequential Multiport Fuel Injection (SFI) system. The fuel system consists of the following components:

Air filter, housing and air intake duct between housing and throttle body
Circuit opening relay (engine compartment fuse and relay box)
EFI main relay (engine compartment fuse and relay box)
Electric fuel pump and fuel level sending unit assembly (located in the fuel tank)
Fuel pressure pulsation damper (at the rear end of the fuel rail on four-cylinder models; at the rear end of left fuel rail on V6 models)
Fuel pressure regulator (at the front end of the fuel rail on four-cylinder models; at the rear end of right fuel rail on V6 models)
Fuel pump relay (engine compartment fuse and relay box)
Fuel rail and fuel injector assembly
Fuel tank
Throttle body assembly

SEQUENTIAL MULTIPORT FUEL INJECTION (SFI) SYSTEM

Intake air is drawn through the air filter housing and air intake duct into the throttle body. Inside the throttle body, the throttle valve controls the amount of air passing into the air intake plenum, then into the cylinders. The intake port for each cylinder is equipped with its own injector. The Sequential Multiport Fuel Injection (SFI) system injects fuel into the intake ports in the same sequence as the firing order of the engine. The injectors are controlled by the Powertrain Control Module (PCM). The PCM constantly monitors the operating conditions of the engine - temperature, speed, load, etc. - and delivers the optimal amount of fuel for these conditions. The amount of fuel delivered by each injector is determined by its on time, which the PCM can vary so quickly that two successive injectors can deliver a different amount of fuel. In other words, the PCM responds instantly to any changes in the operating conditions of the engine.

FUEL PUMP AND LINES

An electric fuel pump/fuel gauge sending unit module is located inside the fuel tank. Fuel is pumped from the fuel tank to the fuel rail/injector assembly through a metal line running along the underside of the vehicle. A fuel pressure pulsation damper, which is located at the connection between the fuel supply line and the fuel rail, dampens fuel pump pulsations. A fuel pressure regulator, which is mounted on the fuel rail, routes all excess fuel back to the fuel tank through a separate return line.

The fuel pump speed is controlled by the fuel pump relay and the fuel pump resistor, both of which are controlled by the circuit opening relay, which is controlled by the Powertrain Control Module (PCM). When the engine is cranked, current flows from the ignition switch to the starter relay coil and to the PCM, which also receives a signal from the Crankshaft Position (CKP) sensor. When the PCM receives these two signals (from the ignition switch and from the CKP sensor), the transistor inside the PCM that controls the circuit opening relay coil allows current to flow to the circuit opening relay, which closes and allows current to flow to the fuel pump relay. As long as the transistor inside the PCM that controls the circuit opening relay continues to receive the CKP sensor signal, it continues to supply current to the circuit opening relay, which continues to supply current to the fuel pump relay. The fuel pump speed is determined by the operating condition of the engine. When the engine is being cranked, the transistor inside the PCM that controls current to the fuel pump relay coil is switched off, so the fuel pump relay closes and battery voltage is applied directly to the fuel pump, which operates at high speed. After the engine has started, the transistor inside the PCM that operates the fuel pump relay is on, so the fuel pump relay opens and battery voltage is supplied to the fuel pump through the fuel pump resistor.

The fuel pump circuit can be opened to stop the pump in the event of an accident because the PCM turns off the circuit opening relay if it detects that an airbag has been deployed from the airbag sensor assembly. After the PCM has deactivated the fuel pump circuit, you can reactivate it by turning the ignition key from OFF to ON if you need to restart the engine.

EXHAUST SYSTEM

The exhaust system consists of the exhaust manifold(s), the front exhaust pipe assembly (two pipes bolted together), the catalytic converters, and the rest of the exhaust pipe system, which includes the muffler and the tail pipe. The catalytic converters are emission control devices added to the exhaust system to reduce pollutants. For more information regarding the catalytic converters, refer to Chapter 6.

2 Fuel pressure relief procedure

➧ Refer to illustration 2.2

⁂ WARNING:

Gasoline is extremely flammable, so take extra precautions when you work on any part of the fuel system. Don't smoke or allow open flames or bare light bulbs near the work area, and don't work in a garage where a gas-type appliance (such as a water heater or clothes dryer) is present. Since gasoline is carcinogenic, wear fuel-resistant gloves when there's a possibility of being exposed to fuel, and, if you spill any fuel on your skin, rinse it off immediately with soap and water. Mop up any spills immediately and do not store fuel-soaked rags where they could ignite. The fuel system is under constant pressure, so, if any fuel lines are to be disconnected, the fuel pressure in the system must be relieved first. When you perform any kind of work on the fuel system, wear safety glasses and have a Class B type fire extinguisher on hand.

1 Remove the fuel filler cap to relieve any pressure inside the fuel tank.

2 Remove the circuit opening relay (C/OPN) from the engine compartment fuse and relay box (see illustration).

➡Note: The location of the circuit opening relay on your vehicle might be different from the location shown in the accompanying photograph. To be sure that you have the correct relay, refer to the guide on the underside of the cover for the engine compartment fuse and relay box.

3 Start the engine and let it run until it stops (it might not start at all). When the engine stops, turn the ignition switch to OFF.

4 Always disconnect the cable from the negative battery terminal (see Chapter 5, Section 1) before working on the fuel system.

5 The fuel system pressure is now relieved. When you're finished working on the fuel system, install the relay and connect the negative cable to the battery.

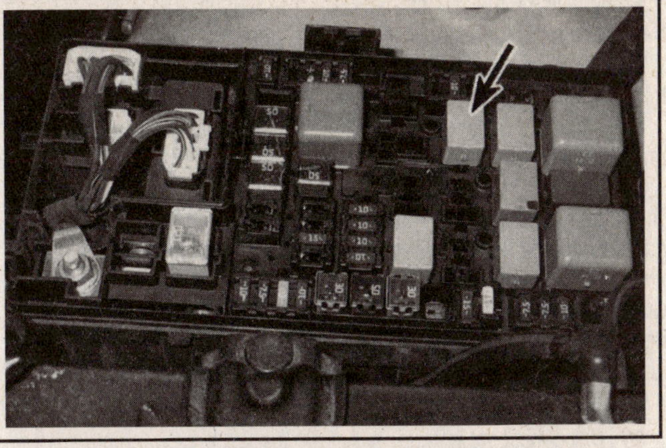

2.2 To relieve fuel pressure, disable the fuel pump by removing the circuit opening relay (be sure to verify the location of the relay by referring to the underside of the fuse/relay box cover)

3 Fuel pump/fuel pressure - check

⁂ WARNING:

Gasoline is extremely flammable, so take extra precautions when you work on any part of the fuel system. See the Warning in Section 2.

GENERAL CHECKS

1 If you suspect insufficient fuel delivery check the following items first:

a) Check the battery and make sure that it's fully charged (see Chapter 5).

b) Inspect all fuel lines. Verify that the problem is not simply a leak in a line.

2 Verify that the fuel pump actually runs. Remove the fuel filler cap, then have an assistant turn the ignition switch to ON while you listen carefully for the sound of the fuel pump operating. You should hear a brief whirring noise (for about two seconds) as the pump comes on and pressurizes the system.

3 If the fuel pump makes no sound, check the fuel pump electrical circuit. Make sure that the electrical connector at the pump is firmly connected and inspect the wiring harness for the pump. To access the fuel pump connector, you'll have to lower the fuel tank (see Section 5). If the fuel pump runs, but a fuel system problem persists, check the fuel pressure.

FUEL PRESSURE CHECK

V6 models

➧ Refer to illustrations 3.4 and 3.7

4 Before proceeding, make sure that you have a fuel pressure gauge capable of reading fuel pressure up to 60 psi. You'll also need a special fitting that will allow you to tee into the quick-connect fitting for the fuel supply hose in the engine compartment and either hose clamps or appropriate quick-connect fittings to connect the gauge to the tee fitting (see illustration).

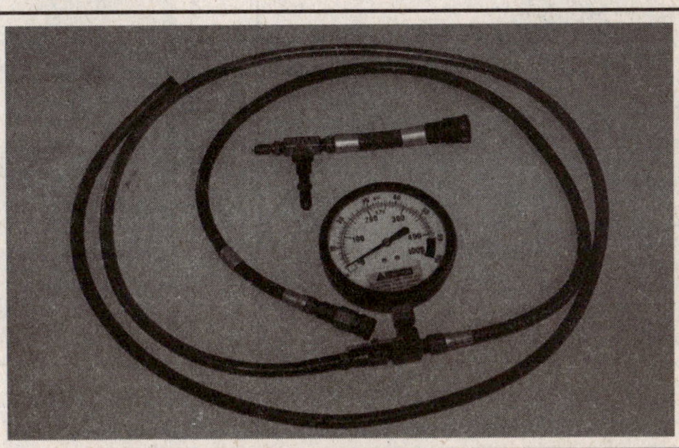

3.4 Here's a typical fuel pressure testing rig that you can use to tee into the fuel supply hose quick-connect fitting in the engine compartment of a V6 model

5 Relieve the fuel system pressure (see Section 2).
6 Locate the fuel supply hose quick-connect fitting in the left side of the engine compartment and disconnect it. If you're unfamiliar with this type of fitting, refer to Section 4.
7 Connect the fuel pressure gauge between the two sides of the quick-connect fitting with the tee fitting (see illustration).
8 Start the engine and let it idle at normal operating temperature. Check the gauge and compare the fuel pressure with the value listed in this Chapter's Specifications. If the pressure is lower than specified, pinch the fuel return line (connected to the fuel pressure regulator) and watch the gauge; if the pressure increases, replace the fuel pressure regulator. If it does not increase, the fuel filter (which is part of the fuel pump module) might be clogged, or the fuel pump might be faulty.
9 If the pressure is higher than specified:

a) *Disconnect the vacuum hose from the fuel pressure regulator and check for the presence of vacuum at the hose. If vacuum is not present, check the hose for a leak or an obstruction.*
b) *If a vacuum signal is present, connect a hand-held vacuum pump to the fitting on the fuel pressure regulator and apply vacuum; if the pressure doesn't decrease when vacuum is applied, the pressure regulator is faulty.*
c) *Turn the engine off and depressurize the fuel system (see Section 2). Detach fuel return line from the fuel pressure regulator and blow through the return line to check it for an obstruction. If there is no obstruction, replace the fuel pressure regulator.*

10 Reconnect the return line, restart the engine, then shut it off. Wait five minutes, then verify that the fuel pressure doesn't drop below the minimum residual pressure listed in this Chapter's Specifications.
11 Relieve the system fuel pressure (see Section 2).
12 Remove the fuel pressure gauge.
13 Reconnect the fuel supply hose quick-connect fitting.
14 Install the quick-connect fitting cover.
15 Reconnect the battery and check the quick-connect fitting for leaks.

3.7 Connect the fuel pressure gauge between the fuel rail and the fuel feed line (V6 models)

Four-cylinder models

➧ Refer to illustrations 3.18a, 3.18b and 3.19

16 Before proceeding, make sure that you have a fuel pressure gauge capable of measuring fuel pressure up to 60 psi. You'll also need a special fitting that will allow you to tee into the quick-connect fitting for the fuel supply hose in the engine compartment and either hose clamps or appropriate quick-connect fittings to connect the gauge to the tee fitting (see illustration 3.4).
17 Relieve the fuel system pressure (see Section 2).
18 Locate the fuel supply hose quick-connect fitting on the left side of the engine compartment, remove the cover and disconnect the fitting (see illustrations).
19 Connect the fuel pressure gauge between the two sides of the quick-connect fitting with the tee fitting (see illustration).
20 Start the engine and let it idle at normal operating temperature. Check the gauge and compare the fuel pressure with the value listed in this Chapter's Specifications. If the pressure is lower than specified, pinch the fuel return line (connected to the fuel pressure regulator) and watch the gauge; if the pressure increases, replace the fuel pressure regulator. If it does not increase, the fuel filter (which is part of the fuel pump module) might be clogged, or the fuel pump might be faulty.

3.18a Remove the cover from this quick-connect fitting on the left side of the engine compartment . . .

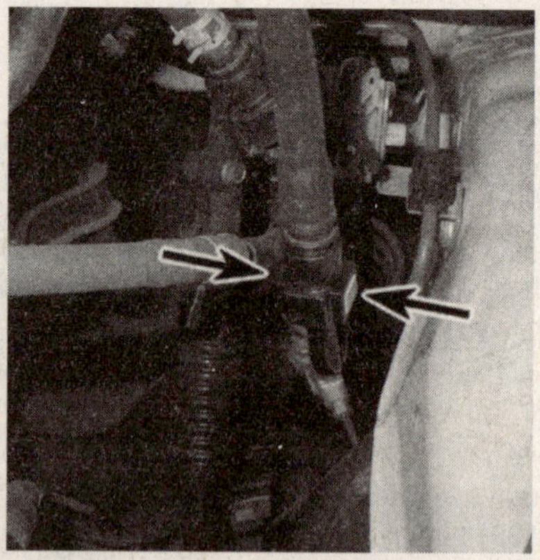

3.18b . . . then depress the two release buttons and disconnect the fitting

3.19 Connect the fuel pressure gauge between the fuel rail and the fuel feed line (four-cylinder models)

21 If the pressure is higher than specified:

a) *Disconnect the vacuum hose from the fuel pressure regulator and check for the presence of vacuum at the hose. If vacuum is not present, check the hose for a leak or an obstruction.*
b) *If a vacuum signal is present, connect a hand-held vacuum pump to the fitting on the fuel pressure regulator and apply vacuum; if the pressure doesn't decrease when vacuum is applied, the pressure regulator is faulty.*
c) *Turn the engine off and depressurize the fuel system (see Section 2). Detach fuel return line from the fuel pressure regulator and blow through the return line to check it for an obstruction. If there is no obstruction, replace the fuel pressure regulator.*

22 Reconnect the return line, restart the engine, then shut it off. Wait five minutes, then verify that the fuel pressure doesn't drop below the minimum residual pressure listed in this Chapter's Specifications.

23 Relieve the system fuel pressure (see Section 2).

24 Remove the fuel pressure gauge.

25 Reconnect the fuel supply hose quick-connect fitting.

26 Install the quick-connect fitting cover.

27 Reconnect the battery and check the quick-connect fitting for leaks.

4 Fuel lines and fittings - general information

WARNING:

Gasoline is extremely flammable, so take extra precautions when you work on any part of the fuel system. See the Warning in Section 2.

1 Always relieve the fuel pressure before servicing fuel lines or fittings (see Section 2).

2 The lines are secured to the underbody with clips. Whenever you raise the vehicle to service anything underneath the vehicle, always inspect the condition of the fuel line clips, and inspect the fuel supply and return lines themselves for dents, kinks and leaks.

3 If evidence of dirt is found in the system or fuel filter during disassembly, disconnect the fuel supply line and blow it out with compressed air. Inspect the inlet fuel strainer on the fuel pump (see Section 8) for damage and deterioration.

STEEL TUBING

4 If replacement of a fuel line or emission line is called for, use welded steel tubing meeting the manufacturer's specifications or its equivalent. Don't use copper or aluminum tubing to replace steel tubing. These materials cannot withstand normal vehicle vibration.

5 Because fuel lines used on fuel-injected vehicles are under high pressure, they require special consideration.

6 Some fuel lines have threaded fittings with O-rings. Any time the fittings are loosened to service or replace components:

a) *Hold the stationary fitting of the fuel line with one wrench while loosening the tube nut with another. This will prevent the line from twisting.*
b) *Check all O-rings for cuts, cracks and deterioration. Replace any that appear hardened, worn or damaged.*
c) *If the lines are replaced, always use original equipment parts, or parts that meet the original equipment standards specified in this Section.*

FLEXIBLE HOSE

CAUTION:

Use only original equipment replacement hoses or their equivalent. Others may fail from the high pressures of this system.

7 Don't route fuel hose within four inches of any part of the exhaust system or within ten inches of the catalytic converter. Metal lines and rubber hoses must never be allowed to chafe against the frame. A minimum of 1/4-inch clearance must be maintained around a hose to prevent contact with the frame.

8 Some models may be equipped with nylon fuel line and quick-connect fittings at the fuel filter and/or fuel pump. The quick-connect fittings cannot be serviced separately. Do not attempt to service these types of fuel lines in the event the retainer tabs or the line becomes damaged. Replace the entire fuel line as an assembly.

REPLACEMENT

CAUTION:

If a fuel line or fuel hose is damaged, replace it with original equipment replacement parts or equivalent. Do not substitute line or hose of inferior quality; it might not be suitable for, and it might fail from, the operating pressure of this system.

9 Relieve the fuel pressure.

10 Remove all clamps and/or clips attaching the lines to the vehicle body. Pay close attention to all clips; they not only secure the fuel lines and hoses, they also route them correctly. The hoses and lines must be reattached to their respective clips when reassembled.

11 On fuel hoses that are clamped onto the metal fuel lines, loosen the clamp and pull the hose off the fitting. If a hose sticks to the metal

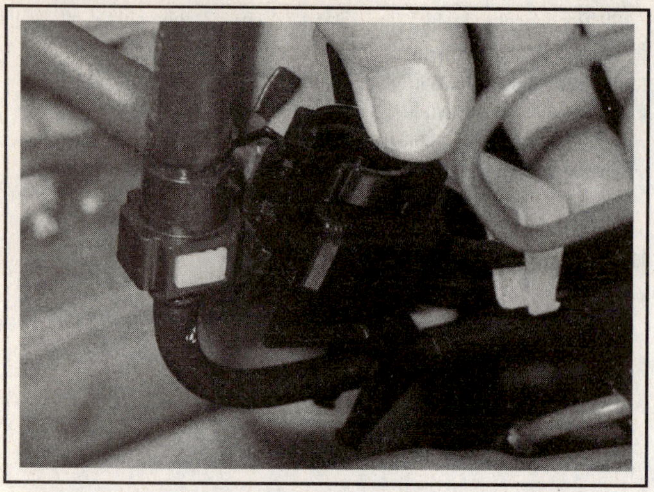

4.12 To remove the protective clamp from a quick-connect fitting, push it sideways, then pull it off

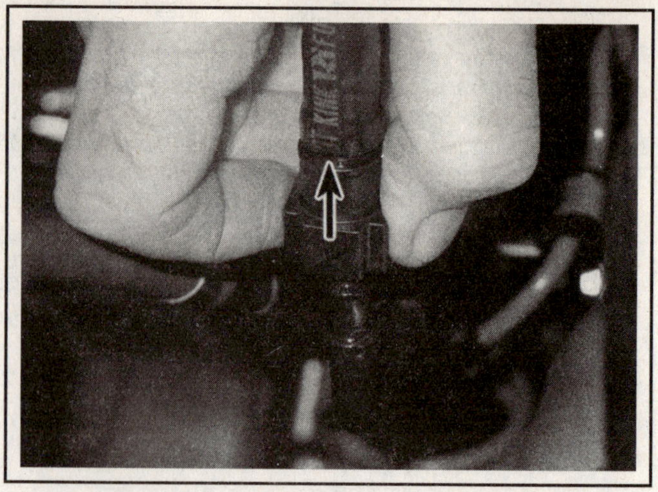

4.13 To disconnect a quick-connect fitting, depress the release buttons and pull the two halves of the connector apart

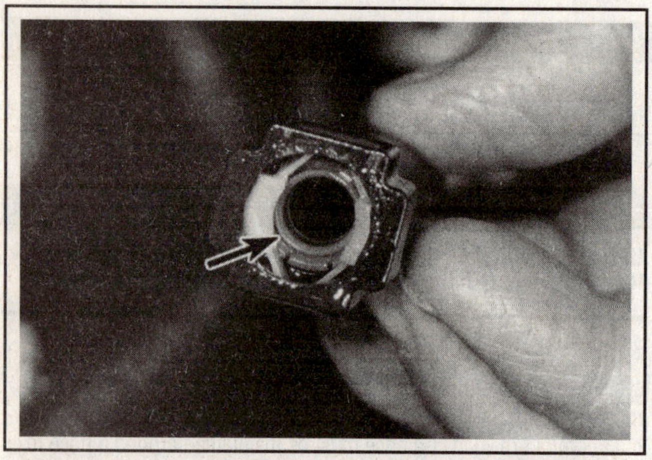

4.14 Whenever you disconnect a quick-connect fitting, inspect the condition of the O-ring inside the fitting. If it's cracked, torn or deteriorated, replace the fitting (O-rings are not available separately)

line, twisting it back and forth will help to loosen it. Before installing old clamps, make sure that they're still tight. Replace any worn clamps.

Quick-connect fittings

➧ Refer to illustrations 4.12, 4.13 and 4.14

➡Note: You'll find quick-connect fittings in the engine compartment and under the vehicle.

12 Remove the protective clamp from the connector (see illustration).

13 Disconnect the quick-connect fitting (see illustration).

14 Inspect the condition of the O-ring inside the fitting (see illustration). If the O-ring is damaged, replace the fitting (O-rings are not available separately). If the fitting has been leaking, replace it even if the O-ring looks okay.

15 To reconnect the fitting, push the male side into the female side until you hear/feel a click. To verify that the fitting is secure, try to pull the two halves of the fitting apart.

16 To install the protective clamp on the quick-connect fitting, push it onto the hose, then push it sideways until it clicks into place.

5 Fuel tank - removal and installation

⁂ WARNING:

Gasoline is extremely flammable, so take extra precautions when you work on any part of the fuel system. See the Warning in Section 2.

REMOVAL

➧ Refer to illustrations 5.5, 5.7a, 5.7b, 5.9a, 5.9b, 5.10a, 5.10b and 5.10c

1 Relieve the fuel system pressure (see Section 2) and remove the fuel tank cap.

2 Disconnect the cable from the negative terminal of the battery (see Chapter 5, Section 1).

3 Raise the vehicle and place it securely on jackstands.

4 Remove the fuel tank rock guard, if equipped.

5 Loosen the lower hose clamp for the fuel tank filler neck hose and disconnect the filler neck hose from the fuel tank (see illustration). If you're unable to disconnect the filler neck hose from the tank at this time, you can disconnect it after you lower the tank slightly in Step 9. Also disconnect the quick-connect fitting for the fuel tank breather line.

6 Unless the fuel level in the tank is very low, siphon any fuel from the tank through the filler neck pipe into an approved container. It's easier to remove the fuel tank when it's nearly empty. But there is no fuel tank drain plug, so if there's still a lot of fuel in the tank, siphon or

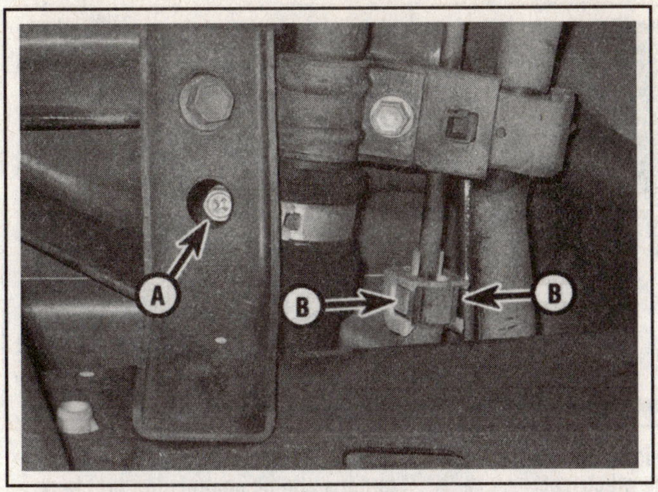

5.5 Loosen the screw (A) for the filler neck hose clamp, slide back the clamp and disconnect the filler neck hose from the fuel tank. Depress the two buttons (B) and disconnect the quick-connect fitting for the fuel tank vent hose

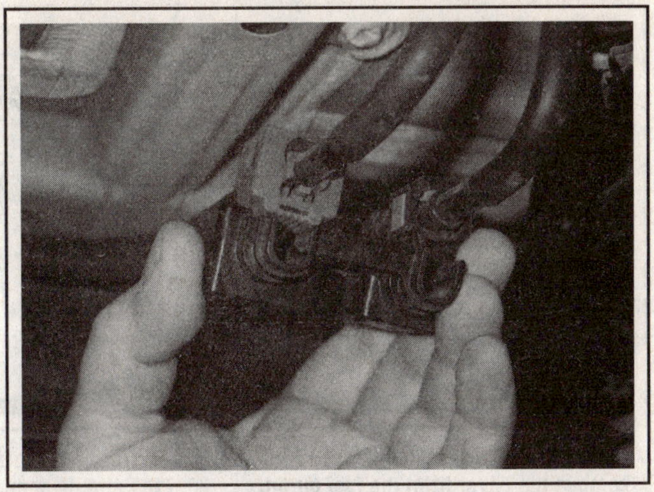

5.7a Remove the clamp that protects both fuel line quick-connect fittings . . .

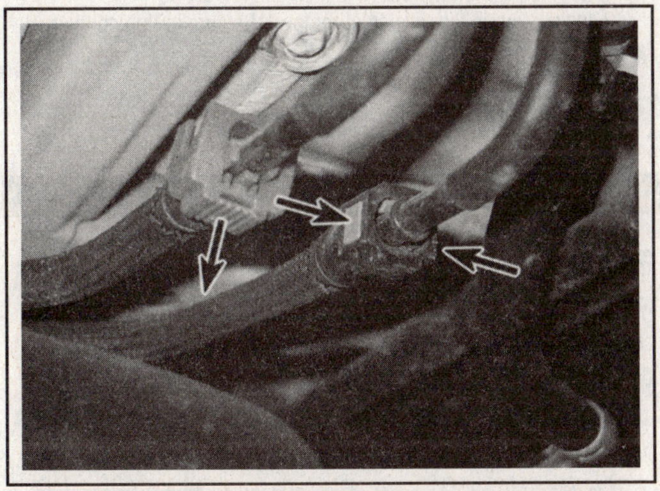

5.7b . . . pull down the orange clamp on the fuel supply line fitting, then depress the two buttons on the side of each fitting and disconnect both fittings

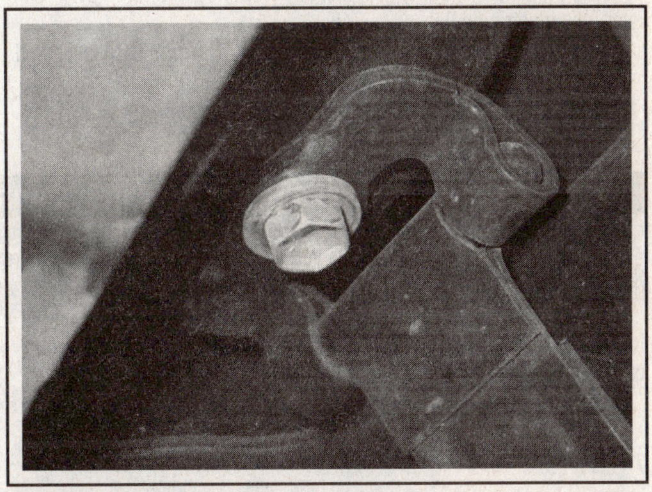

5.9a Remove the two fuel tank strap bolts . . .

hand-pump the remaining fuel from the tank through the tank filler neck pipe.

WARNING:

Don't start the siphoning action by mouth! Use a siphoning kit (available at most auto parts stores).

7 Disconnect the fuel supply and fuel return line quick-connect fittings at the front of the fuel tank (see illustrations).

8 Support the fuel tank.

9 Locate and remove the two fuel tank strap bolts on the left side of the tank, then remove the strap pivot pins on the right side of the tank and remove both fuel tank straps (see illustrations).

10 Carefully lower the tank slightly, remove the fuel pump module cover, disconnect the electrical connector from the fuel pump/fuel gauge

5.9b . . . remove the retaining clips and fuel tank strap pins (other pin not shown), then remove the fuel tank straps

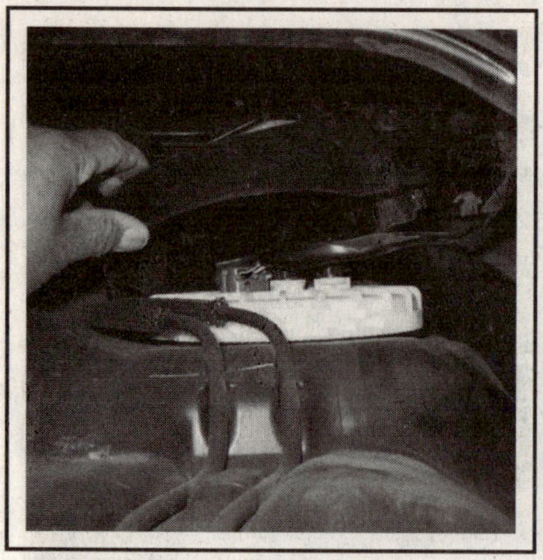

5.10a Lower the tank slightly, remove the cover from the fuel pump/fuel gauge sending unit module . . .

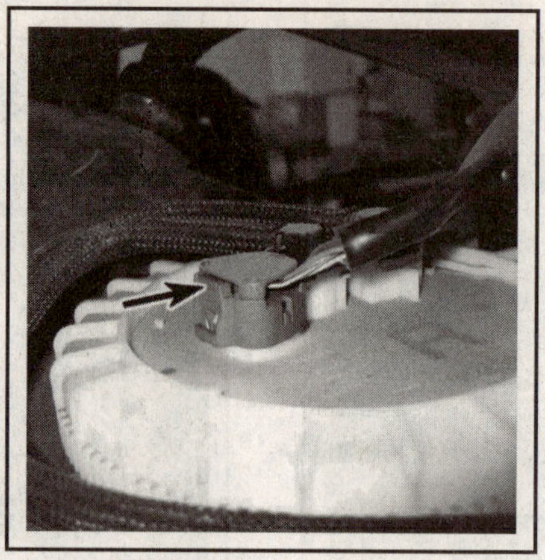

5.10b . . . depress this release tab and disconnect the electrical connector from the fuel pump/fuel gauge sending unit module . . .

5.10c . . . then depress these two buttons and disconnect the fuel tank vent hose quick-connect fitting from the EVAP canister, which is located behind and above the fuel tank

sending unit and disconnect the fuel tank vent hose from the EVAP canister (see illustrations).

11 Make sure that there are no other hoses or wiring harnesses still connected, then carefully lower the fuel tank to the floor.

12 Installation is the reverse of removal. Be sure to tighten the fuel tank strap bolts securely.

6 Fuel tank cleaning and repair - general information

1 The fuel tanks installed in the vehicles covered by this manual are not repairable. If the fuel tank becomes damaged, it must be replaced.

2 Cleaning the fuel tank (due to fuel contamination) should be performed by a professional with the proper training to carry out this critical and potentially dangerous work. Even after cleaning and flushing, explosive fumes may remain inside the fuel tank.

3 If the fuel tank is removed from the vehicle, it should not be placed in an area where sparks or open flames could ignite the fumes coming out of the tank. Be especially careful inside a garage where a gas-type appliance is located.

7 Fuel pump/fuel gauge sending unit module - removal and installation

➧ Refer to illustrations 7.3a, 7.3b, 7.5, 7.6, 7.7, 7.10a and 7.10b

1 Relieve the system fuel pressure (see Section 2).

2 Remove the fuel tank (see Section 5).

3 Disconnect the fuel supply and return line quick-connect fittings from the fuel pump/fuel gauge sending unit module (see illustrations).

4 Note how the fuel and supply lines are routed across the top of the fuel tank, then detach them from the tank.

5 Using a suitable tool (available from automotive special tool suppliers and auto parts stores) unscrew the retainer ring (see illustration).

⁂ CAUTION:

As you unscrew the retainer ring, push each lock tab out (toward the circumference of the retainer) as it approaches each lug so that it clears the lug.

6 Carefully withdraw the fuel pump/fuel gauge sending unit assembly from the fuel tank (see illustration). Make sure you don't damage the

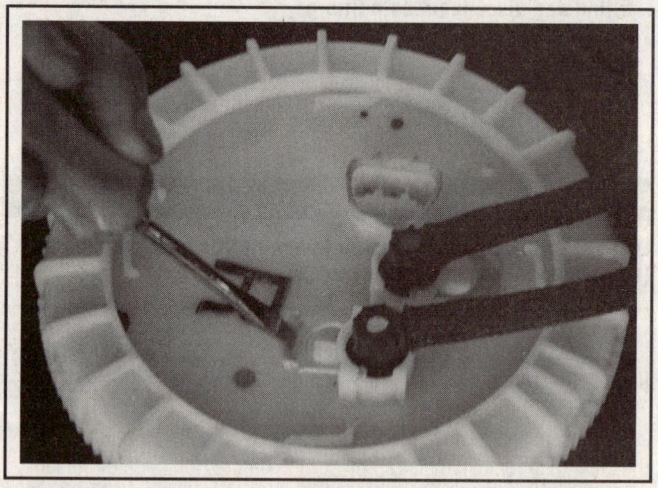

7.3a To disconnect the fuel supply and return line fittings from the fuel pump/fuel gauge sending unit module, slide back the yellow locking retainers . . .

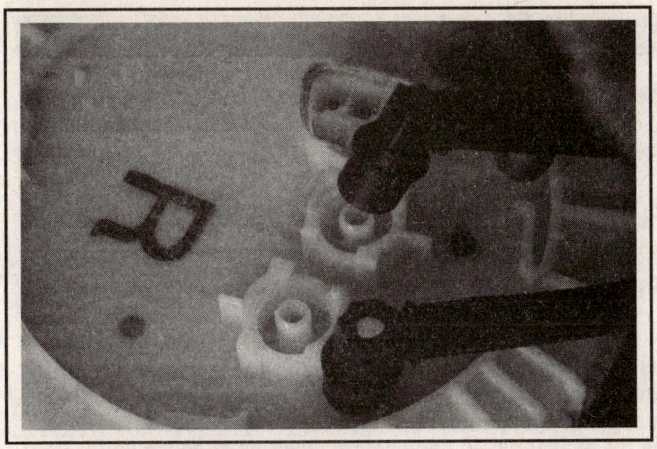

7.3b . . . and pull up on the fittings

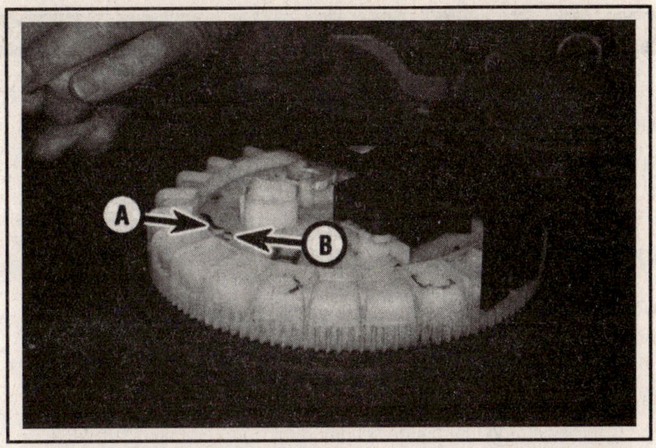

7.5 Carefully unscrew the fuel pump/fuel gauge sending unit retainer ring. As you turn the ring, bend each locking tab (A) to clear the locking lugs (B)

fuel pump filter or bend the sending unit float arm.

7 Remove and inspect the fuel pump/fuel gauge sending unit O-ring (see illustration). If the O-ring is cracked, torn or deteriorated, replace it (it's a good idea to replace this O-ring anytime that you remove the fuel pump/fuel gauge sending unit).

8 If you want to replace either the fuel pump or the fuel gauge sending unit, refer to Section 8.

9 If you have been experiencing fuel delivery problems, disassemble the pump assembly while the pump is removed and inspect the pump inlet strainer (see Section 8). Make sure that it's not clogged or damaged. If the strainer is dirty, try washing it in clean solvent. If it's still clogged, replace the pump assembly (the inlet strainer was not available separately).

10 When installing the fuel pump/fuel gauge sending unit, make sure that the locator lug on the edge of the pump mounting flange is aligned with the notch in the pump mounting ring (see illustration). When tightening the retainer ring, screw it down a couple of turns, then turn it so that the alignment mark on the ring is positioned between the two arrows on the fuel tank (see illustration). Installation is otherwise the reverse of removal.

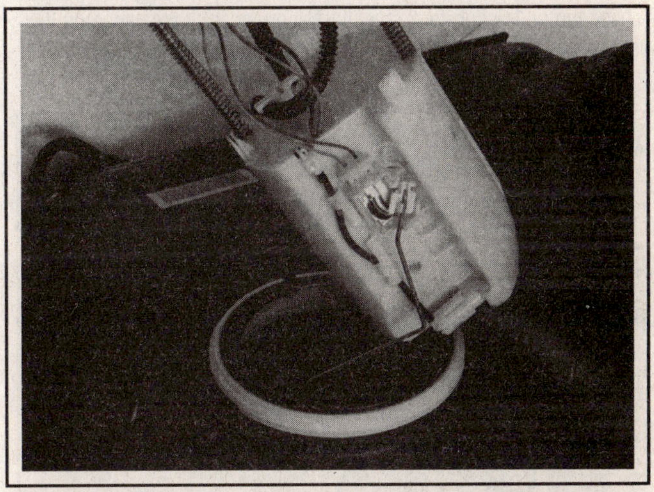

7.6 Carefully pull the fuel pump/fuel gauge sending unit module out of the fuel tank. Angle the pump module as necessary to keep the sending unit float arm from bending

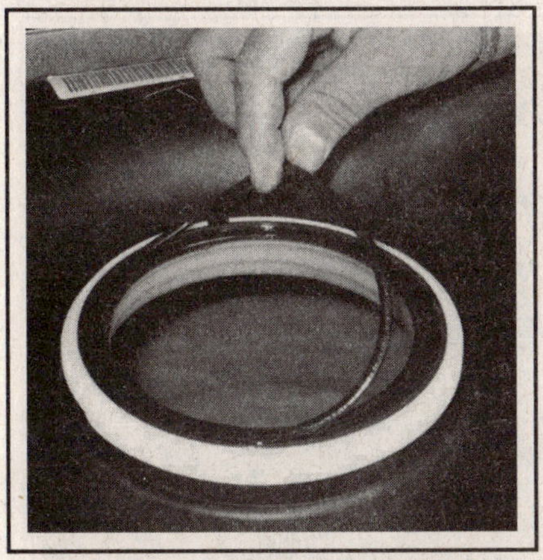

7.7 Remove and inspect the large O-ring that seals the connection between the fuel pump/fuel gauge sending unit module and the fuel tank. If it's cracked, torn or deteriorated, replace it

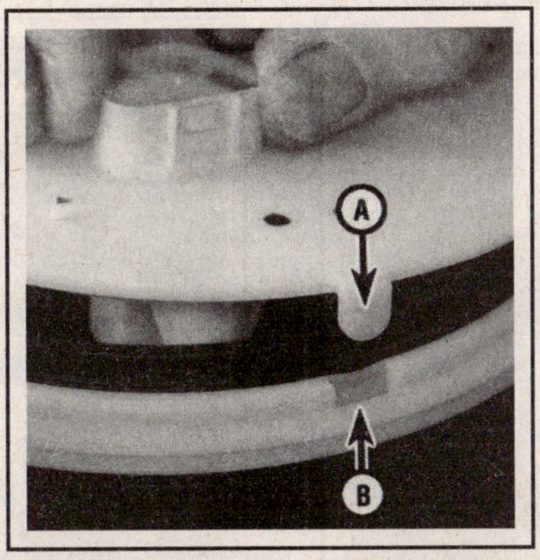

7.10a When installing the fuel pump/ fuel gauge sending unit, make sure that the lug (A) on the pump flange is aligned with the notch (B) in the pump mounting ring

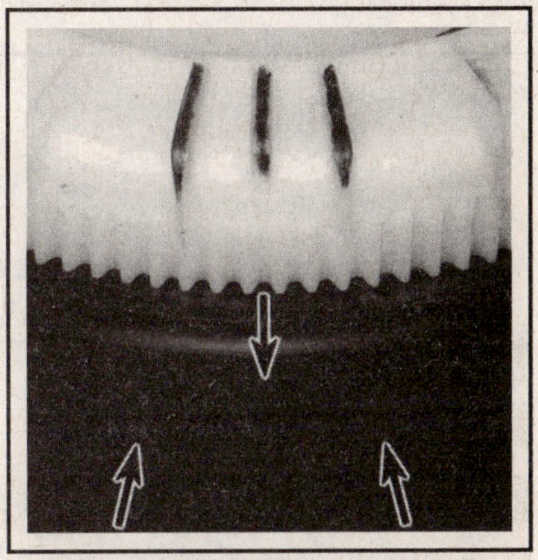

7.10b When tightening the retainer ring, screw it down a couple of turns, then turn it so that this alignment mark on the ring is positioned between the two arrows on the fuel tank

8 Fuel pump/fuel gauge sending unit module - component replacement

1 Remove the fuel tank (see Section 5) and remove the fuel pump/fuel gauge sending unit from the tank (see Section 7).

FUEL GAUGE SENDING UNIT

➧ Refer to illustrations 8.2, 8.3 and 8.4

➡Note: Check on the availability of parts before disassembling the unit.

2 Disconnect the fuel gauge sending unit electrical connector (see illustration).

3 To detach the fuel gauge sending unit from the pump sub-tank, release the lock tab (see illustration) and the sending unit down (toward the inlet strainer).

4 To install the fuel gauge sending unit on the sub-tank, align the T-shaped recess in the underside of the fuel gauge sending unit with the T-shape lug on the sub-tank (see illustration), then fit the gauge on the sub-tank and push down (toward the lower end of the sub-tank) until the lock tab snaps into place.

5 Install the fuel pump/fuel gauge sending unit into the tank (see Section 7) and install the fuel tank (see Section 5).

FUEL PUMP MODULE

➧ Refer to illustration 8.7

➡Note: The fuel pump module consists of the upper mounting flange, the sub-tank (submersible tank that houses all the following components), the fuel filter housing, the fuel pump, the fuel pump inlet strainer and the fuel pressure regulator. Check on the availability of parts before disassembling the unit.

6 Disconnect the upper end of the fuel pump harness from the underside of the fuel pump module mounting flange.

7 Using three screwdrivers, disengage the three claw fittings that secure the sub-tank to the lower end of the fuel pump module mounting flange assembly, then remove the fuel filter assembly from the fuel sub-tank (see illustration).

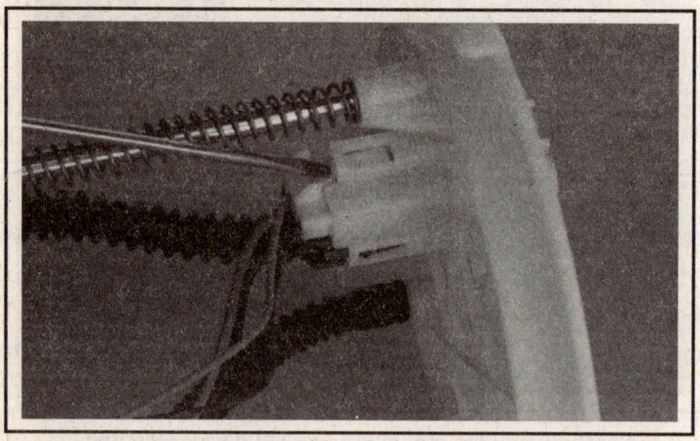

8.2 To disconnect the fuel gauge sending unit electrical connector from its terminal on the underside of the pump module mounting flange, depress this release tab with a small screwdriver and pull out the connector

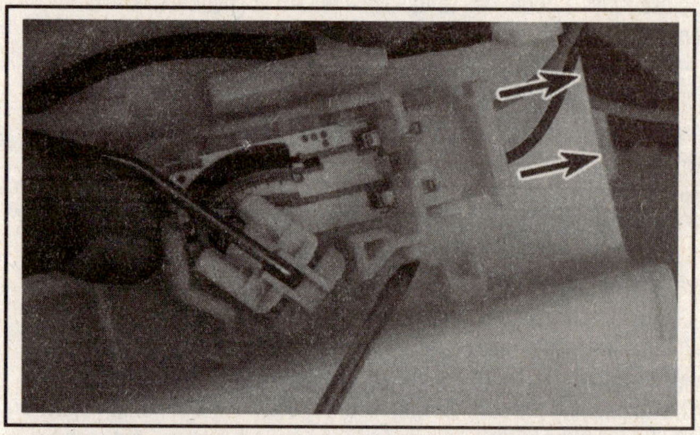

8.3 To detach the fuel gauge sending unit from the sub-tank, release this lock tab with a small screwdriver and slide the sending unit toward the upper end of the sub-tank

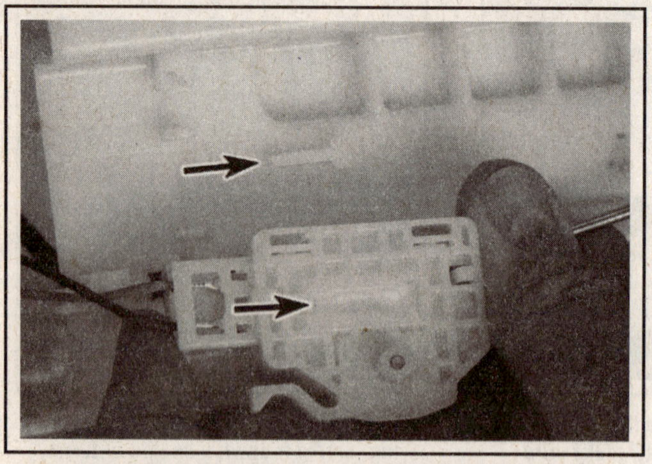

8.4 To install the fuel gauge sending unit on the sub-tank, align the T-shaped recess in the mounting surface of the sending unit with the T-shaped lug on the sub-tank

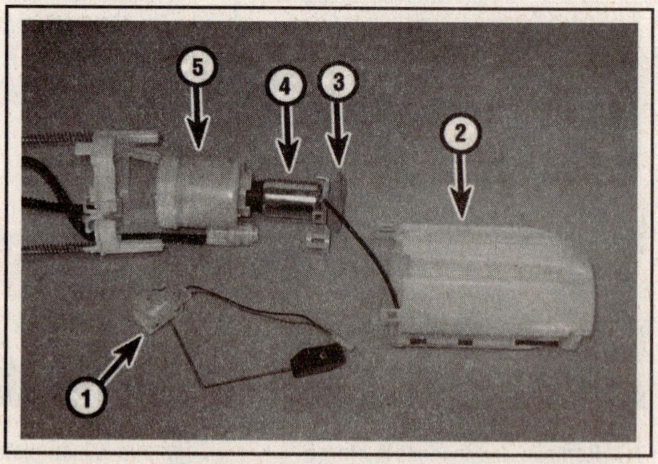

8.7 Fuel pump/fuel gauge sending unit module components:

1 *Fuel gauge sending unit*
2 *Sub-tank (submersible tank)*
3 *Fuel pump inlet strainer*
4 *Fuel pump*
5 *Fuel filter housing*

8 Using screwdrivers, disengage the two claw fittings that secure the fuel filter housing/fuel pump/fuel pressure regulator assembly to the wall of the sub-tank, pull the filter/pump/pressure regulator out of the sub-tank and disconnect the electrical connector from the fuel pump.

9 Using a screwdriver, disengage the three claw fittings that secure the inlet strainer to the fuel filter housing and the two claw fittings that secure the strainer to the fuel pressure regulator, then pull the inlet strainer off the lower end of the fuel pump.

10 Pull the fuel pump out of the fuel filter housing, pull the harness through the upper end of the filter housing, then disconnect the harness connector from the upper end of the fuel pump.

11 To remove the fuel pressure regulator from the fuel filter housing, simply pull it out. Inspect the condition of the two pressure regulator O-rings.

12 Reassembly is the reverse of disassembly.

9 Air filter housing - removal and installation

V6 MODELS

Air intake duct

➧ **Refer to illustration 9.1**

1 Disengage the ground wire from the clip on the air intake duct (see illustration).

2 Loosen the hose clamp screws (see illustration 9.1) and disconnect the air intake duct from the fresh air inlet and from the air filter housing.

3 Installation is the reverse of removal.

9.1 On V6 models, disengage the ground wire from this clip (A) on the air intake duct, then loosen these two hose clamp screws (B) and disconnect the duct from the fresh air inlet and the air filter housing

Air filter housing

➧ **Refer to illustrations 9.6, 9.7, 9.8, 9.9a, 9.9b and 9.10**

4 Remove the engine cover (see Chapter 2B).

5 Remove the air intake duct (see Steps 1 and 2).

6 Disconnect the electrical connector from the Mass Air Flow/Intake Air Temperature (MAF/IAT) sensor (see illustration 9.2a in Chapter 6). Then detach the MAF/IAT sensor harness clip from the air filter housing (see illustration).

7 Loosen the spring-type clamp and disconnect the Positive Crankcase Ventilation (PCV) fresh air inlet hose from the air filter housing (see illustration).

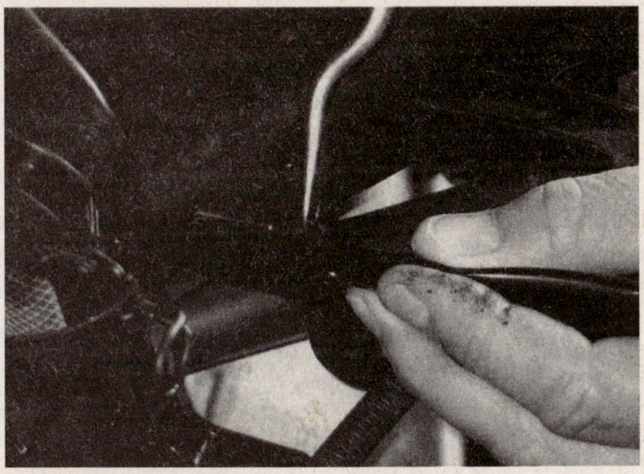

9.6 After disconnecting the MAF/IAT sensor electrical connector, detach the sensor harness clip from the air filter housing with a trim removal tool or a screwdriver (V6 models)

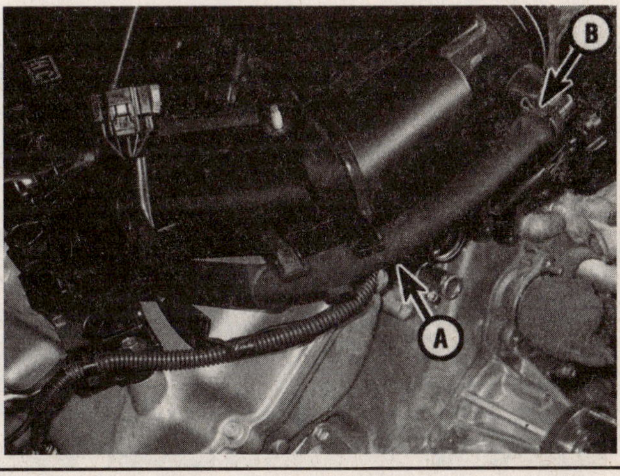

9.7 To disconnect the PCV fresh air inlet hose (A) from the air filter housing, loosen the spring-type hose clamp (B) and disconnect the hose from the pipe on the filter housing (V6 models)

9.8 Loosen the clamp securing the air filter housing to the throttle body

9.9a Remove these two bolts securing the housing to the engine . . .

9.9b . . . then lift up the air filter housing and disconnect the vacuum hose from the backside of the housing (V6 models)

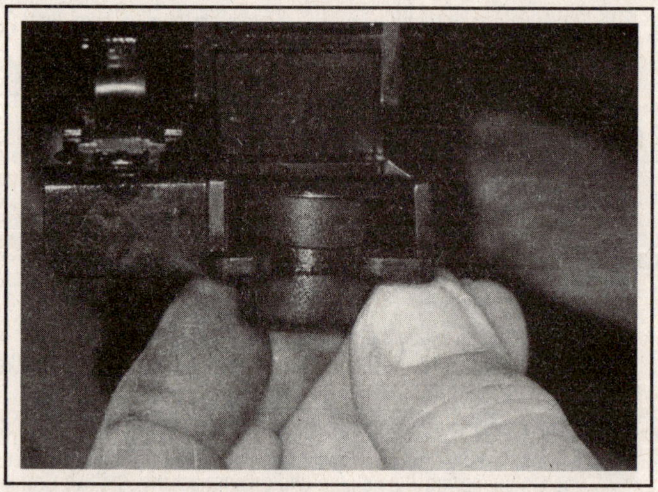

9.10 Inspect the condition of the air filter housing grommets. If either grommet is cracked, torn or deteriorated, replace it (V6 models)

9.12 To remove the air intake duct on four-cylinder models, loosen these two hose clamp screws, then work the duct off the air filter housing and the intake air connector (four-cylinder models)

8 Loosen the screw on the hose clamp that secures the air filter housing to the throttle body (see illustration).

9 Remove the two air filter housing mounting bolts (see illustration), pull the air filter housing forward, disconnect the vacuum hose from the backside of the housing (see illustration) and remove the air filter housing.

10 Inspect the two air filter housing insulator grommets (see illustration). If a grommet is cracked, torn or deteriorated, replace it.

11 Installation is the reverse of removal.

FOUR-CYLINDER MODELS

Air intake duct

➧ **Refer to illustration 9.12**

12 Loosen the hose clamp screws at the air filter housing and at the intake air connector (see illustration), then carefully work the air intake duct off the air filter housing and the intake air connector and remove it.

9.16 Depress the release tab (A) and disconnect the electrical connector, then detach the MAF/IAT sensor harness clip (B) from the filter housing (four-cylinder models)

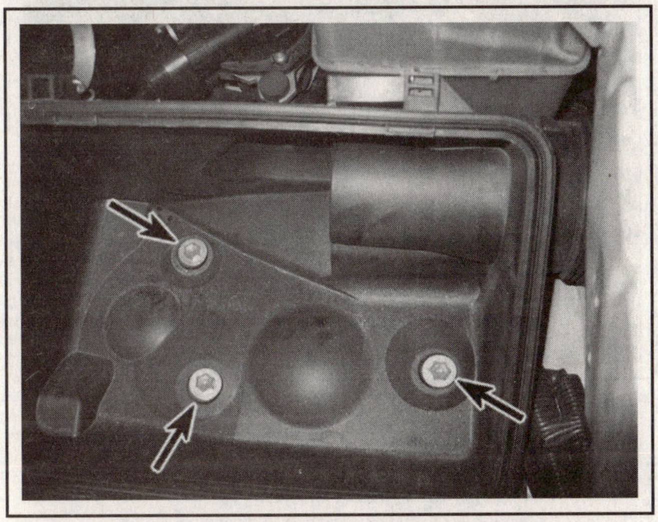

9.19 To detach the air filter housing from the vehicle, remove these three bolts (four-cylinder models)

13 Inspect the condition of the air intake duct. If it's cracked, torn or otherwise deteriorated, replace it.
14 Installation is the reverse of removal.

Air filter housing

➧ **Refer to illustrations 9.16, 9.18 and 9.19**

15 Remove the air intake duct (see Step 12).
16 Disconnect the electrical connector from the Mass Air Flow/Intake Air Temperature (MAF/IAT) sensor, then detach the MAF/IAT harness clip from the filter housing (see illustration) and set the harness aside.
17 Remove the air filter element (see Chapter 1).

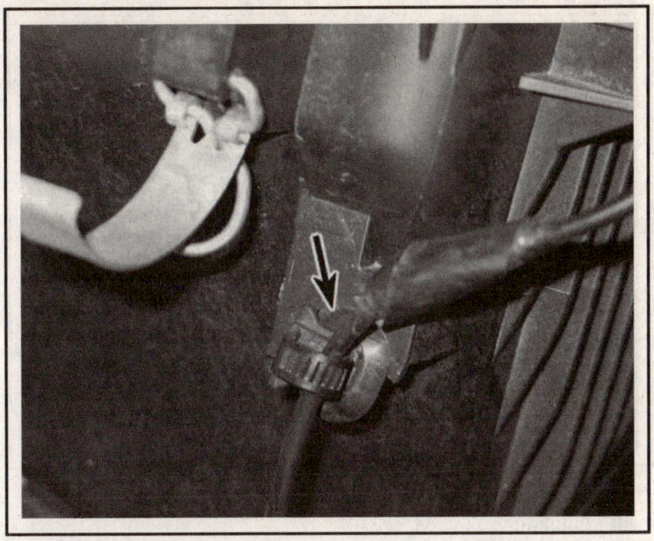

9.18 Detach this ground wire clip from the air filter housing (four-cylinder models)

9.22 To disconnect the electrical connector from the air pressure sensor, depress this release tab (four-cylinder models)

18 Detach the ground wire harness clip from the air filter housing (see illustration).
19 Remove the three air filter housing mounting bolts (see illustration) and remove the air filter housing.
20 Installation is the reverse of removal.

Intake air connector

➧ **Refer to illustrations 9.22, 9.23a, 9.23b, 9.24, 9.25a, 9.25b, 9.26a and 9.26b**

21 Remove the air intake duct (see Step 12).
22 Disconnect the electrical connector from the air injection system air pressure sensor (see illustration).

9.23a Disconnect the vacuum hose from the intake air connector . . .

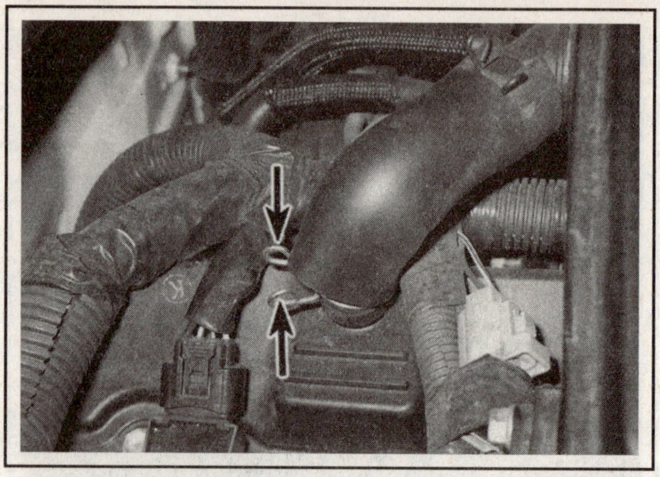

9.23b . . . then squeeze the clamp and disconnect the PCV fresh air inlet hose from the valve cover (four-cylinder models)

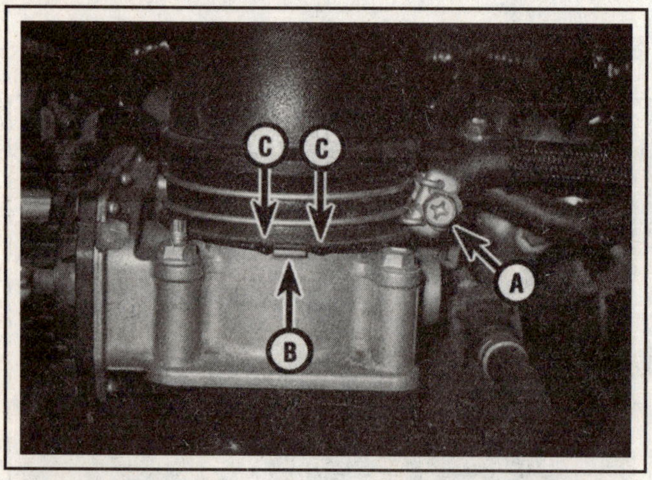

9.24 Loosen the hose clamp screw (A) that secures the short elbow-shaped intake duct to the throttle body. When installing the intake air connector, make sure that the lug (B) on the throttle body is centered between the two lugs (C) on the air duct (four-cylinder models)

9.25a Carefully detach the clip (A) for the electrical harness between the air injection control valve and the air pressure sensor, then detach the control valve-to-pressure sensor vacuum hose (B) (four-cylinder models)

9.25b Detach the MAF/IAT sensor harness clip from the intake air connector (four-cylinder models)

23 Disconnect the vacuum hose (see illustration) and the PCV fresh air inlet hose (see illustration).

24 Loosen the hose clamp that secures the short elbow-shaped intake duct to the throttle body (see illustration).

25 Detach the electrical harness and the vacuum hose for the air pressure sensor (see illustration) and detach the MAF/IAT sensor harness from the intake air connector (see illustration).

26 Remove the three intake air connector mounting bolts (see illustrations) and remove the intake air connector.

27 Installation is the reverse of removal.

9.26a To detach the intake air connector, remove these three mounting bolts and lift off the connector . . .

9.26b . . . then turn the intake air connector upside down and disconnect the vacuum hose from the air pressure sensor (four-cylinder models)

10 Sequential Multiport Fuel Injection (SFI) system - general information

The Sequential Multiport Fuel Injection (SFI) system consists of three sub-systems: air intake, electronic control and fuel delivery. The SFI system uses a Powertrain Control Module (PCM) and various information sensors to calculate the correct air/fuel ratio under all operating conditions. The SFI system and the emission control systems are closely linked. For more information on the emission control systems, refer to Chapter 6.

AIR INTAKE SYSTEM

The air intake system consists of the air filter housing, the air intake duct, the throttle body and the intake manifold. V6 models are equipped with a two-piece (upper and lower) intake manifold. For removal and installation procedures for the intake manifold, see Chapter 2A (four-cylinder models) or Chapter 2B (V6 models).

The throttle body is a single-barrel, side-draft design. The throttle body is heated by engine coolant to prevent icing in cold weather. The throttle body is fully electronic; there is no accelerator cable. The accelerator pedal position sensor is located at, and is an integral component of, the accelerator pedal. For more information on the accelerator pedal position sensor, refer to Chapter 6.

ELECTRONIC CONTROL SYSTEM

For information on the electronic control system, the information sensors, the Powertrain Control Module (PCM) and the output actuators, refer to Chapter 6.

FUEL DELIVERY SYSTEM

The fuel delivery system consists of the fuel pump module, the fuel supply line connecting the fuel tank to the fuel rail, the fuel pulsation damper, the fuel rail and injectors, the fuel pressure regulator and the fuel return line between the fuel rail and the fuel tank.

The electric fuel pump module is located inside the fuel tank. Fuel is drawn through an inlet strainer into the pump, from which it flows through a fuel supply line up to the fuel rail, from which it's sprayed by the injectors into the intake ports. The fuel pulsation damper, which is located at the connection between the fuel supply line and the fuel rail, dampens the pressure pulses from the fuel pump. The vacuum-operated fuel pressure regulator, which is located at the connection between the fuel rail and the fuel return line, maintains a constant fuel pressure to the injectors. When the engine is accelerating or cruising at a steady speed with no load, there is little manifold vacuum, so the pressure regulator is closed, which maintains higher fuel pressure inside the fuel rail and plenty of fuel for the injectors. When the engine is decelerated, intake manifold vacuum is high, which opens the regulator and allows fuel to return to the fuel tank. A smaller supplemental fuel pressure regulator, which is controlled by a spring, is an integral component of the fuel pump/fuel gauge sending unit module; if fuel pressure rises too quickly for the system to handle it, this regulator opens when the pressure hits a specified level and dumps fuel back into the fuel tank.

Each injector consists of a solenoid coil, a pintle valve and the housing. When current is applied to the solenoid by the PCM, the pintle valve rises off its seat and the pressurized fuel inside the housing squirts out the nozzle. The amount of fuel injected is determined by the length of time that the pintle valve is open, which is determined by the length of time during which current is supplied to the solenoid. Because the injector on-time determines the air-fuel mixture ratio, injector timing must be very precise.

11 Fuel injection system - check

Refer to illustration 11.7

WARNING:

Gasoline is extremely flammable, so take extra precautions when you work on any part of the fuel system. See the Warning in Section 2.

1 Inspect all system electrical connectors, especially the ground connections. Loose connectors and poor grounds are a common cause of many engine control system problems.

2 Verify that the battery is fully charged. The Powertrain Control Module (PCM) and sensors don't operate correctly without adequate supply voltage.

3 Inspect the air filter element (see Chapter 1). A dirty or partially blocked filter reduces performance and economy.

4 Check fuel pump operation (see Section 3). If the fuel pump fuse is blown, replace it and note whether it blows again. If it does, look for a short in the wiring harness to the fuel pump (see the wiring diagrams in Chapter 12).

5 Inspect all vacuum hoses connected to the intake manifold for damage, deterioration and leakage.

6 Disconnect the air filter housing (V6 models) or the intake air connector (four-cylinder models) from the throttle body and look for dirt, carbon, varnish, or other residue inside the throttle body bore, particularly around the throttle plate. If it's dirty, inspect the PCV system for excessive residue (see Chapter 6).

7 On four-cylinder models, use a stethoscope to listen to the injectors while the engine is running (see illustration). Touch the stethoscope to each injector, one at a time, and listen for a clicking sound that indicates operation.

Note: It's difficult to access all of the injectors on V6 models.

8 Measure the resistance of the injectors. To do so on V6 models, remove the air filter housing (see Section 9) and, if you're going to measure the resistance of the injectors for the left cylinder bank, remove the intake manifold as well (see Chapter 2B). On four-cylinder models, remove the intake air connector (see Section 9).

9 Disconnect the electrical connectors from the fuel injectors (see Section 15). With the engine off and the fuel injector electrical connectors disconnected, measure the resistance of each injector with an ohmmeter and compare your readings with the injector resistance listed in this Chapter's Specifications. If the indicated resistance is outside the specified range of resistance, replace the injector (see Section 15).

11.7 Use a stethoscope to listen for the clicking sound that indicates that each injector is working correctly; the clicking sound should rise and fall with changes in engine speed (on V6 models, the injectors aren't accessible)

12 Throttle body - removal and installation

WARNING:

Wait until the engine is completely cool before beginning this procedure.

1 Disconnect the cable from the negative terminal of the battery (see Chapter 5, Section 1).

V6 MODELS

Refer to illustrations 12.4, 12.6 and 12.8

2 Remove the engine cover (see Chapter 2B).

3 Remove the air filter housing (see Section 9).

4 Disconnect the electrical connector from the throttle body (see illustration).

12.4 Depress the release tab and disconnect the electrical connector from the throttle body (V6 models)

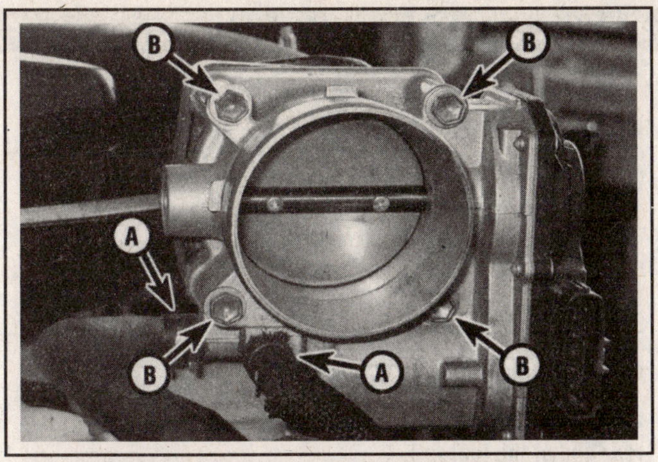

12.6 Loosen the two hose clamps (A) and disconnect the coolant bypass hoses from the throttle body, then remove the four throttle body mounting bolts (B) (V6 models)

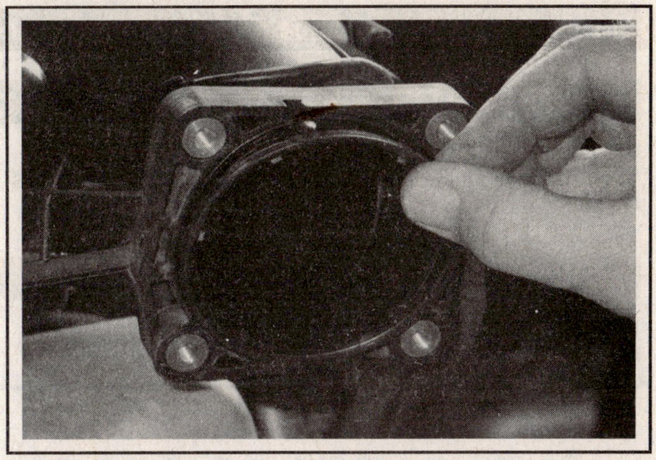

12.8 Remove the O-ring type gasket and inspect it for damage (V6 models)

12.11 Disconnect the electrical connector from the throttle body (four-cylinder models)

12.12 Loosen the two hose clamps and disconnect the coolant bypass hoses from the throttle body, then remove the throttle body mounting fasteners (four-cylinder models)

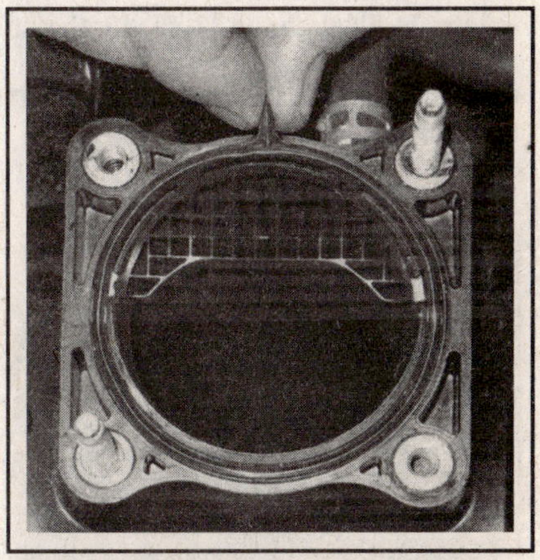

12.14 Remove and discard the old throttle body gasket (four-cylinder models)

5 Clamp-off the coolant hoses connected to the throttle body.

6 Loosen the hose clamps and disconnect both coolant bypass hoses from the throttle body (see illustration).

7 Remove the four throttle body mounting bolts and remove the throttle body.

8 Remove the old throttle body O-ring type gasket (see illustration). Inspect the condition of the gasket. If it's cracked, torn or deteriorated, replace it.

➡Note: To ensure that an air leak doesn't develop between the throttle body and the intake manifold, it's a good idea to simply replace this gasket anytime that you remove the throttle body.

9 Installation is the reverse of removal. Be sure to tighten the throttle body mounting bolts to the torque listed in this Chapter's Specifications. Check the coolant level, adding as necessary (see Chapter 1).

FOUR-CYLINDER MODELS

➧ Refer to illustrations 12.11, 12.12 and 12.14

10 Remove the intake air connector (see Section 9).

11 Disconnect the electrical connector from the throttle body (see illustration).

12 Loosen the hose clamps and disconnect both coolant bypass hoses from the throttle body (see illustration).

13 Unscrew the fasteners and remove the throttle body.

14 Remove the old throttle body O-ring type gasket (see illustration). Inspect the condition of the gasket. If it's cracked, torn or deteriorated, replace it.

➡Note: To ensure that an air leak doesn't develop between the throttle body and the intake manifold, it's a good idea to simply replace this gasket anytime that you remove the throttle body.

15 Installation is the reverse of removal. Be sure to tighten the throttle body mounting bolts to the torque listed in this Chapter's Specifications. Check the coolant level, adding as necessary (see Chapter 1).

13 Fuel pressure pulsation damper - removal and installation

WARNING 1:

Gasoline is extremely flammable, so take extra precautions when you work on any part of the fuel system. See the Warning in Section 2.

WARNING 2:

Wait until the engine is completely cool before beginning this procedure.

1 Relieve the fuel system pressure (see Section 2).

2 Disconnect the cable from the negative terminal of the battery (see Chapter 5, Section 1).

V6 MODELS

Refer to illustrations 13.4a, 13.4b and 13.5

Note: The fuel pressure pulsation damper is located at the rear end of the left fuel rail.

3 Remove the engine cover.

4 Remove the fuel pressure pulsation damper retaining clip and pull the pulsation damper out of the fuel rail (see illustrations).

5 Remove the old O-ring from the fuel pressure pulsation damper (see illustration) and replace it with a new one. Coat the new O-ring with a little clean engine oil before installing it.

6 Installation is the reverse of removal.

FOUR-CYLINDER MODELS

Refer to illustrations 13.7 and 13.8

Note: The fuel pressure pulsation damper is located at the rear end of the fuel rail.

7 Disconnect the fuel supply line from the fuel rail (see illustration)

8 Remove the two fuel pressure pulsation damper mounting bolts (see illustration).

9 Remove and discard the old pulsation damper O-ring. Always replace the O-ring before installing an old or a new fuel pressure pulsation damper. Coat the new O-ring with a little clean engine oil before installing it.

10 Installation is the reverse of removal. Be sure to tighten the fuel pressure pulsation damper mounting bolts to the torque listed in this Chapter's Specifications.

13.4a To remove the fuel pressure pulsation damper from the fuel rail on a V6 model, remove this retainer clip . . .

13.4b . . . then pull the pulsation damper out of the fuel rail

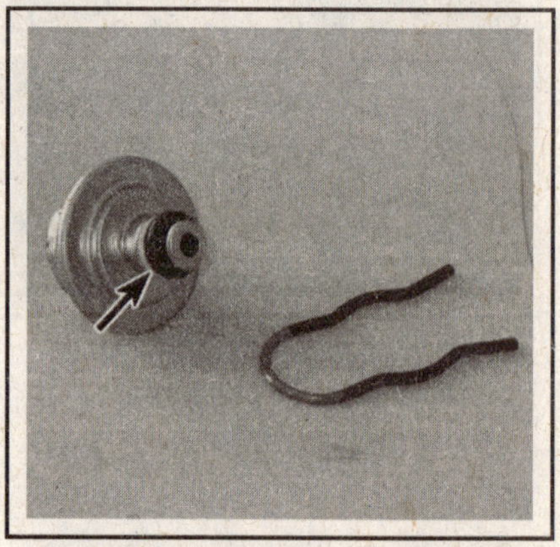

13.5 Remove the old O-ring from the fuel pressure pulsation damper and replace it (V6 models)

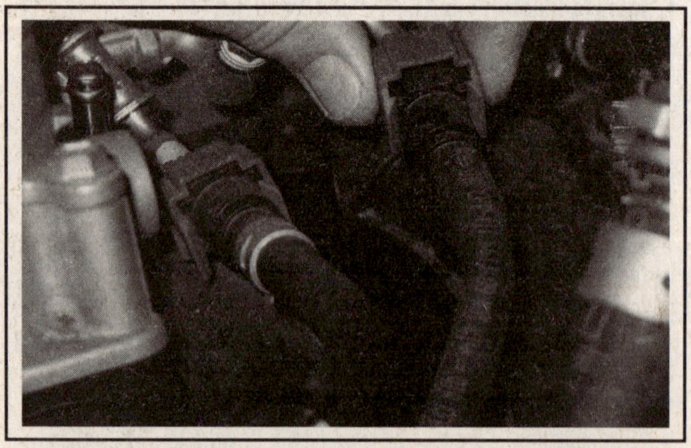

13.7 Pull up the clamp and disconnect the fuel supply line from the fuel pressure pulsation damper (four-cylinder models)

13.8 To detach the fuel pressure pulsation damper from the fuel rail, remove these two bolts (four-cylinder models)

14 Fuel pressure regulator - removal and installation

WARNING:

Gasoline is extremely flammable, so take extra precautions when you work on any part of the fuel system. See the Warning in Section 2.

1 Relieve the fuel system pressure (see Section 2).
2 Disconnect the cable from the negative terminal of the battery (see Chapter 5, Section 1).

V6 MODELS

Refer to illustrations 14.6 and 14.7

Note: The fuel pressure regulator is located at the rear end of the right fuel rail.

3 Remove the engine cover (see Chapter 2B).
4 Remove the air filter housing (see Section 9).
5 Remove the fuel rail and injectors (see Section 15).
6 Remove the fuel pressure regulator mounting bolts (see illustration) and remove the pressure regulator.
7 Remove and discard the old pressure regulator O-ring (see illustration). Always use a new O-ring when installing the old fuel pressure regulator or a new unit. Coat the new O-ring with some clean engine oil and install it on the pressure regulator.
8 Installation is otherwise the reverse of removal. Be sure to tighten the pressure regulator mounting bolts to the torque listed in this Chapter's Specifications.

FOUR-CYLINDER MODELS

Refer to illustrations 14.10, 14.11 and 14.13

Note: The fuel pressure regulator is located at the front end of the fuel rail.

9 Remove the intake air connector (see Section 9).
10 Disconnect the vacuum hose from the fuel pressure regulator (see illustration).
11 Disconnect the fuel return line from the fuel pressure regulator pipe (see illustration).

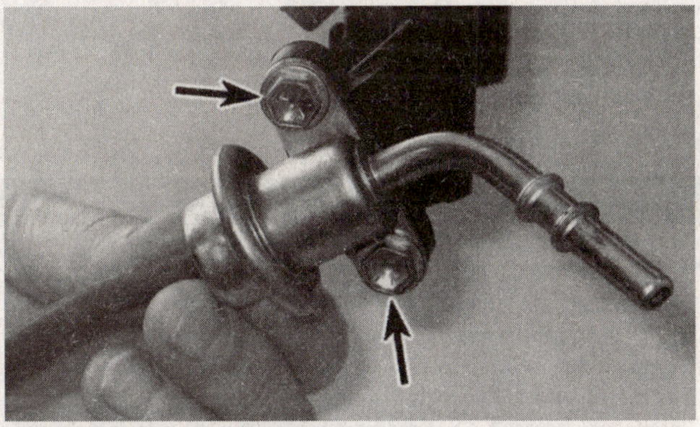

14.6 To detach the fuel pressure regulator from the fuel rail on a V6 model, remove these two bolts

14.7 Remove and discard the old pressure regulator O-ring (V6 models)

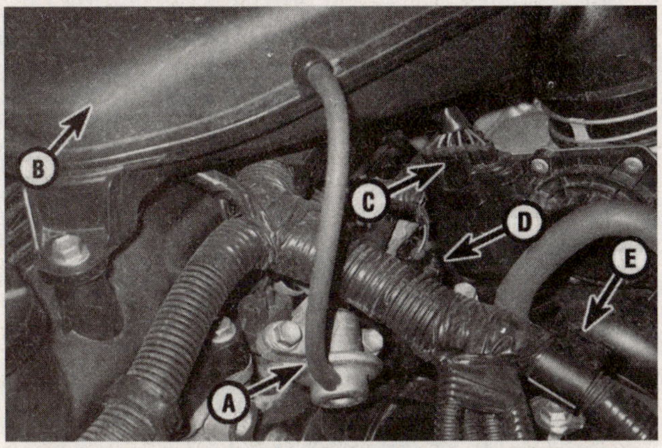

14.10 Disconnect the vacuum hose (A) from the fuel pressure regulator, remove the intake air connector (B), disconnect the electrical connectors from the throttle body (C) and the EGR purge valve (D), then use a screwdriver to disengage the harness clamp (E) (four-cylinder models)

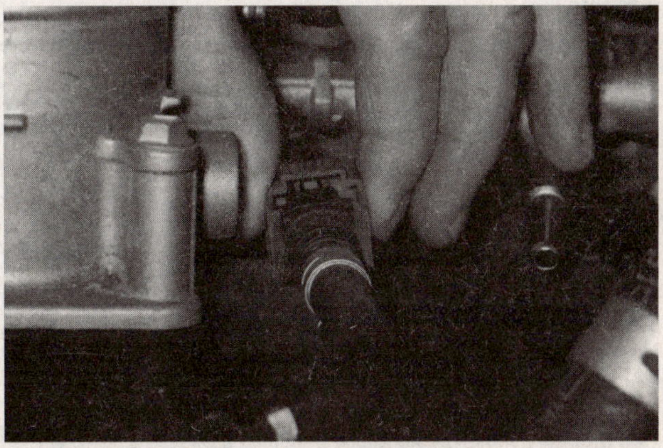

14.11 Pull up the clamp and disconnect the fuel return line from the fuel pressure regulator pipe (four-cylinder models)

14.13 To detach the fuel pressure regulator, remove these three bolts (four-cylinder models)

12 Disconnect the electrical connectors from the throttle body and from the EGR purge valve, then detach the engine wiring harness clamp, remove the bolt that secures the harness clamp bracket and push the harness aside.

13 Remove the two bolts that secure the fuel pressure regulator to the fuel rail, remove the bolt that secures the rear end of the pressure regulator return pipe (see illustration) and remove the regulator.

14 Remove and discard the old fuel pressure regulator O-ring. Always use a new O-ring when installing the old fuel pressure regulator or a new unit. Coat the new O-ring with clean engine oil and install it on the pressure regulator.

15 Installation is otherwise the reverse of removal. Be sure to tighten the fuel pressure regulator bolts to the torque listed in this Chapter's Specifications.

15 Fuel rail and injectors - removal and installation

⁂ WARNING 1:

Gasoline is extremely flammable, so take extra precautions when you work on any part of the fuel system. See the Warning in Section 2.

⁂ WARNING 2:

Wait until the engine is completely cool before beginning this procedure.

1 Relieve the fuel system pressure (see Section 2).

2 Disconnect the cable from the negative terminal of the battery (see Chapter 5, Section 1).

V6 MODELS

➧ Refer to illustration 15.6, 15.7a, 15.7b, 15.9, 15.10, 15.11, 15.12 and 15.13

3 Remove the engine cover (see Chapter 2B) and the air filter housing (see Section 9).

4 Drain the engine coolant (see Chapter 1).

5 Remove the intake manifold (see Chapter 2B).

6 Disconnect the vacuum signal hose from the fuel pressure regulator (see illustration).

7 Remove the clamp from the fuel supply line quick-connect fitting and disconnect the fitting (see illustrations).

8 Remove the clamp from the fuel return line quick-connect fitting and disconnect the fitting.

9 Disconnect the electrical connectors from the fuel injectors (see illustration).

10 Remove the six fuel rail mounting bolts (see illustration).

11 Remove the fuel rail and injectors as a single assembly (see illustration).

12 Remove the injectors from the fuel rail (see illustration).

15.6 Disconnect the vacuum signal hose from the fuel pressure regulator (V6 models)

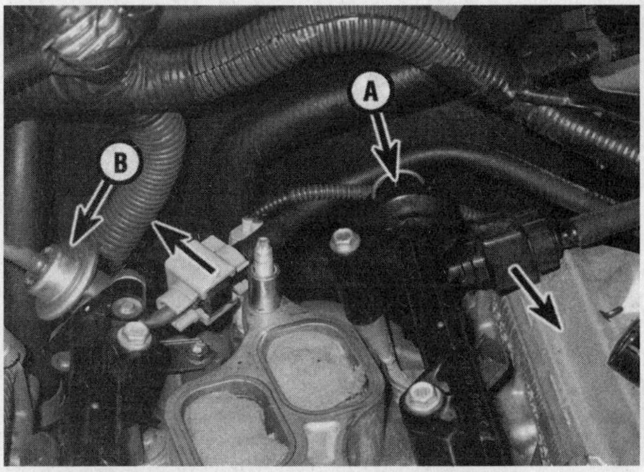

15.7a The fuel supply line connection at the pulsation damper (A) and the fuel return line connection at the fuel pressure regulator (B) are protected by plastic clamps. To remove a clamp, simply pull it off (V6 models)

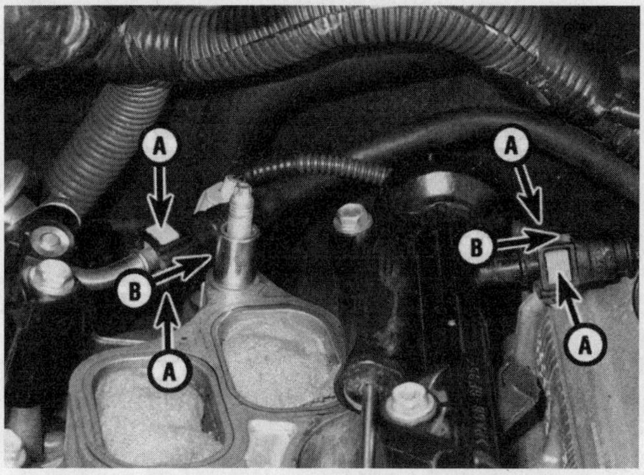

15.7b To disconnect the fuel supply and return line fittings, depress the two release buttons (A) on each fitting and pull the fitting (B) off the pipe (V6 models)

15.9 Disconnect the electrical connectors from the fuel injectors (V6 models)

15.10 To detach the fuel rail from a V6 engine, remove these six bolts

15.11 Remove the fuel rail and injectors as a single assembly (V6 models)

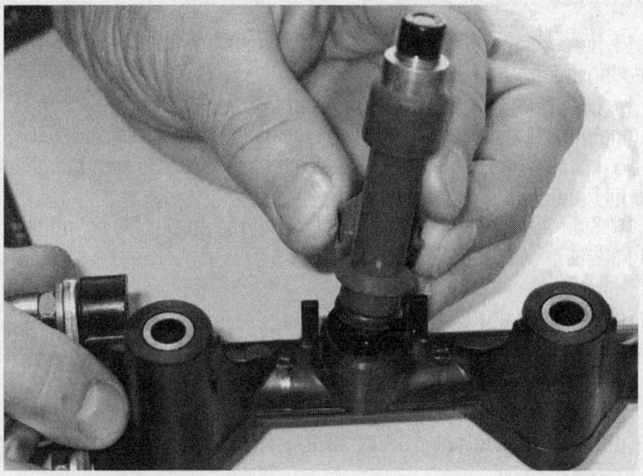

15.12 Remove each injector from the fuel rail (V6 models)

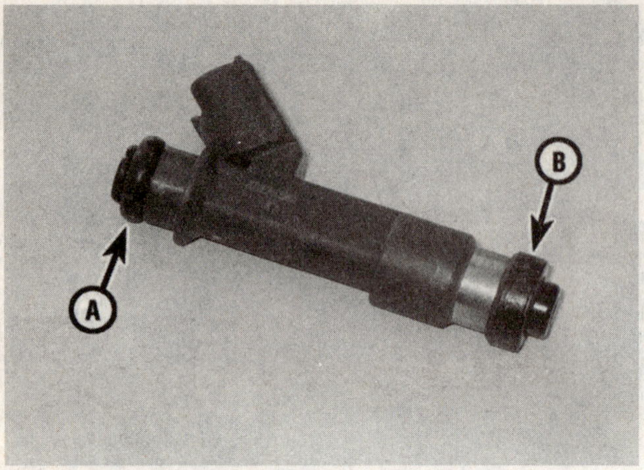

15.13 Remove the O-ring (A) and insulator (B) from each injector and discard them. Always use a new O-ring and insulator when installing an injector, whether it's a new injector or an old unit (V6 models)

15.20 To disengage an engine harness clamp from its retaining bracket, carefully pry up this lock and pull the clamp off the bracket (four-cylinder models)

15.21 To disconnect the electrical connectors from the fuel injectors, depress the release tab on the forward end of each connector and pull up (four-cylinder models)

15.22 To detach the fuel rail from the cylinder head, remove these two bolts (four-cylinder models)

13 Remove and discard the O-ring and insulator from each injector (see illustration).

14 Coat each new injector O-ring and insulator with a little clean engine oil and install the O-rings and insulators on each injector.

15 Installation is otherwise the reverse of removal. Be sure to tighten the fuel rail mounting bolts securely.

16 Start the engine and check for leaks at the quick-connect fittings and at the injectors.

FOUR-CYLINDER MODELS

Removal

➧ Refer to illustrations 15.20, 15.21, 15.22, 15.24, 15.26 and 15.27

17 Remove the intake air connector (see Section 9).

18 Remove the throttle body (see Section 12).

19 Disconnect the fuel supply line (see illustration 13.7) and the fuel return line (see illustration 14.11).

20 Disengage all engine harness clamps from their brackets as necessary (see illustration).

21 Disconnect the electrical connectors from the fuel injectors (see illustration).

22 Remove the two fuel rail mounting bolts (see illustration).

23 Pull straight up on the fuel rail while wiggling the injectors free of their bores in the intake manifold, then remove the fuel rail and injectors as a single assembly.

24 Remove the two fuel rail mounting bolt spacers (see illustration).

25 Remove the four injector spacers.

26 Remove each injector from the fuel rail (see illustration).

27 Remove the O-ring and vibration insulator from each injector (see illustration) and discard them.

Installation

28 Apply a light coat of clean engine oil to each new injector O-ring,

15.24 Remove the two fuel rail mounting bolt spacers (A) and the four injector spacers (B) (four-cylinder models)

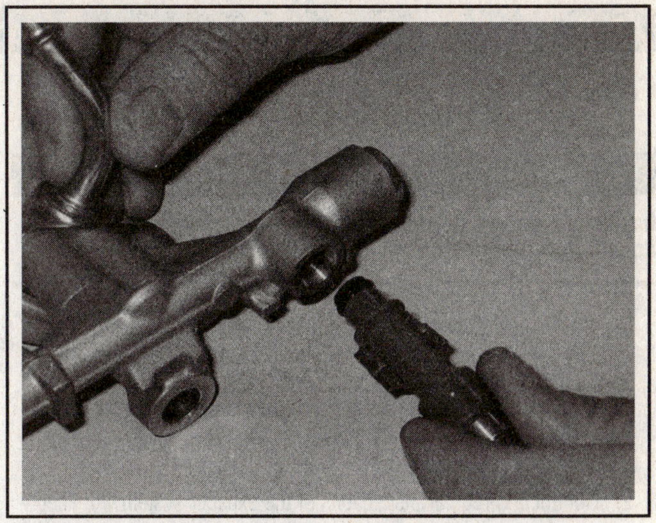

15.26 To remove each injector from the fuel rail, carefully pull on it and wiggle it side-to-side (four-cylinder models)

then install a new O-ring into its groove in the upper end of each injector.

29 Coat the O-rings and the contact surfaces of the injector bores in the fuel rail with a little clean engine oil, then insert the injectors into the fuel rail. Carefully rotate each injector back and forth and work it into the fuel rail. Verify that each injector rotates freely when it's fully seated. If it doesn't, pull it out, replace the O-ring and re-install the injector.

30 Install a new injector vibration insulator onto the lower end of each injector.

31 Apply a light coat of clean engine oil to a new O-ring for each new spacer, then install a new O-ring into the groove in each spacer.

32 Install the four new injector spacers, with their new O-rings, into the injector bores in the cylinder head.

33 Install the two fuel rail mounting bolt spacers.

34 Install the fuel rail and injectors as a single assembly. Push the injectors back into their bores in the cylinder head until they're fully seated.

35 Install the fuel rail mounting bolts and tighten them securely.

36 Installation is otherwise the reverse of removal.

37 Start the engine and check for leaks at the quick-connect fittings and at the fuel injectors.

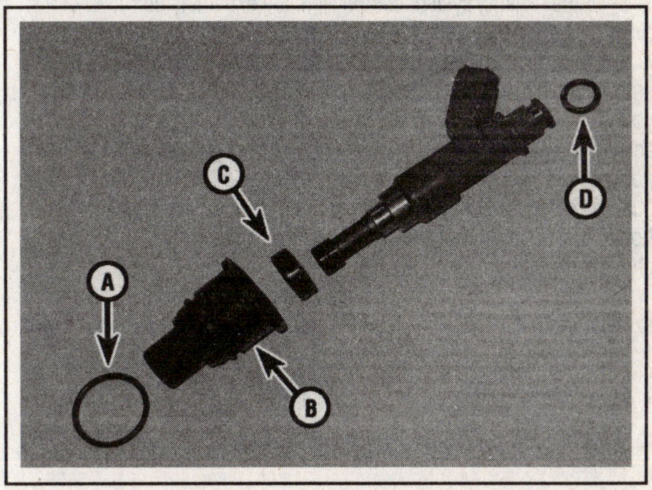

15.27 Fuel injector assembly details:

A Spacer O-ring
B Spacer
C Injector vibration insulator
D Injector O-ring

16 Exhaust system servicing - general information

➧ **Refer to illustration 16.1**

⁂ WARNING:

Inspection and repair of exhaust system components should be done only after the system components have cooled completely.

1 The exhaust system consists of the exhaust manifolds, the catalytic converters, the muffler, the tailpipe and all connecting pipes, brackets, hangers and clamps. The exhaust system is attached to the body with mounting brackets and rubber hangers (see illustration). If any of these parts are damaged or deteriorated, excessive noise and vibration will be transmitted to the body.

2 Conducting regular inspections of the exhaust system will keep it safe and quiet. Look for any damaged or bent parts, open seams, holes, loose connections, excessive corrosion or other defects which could allow exhaust fumes to enter the vehicle. Deteriorated exhaust system components should not be repaired - they should be replaced with new parts.

3 If the exhaust system components are extremely corroded or rusted together, they will probably have to be cut from the exhaust system. The convenient way to accomplish this is to have a muffler repair shop remove the corroded sections with a cutting torch. If, however, you want to save money by doing it yourself and you don't have an oxy/acetylene welding outfit with a cutting torch, simply cut off the old components with a hack-saw. If you have compressed air, special pneumatic cutting chisels can also be used. If you do decide to tackle the job at home, be sure to wear eye protection to protect your eyes from metal chips and work gloves to protect your hands.

4 Here are some simple guidelines to apply when repairing the exhaust system:

a) *Work from the back to the front when removing exhaust system components.*
b) *Apply penetrating oil to the exhaust system component fasteners to make them easier to remove.*
c) *Use new gaskets, hangers and clamps when installing exhaust system components.*
d) *Apply anti-seize compound to the threads of all exhaust system fasteners during reassembly. Be sure to allow sufficient clearance between newly installed parts and all points on the underbody to avoid overheating the floor pan and possibly damaging the interior carpet and insulation. Pay particularly close attention to the catalytic converter and its heat shield.*

⁂ WARNING:

The catalytic converter operates at very high temperatures and takes a long time to cool. Wait until it's completely cool before attempting to remove the converter. Failure to do so could result in serious burns.

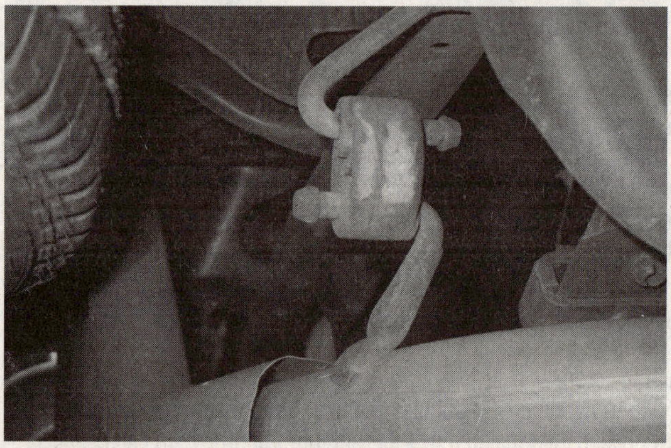

16.1 A typical rubber exhaust hanger

Specifications

Fuel system

Fuel system pressure (all models)	40.8 to 41.7 psi (281 to 287 kPa)
Residual fuel system pressure	
(five minutes after engine turned off)	21 psi (147 KPa) minimum
Injector resistance (approximate, all models)	11.6 to 12.4 ohms, at 68 degrees F (20 degrees C)

Torque specifications

➡Note: One foot-pound (ft-lb) of torque is equivalent to 12 inch-pounds (in-lbs) of torque. Torque values below approximately 15 ft-lbs are expressed in inch-pounds, since most foot-pound torque wrenches are not accurate at these smaller values.

Fuel pressure regulator retaining bolts		
V6 engines	80 in-lbs	9 Nm
Four-cylinder engines	75 in-lbs	8.5 Nm
Fuel pressure pulsation damper		
(four-cylinder engines)	75 in-lbs	8.5 Nm
Throttle body mounting bolts/nuts		
V6 engines	108 in-lbs	12 Nm
Four-cylinder engines	80 in-lbs	9 Nm

Notes

5

ENGINE ELECTRICAL SYSTEMS

Section

Reference to other Chapters

Battery check, maintenance and charging - See Chapter 1
Camshaft Position (CMP) sensor - replacement - See Chapter 6
CHECK ENGINE light - See Chapter 6
Drivebelt check and replacement - See Chapter 1
Spark plug replacement - See Chapter 1
Spark plug wire check and replacement - See Chapter 1

1 General information, precautions and battery disconnection

The engine electrical systems include all ignition, charging and starting components. Because of their engine-related functions, these components are discussed separately from body electrical devices such as the lights, the instruments, etc. (which are included in Chapter 12).

PRECAUTIONS

Always observe the following precautions when working on the electrical system:

a) *Be extremely careful when servicing engine electrical components. They are easily damaged if checked, connected or handled improperly.*
b) *Never disconnect the battery cables while the engine is running.*
c) *Maintain correct polarity when connecting battery cables from another vehicle during jump starting - see the "Booster battery (jump) starting" section at the front of this manual.*
d) *Always disconnect the negative battery cable from the battery before working on the electrical system, but read the following battery disconnection procedure first.*

It's also a good idea to review the safety-related information regarding the engine electrical systems located in the "Safety first!" Section at the front of this manual, before beginning any operation included in this Chapter.

BATTERY DISCONNECTION

Several systems on the vehicle require battery power to be available at all times, either to ensure their continued operation (such as the radio, alarm system, power door locks, windows, etc.) or to maintain control unit memories (such as that in the engine management system's Powertrain Control Module [PCM]) which would be lost if the battery were to be disconnected. Therefore, whenever the battery is to be disconnected, first note the following to ensure that there are no unforeseen consequences of this action:

a) *The engine management system's PCM will lose the information stored in its memory when the battery is disconnected. This includes idling and operating values, any fault codes detected and system monitors required for emissions testing. Whenever the battery is disconnected, the computer may require a certain period of time to re-learn the operating values.*
b) *On any vehicle with power door locks, it is a wise precaution to remove the key from the ignition and to keep it with you, so that it does not get locked inside if the power door locks should engage accidentally when the battery is reconnected!*

Devices known as "memory-savers" can be used to avoid some of the above problems. Precise details vary according to the device used. Typically, it is plugged into the cigarette lighter and is connected by its own wires to a spare battery; the vehicle's own battery is then disconnected from the electrical system, leaving the memory-saver to pass sufficient current to maintain audio unit security codes and ECM memory values, and also to run permanently live circuits such as the clock and radio memory, all the while isolating the battery in the event of a short-circuit occurring while work is carried out.

⁂ WARNING 1:

Some of these devices allow a considerable amount of current to pass, which can mean that many of the vehicle's systems are still operational when the main battery is disconnected. If a memory-saver is used, ensure that the circuit concerned is actually dead before carrying out any work on it!

⁂ WARNING 2:

If work is to be performed around any of the airbag system components, the battery must be disconnected and no memory saver can be used. If a memory-saver device is used, power will be supplied to the airbag and personal injury may result if the airbag is accidentally deployed.

The battery on all vehicles is located in the front left corner of the engine compartment. To disconnect the battery for service procedures requiring power to be cut from the vehicle, loosen the negative cable clamp nut and detach the negative cable from the negative battery post. Isolate the cable end to prevent it from accidentally coming into contact with the battery post.

2 Battery - emergency jump starting

Refer to the *Booster battery (jump) starting* procedure at the front of this manual.

3 Battery - check and replacement

⁂ WARNING:

Hydrogen gas is produced by the battery, so keep open flames and lighted cigarettes away from it at all times. Always wear eye protection when working around a battery. Rinse off spilled electrolyte immediately with large amounts of water.

CHECK

▸ Refer to illustrations 3.1a, 3.1b and 3.1c

1 A battery cannot be accurately tested until it is at or near a fully charged state. Disconnect the negative battery cable from the battery

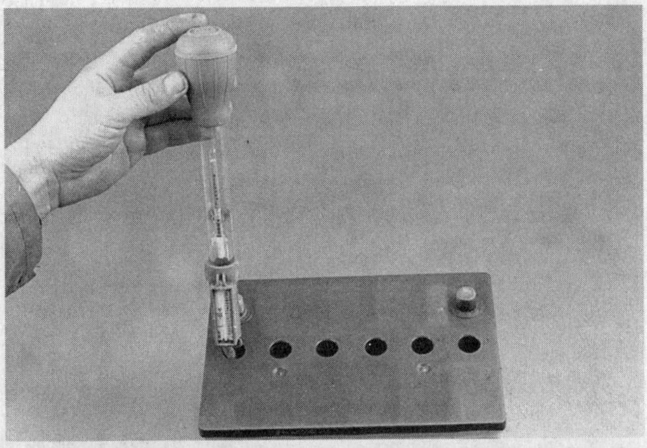

3.1a Checking the specific gravity of the battery electrolyte with a hydrometer; this hydrometer is equipped with a thermometer to make temperature corrections

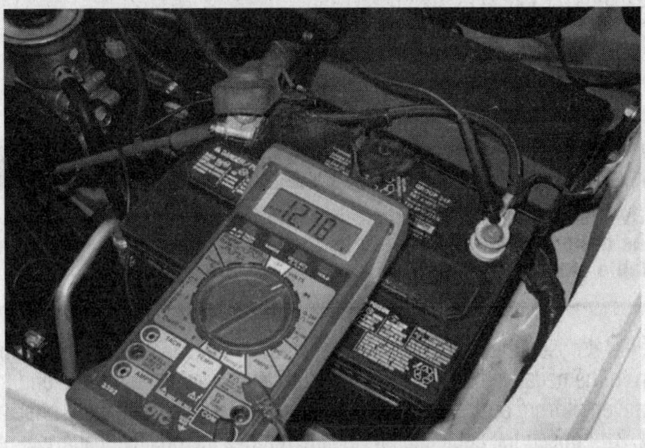

3.1b To test the open circuit voltage of the battery, connect the black probe of the voltmeter to the negative terminal and the red probe to the positive terminal of the battery - a fully charged battery should indicate at least 12.6 volts (depending on the outside air temperature)

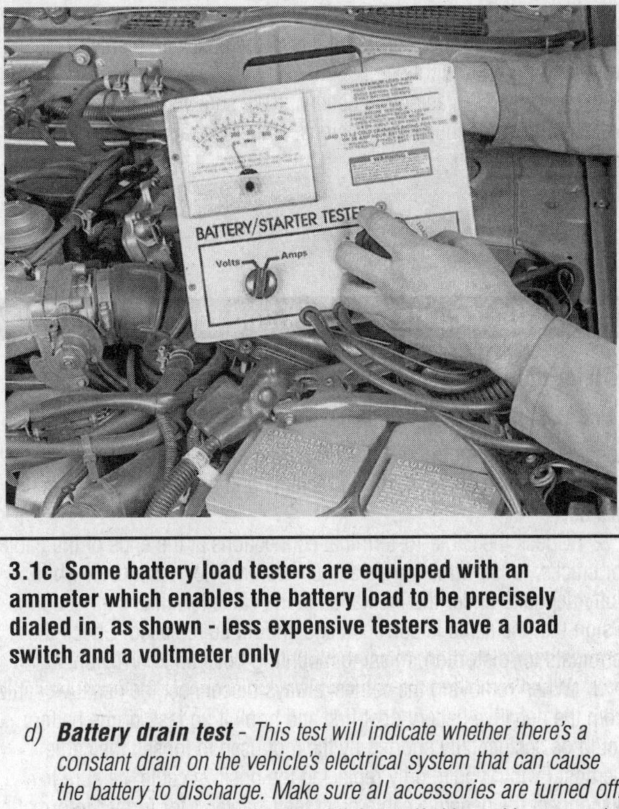

3.1c Some battery load testers are equipped with an ammeter which enables the battery load to be precisely dialed in, as shown - less expensive testers have a load switch and a voltmeter only

and perform the following tests:

a) ***Battery state of charge test*** - *Visually inspect the indicator eye (if equipped) on the top of the battery. If the indicator eye is dark in color, charge the battery as described in Chapter 1. If the battery is equipped with removable caps, check the battery electrolyte. The electrolyte level should be above the upper edge of the plates. If the level is low, add distilled water. DO NOT OVERFILL. The excess electrolyte may spill over during periods of heavy charging. Test the specific gravity of the electrolyte using a hydrometer (see illustration). Remove the caps and extract a sample of the electrolyte and observe the float inside the barrel of the hydrometer. Follow the instructions from the tool manufacturer and determine the specific gravity of the electrolyte for each cell. A fully charged battery will indicate approximately 1.270 (green zone) at 68-degrees F (20-degrees C). If the specific gravity of the electrolyte is low (red zone), charge the battery as described in Chapter 1.*

b) ***Open circuit voltage test*** - *Using a digital voltmeter, perform an open circuit voltage test (see illustration). Connect the negative probe of the voltmeter to the negative battery post and the positive probe to the positive battery post. The battery voltage should be greater than 12.5 volts. If the battery is less than the specified voltage, charge the battery before proceeding to the next test. Do not proceed with the battery load test until the battery is fully charged.*

c) ***Battery load test*** - *An accurate check of the battery condition can only be performed with a load tester (available at most auto parts stores). This test evaluates the ability of the battery to operate the starter and other accessories during periods of heavy amperage draw (load). Connect a battery load testing tool to the battery terminals (see illustration). Load test the battery according to the tool manufacturer's instructions. This tool increases the load demand (amperage draw) on the battery. Maintain the load on the battery for 15 seconds and observe that the battery voltage does not drop below 9.6 volts. If the battery condition is weak or defective, the tool will indicate this condition immediately.*

➡Note: Cold temperatures will cause the minimum voltage reading to drop slightly. Follow the chart given in the tool manufacturer's instructions to compensate for cold climates. Minimum load voltage for freezing temperatures (32 degrees F/0-degrees C) should be approximately 9.1 volts.

d) ***Battery drain test*** - *This test will indicate whether there's a constant drain on the vehicle's electrical system that can cause the battery to discharge. Make sure all accessories are turned off. If the vehicle has an underhood light, verify it's working properly, then disconnect it. Connect one lead of a digital ammeter to the disconnected negative battery cable clamp and the other lead to the negative battery post. A drain of approximately 100 milliamps or less is considered normal (due to the Powertrain Control Module, digital clocks, digital radios and other components which normally cause a key-off battery drain). An excessive drain (approximately 500 milliamps or more) will cause the battery to discharge. The problem circuit or component can be located by removing the fuses, one at a time, until the excessive drain stops and normal drain is indicated on the meter.*

REPLACEMENT

➧ **Refer to illustration 3.2**

⁂ CAUTION:

Always disconnect the negative cable first and hook it up last or the battery might be shorted by the tool being used to loosen the cable clamps.

2 Loosen the cable clamp nut and remove the negative battery cable from the negative battery post (see illustration). Isolate the cable end to prevent it from accidentally coming into contact with the battery post.
3 Loosen the cable clamp nut and remove the positive battery cable from the positive battery post.
4 Remove the battery hold-down strap.
5 Lift out the battery. Be careful - it's heavy.

➡**Note: Battery straps and handlers are available at most auto parts stores for a reasonable price. They make it easier to remove and carry the battery.**

6 While the battery is out, inspect the battery tray for corrosion. If corrosion exists, clean the deposits with a mixture of baking soda and water to prevent further corrosion. Flush the area with plenty of clean water and dry thoroughly.
7 If you are replacing the battery, make sure you replace it with a battery with the identical dimensions, amperage rating, cold cranking rating, etc.

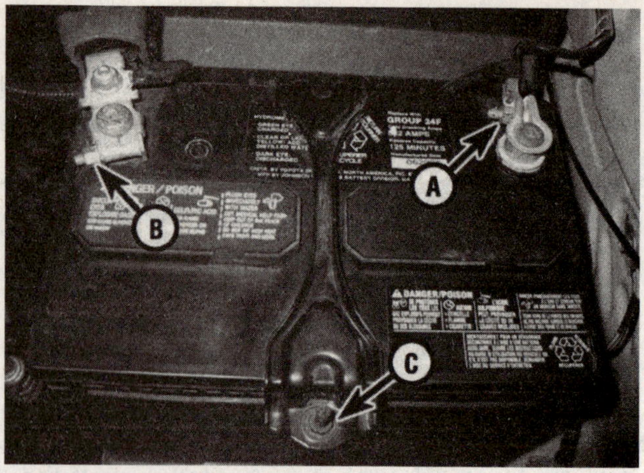

3.2 Battery mounting details

A Negative cable clamp (remove first, connect last)
B Positive cable clamp (remove last, connect first)
C Hold-down strap nut

8 When installing the battery, tighten the hold-down strap nut securely, but don't over-tighten it.
9 When reconnecting the cables, connect the positive cable first and the negative cable last.

4 Battery cables - check and replacement

➧ **Refer to illustrations 4.4a, 4.4b and 4.4c**

1 Periodically inspect the entire length of each battery cable for damage, cracked or burned insulation and corrosion. Poor battery cable connections can cause starting problems and decreased engine performance.
2 Check the cable-to-terminal connections at the ends of the cables for cracks, loose wire strands and corrosion. The presence of white, fluffy deposits under the insulation at the cable terminal connection is a sign that the cable is corroded and should be replaced. Check the terminals for distortion, missing mounting bolts and corrosion.
3 When removing the cables, always disconnect the negative cable from the negative battery post first and hook it up last, or the battery could be accidentally shorted by the tool used to loosen the cable clamps. Even if you're only replacing the positive cable, be sure to disconnect the negative cable first (see Chapter 1 for further information regarding battery cable maintenance).
4 Disconnect the old cables from the battery first (see illustration 3.2). You can disconnect the opposite ends of the cables from whatever they're bolted to (see illustrations). Typically, a smaller ground cable is connected to the vehicle body right next to the underhood fuse/relay box and a larger ground cable is attached to the engine block or to a bellhousing bolt. Note the routing of each cable to ensure correct installation.

➡**Note: The accompanying photos depict typical battery cable connections on four-cylinder and V6 models. The actual connections on your specific model might differ somewhat from the connections shown in these photos.**

5 If you are replacing either or both of the battery cables, take them with you when buying new cables. It is vitally important that you replace the cables with identical parts.
6 Clean the threads of the starter solenoid or ground connection with a wire brush to remove rust and corrosion. Apply a light coat of

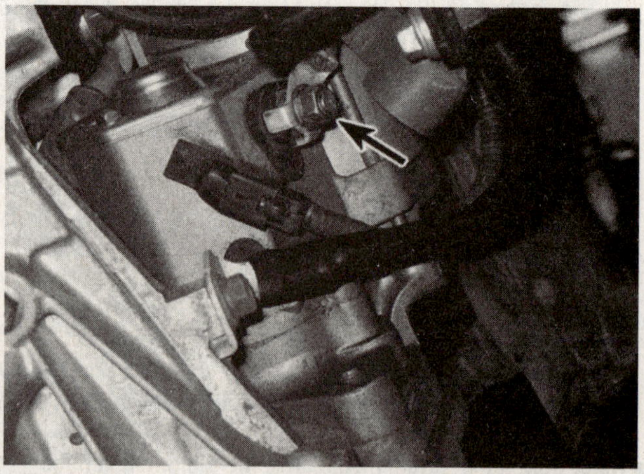

4.4a The positive battery cable is connected to this terminal on the starter motor solenoid (V6 model shown, four-cylinder models similar)

4.4b The positive battery cable harness is also secured to the alternator battery (B+) stud-type terminal by a large nut (under the rubber boot) (V6 model shown, four-cylinder models similar)

4.4c The battery ground cable is grounded to the vehicle adjacent to the battery, and to the engine block

battery terminal corrosion inhibitor or petroleum jelly to the threads to prevent future corrosion.

7 Attach the cable to the terminal and tighten the mounting nut/bolt securely.

8 Before connecting a new cable to the battery, make sure that it reaches the battery post without having to be stretched.

9 When reconnecting the cables to the battery, connect the positive cable first and the negative cable last.

5 Ignition system - general information and precautions

1 All models are equipped with a Direct Ignition System (DIS) which has no distributor. The DIS system includes the Powertrain Control Module (PCM), the Camshaft Position (CMP) sensor, the Crankshaft Position (CKP) sensor, the Variable Valve Timing (VVT) sensors, the ignition coils and the spark plugs. For more information on the PCM and the CMP, CKP and VVT sensors, refer to Chapter 6.

2 Four-cylinder models have four ignition coils; V6 models have six ignition coils. The ignition coils are mounted on the valve covers, directly over the spark plugs. There are no spark plug wires. Each ignition coil has an integral igniter.

WARNING:

Never touch an ignition coil while the engine is running or being cranked-over. The voltage produced by the coils is potentially lethal.

6 Ignition system - check

Refer to illustration 6.2

WARNING:

Because of the high voltage generated by the ignition system, use extreme care when performing a procedure involving ignition components. Never touch an ignition coil while the engine is running or being cranked over. The voltage produced by the coils is potentially lethal.

1 If a malfunction occurs in the ignition system, check the following items first:

a) *Make sure that the cable clamps at the battery terminals are clean and tight.*

b) *Test the condition of the battery (see Section 3). If it doesn't pass all of the tests, replace it.*

c) *Check the ignition coil connections.*

d) *Check any relevant fuses in the engine compartment fuse and relay box (see Chapter 12). If any are burned, determine the cause and repair the circuit.*

e) *Check to see if any trouble codes are stored in the PCM (see Chapter 6).*

2 If the engine turns over but won't start, disconnect an ignition coil from a spark plug (see Section 7), reconnect the electrical connector to the coil, then attach a spark tester between the coil and the spark plug (see illustration). Spark testers are available at most auto parts stores.

3 If the tester flashes during cranking, the coil is delivering sufficient voltage to the spark plug to fire it. Repeat this test for each cylinder to verify that the other coils are OK.

4 If the tester doesn't flash, remove a coil from another cylinder and swap it for the one being tested. If the tester now flashes, you know that the original coil is bad. If the tester still doesn't flash, the PCM or wiring harness is probably defective. Have the PCM checked out by a dealership service department or other qualified repair shop (testing the PCM is beyond the scope of the do-it-yourselfer because it requires expensive special tools).

5 If the tester flashes during cranking but a misfire code (related to the cylinder being tested) has been stored, the spark plug could be fouled or defective.

6.2 Remove the coil and install the tester inline between the coil high tension terminal and the spark plug, then crank the engine. If there's enough power to fire the plug, the tester will flash

7 Ignition coils - replacement

1 Disconnect the cable from the negative battery terminal (see Section 1).

V6 MODELS

➧ Refer to illustration 7.5

2 Remove the engine cover (see Chapter 2B).

3 Remove the air filter housing (see Chapter 4).

4 If you're removing a coil from the left cylinder bank, remove the upper intake manifold (see Chapter 2B).

5 Disconnect the electrical connector from the ignition coil (see illustration).

6 Remove the ignition coil mounting bolt and remove the coil.

7 Installation is the reverse of removal.

FOUR-CYLINDER MODELS

➧ Refer to illustration 7.9

8 Remove the intake air connector (see *Air filter housing - removal and installation* in Chapter 4).

9 Disconnect the electrical connector from the ignition coil (see illustration).

10 Remove the ignition coil mounting bolt, then remove the coil by pulling it straight up.

11 Installation is the reverse of removal.

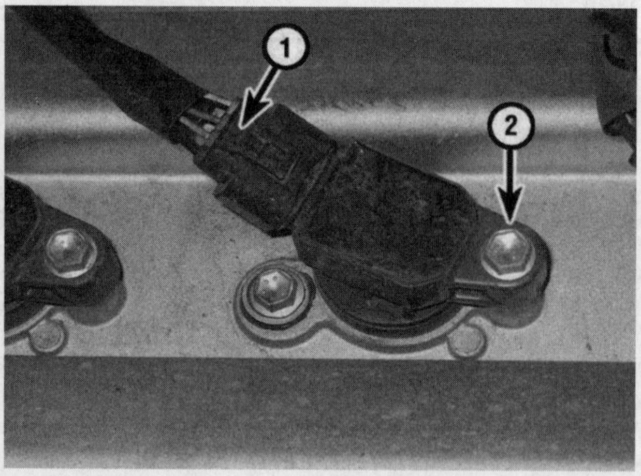

7.5 To remove an ignition coil from a V6 engine, depress the release tab (1) and disconnect the electrical connector, then remove the coil mounting bolt (2)

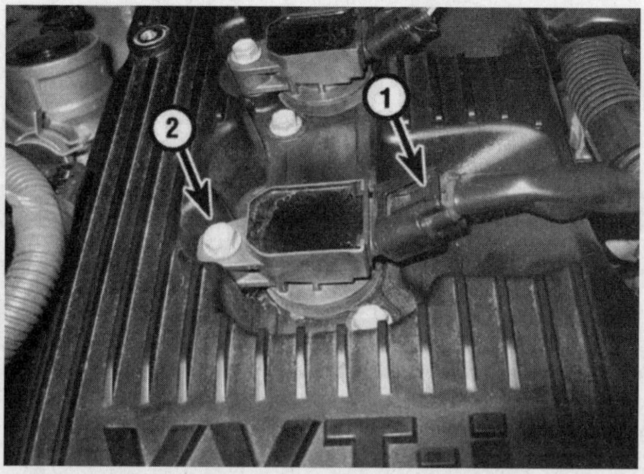

7.9 To remove an ignition coil from a four-cylinder engine, depress the release tab (1) and disconnect the electrical connector from the coil, then remove the coil mounting bolt (2) and pull the coil straight up

8 Charging system - general information and precautions

The charging system includes the alternator, an internal voltage regulator, a charge indicator, the battery and the wiring between all the components. The charging system supplies electrical power for the ignition system, the lights, the radio, etc. The alternator is driven by a drivebelt at the front of the engine.

The purpose of the voltage regulator is to limit the alternator's voltage to a preset value. This prevents power surges, circuit overloads, etc., during peak voltage output.

The charging system doesn't ordinarily require periodic maintenance. However, the drivebelt, battery and wires and connections should be inspected at the intervals outlined in Chapter 1.

The dashboard warning light should come on when the ignition key is turned to START, then should go off immediately. If it remains on, there is a malfunction in the charging system. These vehicles are also equipped with a voltage gauge. If the voltage gauge indicates abnormally high or low voltage, check the charging system (see Section 9).

Be very careful when making electrical circuit connections to a vehicle equipped with an alternator and note the following:

a) *When reconnecting wires to the alternator from the battery, be sure to note the polarity.*
b) *Don't disconnect the battery while the engine is running.*
c) *Before using arc welding equipment to repair any part of the vehicle, disconnect the wires from the alternator and the battery terminals.*
d) *Never start the engine with a battery charger connected.*
e) *Always disconnect both battery leads before using a battery charger.*
f) *The alternator is driven by an engine drivebelt which could cause serious injury if your hand, hair or clothes become entangled in it with the engine running.*
g) *Because the alternator is connected directly to the battery, it could arc or cause a fire if overloaded or shorted out.*

9 Charging system - check

Refer to illustration 9.2

1 If a malfunction occurs in the charging circuit, do not immediately assume that the alternator is causing the problem. First, check the following items:

a) *Make sure the battery cable clamps, where they connect to the battery, are clean and tight.*
b) *Test the condition of the battery (see Section 3). If it does not pass all the tests, replace it with a new battery.*
c) *Check the external alternator wiring and connections.*
d) *Check the drivebelt condition and tension (see Chapter 1).*
e) *Check the alternator mounting bolts for tightness.*
f) *Run the engine and check the alternator for abnormal noise.*
g) *Check the charge light on the dash. It should illuminate when the ignition key is turned ON (engine not running). If it does not, check the circuit from the alternator to the charge light on the dash.*
h) *Check all the fuses that are in series with the charging system circuit. The location of these fuses may vary from year and model but the designations are generally the same. Refer to the wiring diagrams at the end of Chapter 12 for additional information.*

2 With the ignition key off, check the battery voltage with no accessories operating (see illustration). It should be at least 12.6 volts. It may be slightly higher if the engine has been operating within the last hour.

3 Start the engine and check the battery voltage again. It should now be greater than the voltage recorded in Step 2, but not more than 15.0 volts. Turn on all the vehicle accessories (air conditioning, rear window defogger, blower motor, etc.) and increase the engine speed to 2,000 rpm - the voltage should not drop below the voltage recorded in Step 2.

4 If the indicated voltage is greater than the specified charging voltage, replace the voltage regulator.

Note: We recommend replacing the alternator/voltage regulator as a complete unit, using either a rebuilt or new alternator.

5 If the indicated voltage reading is less than the specified charging voltage, the alternator is probably defective. Have the charging system checked at a dealer service department or other properly equipped repair facility.

Note: Many auto parts stores will bench test an alternator off the vehicle. Refer to your local auto parts store regarding their policy; many will perform this service free of charge.

9.2 Connect a voltmeter to the battery terminals, then check the battery voltage with the engine off and again with the engine running

10 Alternator - removal and installation

➡Note: Many new/rebuilt alternators do not have a pulley installed, so you may have to switch the pulley from the old unit to the new/rebuilt one. When buying an alternator, find out the shop's policy regarding installation of pulleys - some shops will perform this service free of charge.

V6 MODELS

➧ Refer to illustrations 10.7, 10.9 and 10.10

➡Note: The alternator is located on the left side of the engine.

1 Remove the engine cover (see Chapter 2B).
2 Remove the battery (see Section 3).
3 Raise the front of the vehicle and place it securely on jackstands.
4 Remove the engine under-cover.
5 Remove the drivebelt (see Chapter 1).
6 Unbolting the air conditioning compressor (see Chapter 3) will give you more room to work.

***** WARNING:**

If you do decide to unbolt the compressor, DON'T disconnect the refrigerant lines from the compressor.

7 Disconnect the electrical connectors from the alternator (see illustration).
8 Remove the bolt that secures the alternator harness bracket to the battery tray and detach the harness bracket from the tray.
9 Remove the two bolts that secure the alternator wire harness bracket to the alternator (see illustration).
10 Remove the two alternator mounting bolts (see illustration) and remove the alternator.
11 Installation is the reverse of removal. Be sure to tighten the alternator mounting fasteners securely.
12 Check the charging voltage to verify that the alternator is operating correctly (see Section 9).

FOUR-CYLINDER MODELS

➧ Refer to illustrations 10.15 and 10.17

➡Note: The alternator is located on the right side of the engine.

13 Disconnect the cable from the negative battery terminal (see Section 1).
14 Remove the drivebelt (see Chapter 1).
15 Remove the rubber weather boot from the B+ terminal (see illustration) and remove the nut that secures the battery cable to the B+ terminal.
16 Disconnect the electrical connector from the alternator.
17 Remove the two alternator mounting bolts (see illustration) and remove the alternator.
18 Installation is the reverse of removal. Be sure to tighten the alternator mounting bolts securely.
19 Check the charging voltage to verify that the alternator is operating correctly (see Section 9).

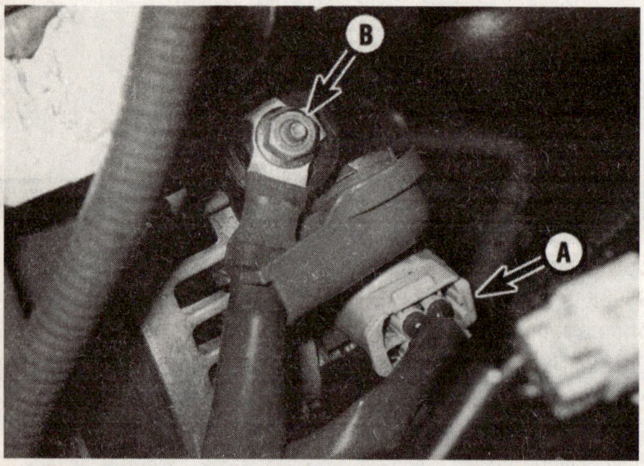

10.7 Depress the release tab (A) and disconnect the electrical connector from the alternator, then remove the nut (B) and disconnect the battery cable from the B+ terminal (V6 models)

10.9 To detach the alternator harness bracket from the alternator, remove these two bolts (V6 models)

10.10 To detach the alternator from the engine, remove these two mounting bolts (V6 models)

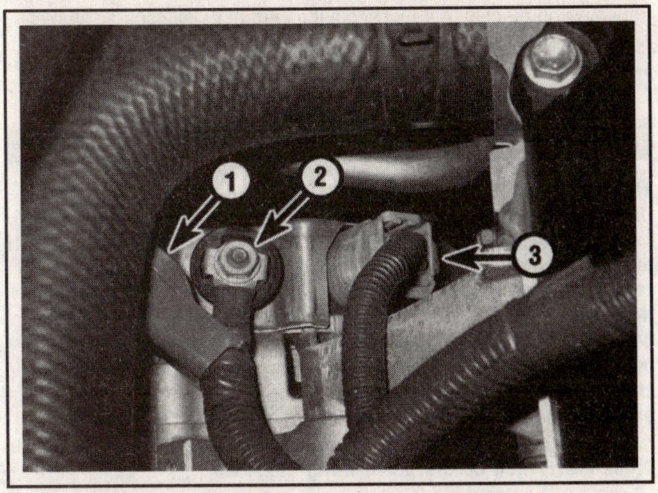

10.15 To disconnect the battery cable from the B+ terminal, pull off the cover (1) and remove this nut (2), then depress the release tab (3) and disconnect the electrical connector (four-cylinder models)

10.17 To detach the alternator, remove these two mounting bolts (four-cylinder models)

11 Starting system - general information and precautions

The starting system consists of the battery, the starter motor, the starter solenoid and the electrical circuit connecting the components. The solenoid is mounted directly on the starter motor. The starter circuit consists of the ignition switch, the starter relay, the MAIN and AM2 fuses, the harness wiring to the solenoid and the heavy gauge wiring to the starter. The manual transmission starting systems include the clutch start switch mounted at the clutch pedal, while automatic transmission systems include the Park/Neutral Position switch mounted on the transmission.

When the ignition key is turned to the START position, the starter solenoid is actuated through the starter control circuit. The starter solenoid then connects the battery to the starter. The battery supplies the electrical energy to the starter motor, which does the actual work of cranking the engine.

The starter motor on a vehicle equipped with a manual transmission can be operated only when the clutch pedal is depressed; the starter on a vehicle equipped with an automatic transmission can be operated only when the transmission selector lever is in Park or Neutral.

Always observe the following precautions when working on the starting system:

a) *Excessive cranking of the starter motor can overheat it and cause serious damage. Never operate the starter motor for more than 15 seconds at a time without pausing to allow it to cool for at least two minutes.*
b) *The starter is connected directly to the battery and could arc or cause a fire if mishandled, overloaded or short circuited.*
c) *Always detach the cable from the negative terminal of the battery before working on the starting system.*

12 Starter motor and circuit - check

➧ Refer to illustrations 12.3 and 12.4

1 If a malfunction occurs in the starting circuit, do not immediately assume that the starter is causing the problem. First, check the following items:

a) *Make sure the battery cable clamps, where they connect to the battery, are clean and tight.*
b) *Check the condition of the battery cables (see Section 4). Replace any defective battery cables with new parts.*
c) *Test the condition of the battery (see Section 3). If it does not pass all the tests, replace it with a new battery.*
d) *Check the starter solenoid wiring and connections. Refer to the wiring diagrams at the end of Chapter 12.*
e) *Check the starter mounting bolts for tightness.*
f) *Check the ignition switch circuit for correct operation (see the wiring diagrams at the end of Chapter 12).*
g) *Check the operation of the Park/Neutral Position switch (automatic transmission) or clutch start switch (manual transmission). Make sure the shift lever is in PARK or NEUTRAL (automatic transmission) or the clutch pedal is pressed (manual transmission). Refer to Chapter 7 for the Park/Neutral Position switch check and adjustment procedure. When performing circuit checks, refer to the wiring diagrams at the end of Chapter 12, if necessary. These systems must operate correctly to provide battery voltage to the ignition solenoid.*
h) *Check the operation of the starter relay. The starter relay is located in the fuse/relay box inside the engine compartment. Refer to Chapter 12 for the testing procedure.*

12.3 To use an inductive ammeter, simply hold the ammeter over the positive or negative cable (whichever is easier in terms of clearance)

2 If the starter does not actuate when the ignition switch is turned to the start position, check for battery voltage to the solenoid. This will determine if the solenoid is receiving the correct voltage signal from the ignition switch. Connect a test light or voltmeter to the starter solenoid positive terminal and while an assistant turns the ignition switch to the start position. If voltage is not available, refer to the wiring diagrams in Chapter 12 and check all the fuses and relays in the starting system. If voltage is available but the starter motor does not operate, remove the starter (see Section 13) and bench test it (see Step 4).

3 If the starter turns over slowly, check the starter cranking voltage and the current draw from the battery. This test must be performed with the starter assembly on the engine. Crank the engine over (for 10 seconds or less) and observe the battery voltage. It should not drop below 8.0 volts on manual transmission models or 8.5 volts on automatic transmission models. Also, observe the current draw using an ammeter (see illustration). It should not exceed 400 amps or drop below 250 amps.

⁂ CAUTION:

The battery cables may be excessively heated because of the large amount of amperage being drawn from the battery. Discontinue the testing until the starting system has cooled down. If the starter motor cranking amp values are not within the correct range, replace it with a new unit. There are several conditions that may affect the starter cranking potential. The battery must be in good condition and the battery cold-cranking rating must not be under-rated for the particular application. Be sure to check the battery specifications carefully. The battery terminals and cables must be clean and not corroded. Also, in cases of extreme cold temperatures, make sure the battery and/or engine block is warmed before performing the tests.

4 If the starter is receiving voltage but does not activate, remove and check the starter/solenoid assembly on the bench (see illustration). Most likely the solenoid is defective. In some rare cases, the engine may be seized so be sure to try and rotate the crankshaft pulley (see Chapter 2A or 2B) before proceeding. With the starter/solenoid assembly mounted in a vise on the bench, install one jumper cable from the negative battery terminal to the body of the starter. Install the other jumper cable from the positive battery terminal to the B+ terminal on the starter. Install a starter switch and apply battery voltage to the solenoid S terminal (for 10 seconds or less) and see if the solenoid plunger, shift lever and overrunning clutch extends and rotates the pinion drive. If the pinion drive extends but does not rotate, the solenoid is operating but the starter motor is defective. If there is no movement but the solenoid clicks, the solenoid and/or the starter motor is defective. If the solenoid plunger extends and rotates the pinion drive, the starter/solenoid assembly is working properly.

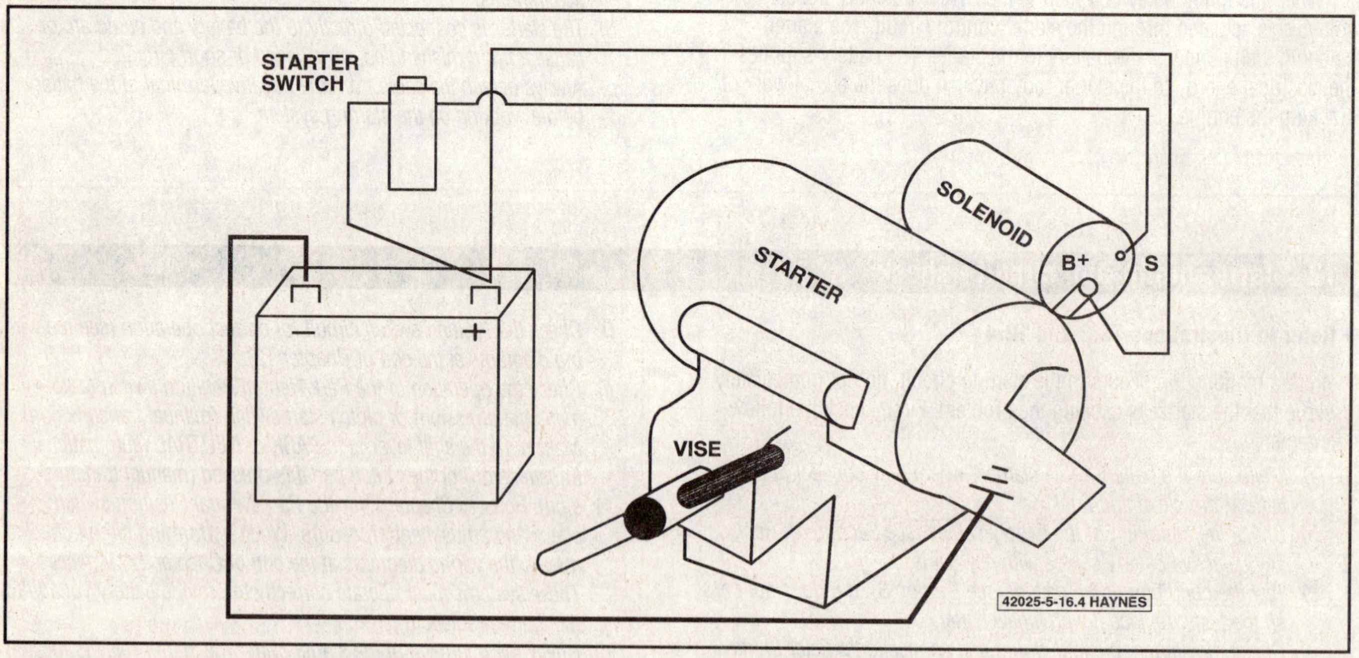

12.4 Starter motor bench testing details

13 Starter motor - removal and installation

1 Disconnect the cable from the negative battery terminal (see Section 1).

2 Raise the front of the vehicle and place it securely on jackstands.

V6 MODELS

Refer to illustrations 13.8 and 13.9

Note: The starter motor is located on the left side of the engine block, just ahead of the transmission.

3 Remove the rear engine under-cover, if equipped.

4 On 2WD models, if applicable, remove the three manifold support bracket bolts and remove the support bracket.

5 On 4WD models, remove the left part of the front exhaust pipe assembly.

6 On 4WD models, disengage the clips that secure the left rear part of the inner fender splash shield and remove the splash shield.

7 On 4WD models, disconnect the steering column intermediate shaft (see Chapter 10).

8 Disconnect the electrical connectors from the starter motor/solenoid assembly (see illustration).

9 Remove the starter motor mounting bolts (see illustration) and remove the starter motor.

10 Installation is the reverse of removal. Be sure to tighten the starter motor mounting bolts securely.

FOUR-CYLINDER MODELS

Refer to illustrations 13.12 and 13.13

Note: The starter motor is located on the left side of the engine block, just ahead of the transmission.

11 Disconnect the cable from the negative battery terminal (see Section 1).

12 Disconnect the electrical connectors from the starter motor (see illustration).

13 Remove the two starter motor mounting bolts (see illustration) and remove the starter.

14 Installation is the reverse of removal. Be sure to tighten the starter motor mounting bolts securely.

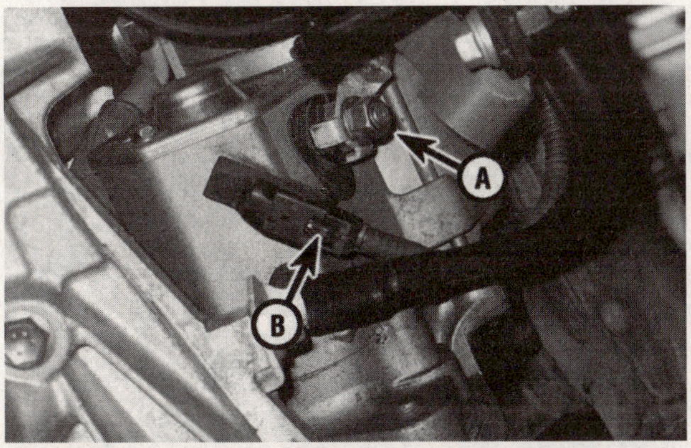

13.8 Remove this nut (A) and disconnect the battery starter cable from the B+ terminal, then disconnect the electrical connector (B) from the starter (V6 models)

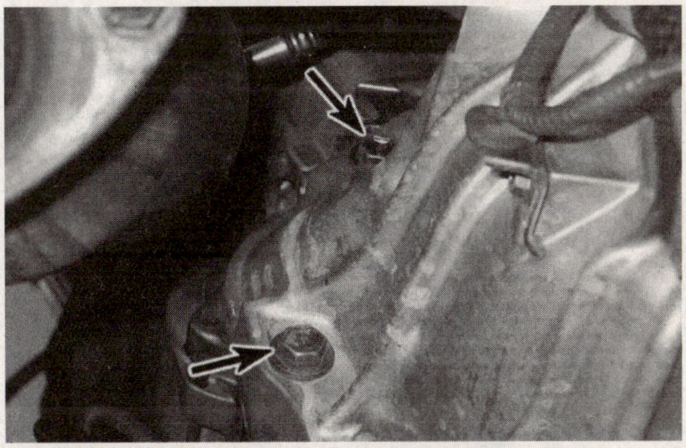

13.9 To detach the starter motor, remove these two mounting bolts (V6 models)

13.12 Remove the cover (1), remove this nut (2) and disconnect the battery starter cable from the B+ terminal, then depress this release tab (2) and disconnect the electrical connector from the starter (four-cylinder models)

13.13 To detach the starter motor, remove these two mounting bolts (four-cylinder models)

Specifications

Charging system

Charging voltage	13.2 to 14.8 volts

6 EMISSIONS AND ENGINE CONTROL SYSTEMS

Section

1 General information

➧ Refer to illustration 1.6

To prevent pollution of the atmosphere from incompletely burned and evaporating gases, and to maintain good driveability and fuel economy, a number of emission control systems are incorporated. They include the:

Acoustic Control Induction System (ACIS) (V6 models)
Air injection system (four-cylinder models)
Catalytic converters
Electronic Throttle Control System-intelligent (ETCS-i)
Evaporative Emissions Control (EVAP) system
On-Board Diagnostics (OBD-II) system
Positive Crankcase Ventilation (PCV) system
Sequential Electronic Fuel Injection (SFI) system
Variable Valve Timing-intelligent (VVT-i) system

The Sections in this Chapter include general descriptions, checking procedures within the scope of the home mechanic and component replacement procedures (when possible) for each of the systems listed above.

Before assuming that an emissions control system is malfunctioning, check the fuel and ignition systems carefully. The diagnosis of some emission control devices requires specialized tools, equipment and training. If checking and servicing become too difficult or if a procedure is beyond your ability, consult a dealer service department or other repair shop. Remember, the most frequent cause of emissions problems is simply a loose or broken wire or vacuum hose, so always check the hose and wiring connections first.

This doesn't mean, however, that emissions control systems are particularly difficult to maintain and repair. You can quickly and easily perform many checks and do most of the regular maintenance at home with common tune-up and hand tools.

➡Note: Because of a federally mandated warranty which covers the emissions control system components, check with your dealer about warranty coverage before working on any emissions-related systems. Once the warranty has expired, you may wish to perform some of the component checks and/or replacement procedures in this Chapter to save money.

Pay close attention to any special precautions outlined in this Chapter. It should be noted that the illustrations of the various systems may not exactly match the system installed on your vehicle because of changes made by the manufacturer during production or from year-to-year.

A Vehicle Emissions Control Information (VECI) label (see illustration) is attached to the underside of the hood. This label contains important emissions specifications and adjustment information. Another label, the Vacuum Hose Routing Diagram, provides a vacuum hose schematic with emissions components identified. When servicing the engine or emissions systems, the VECI label and the vacuum hose routing diagram in your particular vehicle should always be checked.

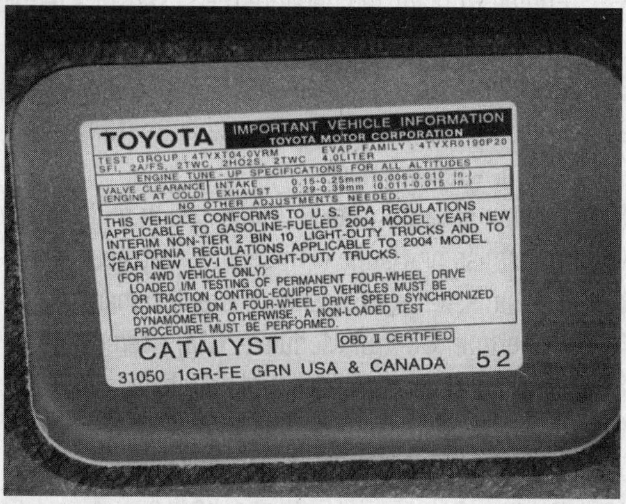

1.6 The Vehicle Emission Control Information (VECI) label contains such essential information as the types of emission control systems installed on the engine and the idle speed and ignition timing specifications (V6 model shown)

2 On-Board Diagnostic (OBD) system and trouble codes

SCAN TOOL INFORMATION

➧ Refer to illustration 2.2

1 Hand-held scanners are handy for analyzing the engine management systems used on late-model vehicles. Because extracting the Diagnostic Trouble Codes (DTCs) from an engine management system is now the first step in troubleshooting many computer-controlled systems and components, even the most basic generic code readers are capable of accessing a computer's DTCs (see illustration). More powerful scan tools can also perform many of the diagnostics once associated with expensive factory scan tools. If you're planning to obtain a generic scan tool for your vehicle, make sure that it's compatible with OBD-II systems. If you don't plan to purchase a code reader or scan tool and don't have access to one, you can have the codes extracted by a dealer service department or by an independent repair shop.

➡Note: Before purchasing an aftermarket generic scan tool, verify that it will work properly with the OBD-II system you want to scan. If necessary, of course, you can always have the codes extracted by a dealer service department or an independent repair shop with a professional scan tool. Some auto parts stores even provide this service for free.

2 With the advent of the Federally mandated emission control system known as On-Board Diagnostics-II (OBD-II), specially designed scanners were developed. Several tool manufacturers have released OBD-II scan tools for the home mechanic (see illustration).

OBD-II SYSTEM

3 All vehicles covered by this manual are equipped with the OBD-II system. This system consists of the on-board computer, known as the Powertrain Control Module (PCM) and information sensors that moni-

2.1 Simple code readers are an economical way to extract trouble codes when the CHECK ENGINE light comes on

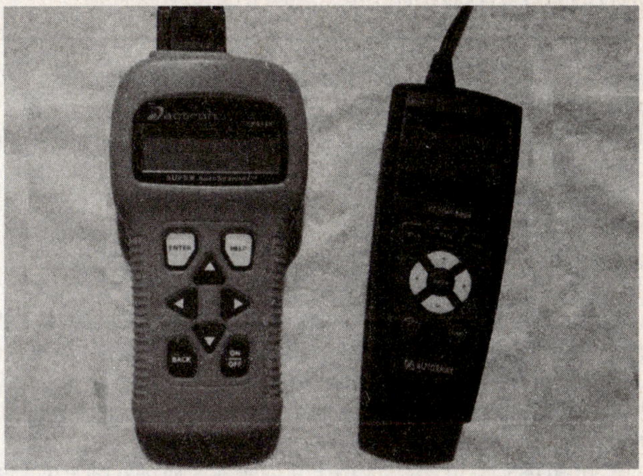

2.2 Scanners like these two units from Actron and AutoXray are powerful diagnostic aids; they can tell you just about anything you want to know about your engine management system

tor various functions of the engine and send a constant stream of data to the PCM during engine operation.

4 The PCM is the brain of the electronically controlled OBD-II system. It receives data from a number of information sensors and switches. Based on the data that it receives from the sensors, the PCM constantly alters engine operating conditions to optimize driveability, performance, emissions and fuel economy. It does so by turning on and off and by controlling various output actuators such as relays, solenoids, valves and other devices. The PCM can only be accessed with an OBD-II scan tool plugged into the 16-pin Data Link Connector (DLC), which is located underneath the driver's end of the dashboard, near the steering column.

5 If your vehicle is still under warranty, virtually every fuel, ignition and emission control component in the OBD-II system is covered by a Federally mandated emissions warranty that is longer than the warranty covering the rest of the vehicle. Vehicles sold in California and in some other states have even longer emissions warranties than other states. Read your owner's manual for the terms of the warranty protecting the emission-control systems on your vehicle. It isn't a good idea to do-it-yourself at home while the vehicle emission systems are still under warranty because owner-induced damage to the PCM, the sensors and/or the control devices might VOID this warranty. So as long as the emission systems are still under warranty, take the vehicle to a dealer service department if there's a problem.

INFORMATION SENSORS

6 **Accelerator pedal position sensor** - The accelerator pedal position sensor is an integral component of the accelerator pedal assembly. There is no accelerator cable. This accelerator pedal position sensor is a variable potentiometer that uses the position of the accelerator pedal as its input. The PCM uses this data to calculate the correct position for the throttle plate and directs the throttle motor inside the throttle body to open and close the throttle plate accordingly.

7 **Air/Fuel Ratio Sensor** - Toyota also refers to the upstream oxygen sensors as air/fuel ratio sensors. But they're essentially the same devices; they're installed upstream in relation to the first catalytic converters and work the same way as oxygen sensors. See Step 13 for more information about the oxygen sensors.

8 **Camshaft Position (CMP) sensor** - The camshaft position sensor produces a signal that the PCM uses to identify the number 1 cylinder and to time the sequential fuel injection. On V6 models, there are two CMP sensors. Toyota refers to the CMP sensors on V6 models as Variable Valve Timing (VVT) sensors. For more information about the VVT sensors, refer to Section 19, Variable Valve Timing-intelligent (VVT-i) system - description and component replacement.

9 **Crankshaft Position (CKP) sensor** - The crankshaft sensor signal provides data on crankshaft position and engine speed to the PCM.

10 **Engine Coolant Temperature (ECT) sensor** - The coolant temperature (ECT) sensor monitors engine coolant temperature and sends the PCM a voltage signal that affects PCM control of the fuel mixture, ignition timing, and EGR operation.

11 **Knock sensors** - The knock sensors monitor engine knock (pre-ignition or detonation), then signal the PCM when knock occurs, so that the PCM can retard ignition timing accordingly. There are two knock sensors on all engines, one for each cylinder bank.

12 **Mass Air Flow/Intake Air Temperature (MAF/IAT) sensor** - The MAF/IAT sensor measures the mass of the intake air by monitoring the volume and weight of the air passing over a hot-wire element.

13 **Oxygen sensors** - Four-cylinder models have one upstream and one downstream oxygen sensor; V6 models have two upstream and two downstream oxygen sensors. The upstream oxygen sensors generate a voltage signal that varies in accordance with the difference between the oxygen content of the exhaust gases and the oxygen in the surrounding air. The downstream oxygen sensors monitor the content of the exhaust gases as they exit the downstream catalytic converter. This information is used by the PCM to predict catalyst deterioration and/or failure.

14 **Transmission Range (TR) sensor** - The TR sensor prevents the engine from being started unless the shift lever is in PARK or NEUTRAL. The TR sensor also sends a signal to the PCM so that it knows what gear the transmission is in. Toyota still refers to the TR sensor as the Park/Neutral Position (PNP) switch. Refer to Chapter 7B for information on replacing and adjusting the PNP switch.

15 **Transmission speed sensors** - The transmission speed sensors tell the PCM the rotational speed of the input shaft and/or output shaft. The PCM uses this information to monitor slippage inside the transmission and to predict failure of critical transmission components.

OUTPUT ACTUATORS

16 **Acoustic Control Induction System (ACIS) Vacuum Switching Valve (VSV)** - The ACIS is used on V6 engines. The ACIS consists of a specially designed intake manifold with an intake air control valve inside the manifold that can direct incoming air through a longer or shorter intake length in accordance with engine speed and throttle valve opening angle. The PCM-controlled ACIS VSV controls the vacuum signal to the vacuum-controlled actuator, which opens and closes the intake air control valve inside the manifold. The ACIS increases power throughout the engine operating range from low speed to high speed.

17 **Air injection control valve** - The PCM-controlled air injection control valve, which is a component in the air injection system employed on four-cylinder models, consists of an air switching valve and a reed valve. When directed to do so by the PCM, the air switching valve directs air from the air pump into the exhaust manifold. The reed valve prevents exhaust gases from back flowing into the air injection control valve. For more information about the air injection system, refer to Section 18, Air injection system - description and component replacement.

18 **Air pump** - The PCM-controlled electric air pump is a component in the air injection system employed on four-cylinder models. The air pump supplies air to the air injection control valve (see above), which directs air to the exhaust manifold. For more information about the air injection system, refer to Section 18, Air injection system - description and component replacement.

19 **EVAP canister vent valve** - The PCM-controlled EVAP canister vent valve opens and closes the ventilation line that brings fresh air into the EVAP system and controls pressure relief if the internal pressure in the fuel tank exceeds the designed threshold. For more information about the EVAP system, refer to Section 15, Evaporative emissions control (EVAP) system - description and component replacement.

20 **EVAP canister purge valve** - The PCM-controlled EVAP canister purge valve is a solenoid valve that meters vapors from the EVAP canister into the intake manifold, from which they're drawn into the combustion chambers and consumed with the incoming air/fuel mixture. For more information about the EVAP system, refer to Section 15, Evaporative emissions control (EVAP) system - description and component replacement.

21 **Fuel injectors** - The PCM opens the fuel injectors sequentially (in firing order sequence). The PCM also controls the pulse width, the interval of time during which each injector is open. The pulse width of an injector (measured in milliseconds) determines the amount of fuel delivered. For more information on the fuel delivery system and the fuel injectors, including injector replacement, refer to Chapter 4.

22 **Ignition coils** - The ignition coils are triggered by the PCM. Refer to Chapter 5 for more information on the ignition coils.

23 **Throttle control motor** - The throttle control motor, which is mounted on the throttle body, opens and closes the throttle plate. The PCM, which uses inputs from the accelerator pedal position sensor, the throttle position sensor and other sensors to calculate the correct throttle angle for the conditions, controls the throttle control motor, which in turn controls the throttle plate.

OBTAINING OBD-II SYSTEM TROUBLE CODES

Refer to illustration 2.25

24 The PCM will illuminate the CHECK ENGINE light (also called the Malfunction Indicator Light) on the dash if it recognizes a component fault for two consecutive drive cycles. It will continue to set the light until the PCM does not detect any malfunction for three or more consecutive drive cycles.

25 The diagnostic codes for the OBD-II system can be extracted from the PCM by plugging a generic OBD-II scan tool (see illustration 2.2) into the PCM's data link connector (see illustration), which is located under the left end of the dash, above the parking brake pedal assembly.

26 Plug the scan tool into the 16-pin data link connector (DLC), and follow the instructions included with the scan tool to extract all the diagnostic codes.

CLEARING DIAGNOSTIC TROUBLE CODES

27 After the system has been repaired, the codes must be cleared from the PCM memory using a scan tool. Do not attempt to clear the codes by disconnecting battery power. If battery power is disconnected from the PCM, the PCM will lose the current engine operating parameters and driveability will suffer until the PCM re-learns the optimum operating parameters after driving the vehicle for a period of time.

28 Always clear the codes from the PCM before starting the engine after a new electronic emission control component is installed. The PCM stores the operating parameters of each sensor and may set a trouble code if a new sensor is allowed to operate before the parameters from the old sensor have been erased.

2.25 The 16-pin Data Link Connector (DLC) is located under the left end of the dash, above the parking brake pedal assembly

DIAGNOSTIC TROUBLE CODES

Code	Code identification
P0010	Camshaft position A actuator circuit (Bank 1)
P0011	Camshaft position A, timing over-advanced or system performance (Bank 1)
P0012	Camshaft position A, timing over-retarded (Bank 1)
P0016	Crankshaft position/camshaft position correlation (Bank 1, sensor A)
P0018	Crankshaft position/camshaft position correlation (Bank 2, sensor A)
P0020	Camshaft position A actuator circuit (Bank 2)
P0021	Camshaft position A, timing over-advanced or system performance (Bank 1)
P0022	Camshaft position A, timing over-retarded (Bank 1)
P0031	Oxygen sensor heater control circuit, low voltage (Bank 1, sensor 1)
P0032	Oxygen sensor heater control circuit, high voltage (Bank 1, sensor 1)
P0037	Oxygen sensor heater control circuit, low voltage (Bank 1, sensor 2)
P0038	Oxygen sensor heater control circuit, high voltage (Bank 1, sensor 2)
P0051	Oxygen sensor heater control circuit, low voltage (Bank 2, sensor 1)
P0052	Oxygen sensor heater control circuit, high voltage (Bank 2, sensor 1)
P0057	Oxygen sensor heater control circuit, low voltage (Bank 2, sensor 2)
P0058	Oxygen sensor heater control circuit, high voltage (Bank 2, sensor 2)
P0100	Mass Air Flow (MAF) sensor, circuit fault
P0101	Mass Air Flow (MAF) sensor, range or performance problem
P0102	Mass Air Flow (MAF) sensor circuit, low input voltage
P0103	Mass Air Flow (MAF) sensor circuit, high input voltage
P0110	Intake Air Temperature (IAT) sensor (in MAF sensor), circuit fault
P0112	Intake Air Temperature (IAT) sensor circuit, low input voltage
P0113	Intake Air Temperature (IAT) sensor circuit, high input voltage
P0115	Engine Coolant Temperature (ECT) sensor, circuit fault
P0116	Engine Coolant Temperature (ECT) sensor, circuit range or performance problem
P0117	Engine Coolant Temperature (ECT) sensor, low input voltage
P0118	Engine Coolant Temperature (ECT) sensor, high input voltage
P0120	Throttle/Accelerator Pedal Position sensor A or circuit fault
P0121	Throttle/Accelerator Pedal Position sensor A, circuit range or performance problem
P0122	Throttle/Accelerator Pedal Position sensor A circuit, low voltage input

DIAGNOSTIC TROUBLE CODES (CONTINUED)

Code	Code identification
P0123	Throttle/Accelerator Pedal Position sensor A circuit, high voltage input
P0125	Insufficient coolant temperature for closed-loop fuel control
P0128	Coolant temperature below thermostat regulating temperature
P0136	Oxygen sensor circuit fault (Bank 1, sensor 2)
P0137	Oxygen sensor circuit, low voltage (Bank 1, sensor 2)
P0138	Oxygen sensor circuit, high voltage (Bank 1, sensor 2)
P0141	Oxygen sensor heater or circuit fault (Bank 1, sensor 2)
P0156	Oxygen sensor circuit malfunction (Bank 2, sensor 2)
P0157	Oxygen sensor circuit, low voltage (Bank 2, sensor 2)
P0158	Oxygen sensor circuit, high voltage (Bank 2, sensor 2)
P0161	Oxygen sensor heater circuit problem (Bank 2, sensor 2)
P0171	Fuel injection system too lean (Bank 1)
P0172	Fuel injection system too rich (Bank 1)
P0174	Fuel injection system lean (Bank 2)
P0175	Fuel injection system rich (Bank 2)
P0220	Throttle/Accelerator Pedal Position sensor B, circuit fault
P0222	Throttle/Accelerator Pedal Position sensor B circuit, low voltage input
P0223	Throttle/Accelerator Pedal Position sensor B circuit, high voltage input
P0230	Fuel pump primary circuit
P0300	Random/multiple cylinder misfire detected
P0301	Cylinder No. 1 misfire detected
P0302	Cylinder No. 2 misfire detected
P0303	Cylinder No. 3 misfire detected
P0304	Cylinder No. 4 misfire detected
P0305	Cylinder No. 5 misfire detected
P0306	Cylinder No. 6 misfire detected
P0325	Knock Sensor No. 1, circuit fault
P0327	Knock Sensor No. 1 circuit, low voltage (Bank 1 or single sensor)
P0328	Knock Sensor No. 1 circuit, high voltage (Bank 1 or single sensor)
P0330	Knock sensor No. 2, circuit fault

Code	Code identification
P0332	Knock Sensor No. 2 circuit, low voltage (Bank 2)
P0333	Knock sensor No. 2 circuit, high voltage (Bank 2)
P0335	Crankshaft Position (CKP) sensor A, circuit fault
P0339	Crankshaft Position (CKP) sensor A, circuit intermittent
P0340	Camshaft Position (CMP) sensor A circuit (Bank 1 or single sensor)
P0341	Camshaft Position (CMP) sensor A circuit, range or performance problem (Bank 1 or single sensor)
P0345	Camshaft Position (CMP) sensor A circuit (Bank 2)
P0346	Camshaft Position (CMP) sensor A circuit, range or performance problem (Bank 2)
P0351	Ignition coil 1 primary/secondary circuit
P0352	Ignition coil 2 primary/secondary circuit
P0353	Ignition coil 3 primary/secondary circuit
P0354	Ignition coil 4 primary/secondary circuit
P0355	Ignition coil 5 primary/secondary circuit
P0356	Ignition coil 6 primary/secondary circuit
P0412	Air injection system air switching valve malfunction
P0418	Air injection system air pump malfunction
P0420	Catalyst system efficiency below threshold (Bank 1)
P0430	Catalyst system efficiency below threshold (Bank 2)
P043E	Evaporative Emission Control (EVAP) system, reference orifice clogged
P043F	Evaporative Emission Control (EVAP) system, reference orifice high flow
P0441	Evaporative Emission Control (EVAP) system, incorrect purge flow
P0442	Evaporative Emission Control (EVAP) system, small leak detected
P0446	EVAP canister vent control valve, circuit fault
P0450	EVAP system vapor pressure sensor (fuel tank pressure sensor)
P0451	EVAP system vapor pressure sensor, range or performance problem
P0452	EVAP system vapor pressure sensor, low voltage input
P0453	EVAP system vapor pressure sensor, high voltage input
P0455	Evaporative Emission Control (EVAP) system, big leak detected
P0456	Evaporative Emission Control (EVAP) system, very small leak detected
P0500	Vehicle Speed Sensor (VSS) A, circuit malfunction
P0503	Vehicle Speed Sensor (VSS) A, intermittent, erratic or and/or high voltage

DIAGNOSTIC TROUBLE CODES (CONTINUED)

Code	Code identification
P0504	Brake switch correlation
P0505	Idle air control system
P0560	System voltage
P0571	Brake switch A, circuit malfunction
P0604	Internal Control Module Random Access Memory (RAM) error
P0606	Powertrain Control Module (PCM) processor
P0607	Powertrain Control Module (PCM) performance
P0617	Starter relay circuit, high voltage
P0630	Vehicle Identification Number (VIN) not programmed or mismatched PCM
P0657	Actuator supply voltage, open circuit
P0705	Transmission Range (TR) sensor circuit malfunction (PRNDL input)
P0710	Transmission fluid temperature sensor A circuit malfunction
P0711	Transmission fluid temperature sensor A performance
P0712	Transmission fluid temperature sensor A circuit, low input voltage
P0713	Transmission fluid temperature sensor A circuit, high input voltage
P0717	Input/turbine speed sensor A circuit, no signal
P0722	Output speed sensor circuit, no signal
P0724	Brake switch B circuit, high voltage
P0748	Pressure control solenoid A electrical (shift solenoid valve SL1)
P0751	Shift solenoid A performance (shift solenoid valve S1)
P0756	Shift solenoid B performance (shift solenoid valve S2)
P0771	Shift solenoid E performance (shift solenoid valve SR)
P0776	Pressure control solenoid B performance (shift solenoid valve SL2)
P0778	Pressure control solenoid B electrical (shift solenoid valve SL2)
P0781	1-2 shift (1-2 shift valve)
P0850	Park/Neutral Position (PNP) switch, input circuit malfunction
P0973	Shift solenoid A control circuit, low voltage (shift solenoid valve S1)
P0974	Shift solenoid A control circuit, high voltage (shift solenoid valve S1)
P0976	Shift solenoid B control circuit, low voltage (shift solenoid valve S2)
P0977	Shift solenoid B control circuit, high voltage (shift solenoid valve S2)
P0985	Shift solenoid E control circuit, low voltage (shift solenoid valve SR)
P0986	Shift solenoid E control circuit, high voltage (shift solenoid valve SR)

3 Accelerator Pedal Position (APP) sensor - replacement

Refer to illustration 3.2

Note: The APP sensor is an integral component of the accelerator pedal assembly. To replace the APP sensor, you must replace the entire accelerator pedal assembly.

1 Disconnect the cable from the negative battery terminal (see Chapter 5, Section 1).

2 Working under the dash with a flashlight, disconnect the electrical connector from the APP sensor (see illustration).

3 Remove the two APP sensor mounting fasteners and remove the APP sensor.

4 Installation is the reverse of removal.

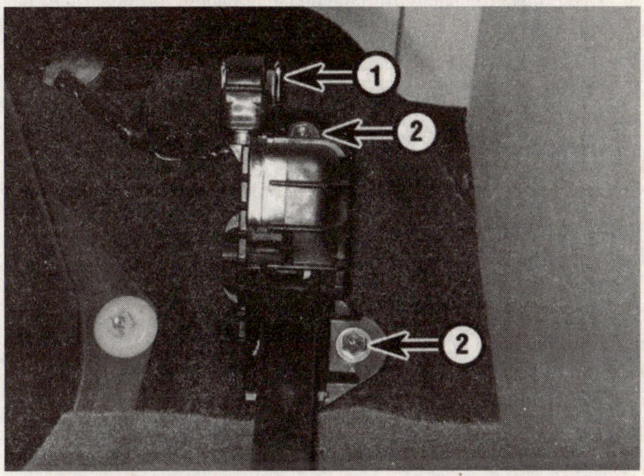

3.2 To remove the APP sensor, depress the release tab (1) and disconnect the electrical connector, then remove the two mounting fasteners (2)

4 Camshaft Position (CMP) sensor - replacement

V6 MODELS

1 On V6 models, there are two CMP sensors, which are located on the inside walls of the cylinder heads, adjacent to the timing rotor installed on the front of each Variable Valve Timing-intelligent (VVT-i) controller. Toyota refers to the CMP sensors on these models as VVT sensors, so they are covered in Section 19.

FOUR-CYLINDER MODELS

Refer to illustration 4.4

Note: The CMP sensor is located on the front of the cylinder head, adjacent to the intake camshaft timing belt sprocket.

2 Disconnect the cable from the negative battery terminal (see Chapter 5, Section 1).

3 Remove the accessory drivebelt (see Chapter 1).

4 Disconnect the CMP sensor electrical connector (see illustration).

5 Remove the CMP sensor retaining bolt.

6 Remove and discard the old CMP sensor O-ring.

7 Installation is the reverse of removal. Be sure to tighten the CMP sensor retaining bolt and stud securely.

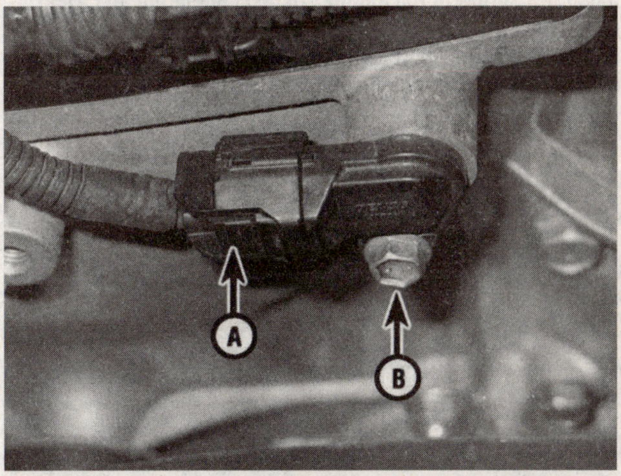

4.4 To disconnect the electrical connector from the CMP sensor, depress the release tab (A) and disconnect the electrical connector, then remove the sensor retaining bolt (B)

5 Crankshaft Position (CKP) sensor - replacement

1 Disconnect the cable from the negative battery terminal (see Chapter 5, Section 1).

2 Raise the vehicle and place it securely on jackstands.

3 Remove the engine under-cover, if equipped.

V6 MODELS

Refer to illustration 5.6

4 Remove the alternator (see Chapter 5).

5 Remove the bolt that secures the air conditioning suction line at the front of the engine, disconnect the electrical connector from the air conditioning compressor and unbolt the compressor (see Chapter 3). Do NOT disconnect the air conditioning hoses from the compressor. Move the compressor aside and support it with wire or rope.

6 Disconnect the CKP sensor electrical connector (see illustration).

7 Remove the CKP sensor retaining bolt.

8 Installation is the reverse of removal. Be sure to tighten the CKP sensor retaining bolt securely.

FOUR-CYLINDER MODELS

Refer to illustration 5.9, 5.10, 5.12 and 5.14

9 Locate the CKP sensor (see illustration).

10 Locate the CKP sensor electrical connector and disconnect it (see illustration).

11 Raise the vehicle and place it securely on jackstands.

12 Unbolt the air conditioning compressor (see Chapter 3), but don't disconnect the air conditioning refrigerant lines. Move the compressor aside and support it with wire or rope. Remove the five air conditioning compressor mounting bracket bolts (see illustration) and remove the air

5.6 On V6 models, the CKP sensor is located at the front lower left corner of the engine. To disconnect the electrical connector (A), slide the white plastic lock to the rear and pull off the connector. To detach the CKP sensor from the engine, remove the sensor mounting bolt (B)

5.9 On four-cylinder models, the CKP sensor (A) is located at the front lower left corner of the engine, beneath the air conditioning compressor mounting bracket. Before you can remove the sensor, you must remove the air conditioning compressor and the compressor mounting bracket. To detach the sensor, remove this bolt (B)

5.10 You'll find the electrical connector for the CKP sensor here, in front of the intake manifold. After disconnecting the two halves of the CKP sensor connector (which is already unplugged in this photo), detach the lower half of the connector from the small bracket to which it's attached

5.12 To detach the air conditioning compressor mounting bracket on four-cylinder models, remove these five bolts

conditioning compressor mounting bracket and drivebelt idler pulley as a single assembly.

13 Remove the CKP sensor retaining bolt (see illustration 5.9).

14 Remove the CKP sensor (see illustration) and remove and discard the old CKP sensor O-ring. Always use a new O-ring when installing the CKP sensor.

15 Installation is the reverse of removal. Be sure to tighten the CKP sensor retaining bolt securely.

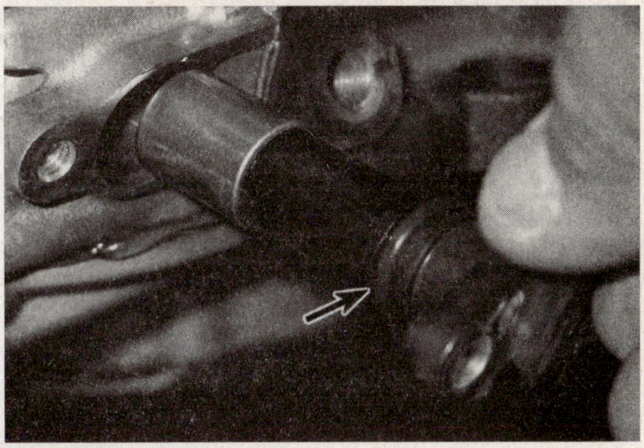

5.14 Be sure to remove and discard the old CKP sensor O-ring

6 Engine Coolant Temperature (ECT) sensor - replacement

⁂ WARNING:

Wait until the engine has cooled completely before beginning this procedure.

⁂ CAUTION:

Handle the Engine Coolant Temperature (ECT) sensor with care. Damage to the ECT sensor will affect the operation of the entire fuel injection system.

1 Drain the engine coolant until the coolant level is below the ECT sensor (see Chapter 1).

V6 MODELS

Refer to illustrations 6.3 and 6.5

2 Remove the engine cover (see Chapter 2B).

3 Disconnect the electrical connector from the ECT sensor (see illustration).

4 Unscrew the ECT sensor from the coolant crossover with a deep socket.

5 Before installing the new ECT sensor, wrap the threads of the sensor with Teflon tape to prevent coolant leakage (see illustration).

6 Installation is otherwise the reverse of removal. Be sure to tighten the ECT sensor to the torque listed in this Chapter's Specifications.

7 Refill the cooling system (see Chapter 1).

6.3 On V6 models, the ECT sensor is located on the rear coolant crossover between the cylinder heads

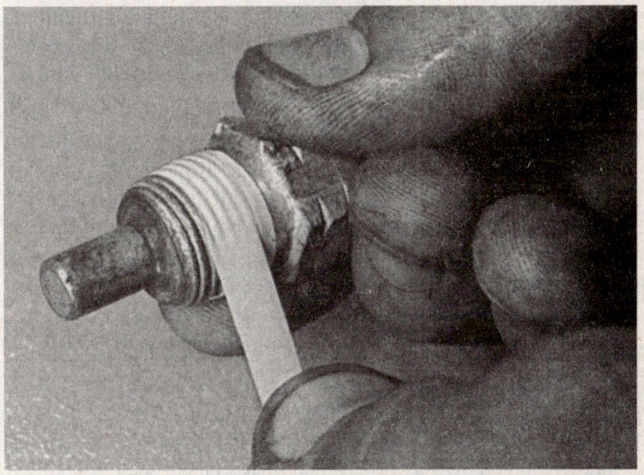

6.5 Wrap the threads of the ECT sensor with Teflon tape to prevent coolant leakage

FOUR-CYLINDER MODELS

➧ Refer to illustration 6.8

8 Disconnect the electrical connector from the ECT sensor (see illustration).

9 Unscrew and remove the ECT sensor from the front coolant crossover with a deep socket.

10 Before installing the new ECT sensor, wrap the threads of the sensor with Teflon tape to prevent coolant leakage (see illustration 6.5).

11 Installation is otherwise the reverse of removal. Be sure to tighten the ECT sensor to the torque listed in this Chapter's Specifications.

12 Refill the cooling system (see Chapter 1).

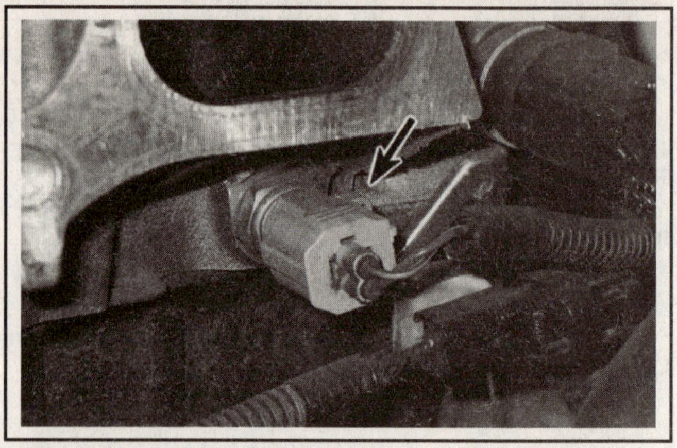

6.8 On four-cylinder models, the ECT sensor is located at the left rear corner of the cylinder head (intake manifold removed for clarity)

7 Intake Air Temperature (IAT) sensor - replacement

The IAT sensor is an integral component of the Mass Air Flow (MAF) sensor. See Section 9.

8 Knock sensor - replacement

⁂ WARNING:

Wait for the engine to cool completely before performing this procedure.

V6 MODELS

➡Note 1: The knock sensors are located on top of the engine block, under the intake manifold.

➡Note 2: Accessing the knock sensors is extremely difficult because you have to remove the cylinder heads, so we recommend that you replace BOTH knock sensors even if only one of them is defective.

1 Drain the engine coolant (see Chapter 1).

2 Remove the cylinder heads (see Chapter 2B).

3 Remove the heater coolant inlet hose and the coolant outlet pipe.

4 Disconnect the electrical connectors from the knock sensors.

5 Carefully note how the knock sensors are oriented, then remove the sensor retaining bolts and remove the sensors.

6 The orientation of the knock sensors is critical because they'll cause clearance problems during installation of the cylinder heads and intake manifold if installed incorrectly. So, when installing the knock sensors, make sure that each sensor is oriented in the same exact position that it was in before removal. When the knock sensors are correctly positioned, tighten the sensor retaining bolts to the torque listed in this Chapter's Specifications.

7 The remainder of installation is the reverse of removal.

8 Refill the engine cooling system (see Chapter 1).

FOUR-CYLINDER MODELS

➧ Refer to illustration 8.10

➡Note: The knock sensor is located on the left side of the engine block, below the intake manifold.

9 Remove the intake manifold (see Chapter 2A).

10 Disconnect the knock sensor electrical connector (see illustration).

8.10 The knock sensor is located on the left side of the engine block, below the intake manifold (intake manifold removed for clarity)

11 Note the orientation of the knock sensor, then remove the knock sensor retaining bolt and remove the sensor.

12 When installing the knock sensor, make sure that it's oriented correctly, with the connector pointing no more than 10 degrees above or below the horizontal.

13 Installation is otherwise the reverse of removal. Be sure to tighten the knock sensor retaining bolt to the torque listed in this Chapter's Specifications.

9 Mass Air Flow/Intake Air Temperature (MAF/IAT) sensor - replacement

➧ Refer to illustration 9.2a and 9.2b

➡Note: The MAF/IAT sensor is located on top of the air filter housing.

1 Make sure the ignition key is in the Off position.

2 Disconnect the MAF/IAT sensor electrical connector (see illustrations).

3 Remove the MAF sensor retaining screws and remove the sensor.

4 Installation is the reverse of removal.

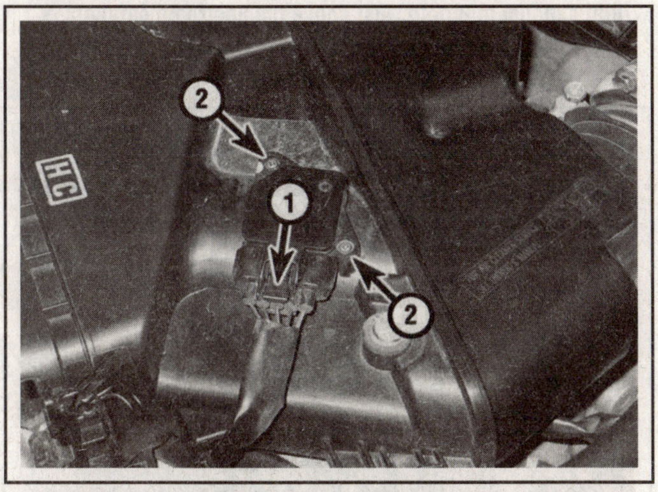

9.2a To detach the MAF/IAT sensor from the air filter housing on a V6 model, depress this release tab (1) and disconnect the electrical connector, then remove the sensor retaining screws (2)

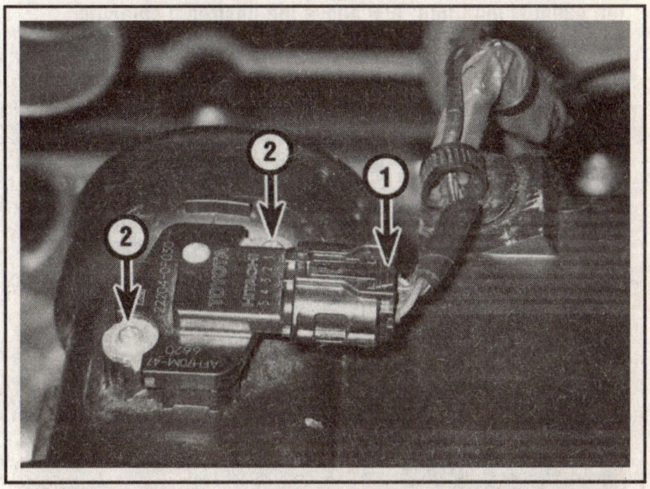

9.2b To detach the MAF/IAT sensor from the air filter housing on a four-cylinder model, depress the release tab (1) and disconnect the electrical connector, then remove the two sensor retaining screws (2)

10 Oxygen sensors - general information and replacement

GENERAL INFORMATION

1 Use special care when servicing an oxygen sensor or air/fuel ratio sensor:

- *a) Oxygen sensors and air/fuel ratio sensors have a permanently attached pigtail and electrical connector which cannot be removed from the sensor. Damage or removal of the pigtail or electrical connector will ruin the sensor.*
- *b) Grease, dirt and other contaminants should be kept away from the electrical connector and the louvered end of the sensor.*
- *c) Do not use cleaning solvents of any kind on an oxygen sensor or air/fuel ratio sensor.*
- *d) Do not drop or roughly handle an oxygen sensor or air/fuel ratio sensor.*
- *e) Be sure to install the silicone boot in the correct position to prevent the boot from melting and to allow the sensor to operate properly.*

REPLACEMENT

➡Note: Because it is installed in an exhaust manifold or exhaust pipe, which contracts when cool, an oxygen sensor or air/fuel ratio sensor might be very difficult to loosen when the engine is cold. Rather than risk damage to the sensor, start and run the engine for a minute or two, then shut it off. Be careful not to burn yourself during the following procedure.

2 Make sure the ignition key is in the OFF position.

3 Raise the vehicle and place it securely on jackstands.

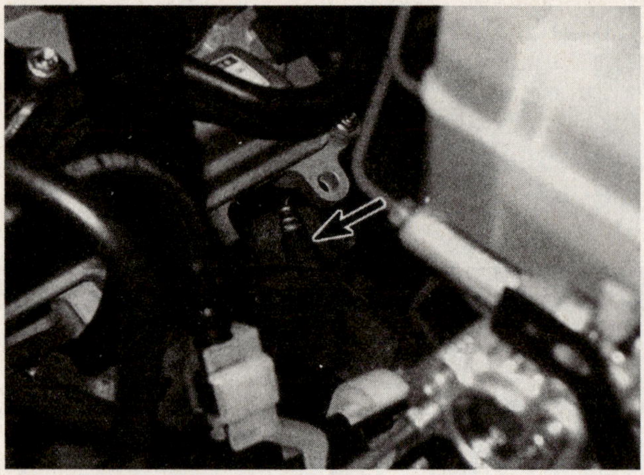

10.4a The left upstream oxygen sensor is located on the left exhaust manifold, just above the upstream catalytic converter (V6 models)

10.4b The right upstream oxygen sensor is located on the right exhaust manifold, just above the upstream catalytic converter (V6 models)

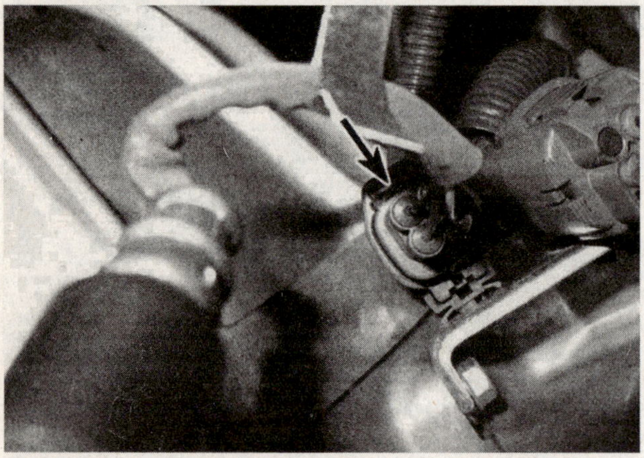

10.4c Trace the electrical lead from the upstream oxygen sensor down to the sensor electrical connector, which is located above the bellhousing, then depress the release tab and disconnect the connector

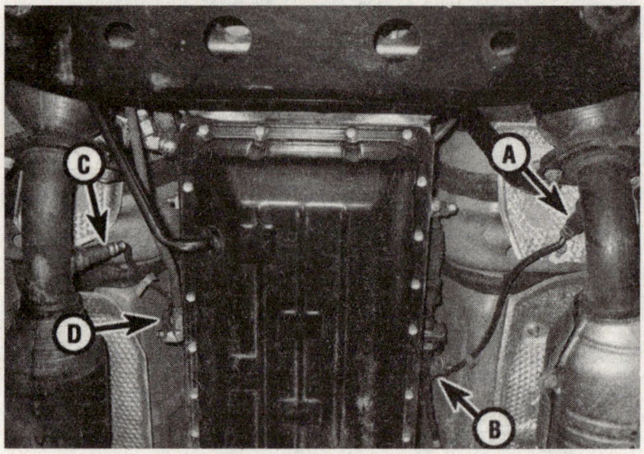

10.9 Downstream oxygen sensor locations (V6 models)

A Left downstream oxygen sensor
B Electrical connector for left sensor
C Right downstream oxygen sensor
D Electrical connector for right sensor

V6 models

➡Note: There are four oxygen sensors, one upstream sensor and one downstream sensor for each cylinder bank.

Upstream oxygen sensor

➧ Refer to illustrations 10.4a, 10.4b and 10.4c

➡Note: The upstream sensors are located at the outlet ends of the exhaust manifolds, just above the integral upstream catalysts.

4 Locate the upstream sensor (see illustrations), then trace the electrical lead to the sensor electrical connector (see illustration) and disconnect it.

5 Unscrew the upstream sensor with a wrench (there isn't room to use an oxygen sensor socket).

6 Apply anti-seize compound to the threads of the sensor to facilitate future removal.

➡Note: Most new sensors already have anti-seize compound applied to the threads.

7 Tighten the oxygen sensor to the torque listed in this Chapter's Specifications.

8 Installation is otherwise the reverse of removal.

Downstream oxygen sensor

➧ Refer to illustrations 10.9 and 10.10

➡Note: The downstream sensors are located in the front exhaust pipe assembly, ahead of the downstream catalysts.

9 Locate the downstream oxygen sensor (see illustration), then trace the electrical lead to the sensor electrical connector and disconnect it.

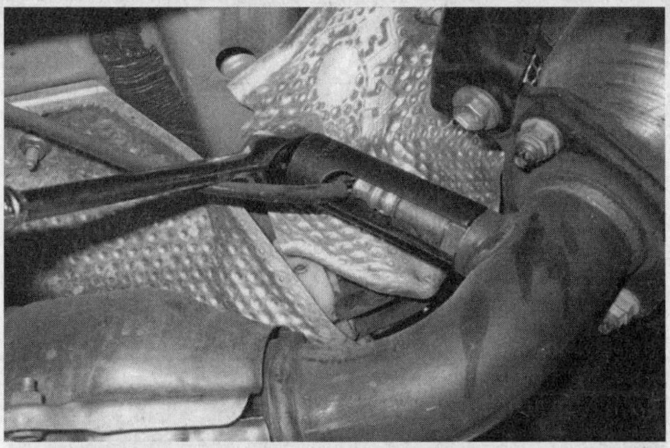

10.10 Use an oxygen sensor socket to unscrew the downstream oxygen sensor

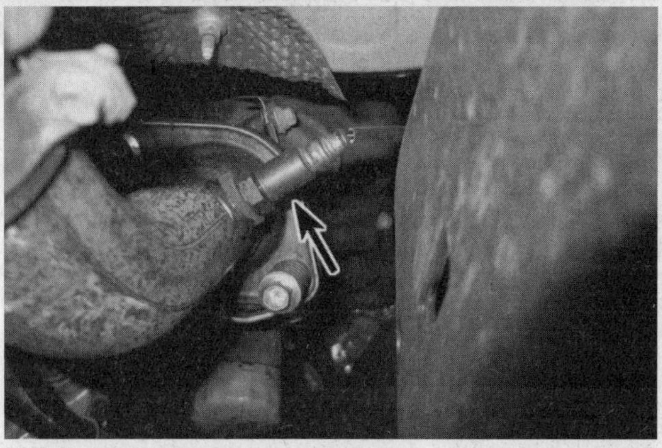

10.14 The upstream oxygen sensor is located on the front exhaust pipe, just below the exhaust manifold flange

10 Unscrew the downstream sensor with an oxygen sensor socket or with a wrench (see illustration).

11 Apply anti-seize compound to the threads of the sensor to facilitate future removal.

➡Note: Most new sensors already have anti-seize compound applied to the threads.

12 Tighten the oxygen sensor to the torque listed in this Chapter's Specifications.

13 Installation is otherwise the reverse of removal.

Four-cylinder models

➡Note: There are two oxygen sensors, one upstream sensor and one downstream sensor.

Upstream oxygen sensor

➧ Refer to illustration 10.14

14 Locate the upstream sensor (see illustration), then trace the electrical lead to the sensor electrical connector and disconnect it.

15 Unscrew the upstream sensor with a wrench (there isn't room to use an oxygen sensor socket).

16 Apply anti-seize compound to the threads of the sensor to facilitate future removal.

➡Note: Most new sensors already have anti-seize compound applied to the threads.

17 Tighten the oxygen sensor to the torque listed in this Chapter's Specifications.

18 Installation is otherwise the reverse of removal.

Downstream oxygen sensor

➧ Refer to illustration 10.19

19 Locate the downstream oxygen sensor (see illustration), then trace the electrical lead to the sensor electrical connector and disconnect it.

20 Unscrew the downstream sensor with an oxygen sensor socket or with a wrench.

21 Apply anti-seize compound to the threads of the sensor to facilitate future removal.

➡Note: Most new sensors already have anti-seize compound applied to the threads.

22 Tighten the oxygen sensor to the torque listed in this Chapter's Specifications.

23 Installation is otherwise the reverse of removal.

10.19 Downstream oxygen sensor location

11 Transmission Range (TR) sensor - replacement

Even though the Society of Automotive Engineers (SAE) has recommended since 1996 that all manufacturers refer to the Park/Neutral Position (PNP) switch as the Transmission Range (TR) sensor, Toyota continues to refer to the TR sensor as the PNP switch. To avoid confusion, we have therefore included the TR sensor/PNP switch in Chapter 7B.

12 Transmission speed sensors - replacement

A340 TRANSMISSION

Refer to illustration 12.2

Note: There are two speed sensors on the A340 transmission: the input speed sensor and the output speed sensor. Both sensors are located on the left side of the transmission. The input speed sensor is located at the front end of the transmission, near the bellhousing. The output speed sensor is located at the rear of the transmission, on the extension housing (2WD models) or the front left corner of the transfer case (4WD models).

1 Raise the vehicle and place it securely on jackstands.
2 Disconnect the electrical connector from the speed sensor that you're replacing (see illustration).
3 Remove the speed sensor retaining bolt and remove the sensor from the transmission.
4 Remove and discard the old NT speed sensor O-ring. Coat the new sensor O-ring with clean automatic transmission fluid (ATF).
5 Installation is the reverse of removal. Be sure to tighten the NT speed sensor retaining bolt securely.

A750E/A750F TRANSMISSIONS

Note: There are two speed sensors on the A750E (2WD)/A750F (4WD) transmissions: the input speed sensor and the output speed sensor. The input speed sensor is located at the left side of the transmission. The output speed sensor is located on the right side of the transmission.

6 Raise the vehicle and place it securely on jackstands.
7 Disconnect the electrical connector from the speed sensor that you're replacing.
8 Remove the speed sensor retaining bolt, then remove the sensor from the transmission.
9 Remove and discard the old speed sensor O-ring. Coat the new O-ring with automatic transmission fluid (ATF).
10 Installation is the reverse of removal. Be sure to tighten the speed sensor retaining bolt securely.

12.2 The input speed sensor (A) is located on the left side of the transmission, near the bellhousing; the output speed sensor (B) is located on the left side of the transmission, either on the extension housing (2WD models, shown) or on the transfer case

13 Powertrain Control Module (PCM) - removal and installation

Refer to illustration 13.3, 13.4a, 13.4b and 13.4c

WARNING:

The models covered by this manual are equipped with Supplemental Restraint Systems (SRS), more commonly known as airbags. Always disable the airbag system before working in the vicinity of any airbag system components to avoid the possibility of accidental deployment of the airbag, which could cause personal injury (see Chapter 12).

CAUTION:

To avoid electrostatic discharge damage to the PCM, handle the PCM only by its case. Do not touch the electrical terminals during removal and installation. If available, ground yourself to the vehicle with an anti-static ground strap, available at computer supply stores.

1 Disconnect the cable from the negative battery terminal (see Chapter 5, Section 1).
2 Remove the right kick panel, the passenger side insulating panel and the glove box (see Chapter 11).
3 Locate the PCM (see illustration) and disconnect all of the electrical connectors from the junction block and from the PCM.

Note: The junction block shares the same mounting brackets with the PCM, so the two components must be removed together.

4 Remove the PCM upper mounting bracket nut and the two lower mounting bracket bolts (see illustrations), then remove the PCM, the junction block and the mounting brackets as a single assembly. As you pull out the assembly, detach the wiring harness clip from the underside of the lower bracket (see illustration).

CAUTION:

Avoid any static electricity damage to the computer by grounding yourself to the body before touching the PCM and using a special anti-static pad to store the PCM on once it is removed.

5 Remove the upper and lower mounting bracket screws (two for each bracket) and the bolt that secures the lower end of the junction block to the lower bracket, then detach both brackets from the PCM. If you're replacing the PCM, install the mounting brackets on the new unit.
6 Installation is the reverse of removal.

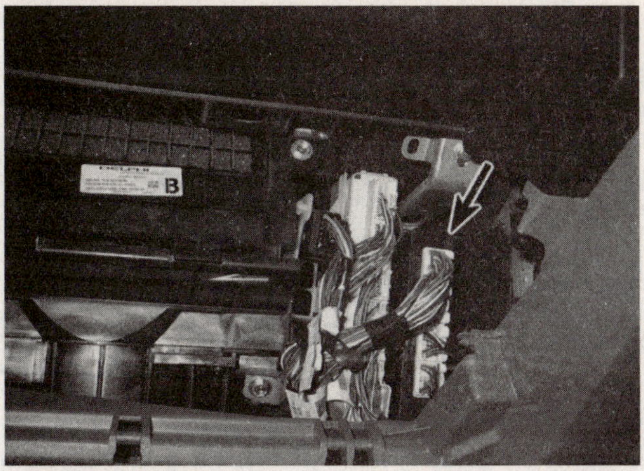

13.3 The PCM is located to the right of the glove box cavity

13.4a To detach the junction block, PCM and mounting brackets, remove the mounting nut from the upper bracket . . .

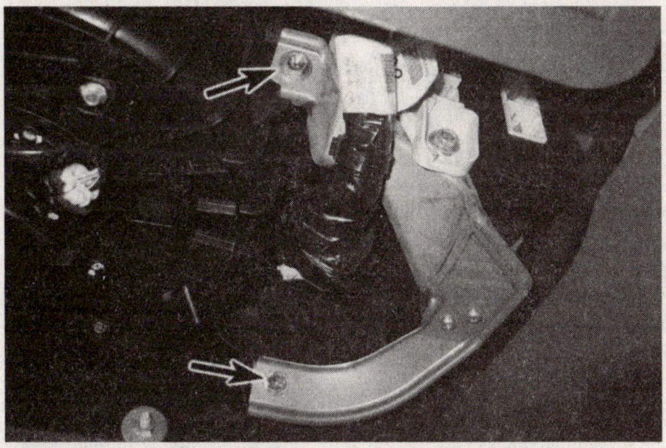

13.4b . . . and remove these two bolts from the lower bracket . . .

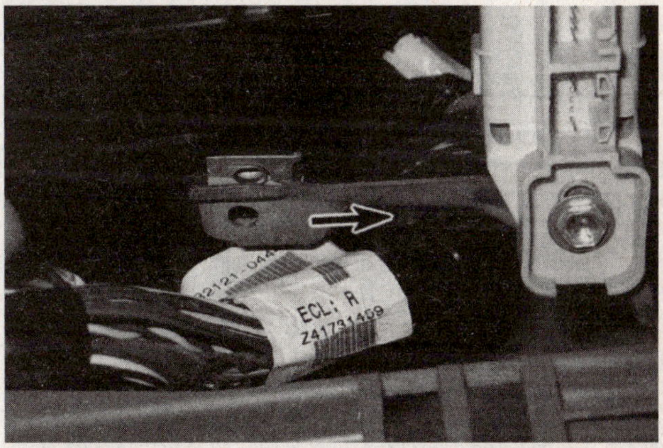

13.4c . . . then pull out the junction block, PCM and mounting brackets as a single assembly and detach this wiring harness clip from the underside of the lower mounting bracket

14 Catalytic converter - description, check and replacement

➡Note: Because of the Federally mandated extended warranty which covers emissions-related components such as the catalytic converter, check with a dealer service department before replacing the converter at your own expense.

DESCRIPTION

1 The catalytic converter is an emission control device added to the exhaust system to reduce pollutants from the exhaust gas stream. There are two types of converters: The oxidation catalyst reduces the levels of hydrocarbon (HC) and carbon monoxide (CO) by adding oxygen to the exhaust stream. The reduction catalyst lowers the levels of oxides of nitrogen (NOx) by removing oxygen from the exhaust gases. These two types of catalysts are combined into a three-way catalyst that reduces all three pollutants.

V6 models

2 There are four catalytic converters, two per cylinder bank. The upstream (warm-up) catalysts are integral parts of the exhaust manifolds. The downstream catalysts are located in the two-piece front exhaust pipe, one per cylinder bank. The two downstream catalysts can be replaced separately.

Four-cylinder models

3 There are two catalytic converters, both of which are integral components of the front exhaust pipe assembly that connects the exhaust manifold to the rest of the exhaust system. This is a one-piece assembly, so if either catalyst fails you must replace the entire assembly.

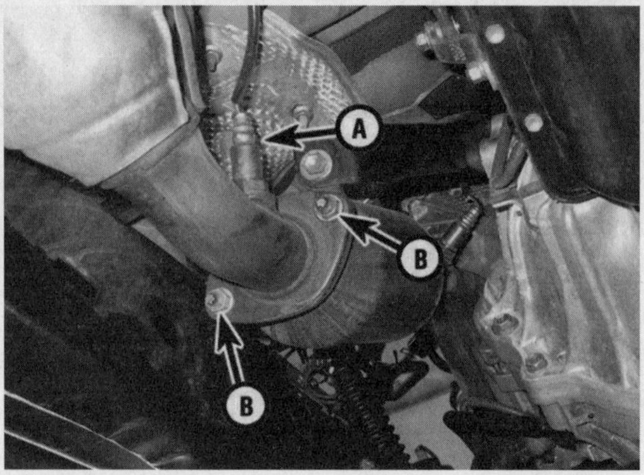

14.12a To detach the left front exhaust pipe from the left exhaust manifold, unscrew and remove the downstream oxygen sensor (A), then remove these two nuts (B)

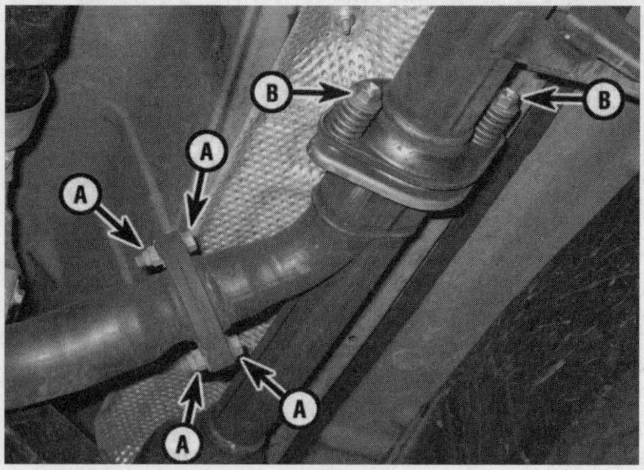

14.12b To detach the left front exhaust pipe from the right front exhaust pipe, remove these two nuts and bolts (A). To detach the right front exhaust pipe from the rest of the exhaust system, remove these two bolts (B)

CHECK

4 The equipment for testing a catalytic converter is expensive. If you suspect that the converter on your vehicle is malfunctioning, take it to a dealer or authorized emissions inspection facility for diagnosis and repair.

5 Whenever the vehicle is raised for servicing underbody components, inspect the converter for leaks, corrosion, dents and other damage. Inspect the welds/flange bolts that attach the front and rear ends of the converter to the exhaust system. If damage is discovered, the converter should be replaced.

6 Although catalytic converters don't break too often, they can become plugged. The easiest way to check for a restricted converter is to use a vacuum gauge to diagnose the effect of a blocked exhaust on intake vacuum.

a) *Connect a vacuum gauge to an intake manifold vacuum source (see Chapter 2C).*
b) *Warm the engine to operating temperature, place the transaxle in Park (automatic) or Neutral (manual) and apply the parking brake.*
c) *Note and record the vacuum reading at idle.*
d) *Quickly open the throttle to near full throttle and release it shut. Note and record the vacuum reading.*
e) *Perform the test three more times, recording the reading after each test.*
f) *If the reading after the fourth test is more than one in-Hg lower than the reading recorded at idle, the exhaust system may be restricted (the catalytic converter could be plugged or an exhaust pipe or muffler could be restricted).*

REPLACEMENT

V6 models

Upstream catalysts

7 To remove or replace the upstream catalysts, refer to *Exhaust manifold - removal and installation* in Chapter 2A.

Downstream catalysts

Refer to illustrations 14.12a, 14.12b and 14.13

Note: The downstream catalysts are located in the two-piece front exhaust pipe assembly. You can remove the downstream catalyst for either cylinder bank separately. Each catalyst and its inlet and outlet pipes are welded into a one-piece assembly. Once you've removed either half, no further disassembly is possible.

8 Raise the vehicle and place it securely on jackstands.

9 Disconnect the electrical connectors from the downstream oxygen sensors (see Section 10).

10 Be sure to spray the nuts on the studs of the exhaust flange and the nuts on the upper flanges of the front exhaust pipe assembly with penetrant. Then spray the nuts and studs at the connection flange between the two halves of the front exhaust pipe assembly and at the rear flange of the front exhaust pipe assembly, where it's connected to the rest of the exhaust system.

11 You can replace either downstream catalyst separately, or you can replace them both by removing the entire front exhaust pipe assembly.

12 To remove the downstream catalyst for the left cylinder bank, remove the nuts from the upper flange, at the lower end of the left exhaust manifold (see illustration), then remove the nuts and bolts from the flange where the left catalyst assembly is connected to the right half of the front exhaust pipe assembly (see illustration).

13 To remove the downstream catalyst for the right cylinder bank, remove the nuts from the upper flange, at the lower end of the right exhaust manifold (see illustration), remove the nuts and bolts from the flange where the right catalyst assembly is connected to the left unit and remove the bolts from the rear flange, where the right catalyst assembly is connected to the rest of the exhaust pipe (see illustration 14.12b).

14 To remove both catalysts, remove the entire front exhaust pipe assembly.

15 Installation is the reverse of removal. Be sure to use new gaskets and fasteners, and tighten all flange nuts securely.

14.13 To detach the right front exhaust pipe from the right exhaust manifold, unscrew and remove the downstream oxygen sensor (A), then remove these two nuts (B)

14.19a To disconnect the upper end of the front exhaust pipe assembly from the exhaust manifold, remove these nuts

Four-cylinder models

Refer to illustrations 14.19a and 14.19b

Note: Both the upstream and downstream catalysts are integral components of the front exhaust pipe assembly, which connects the exhaust manifold to the rest of the exhaust system. If either catalyst fails, you must replace the front exhaust pipe assembly.

16 Raise the vehicle and place it securely on jackstands.

17 Disconnect the electrical connectors from the upstream and downstream oxygen sensors (see Section 10).

18 Be sure to spray the exhaust manifold flange nuts and bolts with penetrant. Then spray the nuts and studs at the rear flange between the front exhaust pipe assembly and the rest of the exhaust system.

19 Remove the nuts from the upper flange, at the lower end of the exhaust manifold, then remove the nuts and bolts from the flange where the front exhaust pipe assembly is connected to the rest of the exhaust system (see illustrations).

20 Installation is the reverse of removal. Be sure to use new gaskets and fasteners and tighten all flange nuts securely.

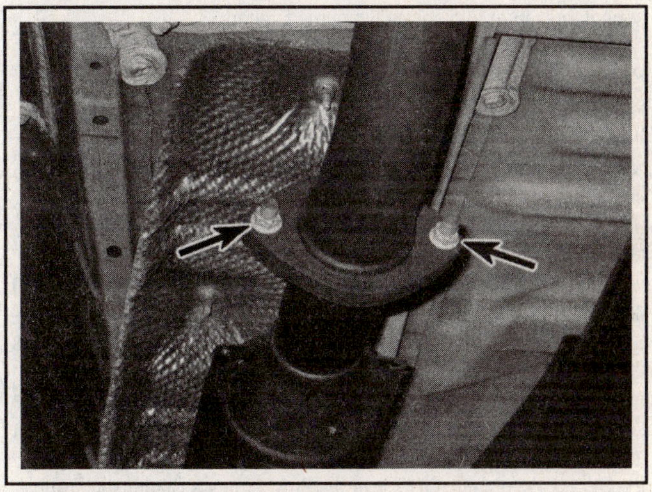

14.19b To disconnect the lower end of the front exhaust pipe assembly from the rest of the exhaust system, remove these nuts

15 Evaporative emissions control (EVAP) system - description and component replacement

DESCRIPTION

1 When the engine is off, gasoline in the fuel tank, and residual fuel in the other components such as the intake manifold, warms up and evaporates, producing unburned hydrocarbon fuel vapors that would waste gas and pollute the air if released into the atmosphere. In an EVAP system, these vapors are routed from the fuel tank, throttle body and intake manifold through a system of hoses to the EVAP system's charcoal canister, where they're stored until the vehicle is operated again. The EVAP canister is able to store these vapors because it's filled with activated charcoal, a substance that can absorb many times its mass in hydrocarbon vapors.

2 The fuel evaporative emission control (EVAP) system absorbs fuel vapors (unburned hydrocarbons) and, during engine operation, releases them into the intake manifold from which they're drawn into the intake ports where they mix with the incoming air-fuel mixture.

3 The EVAP system consists of the air filter, the EVAP canister, the pump module (includes the canister vent valve, vacuum pump and vapor pressure sensor), the refueling valve and the EVAP canister purge valve. All components of the system except the canister purge valve are on or near the EVAP canister, which is located underneath the vehicle, behind the fuel tank. The EVAP canister purge valve is located in the engine compartment, on the intake manifold.

4 The refueling valve, which is located at the EVAP canister, regulates the rate at which fuel vapors flow from the fuel tank to the canister during purging and refueling. During refueling, the vapor pressure inside the fuel tank rises as the fuel level rises; this pressure causes the refueling valve to open, which allows vapors to enter the canister. When the EVAP canister is purged or the EVAP system is leak-checked by the EVAP monitor, vacuum is produced inside the fuel tank. A restrictor

valve inside the refueling valve prevents this vacuum from drawing fuel vapors into the canister to prevent an excessively rich fuel mixture during engine operation; when the PCM pulls down an intentional vacuum to leak-check the EVAP system, the restrictor valve seals off the system to maintain the vacuum.

5 Fresh air, when it's needed, is drawn from the atmosphere near the filler neck cap and is directed through the fresh air line, through the air filter, through the refueling valve and/or the canister vent valve, then into the EVAP canister. The air filter, which is a separate component located to the left of the EVAP canister, prevents dust and debris from contaminating the EVAP system.

6 The vapor pressure sensor inside the pump module monitors the pressure inside the fuel tank and the rest of the EVAP system. The vapor pressure sensor is an integral component of the pump module on the EVAP canister; you cannot replace the vapor pressure sensor separately from the pump module. When the vapor pressure sensor detects excessive pressure inside the fuel tank, it sends a signal to the PCM, which commands the canister purge valve to open. As the vapors are routed from the canister, the open canister vent valve prevents the creation of a relative vacuum inside the tank because it allows outside air (atmospheric pressure) to displace the purged vapors.

7 When the engine is running and the conditions are right, the PCM commands the purge valve to open, and intake vacuum pulls the vapors stored in the EVAP canister out of the canister, through a purge line, through the canister purge valve and into the intake manifold, where they're consumed by the engine. Fresh air is pulled from the atmosphere, through the air filter, through the canister vent valve, then into the EVAP canister, to displace the purged vapors inside the fuel tank. The PCM controls the volume of vapor flow into the manifold by varying the duty cycle of the purge valve in accordance with the driving conditions.

8 As part of the vehicle's OBD-II monitoring system, the PCM uses the vacuum pump inside the pump module to draw down a predetermined vacuum inside the EVAP system, then it uses the vapor pressure sensor to measure the system's ability to hold this vacuum for a certain amount of time. It does this regularly when the vehicle is operated. If the EVAP system fails a leak test, the PCM stores a Diagnostic Trouble Code (DTC).

COMPONENT REPLACEMENT

WARNING:

Gasoline and gasoline vapors are extremely flammable, so take extra precautions when you work on any part of the fuel system. Don't smoke or allow open flames or bare light bulbs near the work area, and don't work in a garage where a gas-type appliance (such as a water heater or clothes dryer) is present. Since gasoline is carcinogenic, wear fuel-resistant gloves when there's a possibility of being exposed to fuel, and, if you spill any fuel on your skin, rinse it off immediately with soap and water. Mop up any spills immediately and do not store fuel-soaked rags where they could ignite. When you perform any kind of work on the fuel system, wear safety glasses and have a Class B type fire extinguisher on hand.

EVAP canister purge valve

Refer to illustrations 15.10a and 15.10b

Note: The EVAP canister purge valve is located on the left side of the intake manifold.

9 On V6 engines, remove the engine cover.

10 Disconnect the electrical connector from the purge valve (see illustrations).

11 Disconnect the EVAP hoses from the purge valve.

12 Remove the purge valve mounting bolt and remove the purge valve.

13 Installation is the reverse of removal.

EVAP canister air filter

Refer to illustrations 15.15, 15.16 and 15.17

Note: The EVAP canister air filter is located near the canister. It filters outside air entering the canister through the fresh air hose, which follows the same route as the fuel filter neck hose and enters the canister through the canister vent valve inside the pump module on the EVAP canister.

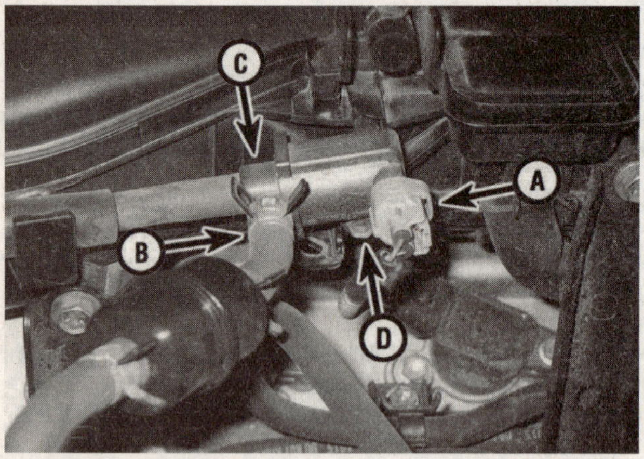

15.10a To remove the EVAP canister purge valve on a V6 model, disconnect the electrical connector (A), disconnect the purge hoses coming from the canister (B) and going to the intake manifold (C), then remove the purge valve mounting bolt (D)

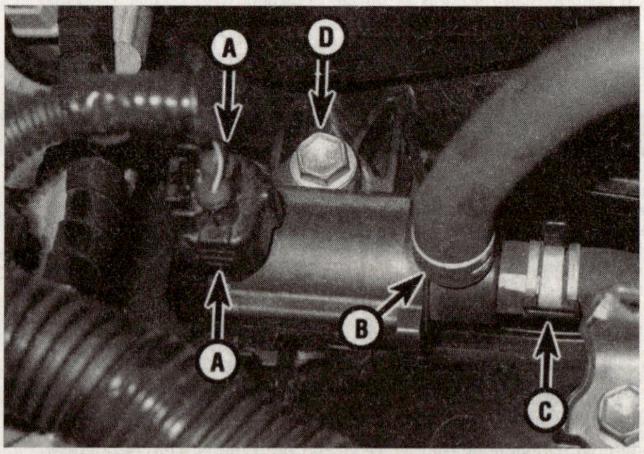

15.10b To remove the EVAP canister purge valve on a four-cylinder model, depress the two release tabs (A) and disconnect the electrical connector, disconnect the purge hoses coming from the canister (B) and going to the intake manifold (C), then remove the purge valve mounting bolt (D)

14 Raise the vehicle and place it securely on jackstands.

15 Detach the ABS harness bracket from the EVAP canister bracket (see illustration).

16 Disconnect the fresh air inlet and outlet hoses from the air filter (see illustration).

17 Remove the air filter from its mounting bracket (see illustration).

18 Installation is the reverse of removal.

EVAP canister and refueling valve

➧ **Refer to illustrations 15.21, 15.22, 15.23a, 15.23b and 15.23c**

19 Raise the vehicle and place it securely on jackstands.

20 Remove the air filter (see Steps 15 through 17).

21 Disconnect the electrical connector from the pump module (see illustration).

22 Disconnect the fuel tank vent line from the refueling valve (see illustration 5.10c in Chapter 4), the fuel tank purge hose from the EVAP canister and the fresh air inlet hose from the pump module (see illustration).

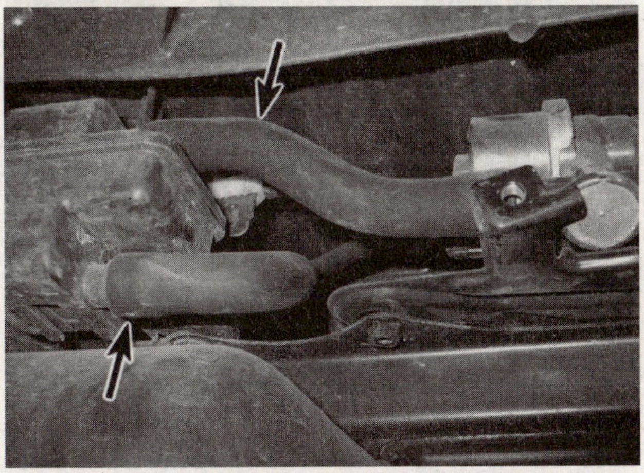

15.16 Disconnect the EVAP hoses from the air filter. The lower hose is the inlet hose drawing outside air into the air filter; the upper hose is the outlet hose that carries filtered fresh air to the pump module, then into the EVAP canister

15.21 Disconnect the electrical connector from the pump module. The connector release tab, which is on the forward side of the connector, is not visible in this photo. If you have difficulty finding the release tab, use a flashlight and a small mirror to locate it

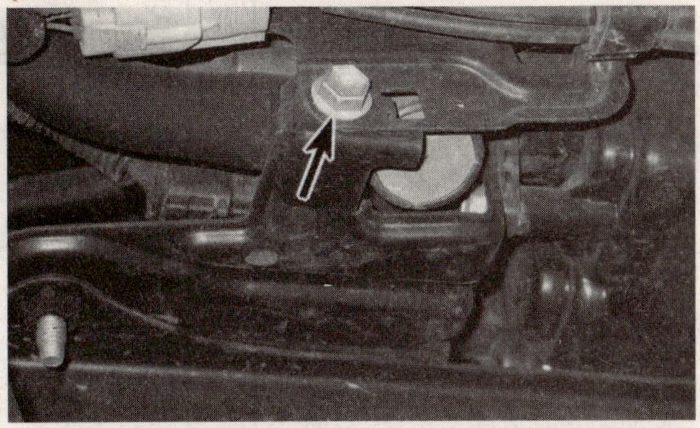

15.15 To detach the ABS harness bracket from the EVAP canister bracket, remove this bolt and carefully pull the ABS harness out of the way

15.17 To disconnect the air filter from its mounting bracket, pry up the locking tab on this mounting bracket and pull off the filter unit

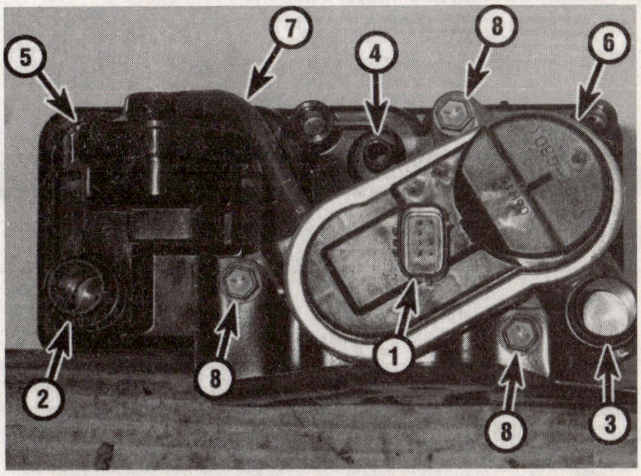

15.22 EVAP canister assembly details:

1 Pump module electrical connector terminal
2 Fuel tank vent line
3 Fresh air inlet hose
4 EVAP canister purge hose
5 Refueling valve
6 Pump module (canister vent valve, vacuum pump and pressure sensor)
7 Refueling valve-to-pump module hose
8 Refueling valve/mounting bracket bolts

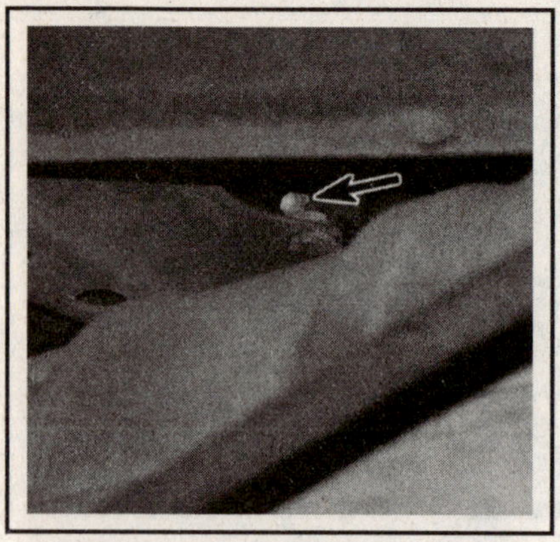

15.23a To detach the EVAP canister, remove this bolt from the front part of the right end of the canister . . .

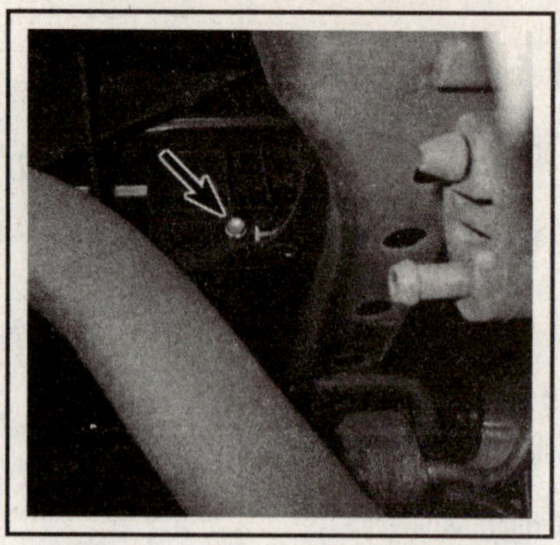

15.23b . . . this bolt from the rear part of the right end . . .

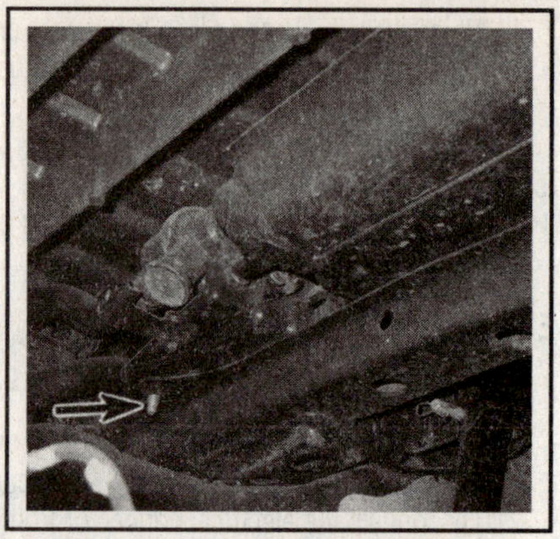

15.23c . . . and this bolt (bolt head, on top, not visible) from the mounting bracket at the left end of the canister

23 Remove the EVAP canister mounting bolts (see illustrations) and remove the canister and mounting bracket as a single assembly.

24 If you're replacing the EVAP canister, unbolt the canister from its mounting bracket (see illustration 15.22), remove the pump module (see Steps 26 through 29) and install the pump module on the new canister.

25 Installation is the reverse of removal.

Pump module

➧ Refer to illustration 15.28

26 Remove the EVAP canister (see Steps 19 through 23).

27 Disconnect the hose that connects the refueling valve to the pump module and remove the pump module/mounting bracket bolts (see illustration 15.22).

28 Remove the pump module from the EVAP canister (see illustration).

29 Remove and discard the old pump module O-ring.

30 Installation is the reverse of removal.

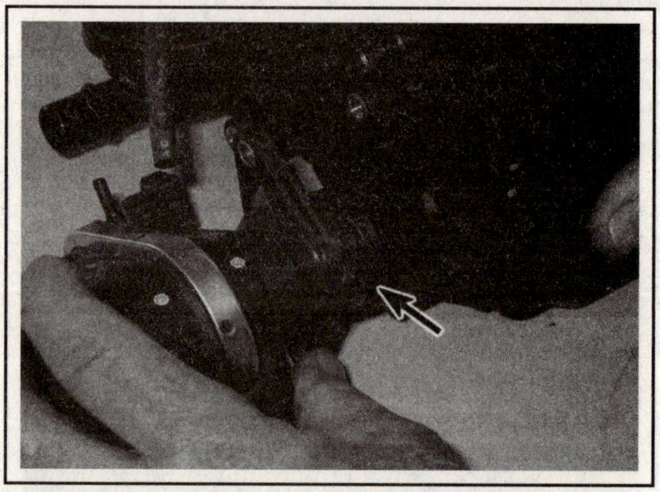

15.28 Remove the pump module from the EVAP canister and remove and discard the old pump module O-ring

16 Positive Crankcase Ventilation (PCV) system - description, check and PCV valve replacement

DESCRIPTION

1 The Positive Crankcase Ventilation (PCV) system reduces hydrocarbon emissions by scavenging crankcase vapors. It does this by circulating fresh air from the air cleaner through the crankcase, where it mixes with blow-by gases, before being drawn through a PCV valve into the intake manifold.

2 The PCV system consists of the PCV valve and two hoses, one of which connects the air filter housing to the crankcase and the other (containing the PCV valve) which connects the crankcase to the intake manifold. On V6 models, the fresh air inlet hose connects the right valve cover to the air filter housing. The PCV valve is located at the rear of the left valve cover, and the PCV hose connects the PCV valve to the intake manifold. On four-cylinder models, the fresh air inlet hose connects the intake air connector to a pipe on the valve cover. The PCV valve is located at the right rear corner of the valve cover and the PCV hose connects the PCV valve to the lower rear part of the intake manifold.

3 To maintain idle quality, the PCV valve restricts the flow when the intake manifold vacuum is high. If abnormal operating conditions (such as piston ring problems) arise, the system is designed to allow excessive amounts of blow-by gases to flow back through the crankcase vent tube into the air cleaner to be consumed by normal combustion.

CHECK

4 There is no scheduled inspection interval for the PCV valve or the PCV system hoses. But, over time the PCV system might become less efficient as an oil residue of sludge builds up inside the PCV valve and the hoses. One symptom of a clogged PCV system is leaking seals. When crankcase vapors can't escape, pressure builds inside the bottom end and eventually causes crankshaft seals to leak. Anytime that you're changing the oil, the air filter, the spark plugs, etc., it's a good idea to pull off the PCV hoses and inspect them and clean them out. If they're cracked, torn or deteriorated, replace them.

16.5a To remove the PCV crankcase ventilation hose (A), slide back the spring-type clamps (B) and disconnect the hose from the PCV valve

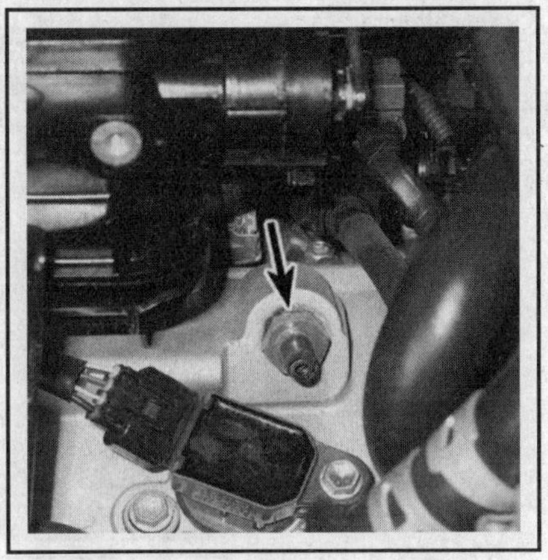

16.5b To remove the PCV valve, unscrew it with a deep socket or wrench

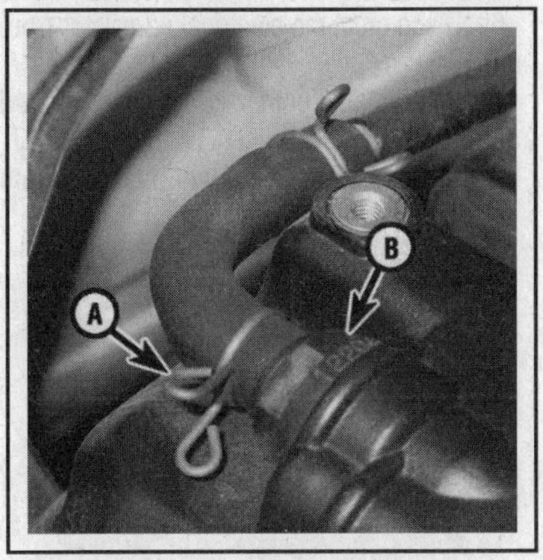

16.8 Slide back the spring-type clamp (A) and disconnect the hose from the PCV valve (B)

PCV VALVE REPLACEMENT

V6 models

➧ Refer to illustrations 16.5a and 16.5b

5 Detach the PCV hose from the PCV valve, then unscrew the PCV valve from the left valve cover (see illustrations).

6 Inspect the condition of the PCV valve. If it's clogged, clean it with solvent, then blow it out with compressed air. If you cannot remove the residue from the inside of the PCV valve, replace it.

7 Installation is the reverse of removal.

Four-cylinder models

➧ Refer to illustration 16.8

8 Detach the PCV hose from the PCV valve (see illustration), then unscrew the PCV valve from the valve cover.

9 Inspect the condition of the hoses. If they're clogged and/or dirty, blow them out with compressed air and wipe them off, then inspect them more carefully. If they're cracked, torn or deteriorated, replace them. Inspect the condition of the PCV valve. If it's clogged, clean it with fresh solvent, then blow it out with compressed air. If you cannot remove the residue from the inside of the PCV valve, replace it.

10 Installation is the reverse of removal.

17 Acoustic Control Induction System (ACIS) - description and component replacement

DESCRIPTION

1 The PCM-controlled Acoustic Control Induction System (ACIS), which is used on V6 models, varies the effective length of the intake manifold runners in response to engine speed and the angle of the throttle plate inside the throttle body. This capability increases efficiency and power at low and high speeds.

2 The ACIS consists of a PCM-controlled Vacuum Switching Valve (VSV), an actuator and an intake air control valve. The VSV is a PCM controlled-device that controls the intake vacuum applied to the actuator. The actuator is a vacuum diaphragm that uses a pushrod and bellcrank to open and close the intake air control valve. The intake air control valve, which is an integral part of the intake manifold, opens and closes to alter the effective length of the intake manifold runners in two stages. On V6 models, there is one large intake air control valve for all six intake runners.

3 At low-to-medium speeds, the PCM activates the VSV, sending vacuum to the actuator diaphragm. The actuator closes the intake air control valve, increasing the length of the intake manifold and improving intake efficiency.

4 At higher speeds, the PCM deactivates the VSV, cutting vacuum to the actuator diaphragm. The actuator opens the intake air control valve, decreasing the length of the intake manifold and improving engine power.

COMPONENT REPLACEMENT

➡Note: The actuator and the intake air control valve are integral components of the intake manifold. If either component fails, replace the intake manifold (see Chapter 2B). The VSV is the only component that you can replace at home.

Vacuum Switching Valve (VSV)

➧ Refer to illustration 17.6

➡Note: The VSV is located at the upper right rear corner of the intake manifold.

5 Remove the engine cover.

6 Disconnect the electrical connector from the VSV (see illustration).

7 Clearly label both vacuum hoses, then disconnect both hoses from the VSV.

8 Remove the VSV mounting bolt and remove the VSV.

9 Installation is the reverse of removal.

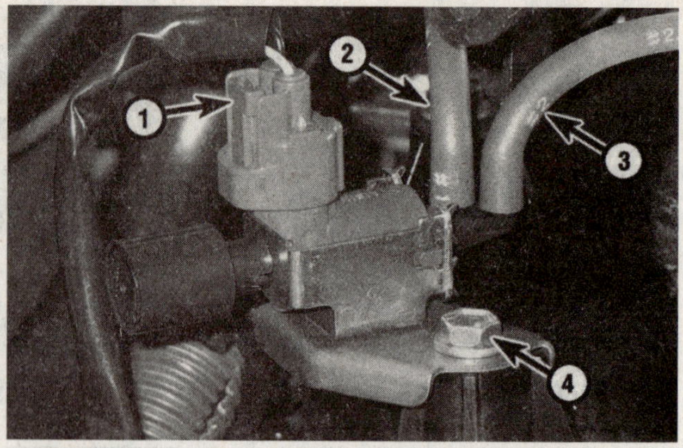

17.6 To remove the Acoustic Control Induction System Vacuum Switching Valve (ACIS VSV) from the intake manifold on a V6 model:

1. *Disconnect the electrical connector*
2. *Disconnect the vacuum hose that connects the VSV to the ACIS actuator*
3. *Disconnect the vacuum hose that directs manifold vacuum to its manifold vacuum source*
4. *Remove the VSV mounting bolt*

18 Air injection system - description and component replacement

DESCRIPTION

1 An air injection system is used on four-cylinder models. Air injection helps the catalytic converters reach their effective operating temperature more rapidly during warm-ups. The system consists of the PCM, an electric air pump, an air injection control valve, an air pressure sensor and an air injection control driver.

2 Using inputs from the Crankshaft Position (CKP) sensor (engine speed), Engine Coolant Temperature (ECT) sensor and Mass Air Flow/Intake Air Temperature (MAF/IAT) sensor, the PCM determines whether the engine is cold (41 to 140 degrees F) or already warmed up (over 140 degrees F). If the engine is already warmed up, the PCM doesn't turn on the air injection system. If the engine is cold or not sufficiently warmed up, the PCM calculates how much air is needed to warm up the catalysts. To do so, it uses input from the MAF/IAT sensor to calculate how long the air injection system should be turned on to bring the catalysts up to their operating temperature (the maximum air injection duration is 80 seconds).

3 When the PCM activates the air injection system, it actuates the air injection control driver, which turns on the electric air pump and the air injection control valve. The air injection control valve consists of an air switching valve (a PCM-controlled solenoid), which switches airflow, and a reed valve that prevents exhaust gases from back flowing into the air injection system. When the air switching valve is turned on by the air injection control driver, it directs air from the air pump through a short pipe into the exhaust ports.

4 When air is pumped into the exhaust manifolds, it helps to burn up any residual unburned fuel vapors, which are always present in the rich mixture used during warm-ups. Without this additional air, the engine exhaust would still warm up, but not as quickly, because the unburned fuel actually cools the exhaust temperature. But when the oxygen in the extra air combines with and promotes the burning of this unburned fuel, it heats up the exhaust gases, and the upstream catalysts, more quickly.

COMPONENT REPLACEMENT

5 Because it is used only during cold-start warm-ups, the air injection system should be trouble-free for years. But if a component of the system fails, you can replace any of the critical components yourself. However, be aware that replacing some of these components - the air pressure sensor, the electric air switching valve and the air pump - is a little more difficult because you'll have to remove the intake manifold to access them.

Air pump assembly

Refer to illustration 18.7

Note: The air pump assembly is located in the right front corner of the engine compartment.

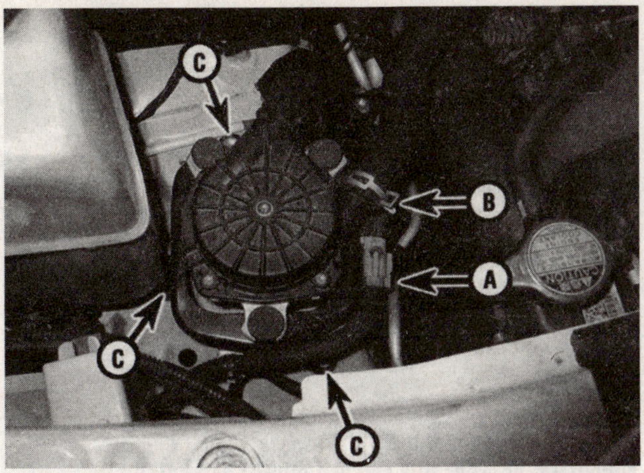

18.7 To disconnect the electrical connector from the air pump, depress the release tab (A) and pull off the connector, release the spring type clamp (B) and disconnect the air injection hose, then remove the three air pump mounting bolts (C)

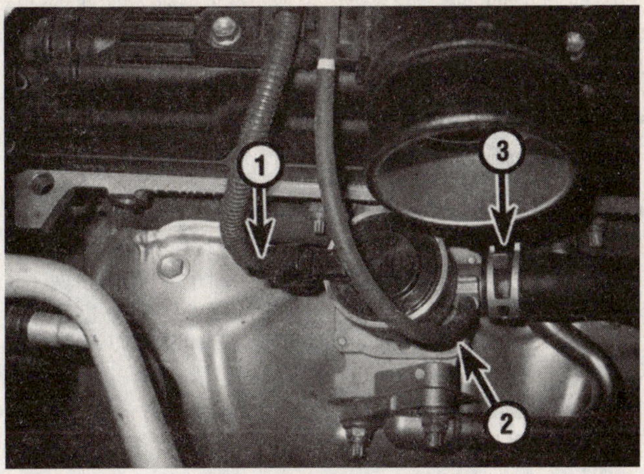

18.12 To disconnect the electrical connector from the air switching valve, depress this release tab (1) and pull off the connector, then disconnect the vacuum hose (2) and the air injection hose (3)

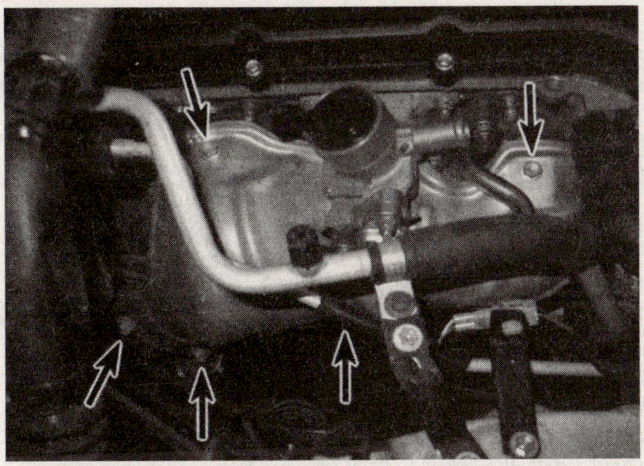

18.13 To detach the heat shield from the exhaust manifold, remove these five bolts

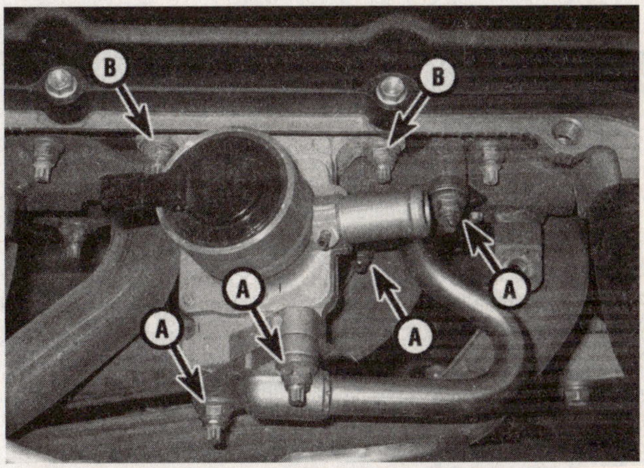

18.14 To detach the air injection pipe, remove these four nuts (A). To remove the air injection valve, remove these two nuts (B)

6 Remove the two windshield washer fluid reservoir mounting bolts and remove the reservoir.

➡**Note: This step is optional; removing the reservoir makes it easier to remove one of the three air pump mounting bolts, but with the right extension you can probably reach this bolt without removing the reservoir.**

7 Disconnect the electrical connector from the air pump and detach the harness clip (see illustration), then detach the other half of the connector (engine harness side) from its bracket and push the connector out of the way. Detach the air pump harness clip from the air pump mounting bracket and set the harness aside.

8 Disconnect the air injection hose from the air pump.

9 Remove the three air pump mounting bolts and remove the pump and mounting bracket as a single assembly.

10 Installation is the reverse of removal.

Air switching valve assembly

➧ **Refer to illustrations 18.12, 18.13 and 18.14**

➡**Note: The air switching valve assembly is located on top of the exhaust manifold.**

11 Remove the air filter housing and the air intake duct (see *Air filter housing - removal and installation* in Chapter 4).

12 Disconnect the electrical connector and vacuum hose from the air switching valve (see illustration).

13 Remove the exhaust manifold heat shield (see illustration).

14 Remove the air injection pipe and gaskets (see illustration).

15 Remove the two air switching valve mounting nuts and remove the air switching valve.

16 Remove and discard the old air switching valve gasket.

17 Installation is the reverse of removal. Be sure to use new gaskets

for the air injection pipe and for the air switching valve and tighten the mounting nuts for the air switching valve and the injection pipe to the torque listed in this Chapter's Specifications.

Air pressure sensor

Refer to illustration 18.18

Note: The air pressure sensor is located on the back of the intake air connector.

18 Disconnect the electrical connector from the air pressure sensor (see illustration).

19 Remove the air pressure sensor mounting bolts and remove the sensor.

20 Installation is the reverse of removal.

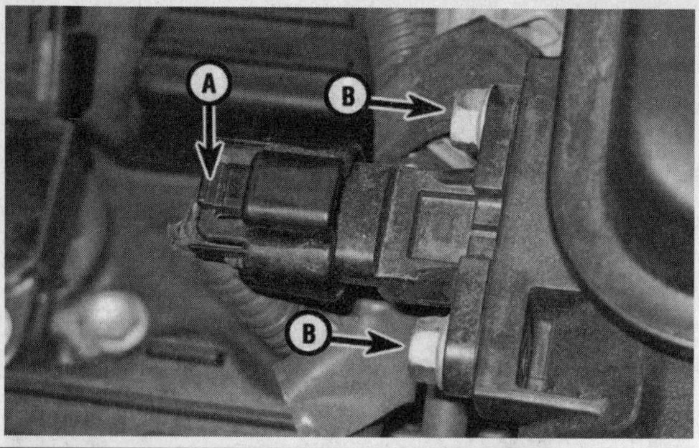

18.18 To disconnect the air pressure sensor electrical connector, depress this release tab (A) and pull off the connector, then remove the two mounting bolts (B) and remove the sensor

19 Variable Valve Timing-intelligent (VVT-i) system - description and component replacement

DESCRIPTION

1 The Variable Valve Timing-intelligent (VVT-i) system is used on all models. The VVT-i system varies intake camshaft timing within a range of 50 degrees (V6 models) or 35 degrees (four-cylinder models) to produce valve timing that is optimized for the driving conditions. The VVT-i system achieves this by using engine oil pressure to advance or retard the controller on the front end of each intake camshaft.

2 The VVT-i system on V6 models consists of two VVT-i (CMP) sensors, the Powertrain Control Module (PCM), two VVT-i oil control valves and two controllers (intake camshaft sprocket/actuator assemblies). The VVT-i system on four-cylinder models is similar, except that it relies on one Camshaft Position (CMP) sensor, one camshaft timing oil control valve and one controller.

Note: Toyota refers to the CMP sensors as "VVT-i sensors" on V6 models and as "Camshaft Position (CMP) sensors" on four-cylinder models.

3 On V6 models, each controller consists of a timing rotor (for the CMP/VVT-i sensor), a housing with an impeller-type vane inside it, a lock pin and the actual timing chain sprocket for the intake camshaft. The vane is fixed on the end of the camshaft. Oil pressure directed into the housing from the advance or retard side of the intake cam causes the vane to rotate in relation to the intake camshaft, advancing or retarding the valve timing. The higher the oil pressure (or flow) the more the actuator assembly will rotate, thereby advancing or retarding the camshaft. When the engine is turned off, the intake cam is in its most retarded position to ensure easy starting. When the engine is first started and no oil pressure has yet been applied to the controller housing, the lock pin locks the housing and vane together to prevent it from making a knocking sound. When oil pressure enters the housing and is applied to the lock pin spring, the spring is compressed and the lock pin retracts, allowing the housing to rotate in relation to the vane.

4 The controllers on four-cylinder models are identical to the controllers used on V6 models.

5 Here's how it works: When oil is applied to the advance side of the vane, the actuator advances the intake camshaft in a clockwise direction. When oil is applied to the retard side of the vane, the actuator rotates the camshaft counterclockwise back to 0 degrees advance, which is the normal position of the actuator during engine operation under no-load or idle conditions. The PCM can also send a signal to the timing oil control valve to stop oil flow to both (advance and retard) passages to hold camshaft advance in its current position. Under light engine loads, the VVT-i system retards the camshaft timing to decrease valve overlap and stabilize engine output. Under medium engine loads, the VVT-i system advances the camshaft timing to increase valve overlap, thereby increasing fuel economy and decreasing exhaust emissions. Under heavy engine loads at low RPM, the VVT-i system advances the camshaft timing to help close the intake valve faster, which improves low to midrange torque. Under heavy engine loads at high RPM, the VVT-i system retards the camshaft timing to slow the closing of the intake valve to improve engine horsepower.

6 The camshaft timing oil control valve is a PCM-controlled device that controls and directs the flow of oil to the advance or retard passages leading to the controller. There are two oil control valves, one for each intake camshaft. A spring-loaded spool valve inside the oil control valve directs oil pumped into the valve toward either the advance outlet port or the retard outlet port, depending on the engine speed, which determines oil pressure. When the engine is turned off, the spring is extended and the spool valve is in its most retarded state. Once the engine is started and oil pressure begins to rise with engine rpm, the spool is slowly pushed against its spring.

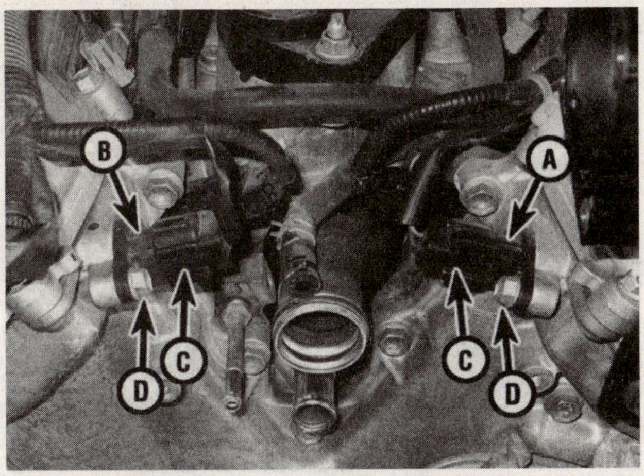

19.11 CMP/VVT sensor details (V6 models):

A Left CMP/VVT sensor
B Right CMP/VVT sensor
C Electrical connector
D Mounting bolt

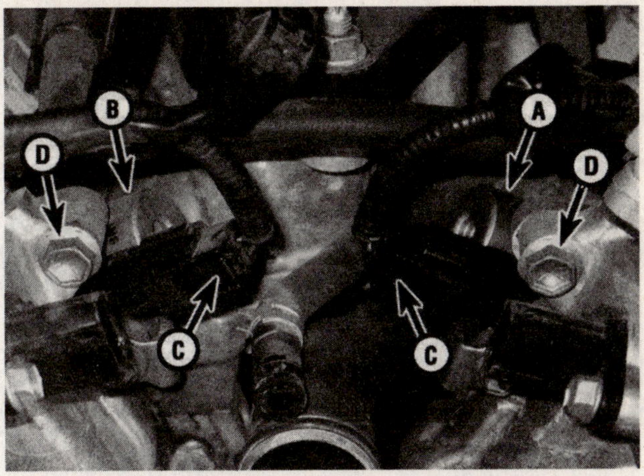

19.17 Camshaft timing oil control valve details (V6 models):

A Left camshaft timing oil control valve
B Right camshaft timing oil control valve
C Electrical connector
D Mounting bolt

COMPONENT REPLACEMENT

V6 models

VVT sensors

➧ Refer to illustration 19.11

WARNING:

Wait until the engine is completely cool before beginning this procedure.

➡Note: There are two VVT (CMP) sensors on V6 models. The sensors are located on the inside walls of the cylinder heads, adjacent to the timing rotor installed on the front of each Variable Valve Timing-intelligent (VVT-i) controller. Toyota refers to these sensors as "CMP/VVT" sensors on earlier models, and simply as VVT sensors on later models. If you're buying a new sensor at a Toyota parts department, use the Toyota terminology.

7 Disconnect the cable from the negative battery terminal (see Chapter 5, Section 1).

8 Remove the engine cover.

9 Remove the air filter housing (see Chapter 4).

10 If you're going to remove the CMP/VVT sensor from the left cylinder head, drain the engine coolant (see Chapter 1) and disconnect the two coolant bypass hoses from the throttle body.

➡Note: As an alternative to draining the coolant, the hoses can be clamped off using locking pliers.

11 Disconnect the electrical connector from the CMP/VVT sensor (see illustration).

12 Remove the CMP/VVT sensor mounting bolt and remove the sensor.

13 Installation is the reverse of removal.

Camshaft timing oil control valve

➧ Refer to illustration 19.17

14 Remove the engine cover.

15 Remove the air filter housing (see Chapter 4).

16 If you're replacing the left camshaft timing oil control valve, remove the upper intake manifold (see Chapter 2B).

17 Disconnect the electrical connector from the camshaft timing oil control valve (see illustration).

18 Remove the camshaft timing oil control valve mounting bolt and remove the oil control valve.

19 Installation is the reverse of removal.

Four-cylinder models

Camshaft Position (CMP) sensor

20 See Section 4.

Camshaft timing oil control valve

➧ Refer to illustrations 19.21 and 19.22

➡Note: The camshaft timing oil control valve is located on the left side of the engine, just ahead of the intake manifold.

21 Disconnect the electrical connector from the camshaft timing oil control valve (see illustration).

22 Remove the camshaft timing oil control valve mounting bolt (see illustration) and remove the camshaft timing oil control valve.

23 Installation is the reverse of removal.

19.21 To disconnect the electrical connector from the camshaft timing oil control valve, depress this release tab

19.22 To detach the camshaft timing oil control valve from the engine block, remove this mounting bolt

Torque specifications	Ft-lbs (unless otherwise noted)	Nm

➡Note: One foot-pound (ft-lb) of torque is equivalent to 12 inch-pounds (in-lbs) of torque. Torque values below approximately 15 ft-lbs are expressed in inch-pounds, since most foot-pound torque wrenches are not accurate at these smaller values.

Air switching valve and air injection pipe mounting nuts	180 in-lbs	20
Engine Coolant Temperature (ECT) sensor (all models)	180 in-lbs	20
Knock sensor retaining bolt (all models)	180 in-lbs	20
Oxygen sensors (all models)	33	45

7A MANUAL TRANSMISSION

Section

1 General information

All vehicles covered by this manual are equipped with a five or six-speed manual transmission or a four or five-speed automatic transmission. Information on the manual transmission and certain procedures common to both transmissions - such as oil seal replacement - can be found in this Part of Chapter 7. Service procedures for the automatic transmission can be found in Part B, and information about the transfer case used on 4WD models can be found in Part C of this Chapter.

Four-cylinder models are equipped with R155 transmissions; V6 models use the RA60 transmission. 4WD transmissions are indicated by an F suffix on the model number.

Depending on the expense involved in having a transmission overhauled, it might be a better idea to consider replacing it with either a new or rebuilt unit. Your local dealer or transmission shop should be able to supply information concerning cost, availability and exchange policy. Regardless of how you decide to remedy a transmission problem, you can still save a lot of money by removing and installing the unit yourself.

2 Rear output shaft oil seal (2WD) - replacement

Refer to illustrations 2.4 and 2.5

Note: This procedure applies to both manual and automatic transmissions.

1 Oil leaks frequently occur due to wear of the extension housing oil seal. Replacement of this seal is relatively easy, since the repair can usually be performed without removing the transmission from the vehicle.

2 The extension housing oil seal is located at the extreme rear of the transmission, where the driveshaft is attached. Raise the vehicle and support it securely on jackstands. If the seal is leaking, transmission lubricant will be built up on the front of the driveshaft and may be dripping from the rear of the transmission.

3 Remove the driveshaft (see Chapter 8).

4 Using a puller or seal removal tool, carefully remove the oil seal out of the rear of the transmission (see illustration). Do not damage the splines on the transmission output shaft.

5 Using a large section of pipe or a very large deep socket as a drift, install the new oil seal (see illustration). Drive it into the bore squarely and make sure it's completely seated.

6 Lubricate the splines of the transmission output shaft and the outside of the driveshaft slip yoke with lightweight grease, then install the driveshaft. Be careful not to damage the lip of the new seal.

7 Check the lubricant level in the transmission, adding as necessary (see Chapter 1).

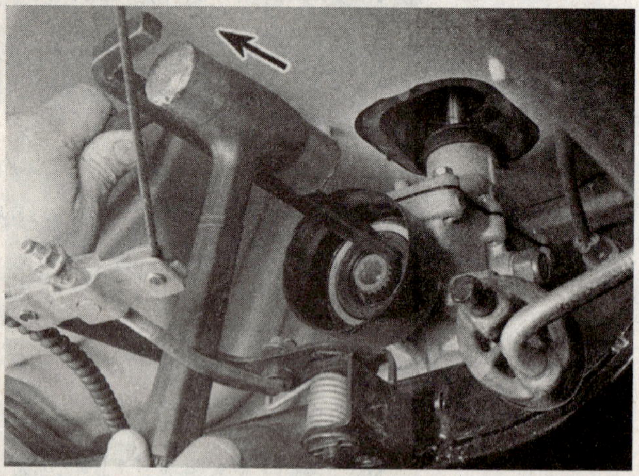

2.4 Using a seal removal tool to remove the seal

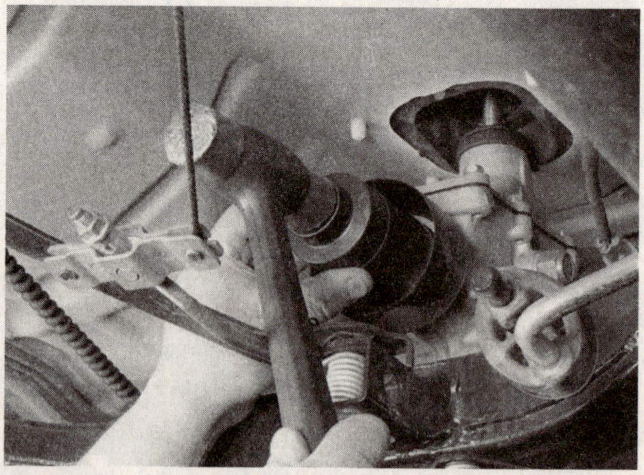

2.5 To install the new seal, tap it into place with a large socket and hammer

3 Transmission mount - check and replacement

CHECK

Refer to illustration 3.1

1 Insert a large screwdriver or pry bar into the space between the transmission and the crossmember and try to pry the transmission up slightly (see illustration).

2 The transmission should not move away from the insulator much. If there is any separation of the rubber, the mount is worn out.

REPLACEMENT

Refer to illustration 3.3 and 3.4

3 Support the transmission with a jack and remove the transmission mount-to-crossmember retaining bolts (see illustration).

4 Raise the transmission slightly with the jack and remove the mount-to-transmission fasteners (see illustration).

Note: On 4WD models, the mount is attached to the transfer case.

5 Installation is the reverse of the removal procedure. Be sure to tighten the fasteners securely.

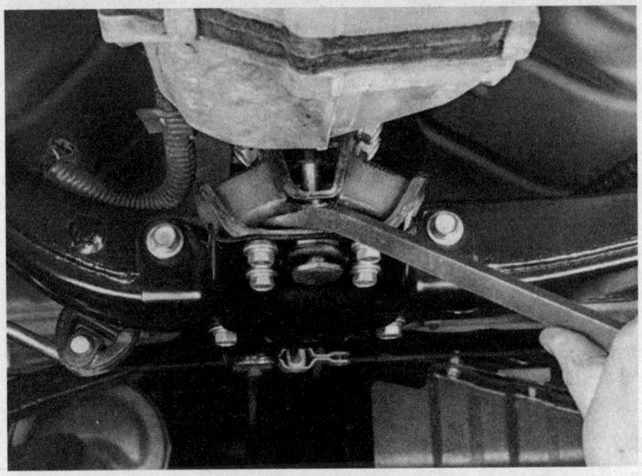

3.1 To check the transmission mount, insert a prybar or large screwdriver between the mount rubber and the mount bracket, then try to pry the transmission up off its mount; if the transmission moves significantly, replace the mount

3.3 Transmission mount-to-crossmember retaining bolts (4WD shown, 2WD similar)

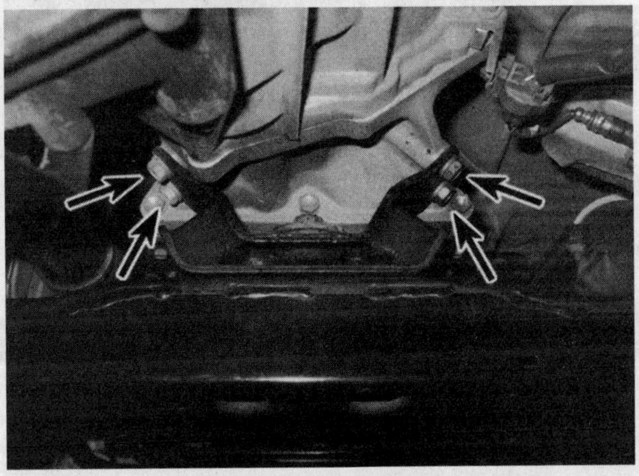

3.4 Remove the four mount-to-transmission fasteners

4 Shift lever - removal and installation

Refer to illustrations 4.4, 4.5a, 4.5b and 4.6

1 Unscrew the shift lever knob(s). Remove the four shift lever boot retaining screws and remove the boot and retainer.

2 Remove the console (see Chapter 11).

3 Remove the rubber shift lever boot.

4 Remove the small dust cover from the shift lever cap (see illustration).

5 Press down on the shift lever cap and rotate it counterclockwise to remove the shift lever (see illustrations).

Note: It's best to first cover the shift lever cap with a rag.

6 Installation is the reverse of removal. Before installing the shift lever, lubricate the friction surfaces with multi-purpose grease (see illustration).

4.4 Remove the dust cover from the shift lever cap

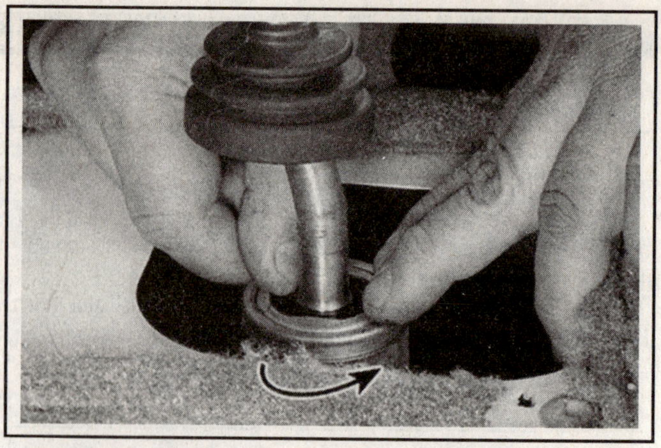

4.5a Push down on the shift lever cap and rotate it counterclockwise . . .

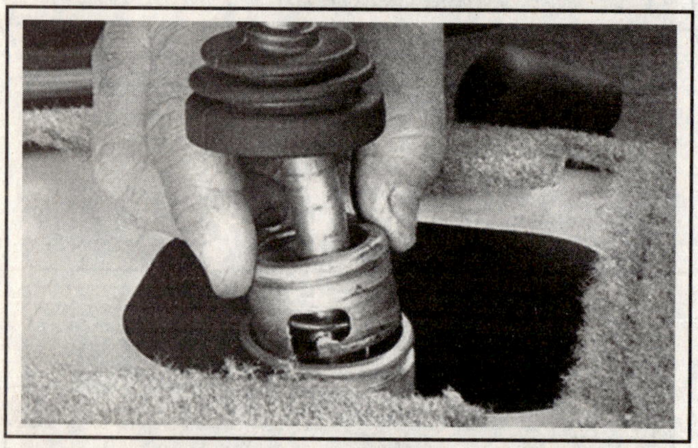

4.5b . . . then remove the shift lever

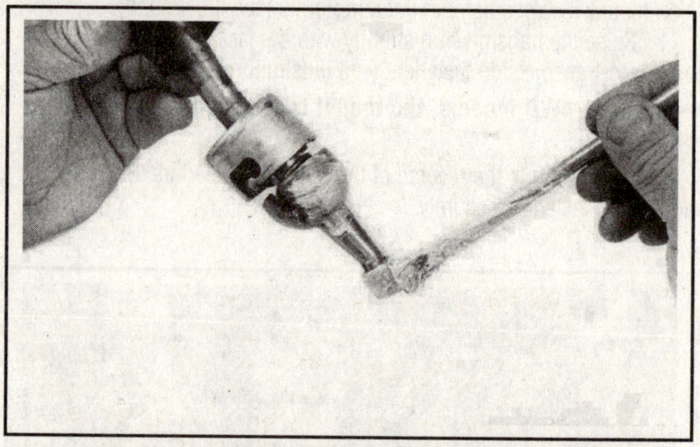

4.6 Lubricate the friction surfaces of the shift lever with multi-purpose grease before installing the shift lever

5 Back-up light switch - removal and installation

1 Raise the vehicle and support it securely on jackstands. The back-up light switch is screwed into the right side of the transmission near the rear.

2 Unplug the back-up light switch electrical connector.

3 Unscrew the back-up light switch from the transmission case and remove the gasket.

4 Install the switch and gasket in the transmission case and tighten it securely.

5 Plug in the electrical connector.

6 Lower the vehicle and check the operation of the back-up lights.

6 Manual transmission - removal and installation

➡Note: On 4WD models, remove the transfer case with the transmission. If necessary, the transfer case can be separated after removal.

REMOVAL

➧ Refer to illustration 6.19

1 Disconnect the cable from the negative terminal of the battery.

2 Drain the lubricant from the transmission (see Chapter 1).

3 Remove the shift lever (see Section 4).

4 Raise the vehicle and support it securely on jackstands.

5 Remove the driveshaft (see Chapter 8).

6 On V6 models, remove the clutch housing cover, then disconnect the fluid lines and remove the clutch fluid accumulator.

7 Unbolt the clutch release cylinder and hydraulic line brackets (see Chapter 8).

⁂ CAUTION:

Don't depress the clutch pedal with the release cylinder removed.

6.19 **Remove the crossmember-to-frame mounting bolts**

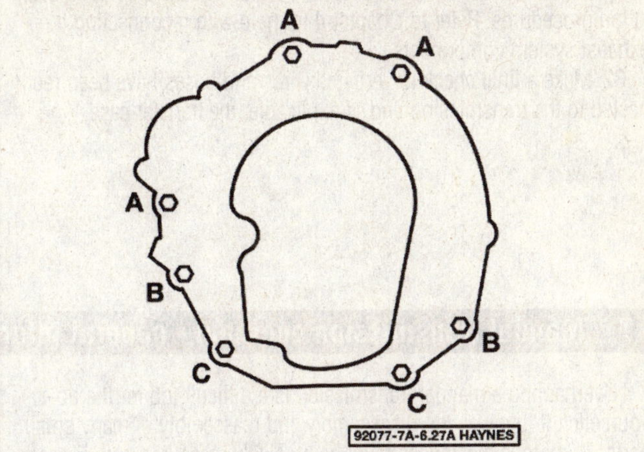

6.27a **Transmission mounting bolt identification - four-cylinder models**

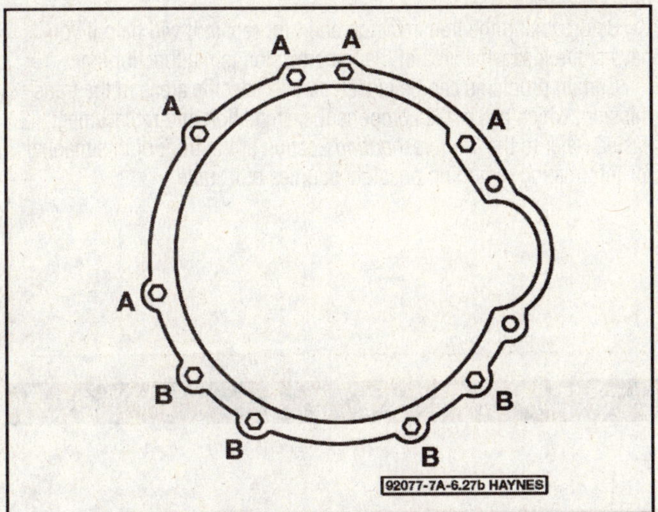

6.27b **Transmission mounting bolt identification - V6 models**

8 Remove the engine lower covers, if so equipped.

9 Remove the front section of exhaust pipe if it interferes. Use wire to support the rear exhaust pipe or remove it as well.

10 Disconnect the electrical connectors for the vehicle speed sensor and back-up light switch. Release the wiring harness from the transmission and move it out of the way. On 4WD models, disconnect the range switches.

11 Remove the front exhaust pipe(s) (see Chapter 4).

12 Remove the starter (see Chapter 5).

13 Remove the manifold brace brackets.

Note: There is one manifold brace on four-cylinder models.

14 Remove the flywheel inspection cover.

15 Support the engine by placing a floor jack and a block of wood under the engine oil pan.

16 Support the transmission with a jack, preferably a special jack made for this purpose. Safety chains will help steady the transmission on the jack. If you don't have a transmission jack, support the transmission with a floor jack.

17 Remove the bolts from the engine rear mount.

18 Remove the rear engine mount.

19 Remove the transmission mount-to-crossmember retaining bolts and the crossmember-to-frame mounting bolts and remove the crossmember (see illustration).

20 Remove the transmission-to-engine bolts.

21 Make a final check that all wires have been disconnected from the transmission (and transfer case, if so equipped).

22 Move the transmission and jack toward the rear of the vehicle until the transmission input shaft is clear of the clutch housing.

CAUTION:

Be careful to avoid tilting the engine so much that the fan contacts the fan shroud.

23 Lower the transmission completely. Remove the transfer case if necessary on 4WD models.

24 Inspect the clutch components. It's a good idea to install new clutch components whenever the transmission is removed (see Chapter 8).

INSTALLATION

Refer to illustrations 6.27a and 6.27b

25 Install the clutch components, if they were removed (see Chapter 8).

26 Apply a light coat of high-temperature grease to the splines and end of the input shaft. With the transmission secured to the jack, raise it into position behind the engine, then carefully slide it forward, engaging the input shaft with the clutch plate hub. Do not use excessive force to install the transmission - if the input shaft won't slide into place, readjust the angle of the transmission so that it's level and/or turn the input shaft so the splines engage properly with the clutch.

27 Once the transmission is fully seated against the engine, install the transmission-to-engine bolts. Tighten the bolts to the torque listed in this Chapter's Specifications (see illustrations).

28 Install the crossmember. Tighten all nuts and bolts securely.

29 Bolt the transmission mount to the crossmember.

30 Remove the jacks supporting the transmission and the engine.

31 Install the various components removed previously. Refer to

Chapter 8 for driveshaft installation and clutch release cylinder installation procedures. Refer to Chapter 4 for help with reconnecting the exhaust system components.

32 Make a final check to verify all wires and hoses have been reconnected to the transmission and, if applicable, the transfer case.

33 If necessary, fill the transmission and or transfer case with lubricant to the proper level (see Chapter 1). Lower the vehicle.

34 Install the shift lever and boot (see Section 4).

35 Connect the negative cable to the negative terminal of the battery. Road test the vehicle and check for leaks.

7 Manual transmission overhaul - general information

Overhauling a manual transmission is a difficult job for the do-it-yourselfer. It involves the disassembly and reassembly of many small parts. Numerous clearances must be precisely measured and, if necessary, changed with select fit spacers and snap-rings. As a result, if transmission problems arise, it can be removed and installed by a competent do-it-yourselfer, but overhaul should be left to a transmission repair shop. Rebuilt transmissions may be available - check with your dealer parts department or a local transmission repair shop. At any rate, the time and money involved in an overhaul is almost sure to exceed the cost of a rebuilt unit.

Nevertheless, it's not impossible for an inexperienced mechanic to rebuild a transmission if the special tools are available and the job is done in a deliberate step-by-step manner so nothing is overlooked.

The tools necessary for an overhaul include internal and external snap-ring pliers, a bearing puller, a slide hammer, a set of pin punches, a dial indicator and possibly a hydraulic press. In addition, a large, sturdy workbench and a vise or transmission stand will be required.

During disassembly of the transmission, make careful notes of how each piece comes off, where it fits in relation to other pieces and what holds it in place.

Before taking the transmission apart for repair, it will help if you have some idea what area of the transmission is malfunctioning.

Certain problems can be closely tied to specific areas in the transmission, which can make component examination and replacement easier. Refer to the *Troubleshooting* Section at the front of this manual for information regarding possible sources of trouble.

Specifications

General

Fluid type and capacity	See Chapter 1

Torque specifications	Ft-lbs	Nm
Crossmember-to-frame bolts	30	40
Transmission-to-engine bolts		
Four-cylinder models (see illustration 6.27a)		
Bolt A	53	72
Bolt B and C	27	37
V6 models (see illustration 6.27b)		
Bolt A	53	72
Bolt B	28	38

7B AUTOMATIC TRANSMISSION

Section

Reference to other Chapters

Automatic transmission fluid and filter change - See Chapter 1
Automatic transmission fluid level check - See Chapter 1

1 General information

Four-cylinder models use a four-speed (three speeds plus overdrive) model A340E automatic transmission. V6 models use a five-speed unit designated as A750E (2WD) or A750F (4WD). All models are equipped with a lock-up torque converter, known as a Torque Converter Clutch (or TCC). The clutch provides a direct connection between the engine and the drive wheels for improved efficiency and fuel economy. All models are electronically controlled; upshifts and downshifts are initiated by a computer, which controls the solenoids mounted on the valve body.

Due to the complexity of the clutches and the hydraulic control system, and because of the special tools and expertise needed to overhaul an automatic transmission, diagnosis and repair of the transmission must be handled by a dealer service department or a transmission repair shop. The procedures in this Chapter are limited to general diagnosis, routine maintenance, adjustment and transmission removal and installation. However, even though the repair work must be done by a transmission specialist, you can save money by removing and installing the transmission yourself. You can also check and adjust the shift linkage, replace the extension housing seal and check and replace the Park/Neutral position switch.

2 Diagnosis - general

➡Note: Automatic transmission malfunctions may be caused by five general conditions: poor engine performance, improper adjustments, hydraulic malfunctions, mechanical malfunctions or malfunctions in the computer or its signal network. Diagnosis of these problems should always begin with a check of the easily repaired items: fluid level and condition (see Chapter 1) and shift linkage adjustment. Next, perform a road test to determine if the problem has been corrected or if more diagnosis is necessary. If the problem persists after the preliminary tests and corrections are completed, additional diagnosis should be done by a dealer service department or transmission repair shop. Refer to the Troubleshooting Section at the front of this manual for information on symptoms of transmission problems.

PRELIMINARY CHECKS

1 Drive the vehicle to warm the transmission to normal operating temperature.

2 Check the fluid level as described in Chapter 1:

a) *If the fluid level is unusually low, add enough fluid to bring the level within the designated area of the dipstick, then check for external leaks (see below).*

b) *If the fluid level is abnormally high, drain off the excess, then check the drained fluid for contamination by coolant. The presence of engine coolant in the automatic transmission fluid indicates that a failure has occurred in the internal radiator walls that separate the coolant from the transmission fluid (see Chapter 3).*

c) *If the fluid is foaming, drain it and refill the transmission, then check for coolant in the fluid or a high fluid level.*

3 Check for any stored trouble codes (see Chapter 6).

➡Note: If the engine is malfunctioning, do not proceed with the preliminary checks until it has been repaired and runs normally.

4 Inspect the shift cable (see Section 3). Make sure it's properly adjusted and operates smoothly.

FLUID LEAK DIAGNOSIS

5 Most fluid leaks are easy to locate visually. Repair usually consists of replacing a seal or gasket. If a leak is difficult to find, the following procedure may help.

6 Identify the fluid. Make sure it's transmission fluid and not engine oil or brake fluid (automatic transmission fluid is a deep red color).

7 Try to pinpoint the source of the leak. Drive the vehicle several miles, then park it over a large sheet of cardboard. After a minute or two, you should be able to locate the leak by determining the source of the fluid dripping onto the cardboard.

8 Make a careful visual inspection of the suspected component and the area immediately around it. Pay particular attention to gasket mating surfaces. A mirror is often helpful for finding leaks in areas that are hard to see.

9 If the leak still cannot be found, clean the suspected area thoroughly with a degreaser or solvent, then dry it.

10 Drive the vehicle for several miles at normal operating temperature and varying speeds. After driving the vehicle, visually inspect the suspected component again.

11 Once the leak has been located, the cause must be determined before it can be properly repaired. If a gasket is replaced but the sealing flange is bent, the new gasket will not stop the leak. The bent flange must be straightened.

12 Before attempting to repair a leak, check to make sure the following conditions are corrected or they may cause another leak.

➡Note: Some of the following conditions cannot be fixed without highly specialized tools and expertise. Such problems must be referred to a transmission repair shop or a dealer service department.

Gasket leaks

13 Check the pan periodically. Make sure the bolts are tight, no bolts are missing, the gasket is in good condition and the pan is flat (dents in

the pan may indicate damage to the valve body inside).

14 If the pan gasket is leaking, the fluid level or the fluid pressure may be too high, the vent may be plugged, the pan bolts may be too tight, the pan sealing flange may be warped, the sealing surface of the transmission housing may be damaged, the gasket may be damaged or the transmission casting may be cracked or porous. If sealant instead of gasket material has been used to form a seal between the pan and the transmission housing, it may be the wrong sealant.

Seal leaks

15 If a transmission seal is leaking, the fluid level or pressure may be too high, the vent may be plugged, the seal bore may be damaged, the seal itself may be damaged or improperly installed, the surface of the shaft protruding through the seal may be damaged or a loose bearing may be causing excessive shaft movement.

16 Make sure the dipstick tube seal is in good condition and the tube is properly seated. Periodically check the area around the speedometer gear or sensor for leakage. If transmission fluid is evident, check the O-ring for damage.

Case leaks

17 If the case itself appears to be leaking, the casting is porous and will have to be repaired or replaced.

18 Make sure the oil cooler hose fittings are tight and in good condition.

Fluid comes out vent pipe or fill tube

19 If this condition occurs, the transmission is overfilled, there is coolant in the fluid, the case is porous, the dipstick is incorrect, the vent is plugged or the drain back holes are plugged.

3 Shift cable or shift rod - removal and installation

WARNING:

These models are equipped with airbags. Always disable the airbag system before working in the vicinity of any airbag system component to avoid the possibility of accidental deployment of the airbag(s), which could cause personal injury (see Chapter 12).

CABLE

Refer to illustrations 3.4, 3.5 and 3.9

1 Disconnect the cable from the negative battery terminal (see Chapter 5, Section 1).

2 Refer to Chapter 11 and remove the center console to expose the shifter assembly.

3 Disconnect the wiring from the voltage inverter under the console. Remove the mounting bolts and remove the inverter.

4 Disengage the cable end from the shifter (see illustration).

5 Pull out the retaining clip and release the cable from the shifter bracket (see illustration).

6 Remove the nuts and release the cable from the floor of the vehicle.

7 Raise the vehicle and place it securely on jackstands.

8 Remove the fasteners and detach the cable guide bracket from the floor pan.

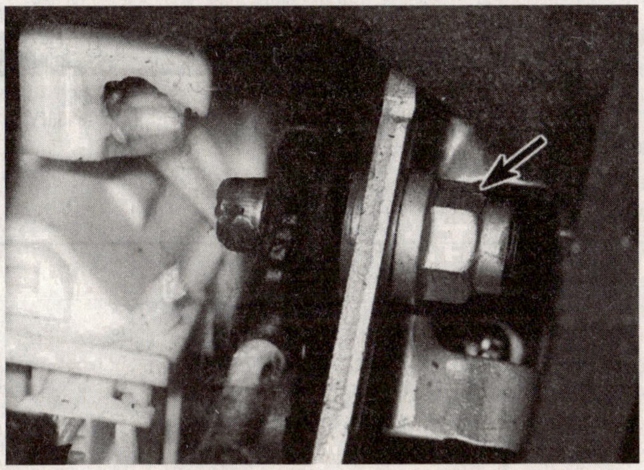

3.4 Remove the fastener securing the cable to the shifter

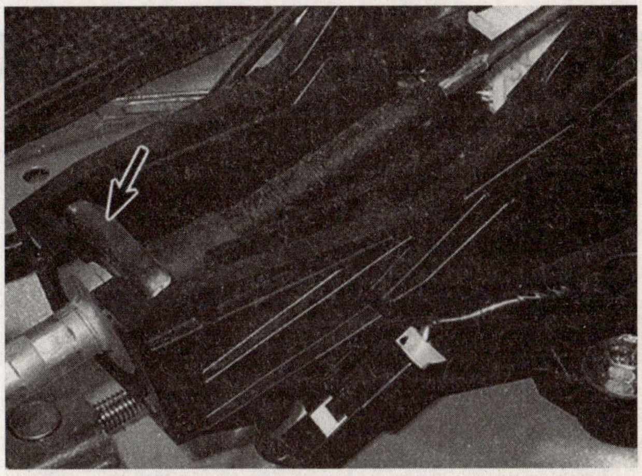

3.5 Detach the retaining clip to release the cable

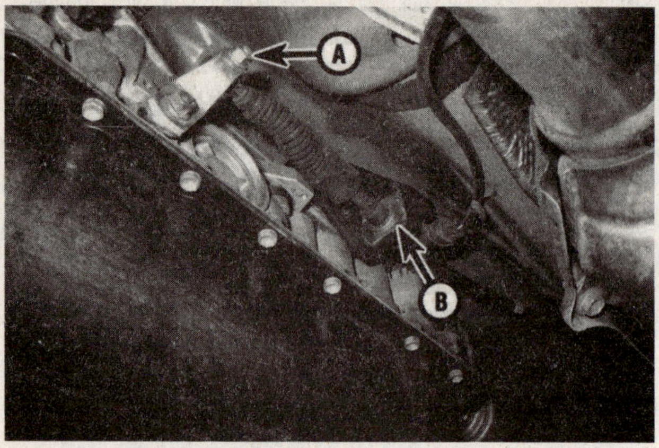

3.9 Remove the fastener securing the cable end (A), then remove the clip securing the cable (B)

3.14 Shift rod attachment points

9 Remove the nut that attaches the end of the shift cable to the manual lever and disconnect the shift cable from the manual lever, then remove the cable retainer clip with pliers (see illustration).

10 Carefully pull the cable through the body.

11 Installation is the reverse of removal. Set the lever on the side of the transmission to Neutral, then set the shift lever to Neutral. The cable should snap into place without moving the lever.

12 Always check the adjustment of the cable to verify that the range indicated on the console is the range the transmission is in. Refer to Section 4 if the vehicle will start in any position other than Neutral or Park.

ROD

➧ Refer to illustration 3.14

13 Raise the vehicle and support it securely on jackstands.

14 The shift rod used on some models is attached with a clevis pin at the transmission end and a bolt at the shifter end (see illustration).

15 If the shifter is correctly adjusted, mark the position of the rod's rear bolt with paint or a marker before removing it.

16 Remove the front cotter pin and slide the clevis out. At the rear, remove the adjusting bolt and nut.

17 Installation is the reverse of removal.

18 To adjust the shifter, loosen the rear bolt. Place the transmission lever and the shifter lever in Neutral.

19 Carefully tighten the adjusting bolt. Always check the adjustment of the rod to verify that the range indicated on the console is the range the transmission is in. Refer to Section 4 if the vehicle will start in any position other than Neutral or Park.

4 Park/Neutral Position (PNP) switch - removal, installation and adjustment

REMOVAL AND INSTALLATION

➧ Refer to illustration 4.5

1 The Park/Neutral Position switch prevents the engine from starting in any gear other than Park or Neutral. If the engine starts in any position other than Park or Neutral, it's either out of adjustment or defective. The switch is mounted on the right side of the transmission near the oil pan.

2 Raise the vehicle and place it securely on jackstands.

3 Unplug the electrical connector from the Park/Neutral Position switch.

4 Pry the lock washer fingers back and remove the nut.

5 Remove the Park/Neutral Position switch retaining bolt (see illustration).

6 Remove the switch.

7 Installation is the reverse of removal. Check and, if necessary, adjust the switch.

ADJUSTMENT

➧ Refer to illustration 4.11

8 Raise the vehicle and place it securely on jackstands.

9 Move the shifter to Neutral.

10 Loosen the switch retaining bolt.

11 Align the groove (inside the nut) and neutral basic line (above it) (see illustration).

12 Holding the switch in this position, tighten the switch retaining bolt.

13 Verify that the engine only starts in Neutral and Park.

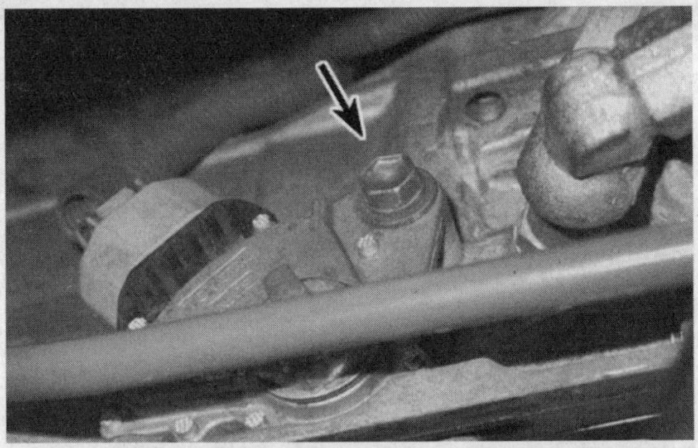

4.5 Loosen the Park/Neutral Position switch mounting bolt

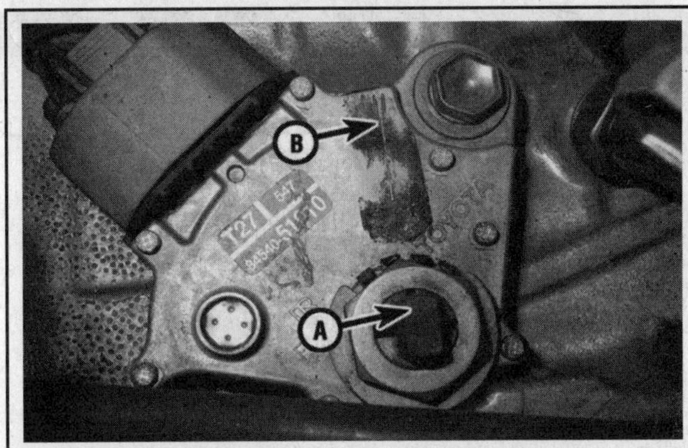

4.11 Align the groove (A) and neutral basic line (B)

5 Transmission oil cooler - removal and installation

Note: This procedure applies to models with an external transmission oil cooler.

1 Remove the grille (see Chapter 11). The cooler is mounted under the horns.

2 Place rags or a drain pan under the cooler.

3 Disconnect the fluid lines from the cooler. Immediately plug the ends to prevent contamination and fluid loss.

4 Remove the mounting bolts and remove the cooler.

5 Installation is the reverse of removal. Tighten all cooler fasteners and cooler lines securely.

6 Check the automatic transmission fluid level and add if necessary (see Chapter 1).

6 Output shaft oil seal - replacement

Refer to illustrations 6.4 and 6.5

1 Oil leaks frequently occur due to wear of the extension housing oil seal. Replacement of this seal is relatively easy, since the repair can usually be performed without removing the transmission from the vehicle.

2 The extension housing oil seal is located at the extreme rear of the transmission, where the driveshaft is attached. Raise the vehicle and support it securely on jackstands. If the seal is leaking, transmission lubricant will be built up on the front of the driveshaft and may be dripping from the rear of the transmission.

3 Remove the driveshaft (see Chapter 8).

4 Using a puller or seal removal tool, carefully remove the oil seal out of the rear of the transmission (see illustration). Do not damage the splines on the transmission output shaft.

5 Using a large section of pipe or a very large deep socket as a drift, install the new oil seal (see illustration). Drive it into the bore squarely and make sure it's completely seated.

6 Lubricate the splines of the transmission output shaft and the outside of the driveshaft slip yoke with lightweight grease, then install the driveshaft. Be careful not to damage the lip of the new seal.

7 Check the lubricant level in the transmission, adding as necessary (see Chapter 1).

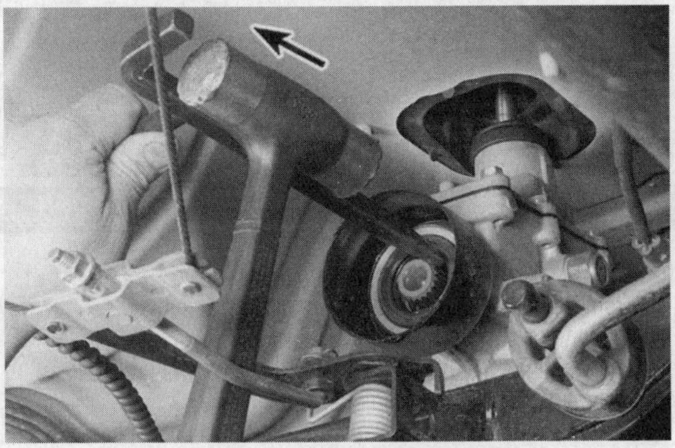

6.4 Using a seal removal tool to remove the seal

6.5 To install the new seal, tap it into place with a large socket and hammer

7 Transmission mount - check and replacement

CHECK

Refer to illustration 7.1

1 Insert a large screwdriver or pry bar into the space between the transmission and the crossmember and try to pry the transmission up slightly (see illustration).

2 The transmission should not move away from the insulator much. If there is any separation of the rubber, the mount is worn out.

REPLACEMENT

Refer to illustrations 7.3 and 7.4

3 Support the transmission with a jack and remove the transmission mount-to-crossmember retaining bolts (see illustration).

4 Raise the transmission slightly with the jack and remove the mount-to-transmission fasteners (see illustration).

5 Remove the mount.

6 Installation is the reverse of the removal procedure. Be sure to tighten the fasteners securely.

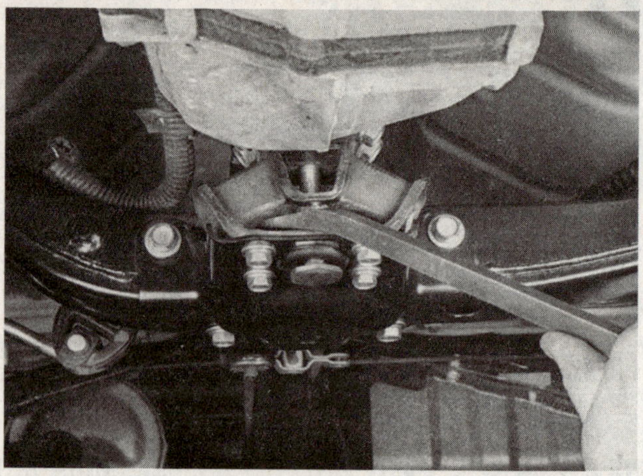

7.1 To check the transmission mount, insert a prybar or large screwdriver between the mount rubber and the mount bracket, then try to pry the transmission up off its mount; if the transmission moves significantly, replace the mount

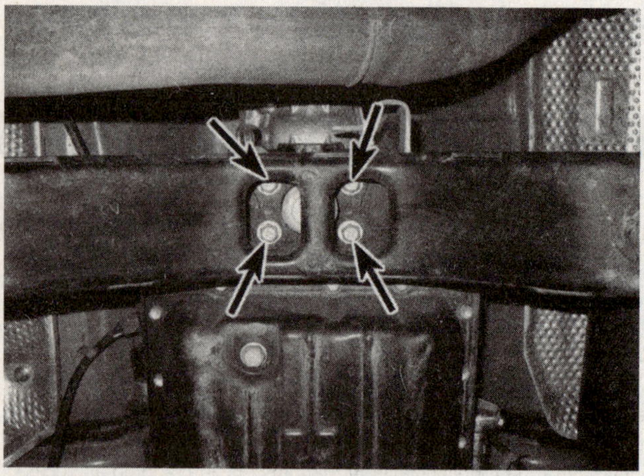

7.3 Transmission mount-to-crossmember retaining bolts

7.4 Remove the mount-to-transmission fasteners (left side shown, right side similar)

8 Automatic transmission - removal and installation

REMOVAL

Note: The transfer case will be removed along with the transmission on 4WD models.

1 Raise the vehicle and support it securely on jackstands. It must be high enough to slide the transmission under the body.

2 Disconnect the cable from the negative battery terminal (see Chapter 5, Section 1).

3 Remove any engine/transmission undercovers that are in the way.

4 Drain the transmission fluid (see Chapter 1).

5 Remove the transmission fluid dipstick tube and bracket.

6 Remove the engine manifold brace(s).

7 Depending which model you are working on, disconnect the shift cable or shift rod at the transmission (see Section 3).

8 Remove the rear driveshaft (see Chapter 8).

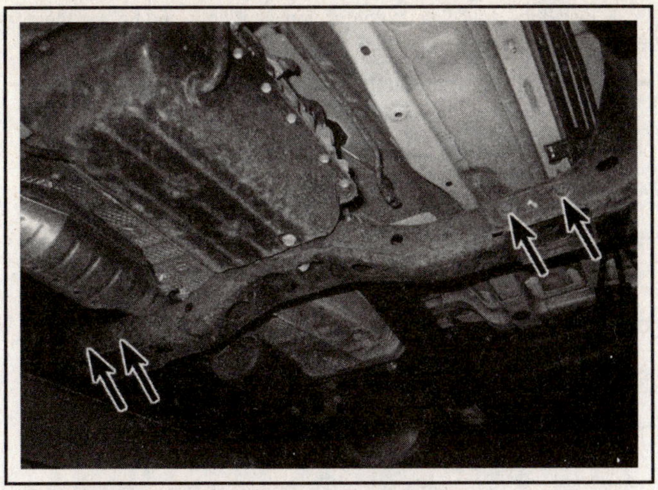

8.13 Crossmember-to-frame mounting bolts

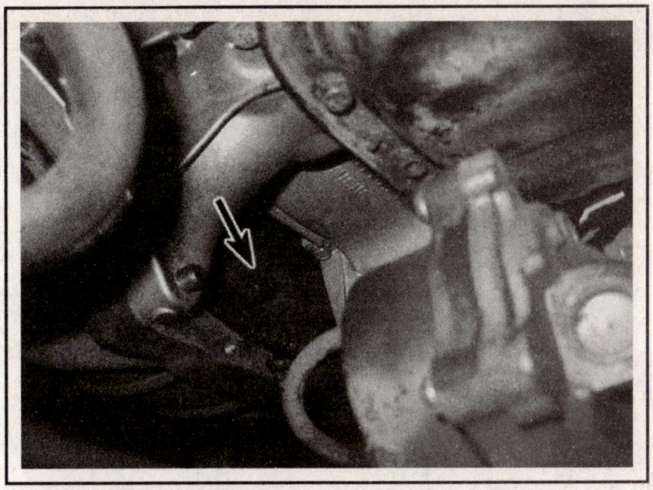

8.18 Remove the driveplate-to-torque converter bolts by turning the crankshaft (in a clockwise direction only, viewed from the front) for access to each bolt

4WD models

9 Disconnect the wiring from the transfer case indicator switches and the shifting wiring harness.

10 Mark the front driveshaft flange and the transfer case flange so they can be installed in the same position. Disconnect the front driveshaft from the transfer case and support it with wire.

All models

➧ Refer to illustrations 8.13 and 8.18

11 Disconnect the transmission fluid cooler lines and secure them out of the way so they won't become bent.

12 Support the transmission with a jack - preferably a jack made for this purpose. Safety chains will help steady the transmission on the jack.

13 Remove the rear crossmember (see illustration). Make sure the jack is solidly supporting the transmission. Raise the transmission enough to allow removal of the crossmember, then remove the nuts/ bolts securing the crossmember to the frame side rails and remove the crossmember and the mount.

14 Remove the starter (see Chapter 5).

15 Lower the rear of the transmission slightly and disconnect the wiring from the park/neutral position sensor, the speed sensors and the temperature sensor. Disconnect the wiring harness clips from the transmission.

16 Remove the torque converter access cover from the bellhousing.

17 Mark the torque converter and the driveplate with a scribe or chalk so they can be installed in the same position.

18 Remove the driveplate-to-torque converter bolts (see illustration). Turn the crankshaft (in a clockwise direction only, viewed from the front) for access to each bolt.

19 Support the engine with another jack. Use a block of wood under the oil pan to spread the load.

20 Remove the transmission-to-engine bolts.

21 Slowly lower the jack until you can remove the upper bolts securing the transmission to the engine. Several long extensions may have to be used to reach the upper bolts.

22 Move the transmission to the rear to disengage it from the engine block dowel pins and make sure the torque converter is detached from the driveplate. Secure the torque converter to the transmission so it won't fall out during removal.

INSTALLATION

➧ Refer to illustrations 8.27a and 8.27b

23 Prior to installation, make sure the torque converter hub is securely engaged in the front pump of the transmission. This can be confirmed by pushing in on the converter and turning it; if it isn't seated completely, it will drop into place as this is done.

24 With the transmission secured to the jack, raise it into position. Be sure to keep it level so the torque converter doesn't slide out.

25 Turn the torque converter until the marks on the converter and driveplate are aligned.

26 Move the transmission forward carefully until the dowel pins engage with the holes in the bellhousing.

27 Install the transmission-to-engine bolts. Tighten them to the torque listed in this Chapter's Specifications (see illustrations).

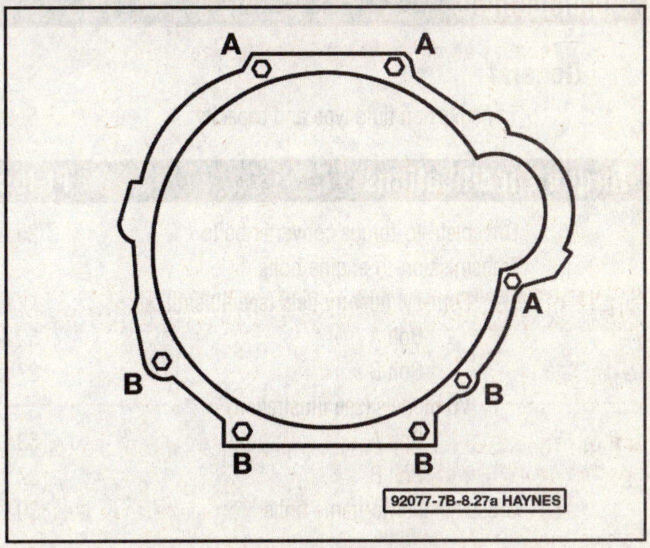

8.27a Transmission mounting bolt identification - four-cylinder models

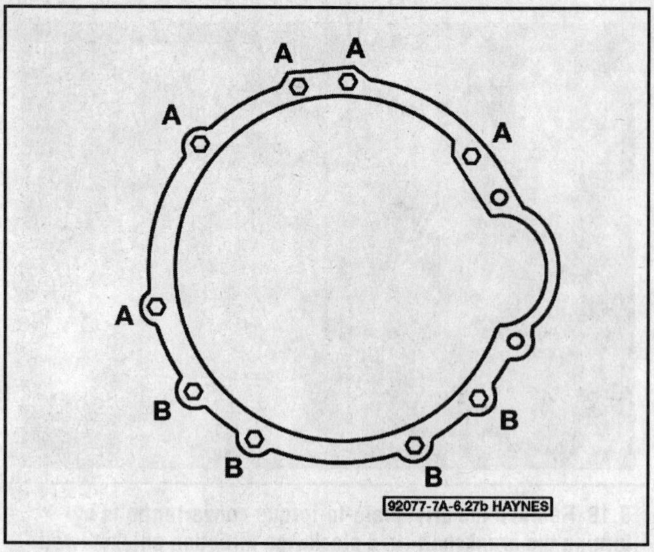

8.27b Transmission mounting bolt identification - V6 models

28 Install the driveplate-to-torque converter bolts and tighten them to the torque listed in this Chapter's Specifications.

➡Note: Install all of the bolts before tightening any of them.

29 Install the crossmember to the frame, then tighten the bolts and nuts to the torque listed in this Chapter's Specifications.

30 Lower the transmission extension housing until the mount is seated on and aligned with the crossmember, then tighten the fasteners securely.

31 The remainder of installation is the reverse of removal.

32 Remove all jacks supporting the transmission and engine and lower the vehicle.

33 Fill the transmission with the specified fluid (see Chapter 1), run the engine and check for fluid leaks.

9 Automatic transmission overhaul - general information

In the event of a fault occurring, it will be necessary to establish whether the fault is electrical, mechanical or hydraulic in nature, before repair work can be contemplated. Diagnosis requires detailed knowledge of the transmission's operation and construction, as well as access to specialized test equipment, and so is deemed to be beyond the scope of this manual. It is therefore essential that problems with the automatic transmission be referred to a dealer service department or other qualified repair facility for assessment.

Note that a faulty transmission should not be removed before the vehicle has been assessed by a knowledgeable technician equipped with the proper tools, as troubleshooting must be performed with the transmission installed in the vehicle.

Specifications

General

Transmission fluid type and capacity	See Chapter 1

Torque specifications	Ft-lbs	Nm
Driveplate-to-torque converter bolts	35	47
Transmission-to-engine bolts		
Four-cylinder models (see illustration)		
Bolt A	52	71
Bolt B	27	37
V6 models (see illustration)		
Bolt A	53	72
Bolt B	27	37
Crossmember-to-frame bolts	30	40

7C
TRANSFER CASE

Section

Reference to other Chapters

Transfer case - lubricant change - See Chapter 1
Transfer case - lubricant level check - See Chapter 1

1 General information

Four-wheel drive (4WD) models are equipped with a transfer case mounted on the rear of the transmission. Drive is transmitted from the engine, through the transmission and the transfer case to the front and rear axles by driveshafts.

We don't recommend trying to rebuild a transfer case at home. It's difficult to overhaul without special tools, and rebuilt units are available for less than it would cost to rebuild your own. However, there are a number of components that you can check, adjust and/or replace - and those are the items covered in this Chapter.

2 Oil seals - replacement

Refer to illustrations 2.4a, 2.4b, 2.5, 2.6 and 2.8

Note: This procedure applies to both the front and rear seals. It's not necessary to remove the transfer case to replace a seal.

1 Raise the vehicle and support it securely on jackstands,

2 Drain the transfer case lubricant (see Chapter 1).

3 If you're replacing the front seal, remove the front driveshaft; if you're replacing the rear seal, remove the rear driveshaft (see Chapter 8).

4 Unstake and remove the companion flange retaining nut (see illustrations).

5 Remove the companion flange (see illustration). If the flange is difficult to remove from the output shaft, use a puller.

6 Pry out the seal with a screwdriver or a seal removal tool (see illustration). Don't damage the seal bore.

7 Lubricate the new seal lip with multi-purpose grease.

8 Drive the seal into place with a large socket (see illustration). The outside diameter of the socket should be slightly smaller than the outside diameter of the seal.

9 The remainder of installation is the reverse of removal. Be sure to tighten the companion flange nut to the torque listed in this Chapter's Specifications.

10 Fill the transfer case with the specified fluid (see Chapter 1), drive the vehicle and check for leaks.

2.4a Unstake the companion flange retaining nut . . .

2.4b . . . then break the nut loose while holding the flange as shown

2.5 Pull the companion flange off the output shaft - it may be necessary to use a small puller

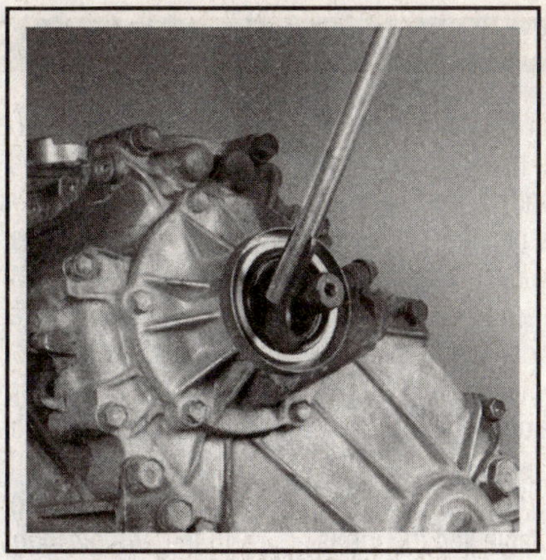

2.6 Pry out the seal with a screwdriver or a seal removal tool

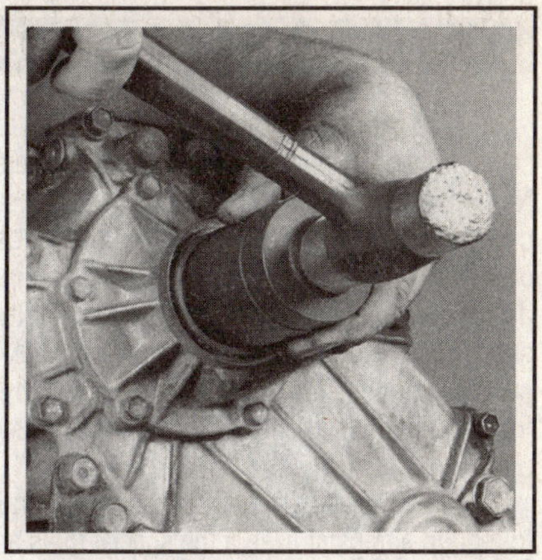

2.8 Drive the seal into place with a seal installer tool or a large socket

3 4WD Shift Selector System switch - replacement

DESCRIPTION

1 The Shift Selector System allows the driver to change the function of the transfer case using a switch on the instrument panel. All 4WD models have a knob that selects 2 wheel drive, 4 wheel drive and low-range 4 wheel drive.

SWITCH REPLACEMENT

2 Remove the center instrument cluster trim panel (see Chapter 11). Disconnect the electrical connectors from the switch.

3 To remove the switch from the center trim panel, depress the release tabs and push the switch out through the front side of the trim panel.

4 Installation is the reverse of removal.

4 Transfer case - removal and installation

➨Note: The manufacturer recommends that the transmission/transfer case assembly be removed before separating them. In some vehicles, it may be possible to simply remove the transfer case with the transmission left attached to the engine. If you decide to attempt removing the transfer case with the transmission left in the vehicle, refer to Chapter 7A or 7B for information about properly supporting the transmission and lowering it for clearance.

1 Raise the vehicle and support it securely on jackstands

2 Drain the transfer case lubricant (see Chapter 1).

3 Remove the front and rear driveshafts (see Chapter 8).

4 Remove any transfer case breather hose from the top of the transmission.

5 If necessary, remove any exhaust pipes that interfere with the removal of the transfer case (see Chapter 4).

6 Unplug all electrical connectors, such as the shift actuator wiring, transfer case position switches and all other connectors or wiring harnesses attached to the transfer case.

7 Refer to Chapter 7A or 7B and remove the transmission/transfer case assembly.

8 With the front of the transmission facing down, remove the bolts securing the transfer case to the transmission. Have an assistant support the transfer case while you lift it away from the rear of the transmission.

9 Installation is the reverse of removal. Be sure to tighten the transfer case-to-transmission bolts to the torque listed in this Chapter's Specifications.

10 Fill the transfer case with the specified fluid (see Chapter 1), drive the vehicle and check for fluid leaks.

5 Transfer case overhaul - general information

1 Overhauling a transfer case is a difficult job for the do-it-yourselfer. It involves the disassembly and reassembly of many small parts. Numerous clearances must be precisely measured and, if necessary, changed with select-fit spacers and snap-rings. As a result, if transfer case problems arise, it can be removed and installed by a competent do-it-yourselfer, but overhaul should be left to a transmission repair shop. Rebuilt transfer cases may be available - check with your dealer parts department and auto parts stores. At any rate, the time and money involved in an overhaul is almost sure to exceed the cost of a rebuilt unit.

2 Nevertheless, it's not impossible for an inexperienced mechanic to rebuild a transfer case if the special tools are available and the job is done in a deliberate step-by-step manner so nothing is overlooked.

3 The tools necessary for an overhaul include internal and external snap-ring pliers, a bearing puller, a slide hammer, a set of pin punches, a dial indicator and possibly a hydraulic press. In addition, a large, sturdy workbench and a vise or transfer case stand will be required.

4 During disassembly of the transfer case, make careful notes of how each piece comes off, where it fits in relation to other pieces and what holds it in place. Noting how they are installed when you remove the parts will make it much easier to get the transfer case back together.

5 Before taking the transfer case apart for repair, it will help if you have some idea what area of the transfer case is malfunctioning. Certain problems can be closely tied to specific areas in the transfer case, which can make component examination and replacement easier. Refer to the *Troubleshooting* Section at the front of this manual for information regarding possible sources of trouble.

Specifications

General

Transfer case - lubricant type	See Chapter 1

Torque specifications

Torque specifications	Ft-lbs	Nm
Driveshaft companion flange nut (front and rear)	87	118
Transfer case-to-transmission bolts	17	23

8 CLUTCH AND DRIVELINE

Section

Reference to other Chapters

Differential lubricant change - See Chapter 1
Differential lubricant level check - See Chapter 1
Driveaxle boot check (4WD models) - See Chapter 1

1 General information

The Sections in this Chapter deal with the components from the rear of the engine to the rear wheels (except for the transmission and transfer case, which are dealt with in Chapter 7) and forward to the front wheels on four-wheel drive (4WD) models. In this Chapter, the components are grouped into three categories: clutch, driveshaft(s) and axle(s). Separate Sections within this Chapter cover checks and repair procedures for components in each of these three groups.

Since nearly all these procedures involve working under the vehicle, make sure it's safely supported on sturdy jackstands or a hoist where the vehicle can be safely raised and lowered.

2 Clutch - description and check

1 All vehicles with a manual transmission have a single dry plate, diaphragm type clutch. The clutch disc has a splined hub which allows it to slide along the splines of the transmission input shaft. The clutch and pressure plate are held in contact by spring pressure exerted by the diaphragm in the pressure plate.

2 The clutch release system is operated by hydraulic pressure. The hydraulic release system consists of the clutch pedal, a master cylinder, the hydraulic line, a release (or slave) cylinder which actuates the clutch release lever and the clutch release (or throwout) bearing.

3 When pressure is applied to the clutch pedal to release the clutch, hydraulic pressure is exerted against the outer end of the clutch release lever. As the lever pivots the shaft fingers push against the release bearing. The bearing pushes against the fingers of the diaphragm spring of the pressure plate assembly, which in turn releases the clutch plate.

4 Terminology can be a problem when discussing the clutch components because common names are in some cases different from those used by the manufacturer. For example, the driven plate is also called the clutch plate or disc, the clutch release bearing is sometimes called a throwout bearing, the release cylinder is sometimes called the slave cylinder.

5 Other than to replace components with obvious damage, some preliminary checks should be performed to diagnose clutch problems.

a) *The clutch release system is supplied with fluid from the brake master cylinder, so the first check should be of the fluid level in the brake master cylinder. If the fluid level is low, add fluid as necessary and inspect the brake and clutch hydraulic systems for leaks. If the reservoir is dry, bleed the system as described in Section 8 and recheck the clutch operation.*

b) *To check clutch spin-down time, run the engine at normal idle speed with the transmission in Neutral (clutch pedal up - engaged). Disengage the clutch (pedal down), wait several seconds and shift the transmission into Reverse. No grinding noise should be heard. A grinding noise would most likely indicate a bad pressure plate or clutch disc.*

c) *To check for complete clutch release, run the engine (with the parking brake applied to prevent vehicle movement) and hold the clutch pedal approximately 1/2-inch from the floor. Shift the transmission between 1st gear and Reverse several times. If the shift is rough, component failure is indicated. Check the release cylinder pushrod travel. With the clutch pedal depressed completely, the release cylinder pushrod should extend substantially (you may have to remove an inspection plug to see the pushrod). If it doesn't, check the fluid level in the clutch master cylinder.*

d) *Visually inspect the pivot bushing at the top of the clutch pedal to make sure there's no binding or excessive play.*

e) *Crawl under the vehicle and make sure the clutch release lever is securely attached to the ballstud.*

3 Clutch master cylinder - removal and installation

1 Clamp-off the fluid feed hose coming from the brake master cylinder.

2 Disconnect the fluid feed hose and the hydraulic line to the release cylinder where they connect to the fittings at the firewall; use a flare-nut wrench to protect the tube nut, and hold the fitting with an open-end wrench. Have rags handy, as some fluid will be lost as the line is removed.

CAUTION:

Don't allow fluid to come into contact with the paint since it will damage the finish. Also have plugs ready and immediately plug the line and hose to prevent leakage and fluid contamination.

3 Working inside the passenger compartment, remove the left door scuff plate, the kick panel and the lower trim panel(s) from the driver's side (see Chapter 11).

4 Remove the fuse and relay box from the left side of the instrument panel (see illustration 11.16 in Chapter 3).

5 Remove the clutch start switch (see Section 10).

6 Remove the clutch pedal mounting fasteners; it's secured by two nuts and one bolt (the bolt is at the top of the pedal bracket).

7 Remove the clutch pedal/master cylinder assembly from the passenger compartment.

8 Disconnect the hydraulic line from the master cylinder.

9 Remove the master cylinder mounting bolts and detach the master cylinder from the pedal bracket.

10 Installation is the reverse of removal.

11 Bleed the clutch system (see Section 8). Add fluid to the brake master cylinder, as necessary (see Chapter 1).

4 Clutch release cylinder and accumulator - removal and installation

1 Clamp-off the fluid feed hose coming from the brake master cylinder.

2 Raise the vehicle and support it securely on jackstands.

RELEASE CYLINDER

3 On V6 models, remove the clutch housing cover.

4 Disconnect the clutch fluid hydraulic line from the release cylinder. Use a flare-nut wrench to protect the tube nut. Have rags handy, as some fluid will be lost as the line is removed. Also have a plug ready and immediately plug the line to prevent leakage.

5 Remove the two release cylinder mounting bolts.

6 Separate the release cylinder from the transmission.

7 Installation is the reverse of removal. Make sure the pushrod dust boot is in good condition and the pushrod is seated correctly in its pocket in the release lever. Tighten the release cylinder mounting bolts to the torque listed in this Chapter's Specifications.

8 Bleed the clutch hydraulic system (see Section 8).

9 Add fluid to the brake master cylinder, as necessary (see Chapter 1).

ACCUMULATOR (V6 MODELS)

10 Remove the clutch housing cover.

11 Disconnect the clutch fluid hydraulic lines from the accumulator. Use a flare-nut wrench to protect the tube nut. Have rags handy, as some fluid will be lost as the line is removed.

CAUTION:

Don't allow fluid to come into contact with the paint - it will damage the finish. Also have a plug ready and immediately plug the line to prevent leakage.

12 Remove the three accumulator mounting bolts and separate the accumulator from the transmission.

13 Installation is the reverse of removal. Tighten the accumulator mounting bolts to the torque listed in this Chapter's Specifications.

14 Bleed the clutch hydraulic system (see Section 8).

15 Add fluid to the brake master cylinder, as necessary (see Chapter 1).

5 Clutch components - removal, inspection and installation

WARNING:

Dust produced by clutch wear and deposited on clutch components is hazardous to your health. DO NOT blow it out with compressed air and DO NOT inhale it. DO NOT use gasoline or petroleum-based solvents to remove the dust. Brake system cleaner should be used to flush the dust into a drain pan. After the clutch components are wiped clean with a rag, dispose of the contaminated rags and cleaner in a covered, marked container.

REMOVAL

Refer to illustrations 5.5a, 5.5b and 5.6

Note: The following procedure is based on the assumption that the transmission is being removed and the engine remains in place. However, anytime the transmission or engine is removed, you should inspect the clutch assembly for wear. Unless the clutch components are new or in near-perfect condition, their relatively low cost - compared to the time and trouble it takes to get to them - warrants their replacement anytime the engine or transmission is removed.

1 Raise the vehicle and support it securely on jackstands.

2 Remove the clutch release cylinder (see Section 4).

3 Remove the transmission (see Chapter 7, Part A). Support the engine while the transmission is out. An engine hoist can be used to support it from above. If you use a jack underneath the engine instead, make sure a piece of wood is positioned between the jack and oil pan to spread the load.

CAUTION:

The pick-up for the oil pump is very close to the bottom of the oil pan. If the pan is bent or distorted in any way, engine oil starvation could occur.

4 The clutch release lever and release bearing can remain attached to the housing for the time being.

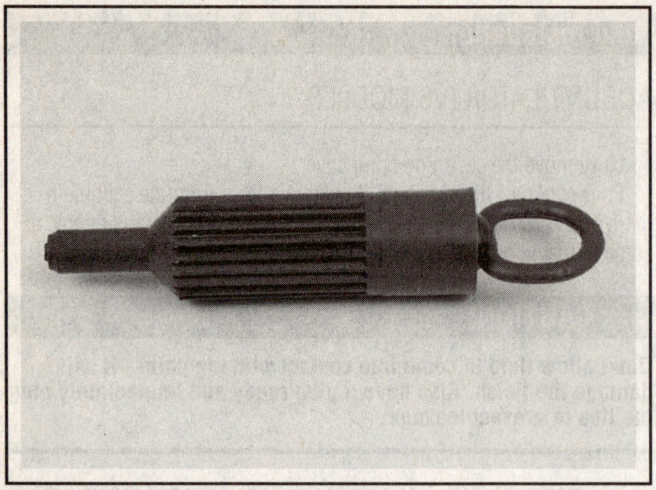

5.5a A clutch alignment tool like this one is available at most auto part stores

5.5b Insert the clutch alignment tool into the clutch disc splines prior to loosening the pressure plate bolts, or to center the clutch disc when reinstalling the pressure plate

5.6 If you're going to re-use the same pressure plate, mark its relationship to the flywheel

5.9 Inspect the surface of the flywheel for cracks, dark-colored areas (signs of overheating) and other obvious defects; resurfacing will correct minor defects

5 To support the clutch disc during removal, install an alignment tool through the clutch disc hub (see illustrations).

6 Carefully inspect the flywheel and pressure plate for indexing marks. The marks are usually an X, an O or a white letter. If they cannot be found, paint a mark so the pressure plate and the flywheel will be in the same alignment during installation (see illustration).

7 Loosen the six pressure plate-to-flywheel bolts in 1/4-turn increments until they can be removed by hand. Work in a criss-cross pattern until all spring pressure is relieved, then hold the pressure plate securely and completely remove the bolts, followed by the pressure plate and clutch disc.

INSPECTION

Refer to illustrations 5.9, 5.11 and 5.13

8 Ordinarily, when a problem occurs in the clutch, it can be attributed to wear of the clutch driven plate assembly (clutch disc). However, all components should be inspected at this time.

Note: If the clutch components are contaminated with oil, there will be shiny, black glazed spots on the clutch disc lining, which will cause the clutch to slip. Replacing clutch components won't completely solve the problem - be sure to check the crankshaft rear oil seal and the transmission input shaft seal for leaks. If it looks like a seal is leaking, be sure to install a new one to avoid the same problem with the new clutch.

9 Check the flywheel for cracks, heat checking, grooves and other obvious defects (see illustration). If the imperfections are slight, a machine shop can machine the surface flat and smooth, which is highly recommended regardless of the surface appearance. Refer to Chapter 2, Part A or B, for the flywheel removal and installation procedure.

10 Inspect the pilot bearing (see Section 7).

11 Check the lining on the clutch disc. There should be at least 1/32-inch of lining above the rivet heads. Check for loose rivets, distortion, cracks, broken springs and other obvious damage (see illustration). As mentioned above, ordinarily the clutch disc is routinely replaced, so if you're in doubt about its condition, replace it.

12 The release bearing should also be replaced along with the clutch disc (see Section 6).

13 Check the machined surfaces and the diaphragm spring fingers of the pressure plate (see illustration). If the surface is grooved or otherwise damaged, replace the pressure plate. Also check for obvious damage, distortion, cracks, etc. Light glazing can be removed with medium-grit emery cloth. If a new pressure plate is required, new and factory-rebuilt units are available.

INSTALLATION

➧ Refer to illustration 5.17

14 Before installation, clean the flywheel and pressure plate machined surfaces with brake system cleaner. It's important that no oil or grease is on these surfaces or the lining of the clutch disc. Handle the parts only with clean hands.

15 Position the clutch disc and pressure plate against the flywheel with the clutch held in place with an alignment tool (see illustration 5.5b). Make sure it's installed properly (most replacement clutch plates will be marked "flywheel side" or something similar - if it's not marked, install the clutch disc with the damper springs toward the transmission).

16 Tighten the pressure plate-to-flywheel bolts only finger-tight, working around the pressure plate.

17 Center the clutch disc by ensuring the alignment tool extends through the splined hub and into the pilot bearing in the crankshaft. Wiggle the tool up, down or from side-to-side as needed to bottom the tool in the pilot bearing. Tighten the pressure plate-to-flywheel bolts a little at a time, working in the order shown (see illustration), to prevent distorting the cover. After all the bolts are snug, tighten them to the torque listed in this Chapter's Specifications. Remove the alignment tool.

18 Using high-temperature grease, lubricate the inner groove of the release bearing. Also apply a thin film of grease on the release lever contact areas and the transmission input shaft bearing retainer.

19 Install the clutch release bearing (see Section 6).

20 Install the transmission, release cylinder and all components removed previously.

5.11 Inspect the clutch plate lining, springs and splines for wear

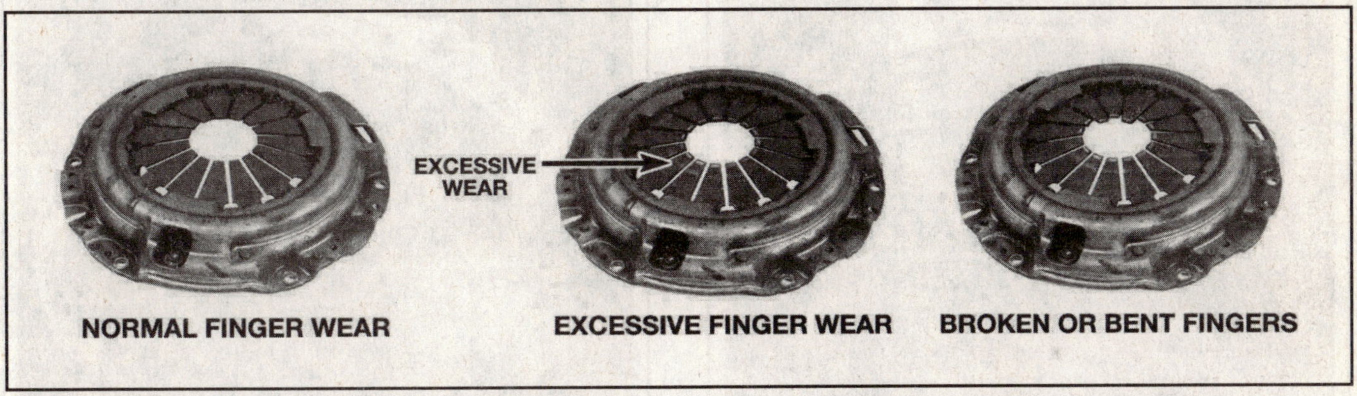

5.13 Replace the pressure plate if excessive wear or damage is noted

5.17 Tighten the pressure plate bolts in the order shown, a little at a time, to the torque listed in this Chapter's Specifications

6 Clutch release bearing - removal, inspection and installation

**** WARNING:**

Dust produced by clutch wear and deposited on clutch components is hazardous to your health. DO NOT blow it out with compressed air and DO NOT inhale it. DO NOT use gasoline or petroleum-based solvents to remove the dust. Brake system cleaner should be used to flush the dust into a drain pan. After the clutch components are wiped clean with a rag, dispose of the contaminated rags and cleaner in a covered, marked container.

REMOVAL

Refer to illustrations 6.4a, 6.4b and 6.4c

1 Remove the release cylinder (see Section 4).
2 Remove the transmission (see Chapter 7, Part A).
3 Disengage the release lever from the ballstud.
4 Note how the release lever fingers are engaged by the wire retainer on the bearing, then disengage the bearing from the lever and slide it off the input shaft (see illustration). Remove the release lever (see illustration). Inspect the release lever boot for cracks or tears. If it's worn or damaged, replace it. Inspect the wire retainer in the lever. If its damaged or distorted, remove it (see illustration) and discard it.

INSPECTION

Refer to illustration 6.5

5 Hold the outer portion of the bearing and rotate the center while applying pressure (see illustration). If the bearing doesn't turn smoothly or if it's noisy, replace it with a new one. Wipe the bearing with a clean rag and inspect it for damage, wear and cracks. Don't immerse the bearing in solvent - it's sealed for life and to do so would ruin it.

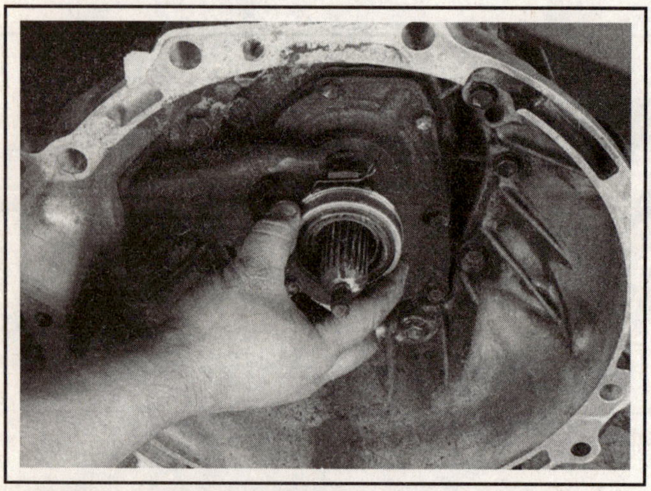

6.4a Note how the release bearing and the release lever fit together, then disengage them and slide them off the bearing - four cylinder model shown

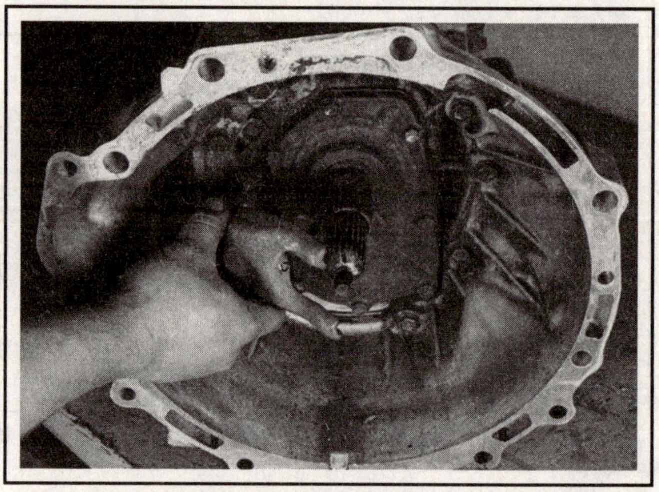

6.4b Remove the release lever

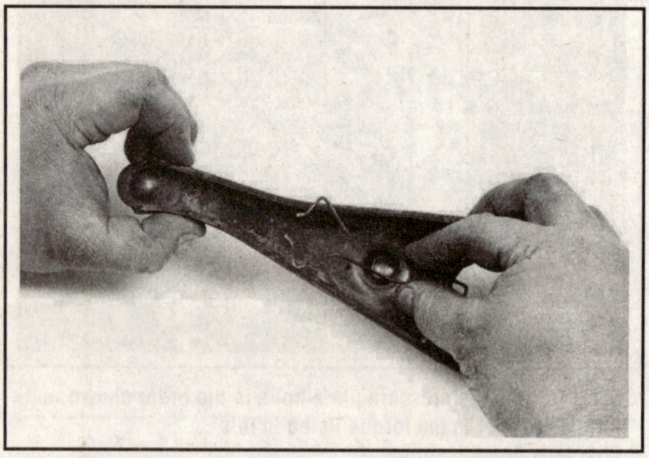

6.4c If the release lever retainer is damaged or distorted, remove and discard it - don't install the release lever with a weak or bent retainer

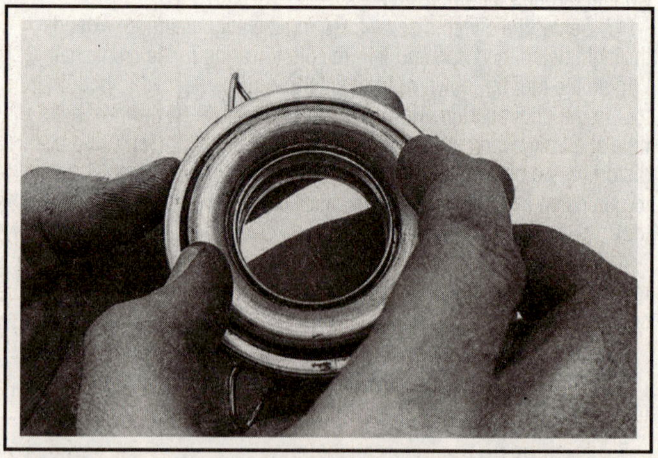

6.5 To check the operation of the release bearing, hold it by the outer race and rotate the inner race while applying pressure; the bearing should turn smoothly - if it doesn't, replace it

INSTALLATION

Refer to illustrations 6.6a, 6.6b, 6.6c, 6.6d, 6.8a and 6.8b

6 Lightly lubricate the clutch release lever at the indicated spots (see illustrations). If you removed the old retainer from the lever, install the new one (see illustration). Also lubricate the groove around the circumference of the release bearing that's engaged by the lever, and the bearing retainer around the input shaft with high-temperature grease (see illustration).

7 Attach the release bearing to the release lever. Make sure the bearing is properly engaged by the retainer clip.

8 Lubricate the clutch release lever ballstud with high-temperature grease, insert the release lever through the boot, slide the release bearing onto the input shaft splines and push the lever onto the ballstud until the lever retainer pops onto the stud (see illustration). Make sure that the release lever pivots freely and the release bearing slides freely on the input shaft splines (see illustration).

9 Apply a light coat of high-temperature grease to the face of the release bearing, where it contacts the pressure plate diaphragm fingers.

10 The remainder of installation is the reverse of the removal procedure. Tighten all transmission-to-engine bolts to the torque listed in the Chapter 7A Specifications.

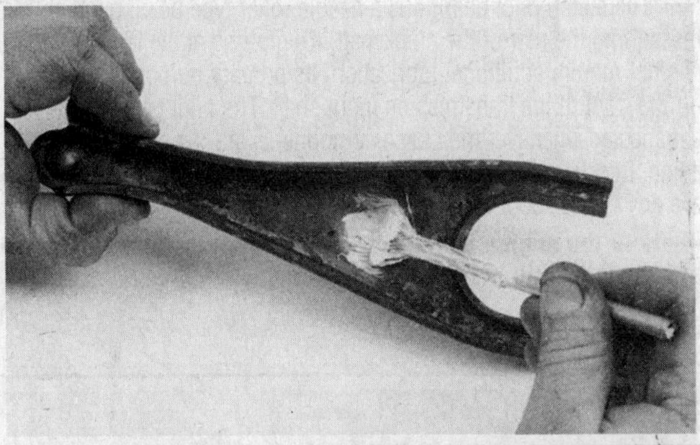

6.6a Lubricate the pocket for the ballstud in the backside of the release lever . . .

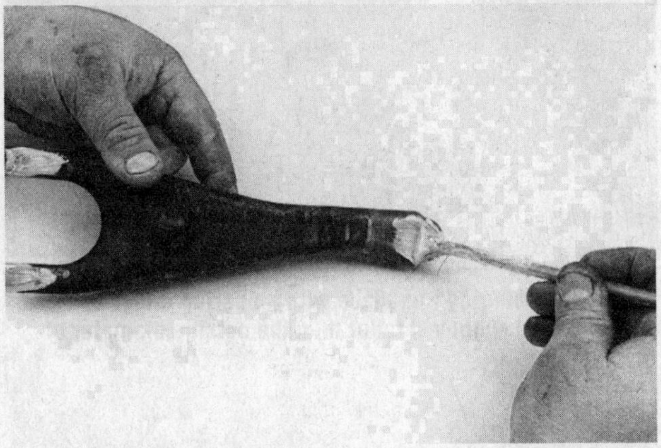

6.6b . . . the release lever fingers and the pocket for the release cylinder pushrod with high-temperature grease

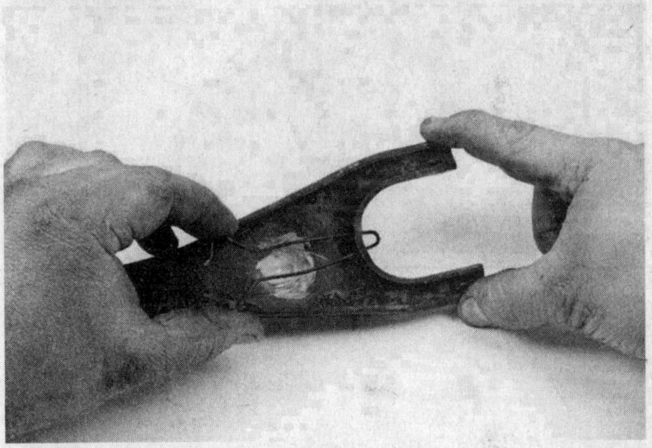

6.6c If you removed the old release lever retainer, install the new one

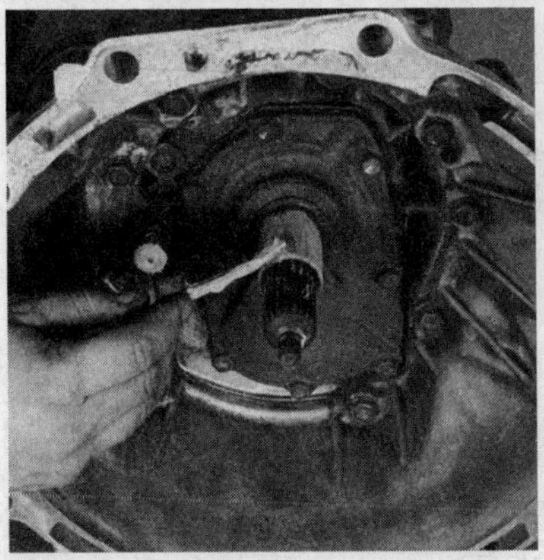

6.6d Lubricate the bearing retainer surrounding the input shaft with high temperature grease

6.8a Install the release lever and release bearing and push the release lever onto the ballstud (you should feel it snap into place when the ballstud pops through the wire retainer)

6.8b Slide the release bearing in and out on the retainer to verify that the release lever is locked onto the ballstud, the bearing slides freely and the two parts are properly engaged

7 Pilot bearing - inspection and replacement

➧ Refer to illustrations 7.5 and 7.6

1 The clutch pilot bearing is a needle roller type bearing which is pressed into the rear of the crankshaft. It's greased at the factory and does not require additional lubrication. Its primary purpose is to support the front of the transmission input shaft. The pilot bearing should be inspected whenever the clutch components are removed from the engine. Because of its inaccessibility, replace it with a new one if you have any doubt about its condition.

➡Note: If the engine has been removed from the vehicle, disregard the following Steps which don't apply.

2 Remove the transmission (see Chapter 7A).

3 Remove the clutch components (see Section 5).

4 Using a flashlight, inspect the bearing for excessive wear, scoring, dryness, roughness and any other obvious damage. If any of these conditions are noted, replace the bearing.

5 Removal can be accomplished with a special puller (see illustration), which is available at most auto parts stores.

6 To install the new bearing, lightly lubricate the outside surface with grease, then drive it into the recess with a seal installer or a socket (see illustration). The bearing seal must face out.

7 Install the clutch components, transmission and all other components removed previously. Tighten all fasteners to the recommended torque.

7.5 A small slide-hammer puller is handy for removing an old pilot bearing

7.6 Tap the bearing into place with a bearing driver or a socket that is slightly smaller than the outside diameter of the bearing

8 Clutch hydraulic system - bleeding

➧ Refer to illustration 8.4

1 The hydraulic system should be bled to remove all air whenever any part of the system has been removed or if the fluid level has been allowed to fall so low that air has been drawn into the master cylinder. The procedure is very similar to bleeding a brake system.

2 Fill the master cylinder with new brake fluid conforming to DOT 3 specifications.

⁂ CAUTION:

Do not re-use any of the fluid coming from the system during the bleeding operation or use fluid which has been inside an open container for an extended period of time.

3 Raise the vehicle and support it securely on jackstands to gain access to the release cylinder, which is located on the left side of the clutch housing.

4 Remove the dust cap which fits over the bleeder valve and push a length of hose over the valve (see illustration). Place the other end of the hose into a clear container with about two inches of brake fluid. The hose end must be in the fluid at the bottom of the container.

8.4 The setup for bleeding the clutch hydraulic system is simple: a length of hose between the bleeder plug and a small container with about two inches of brake fluid in it; make sure the hose is submerged in the fluid

5 Have an assistant depress the clutch pedal and hold it. Open the bleeder valve on the release cylinder, allowing fluid to flow through the hose. Close the bleeder valve when the flow of fluid and air bubbles stops. Once closed, have your assistant release the pedal.

6 Continue this process until all air is evacuated from the system, indicated by a solid stream of fluid being ejected from the bleeder valve each time with no air bubbles in the hose or container. Keep a close watch on the fluid level inside the clutch master cylinder reservoir - if the level drops too low, air will be sucked back into the system and the process will have to be started all over again.

7 Install the dust cap and lower the vehicle. Check carefully for proper operation before placing the vehicle in normal service.

9 Clutch pedal - adjustment

➧ Refer to illustration 9.2

1 Remove any floor mats or carpeting to expose the sound deadener material.

2 With the clutch pedal at its normal (released) height, measure the distance from the floor to the pedal (see illustration) and compare your measurement to the pedal height listed in this Chapter's Specifications.

3 If the pedal height is incorrect, remove the lower finish panel(s) (see Chapter 11) and adjust pedal height by loosening the locknut and turning the adjustment bolt located at the top of the pedal. Once the pedal height is correct, tighten the locknut securely.

4 Depress the clutch pedal to the point at which you feel initial resistance, measure the distance from the floor to the pedal and compare your measurement to the pedal freeplay listed in this Chapter's Specifications.

5 If the pedal freeplay is incorrect, remove the lower finish panel(s) (see Chapter 11) and adjust pedal freeplay by loosening the pushrod locknut and turning the pushrod until pedal freeplay is correct. Once the pedal freeplay is correct, tighten the locknut securely.

6 After adjusting pedal freeplay, check pedal height again and make sure that it's still within the specified range of adjustment.

7 Check and, if necessary, adjust the clutch start switch (see Section 10).

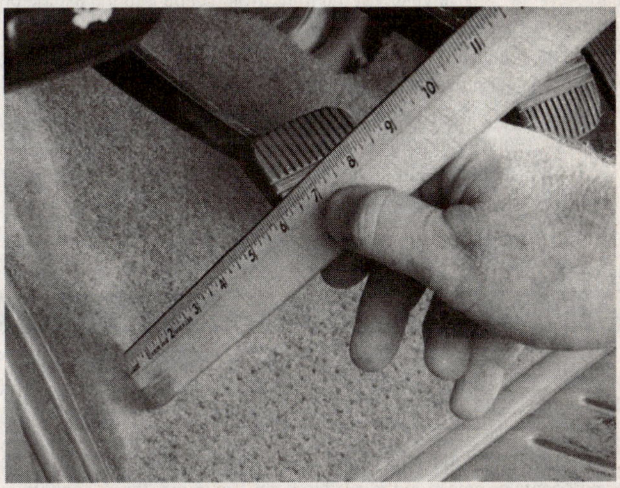

9.2 Measure the clutch pedal height with the pedal in the released position

10 Clutch start switch - check, adjustment and replacement

CHECK

1 The clutch start switch, which is mounted on the clutch pedal bracket, prevents the engine from being started unless the clutch pedal is depressed.

2 To test the switch, verify that the engine will not start if the clutch pedal is depressed, and that it does start when the pedal is depressed.

3 If the engine starts without depressing the clutch pedal, the switch is probably bad, but check the switch circuit first. Make sure there's voltage to the power side of the switch.

4 If there's voltage to the switch, verify that there's only voltage on the ground side of the switch when the clutch pedal is depressed.

5 If there's a short or open in the circuit on either side of the switch, repair it and retest.

6 If the circuit is okay, check the continuity of the switch with an ohmmeter. When the plunger is depressed, there should be continuity; when the plunger is released (free), there should be no continuity. If the switch does not perform as described, replace the switch.

ADJUSTMENT

7 Loosen the switch locknut and screw the switch in or out so that the switch plunger protrudes past the locknut. Press the clutch pedal and check to see that the switch contacts the clutch pedal stop (ON) and releases completely (OFF).

REPLACEMENT

8 Remove the lower finish panel(s) (see Chapter 11).

9 Unplug the electrical connector from the switch, remove the switch locknut and remove the switch.

10 Installation is the reverse of removal. Adjust the switch.

11 Driveshaft(s) and universal joints - general information

The driveshaft is of tubular construction and designed as a one-section type (regular cab models) or two-section type (extended cab models).

The driveshaft is finely balanced during manufacture and it is recommended that care be used when universal joints are replaced to help maintain this balance. It is sometimes better to have the universal joints replaced by a dealership or shop specializing in this type of work. If you replace the joints yourself, mark each individual yoke in relation to the one opposite in order to maintain the balance. Do not drop the assembly during servicing operations.

Driveshafts on 2WD models employ a splined yoke, known as a "slip yoke" or "sleeve yoke," at the front, which slips into the extension housing of the transmission. This arrangement allows the driveshaft to slide back-and-forth within the transmission during vehicle operation. An oil seal prevents leakage of fluid at this point and keeps dirt from entering the transmission. If leakage is evident at the front of the driveshaft, replace the oil seal (see Chapter 7, Part A).

On 4WD models, each driveshaft is attached to the transfer case by a flange yoke. Once a front or rear driveshaft has been removed, either companion flange can be removed from the transfer case to replace the companion seal(s) (each companion flange uses two seals: one seal between the companion flange and the transfer case, the other, smaller, seal inside the companion flange itself). Refer to Chapter 7C for the transfer case seal replacement procedure.

Since the driveshaft is a balanced unit, it's important that no undercoating, mud, etc. be allowed to stay on it. When the vehicle is raised for service, it's a good idea to clean the driveshaft and inspect it for any obvious damage. Also, make sure the small weights used to originally balance the driveshaft are in place and securely attached. Whenever the driveshaft is removed, it must be reinstalled in the same relative position to preserve the balance.

Problems with the driveshaft are usually indicated by a noise or vibration while driving the vehicle. A road test should verify if the problem is the driveshaft or another vehicle component. Refer to the *Troubleshooting* Section at the front of this manual. If you suspect trouble, inspect the driveline (see Section 12).

12 Driveline inspection

1 Raise the rear of the vehicle and support it securely on jackstands. Block the front wheels to keep the vehicle from rolling off the stands.

2 Crawl under the vehicle and visually inspect the driveshaft. Look for any dents or cracks in the tubing. If any are found, the driveshaft must be replaced.

3 Check for oil leakage at the front and rear of the driveshaft. Leakage where the driveshaft enters the transmission or transfer case indicates a defective transmission/transfer case seal (see Chapter 7). Leakage where the driveshaft joins the differential indicates a defective pinion seal (see Section 18).

4 While under the vehicle, have an assistant rotate a rear wheel so the driveshaft will rotate. As it does, make sure the universal joints are operating properly without binding, noise or looseness. Listen for any noise from the center bearing (if equipped), indicating it's worn or damaged. Also check the rubber portion of the center bearing for cracking or separation, which will necessitate replacement.

5 The universal joint can also be checked with the driveshaft motionless, by gripping your hands on either side of the joint and attempting to twist the joint. Any movement at all in the joint is a sign of considerable wear. Lifting up on the shaft will also indicate movement in the universal joints.

6 Finally, check the driveshaft mounting bolts at the ends to make sure they're tight.

7 In addition, check for grease leakage around the sleeve yoke, indicating failure of the yoke seal.

8 Check for leakage where the driveshafts connect to the transfer case and front differential. Leakage indicates worn oil seals.

9 At the same time, on 4WD models, check for looseness in the joints of the front driveaxles. Also check for grease or oil leakage from around the driveaxles by inspecting the rubber boots and both ends of each axle. Oil leakage at the differential junction indicates a defective side oil seal. Leakage at the wheel side indicates a defective front hub seal, while leakage at the boots means a damaged rubber boot. For servicing of these components, see the appropriate Sections.

13 Driveshaft - removal and installation

1 Raise the vehicle and support it securely on jackstands. Place the transmission in Neutral with the parking brake off. Block the front wheels to prevent the vehicle from rolling.

REAR DRIVESHAFT

Removal

➧ Refer to illustrations 13.2, 13.3 and 13.5

2 Make matchmarks on the driveshaft and the differential flange in line with each other (see illustration). This is to make sure the driveshaft is reinstalled in the same position to preserve the balance.

3 Remove the bolts securing the flange yoke to the rear differential (see illustration). Turn the driveshaft (or wheels) as necessary to bring the bolts into the most accessible position.

4 Remove the center bearing protector, if equipped, then remove the bolts from the center support bearing.

5 Lower the rear of the driveshaft. Slide the front of the driveshaft out of the transmission on 2WD models. On 4WD models, separate the flange at the transfer case (see illustration).

6 On 2WD models, wrap a plastic bag over the transmission exten-

13.2 Mark the relationship of the driveshaft flange yoke to the companion flange on the differential - 2WD model shown

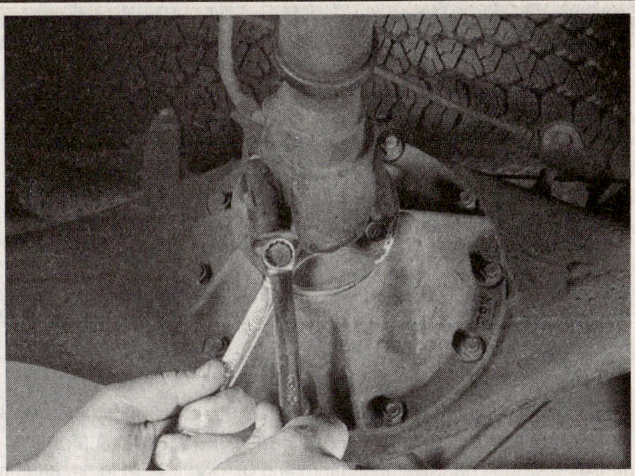

13.3 Using a backup wrench to hold each bolt, break loose all four nuts securing the flange yoke to the differential

sion housing and hold it in place with a rubber band. This will prevent loss of fluid and protect against contamination while the driveshaft is out.

7 If you're overhauling a U-joint, refer to Section 14. If you're replacing the center bearing on a two-piece driveshaft, see Section 15. If you're replacing the pinion seal, see Section 18. If you're replacing the extension housing seal, you'll find the procedure in Chapter 7A. If you're replacing the rear seal in the transfer case on a 4WD model, refer to Chapter 7C.

Installation

8 Remove the plastic bag from the transmission or transfer case and wipe the area clean. Slide the front of the driveshaft into the transmission (2WD models) or bolt the U-joint flange yoke to the companion flange, installing the fasteners finger-tight (4WD models).

9 Raise the center bearing (if equipped) into place and screw the retaining bolts in a few turns. Raise the rear of the driveshaft into position, checking to be sure the marks are in alignment. If not, turn the rear wheels to match the pinion flange and the driveshaft.

10 Tighten all fasteners to the torque listed in this Chapter's Specifications.

13.5 On 4WD models, mark the relationship of the driveshaft U-joint flange yoke to the companion flange, then remove the bolts securing the flange yoke to the transfer case

FRONT DRIVESHAFT (4WD MODELS)

Removal

11 Remove the exhaust pipe assembly (see Chapter 4),

12 Remove any skid plates or splash shields, if equipped.

13 Mark the relationship of the driveshaft to the front differential and the transfer case companion flanges using paint.

14 Remove the nuts, bolts and washers from the differential and transfer case companion flanges and remove the front driveshaft.

15 If you're overhauling a U-joint, refer to Section 14. If you're replacing the pinion seal in the front differential, refer to Section 23. If you're replacing either front seal in the transfer case on a 4WD model, see Chapter 7C.

Installation

16 Attach the front end of the driveshaft to the front differential companion flange and install the nuts and bolts finger-tight.

17 Extend or compress the driveshaft as necessary, attach the rear end to the transfer case flange, install the washers, nuts and bolts and tighten the nuts to the torque listed in this Chapter's Specifications.

18 Install the skid plate and/or splash shields.

19 Remove the jackstands and lower the vehicle.

14 Universal joints - replacement

Refer to illustration 14.3, 14.4 and 14.5

Note: A press or large vise will be required for this procedure. It may be a good idea to take the driveshaft to a repair or machine shop where the U-joints can be replaced for you, usually at a reasonable charge.

1 Remove the driveshaft (see Section 13).

2 Place the driveshaft on a bench equipped with a vise.

3 Mark the shaft and yoke for proper reassembly, then remove the snap-rings from the U-joint (see illustration).

4 Place a piece of pipe or a large socket with an inside diameter slightly larger than the bearing cup over one of the bearing cups. Position a socket which is of slightly smaller diameter than the cup on the opposite bearing cup (see illustration) and use the vise to force the cup out (inside the pipe or large socket), stopping just before it comes completely out of the yoke.

5 Use the vise or large pliers to work the cup the rest of the way out (see illustration).

6 Transfer the sockets to the other side and press the opposite bearing cup out in the same manner.

7 After the bearing cups have been removed, lift the U-joint from the yoke and thoroughly clean all dirt and debris from the yokes on both ends of the driveshaft. Be sure to remove any metal burrs from the yoke bores.

8 Pack the new U-joint bearing cups with grease, this will allow the needle bearings to be held in place while your installing the bearing cups. Ordinarily, specific instructions for lubrication will be included with the U-joint servicing kit and should be followed carefully.

9 Position the U-joint body in the yoke and partially install one bearing cup in the yoke. If the U-joint is equipped with a grease fitting, be sure it points in the same direction as the grease fitting on the opposite end of the driveshaft.

10 Start the U-joint body into the bearing cup and partially install the other cup. Align the U-joint body between the bearing cups and press the bearing cups into position, being careful not to damage the dust seals.

11 Install the snap-rings. If difficulty is encountered in seating the snap-rings, strike the driveshaft yoke sharply with a hammer. This will spring the yoke ears slightly and allow the snap-rings to seat in the groove. This should also be done to center the U-joint after assembly.

Note: If you still have difficulty seating the snap-rings, one of the small needle bearings may have become stuck between the bearing cup and the end of the spider. Disassemble and inspect the joint.

12 Install the driveshaft (see Section 13).

13 If the U-joint is equipped with a grease fitting, lubricate it as described in Chapter 1.

14 Remove the jackstands and lower the vehicle.

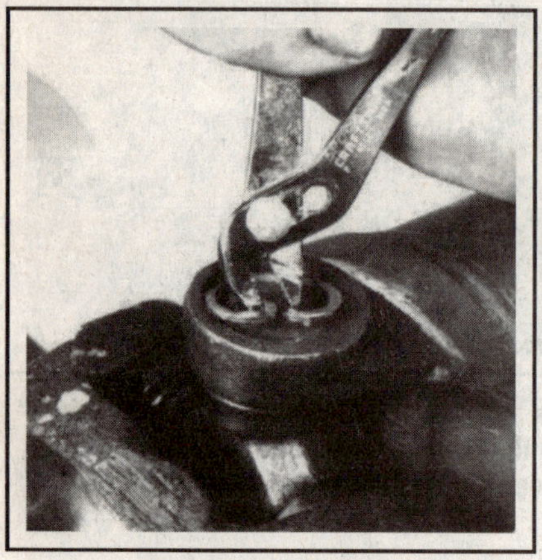

14.3 Use a small pair of pliers to remove the U-joint snap-rings

14.4 To remove the U-joint from the driveshaft, use a vise as a press; the small socket will push the cross and bearing cup into the large socket

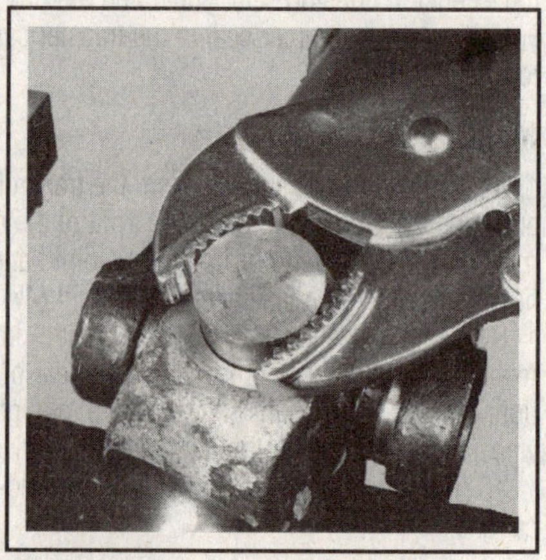

14.5 Grip the bearing cup with locking pliers and remove it from the yoke

15 Center bearing - removal and installation

Note: Obtain a new yoke nut before performing this procedure.

1 Raise the vehicle and support it securely on jackstands.

2 Remove the driveshaft (see Section 13).

3 Clamp the driveshaft securely into a bench vise.

4 Mark the center U-joint yoke and driveshaft, then disassemble the center U-joint (see Section 14).

5 Mark the relationship of the intermediate shaft to the yoke or flange using paint.

6 Unstake and remove the nut, then remove the yoke from the intermediate shaft.

7 Remove the center bearing from the intermediate shaft.

Note: This usually requires a puller or a punch and a hammer to separate the center bearing from the intermediate shaft.

8 Installation is the reverse of removal. Be sure to install a new yoke nut and tighten it to the torque listed in this Chapter's Specifications, then stake the collar of the nut into the slot in the driveshaft.

16 Axles - description and check

DESCRIPTION

1 The rear axle assembly is a hypoid, semi-floating type (the centerline of the pinion gear is below the centerline of the ring gear). When the vehicle goes around a corner, the differential allows the outer rear wheel to turn more quickly than the inner wheel. The axleshafts are splined to the differential side gears, so when the vehicle goes around a corner, the inner wheel, which turns more slowly than the outer wheel, turns its side gear more slowly than the outer wheel turns its side gear. The differential pinion gears roll around the slower side gear, driving the outer side gear - and wheel - more quickly. The steel axle tubes are pressed into and welded to the carrier.

2 Some models are equipped with an electronic locking rear differential. The Differential Lock System (Auto LSD System) allows the differential to operate in two different modes. With the switch OFF, the differential operates normally, but when the switch is turned ON, the differential locks, allowing the axles to rotate equally with no slip. This mode is selected in mud, deep snow or other conditions when one of the wheels starts to spin. When activated, the rear differential lock actuator slides the differential lock sleeve into place, locking the axles together.

3 On 4WD models, a fully independent front axle assembly is used. This consists of a differential and a pair of driveaxles. Each driveaxle has an inner and outer constant velocity (CV) joint.

CHECK

4 Often, a suspected axle problem lies elsewhere. Do a thorough check of other possible causes before assuming the axle is the problem.

5 The following noises are those commonly associated with axle diagnosis procedures:

- a) *Road noise is often mistaken for mechanical faults. Driving the vehicle on different surfaces will show whether the road surface is the cause of the noise. Road noise will remain the same if the vehicle is under power or coasting.*
- b) *Tire noise is sometimes mistaken for mechanical problems. Tires which are worn or low on pressure are particularly susceptible to emitting vibrations and noises. Tire noise will remain about the same during varying driving situations, where axle noise will change during coasting, acceleration, etc.*
- c) *Engine and transmission noise can be deceiving because it will travel along the driveline. To isolate engine and transmission noises, make a note of the engine speed at which the noise is most pronounced. Stop the vehicle and place the transmission in Neutral and run the engine to the same speed. If the noise is the same, the axle is not at fault.*

6 Because of the special tools needed, overhauling the differential isn't cost effective for a do-it-yourselfer. The procedures included in this Chapter describe axleshaft removal and installation, axleshaft oil seal replacement, axleshaft bearing replacement and removal of the entire unit for repair or replacement. Any further work should be left to a dealer service department or other qualified repair shop.

17 Axleshaft, bearing and oil seals (rear) - removal and installation

➧ Refer to illustrations 17.7, 17.8, 17.9 and 17.10

1 Release the parking brake. Raise the rear of the vehicle, support it securely on jackstands and block the front wheels. Remove the wheels and tires.

2 Remove the brake drum (see Chapter 9).

3 Remove the ABS sensor (see Chapter 9).

4 Disconnect the parking brake cables from the brake shoe and backing plate (see Chapter 9).

5 Remove the brake shoes (see Chapter 9).

6 Drain the brake fluid and remove the brake lines from the wheel cylinders (see Chapter 9).

7 Remove the four backing plate mounting nuts (see illustration).

8 Pull the axleshaft out of the rear axle housing along with the backing plate (see illustration).

17.7 To detach the axleshaft from the rear axle housing, remove the four backing plate nuts

17.8 Extract the axleshaft very carefully from the axle housing, especially if you don't want to replace the axleshaft seal

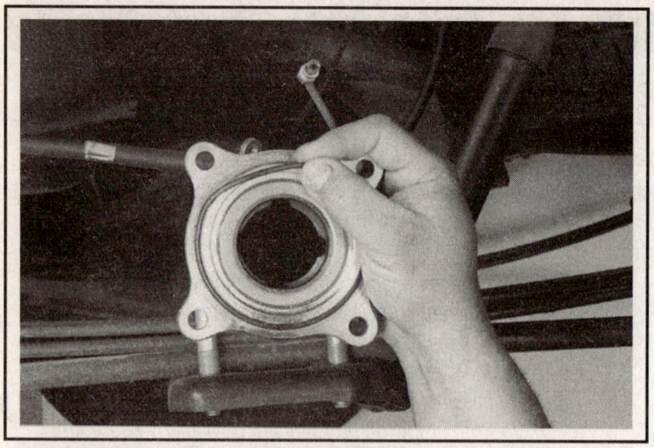

17.9 Remove this O-ring from the rear axle housing; install a new one before installing the axleshaft

17.10 Use a seal removal tool or a big screwdriver to pry out the old axleshaft seal; use a seal installer or a big socket to install the new seal

9 Remove the O-ring from the rear axle housing (see illustration).

10 Remove the axleshaft inner oil seal from the axle housing with a seal removal tool or a big screwdriver (see illustration).

11 Further disassembly of the axleshaft assembly requires special tools and a hydraulic press. If the axleshaft, bearing or outer oil seal needs to be replaced, take the axleshaft assembly to an automotive machine shop.

12 Drive a new axleshaft inner seal into the end of the axle tube with a seal installer or a big socket. Coat the lip of the seal with clean oil or multi-purpose grease.

13 Install a new axle housing O-ring. Apply a light coat of oil to the new O-ring.

14 Make sure the axleshaft is clean and there are no burrs or metal splinters on it. Deburr any surface irregularities so the axleshaft doesn't damage the seal during installation. Lightly coat the axleshaft with clean oil, then insert it into the axle housing. Make sure the splined inner end of the axleshaft doesn't damage the lip of the new axleshaft seal.

15 Installation is the reverse of removal. Tighten the four backing plate mounting nuts to the torque listed in this Chapter's Specifications.

18 Differential pinion oil seal - replacement

Refer to illustrations 18.3, 18.4, 18.6, 18.8, 18.9 and 18.10

Note: This procedure applies to the rear pinion seal on all vehicles and the front pinion seal on 4WD models.

1 Loosen the wheel lug nuts, raise the vehicle and support it securely on jackstands. Block the wheels at the opposite end to keep the vehicle from rolling off the stands. Remove the rear wheels (this will allow you to obtain a more accurate pinion shaft preload reading).

2 Disconnect the driveshaft from the differential (see Section 13) and position it off to the side using rope or mechanics wire.

3 Use an inch-pound torque wrench to check the torque required to rotate the pinion (see illustration).

4 Make alignment marks on the pinion shaft and flange (see illustration). Unstake the pinion nut.

5 Count the number of threads visible between the end of the nut and the end of the pinion shaft and jot it down for later use.

6 A special flange holding tool is the best way to keep the companion flange from moving while the pinion nut is loosened. If you're unable to obtain a flange holding tool, immobilize the flange by inserting a big screwdriver through one of the U-joint bolt holes in the flange and wedge it against a bracket (see illustration) or reinforcement rib on the differential carrier.

7 Remove the pinion nut.

8 Withdraw the companion flange. It may be necessary to use a puller to draw it out (see illustration). Do NOT pry behind the flange or hammer on the flange or the end of the pinion shaft.

9 Pry out the old seal (see illustration) and discard it.

10 Lubricate the lips of the new seal with high-temperature grease and tap it evenly into position with a seal installation tool or a large socket. Make sure it enters the housing squarely and is tapped in to its full depth (see illustration).

11 Align the mating marks made before disassembly and install the companion flange. If necessary, tighten the pinion nut to draw the flange into place. Do not try to hammer the flange into position.

12 Apply non-hardening sealant to the ends of the splines visible in the center of the flange so oil will be sealed in.

13 Install the washer (if equipped) and a new pinion nut. Tighten the nut carefully, until the original number of threads are exposed.

14 Using an inch-pound torque wrench, measure the torque required to rotate the pinion and tighten the nut in small increments until it matches the pinion shaft bearing preload recorded in Step 3. In order to compensate for the drag of the new seal, the nut should be tightened more until the rotational torque of the pinion exceeds what was recorded earlier, but not by more than 5 in-lbs.

15 Once the proper preload is reached, stake the collar of the nut into the slot in the pinion shaft.

16 Reconnect the driveshaft to the pinion flange (see Section 4). Check the differential lubricant level and add some, if necessary, to bring it to the appropriate level (see Chapter 1).

17 Install the wheels and lower the vehicle. Tighten the lug nuts to the torque listed in the Chapter 1 Specifications.

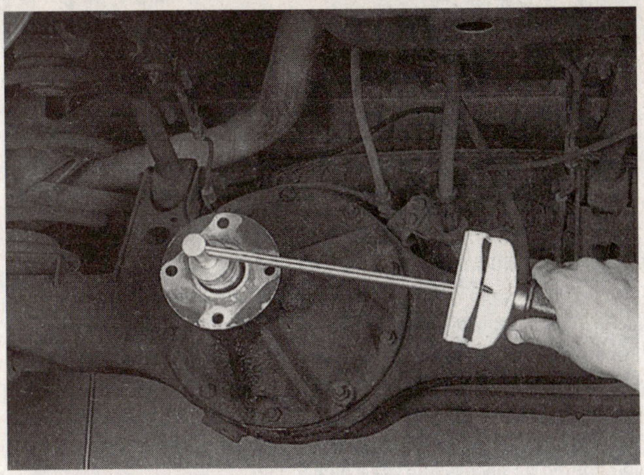

18.3 Use an inch-pound torque wrench to check the torque necessary to rotate the pinion shaft

18.4 Mark the relative positions of the pinion shaft and flange before removing the nut

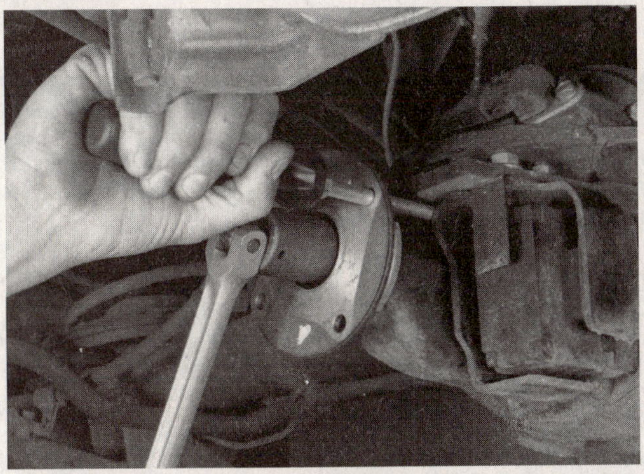

18.6 If you don't have a flange holding tool, lock the flange by jamming a large screwdriver through a bolt hole in the flange and wedge it underneath a bracket as shown, or under a reinforcement rib on the differential carrier

18.8 If you can't pull off the pinion flange by hand, remove it with a small puller

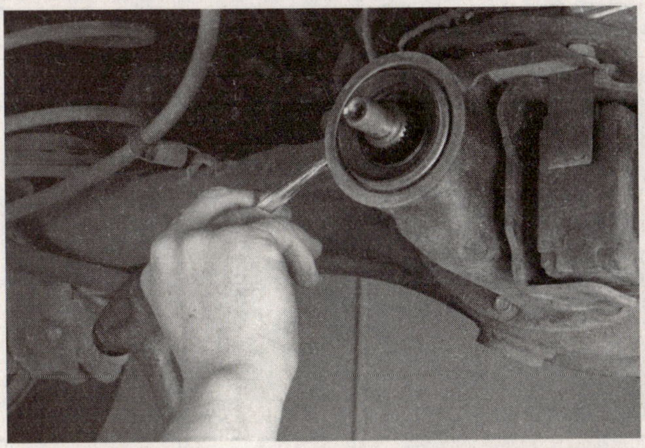

18.9 Pry out the old pinion seal with a seal removal tool or a big screwdriver, or tap it out with a small punch

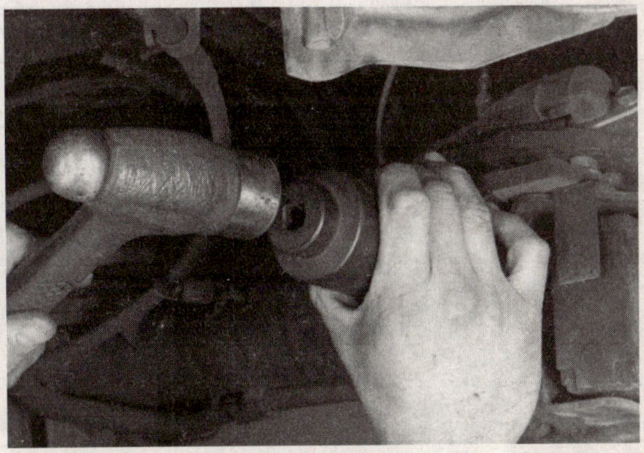

18.10 Lubricate the lips of the new pinion seal and seat it squarely in the bore, then drive it into the carrier with a seal driver or a large socket

19 Axle (rear) - removal and installation

1 Loosen the rear wheel lug nuts, raise the rear of the vehicle and support it securely on jackstands placed under the frame (not under the axle). Block the front wheels to keep the vehicle from rolling off the stands. Remove the rear wheels.

2 Position a floor jack under the rear axle differential housing.

3 Disconnect the driveshaft from the rear axle pinion flange (see Section 13). Fasten the driveshaft out of the way with rope or wire; don't let it hang unsupported.

4 Disconnect the left and right ABS wheel speed sensors (see Chapter 9).

5 Detach all brake hoses and/or lines from the axle housing, then plug them to prevent fluid leakage.

6 Disconnect the parking brake cables from the brake assemblies (see Chapter 9).

7 Detach the vent hose from the axle housing.

8 Disconnect the shock absorbers from the axle brackets (see Chapter 10).

9 If equipped, remove the stabilizer bar.

10 Remove the nuts from the leaf spring U-bolts (see Chapter 10).

11 If you're working on a 4WD model, lower the jack under the differential, then remove the rear axle assembly from under the vehicle.

12 If you're working on a 2WD model, unbolt the front ends of the leaf springs and allow them to swing down, then lower the axle and guide it out.

13 Installation is the reverse of removal. Be sure to tighten all suspension fasteners to the torque listed in the Chapter 10 Specifications.

14 Bleed the brakes (see Chapter 9). Check the differential lubricant level, adding as necessary (see Chapter 1).

20 Driveaxle (4WD models) - removal and installation

Refer to illustrations 20.2, 20.3a and 20.3b

1 Loosen the front wheel lug nuts, raise the front of the vehicle and support it securely on jackstands. Block the rear wheels to keep the vehicle from rolling off the stands. Remove the wheel(s). Drain the differential lubricant (see Chapter 1).

2 Remove the grease cap (see illustration).

3 Remove the cotter pin (see illustration) and nut lock, place a prybar or large screwdriver between the wheel studs to hold the driveaxle and break the driveaxle/hub nut loose with a large breaker bar (see illustration), or have an assistant apply the brakes. Remove the nut.

4 Remove the ABS wheel speed sensor(s) (see Chapter 9) and disconnect the outer tie rod ends from the steering knuckle (see Chapter 10).

5 Unbolt the lower balljoint bracket from the steering knuckle (see Chapter 10).

6 Knock the driveaxle loose from the steering knuckle with a brass drift and hammer. Do NOT use a steel punch or strike the end of the driveaxle with a steel hammer; a steel punch or hammer will damage the threads on the end of the driveaxle.

7 Swing the steering knuckle outward and pull the driveaxle assembly out of the steering knuckle, then detach the driveaxle from the differential. If you're removing the right driveaxle, tap the inner CV joint out of the differential with a hammer and a brass drift; if you're removing the left driveaxle, a slide hammer with a special hooked adapter (available at most auto parts stores) will be needed to pull the inner CV joint from the differential.

8 Installation is the reverse of removal. Be sure to tighten the driveaxle/hub nut to the torque listed in this Chapter's Specifications, then install the nut lock and a new cotter pin. Tighten the lug nuts to the torque listed in the Chapter 1 Specifications. Tighten all suspension fasteners to the torque listed in the Chapter 10 Specifications.

9 Refill the differential with the proper lubricant (see Chapter 1).

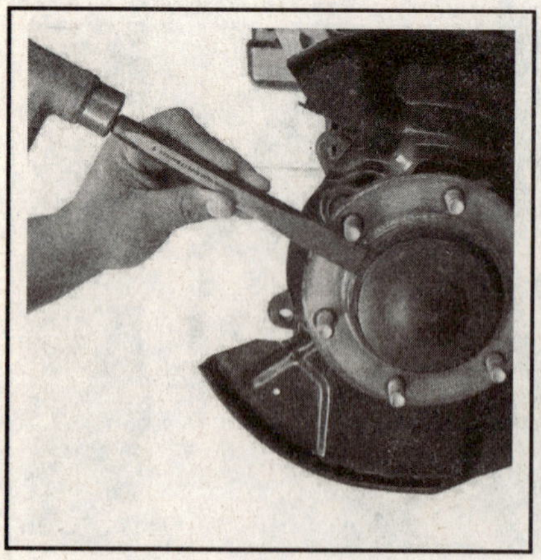

20.2 Using a hammer and chisel, tap the grease cap from the hub

20.3a Remove the cotter pin and the nut lock

20.3b Place a prybar between two of the wheel studs, then loosen the driveaxle/hub nut

21 Driveaxle (4WD models) - boot replacement

1 Remove the driveaxle (see Section 20).

DISASSEMBLY

Refer to illustrations 21.3, 21.4, 21.6 and 21.7

Note: If the CV joint boots must be replaced, explore all options before beginning the job. Complete rebuilt driveaxles are available on an exchange basis, which eliminates much time and work. Whichever route you choose to take, check on the cost and availability of parts before disassembling the vehicle.

2 Mount the driveaxle in a vise with wood lined jaws (to prevent damage to the axleshaft). Check the CV joint for excessive play in the radial direction, which indicates worn parts. Check for smooth operation throughout the full range of motion for each CV joint. If a boot is torn, disassemble the joint, clean the components and inspect for damage due to loss of lubrication and possible contamination by foreign matter.

3 Using a small screwdriver, pry the retaining tabs of the clamps up to loosen them and slide them off (see illustration).

4 Using a screwdriver, carefully pry up on the edge of the outer boot and push it away from the CV joint. Old and worn boots can be cut off. Pull the inner CV joint boot back from the housing and slide the housing off the tripod (see illustration).

5 Mark the tripod and axleshaft to ensure that they are reassembled properly.

6 Remove the tripod joint snap-ring with a pair of snap-ring pliers (see illustration).

7 Use a hammer and a brass punch to drive the tripod joint from the driveaxle (see illustration).

8 If you haven't already cut them off, remove both boots.

Note: Do NOT disassemble the outboard CV joint.

CHECK

9 Thoroughly clean all components, including the outer CV joint assembly, with solvent until the old CV joint grease is completely removed. Inspect all visible bearing surfaces for cracks, pitting, scoring and other signs of wear. If the inner CV joint is worn, you can buy a new

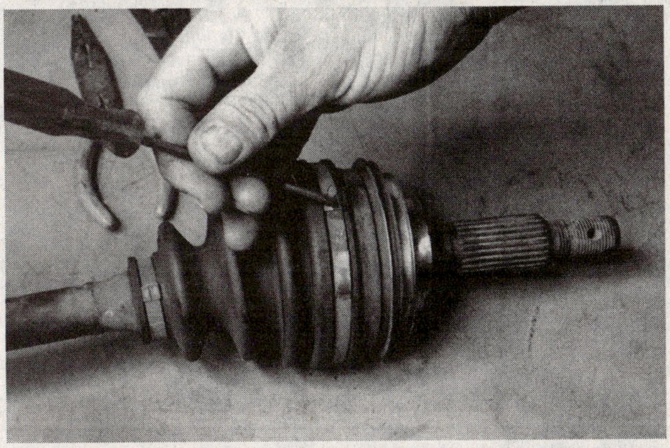

21.3 Lift the tabs on the boot clamps with a small screwdriver, then open the clamps

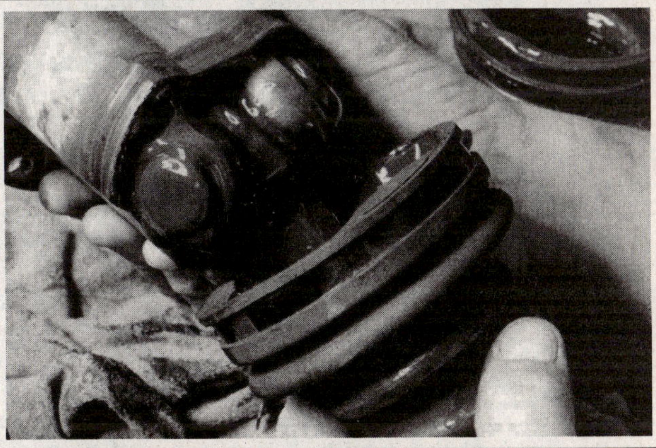

21.4 Remove the boot from the inner CV joint and slide the joint housing from the tripod

21.6 Remove the snap-ring with a pair of snap-ring pliers

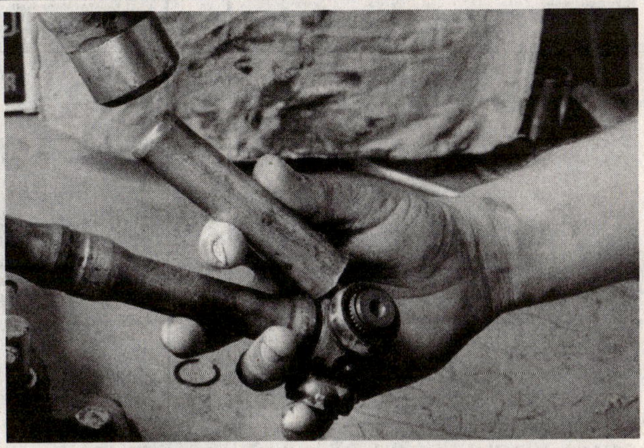

21.7 Drive the tripod joint from the driveaxle with a brass punch and hammer: be careful not to damage the bearing surfaces or the splines of the shaft

inner CV joint and install it on the old axleshaft; if the outer CV joint is worn, you'll have to purchase a new outer CV joint and axleshaft (they're sold preassembled).

REASSEMBLY

➧ Refer to illustrations 21.10a, 21.10b, 21.10c, 21.10d, 21.12a, 21.12b, 21.12c and 21.12d

10 Wrap the splines on the inner end of the axleshaft with electrical or duct tape to protect the boots from the sharp edges of the splines (see illustration). Slide the clamps and boot(s) onto the axleshaft, then place the tripod on the shaft. Apply grease to the tripod assembly and inside the housing. Insert the tripod into the housing and pack the remainder of the grease around the tripod (see illustrations).

11 Slide the boot into place, making sure both ends seat in their grooves. Position the inner CV joint midway through its travel.

12 Equalize the pressure in the boot, then tighten and secure the boot clamps (see illustrations).

13 Install the driveaxle assembly (see Section 20).

21.10a Wrap the splined area of the axleshaft with tape to prevent damage to the boots when installing them

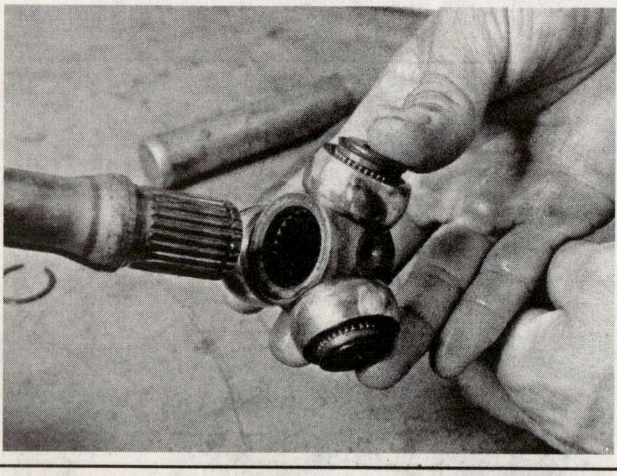

21.10b Install the tripod with the recessed portion of the splines facing the axleshaft

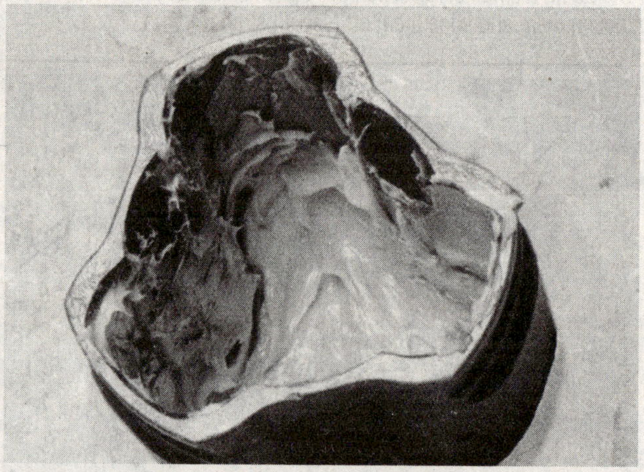

21.10c Place grease at the bottom of the CV joint housing

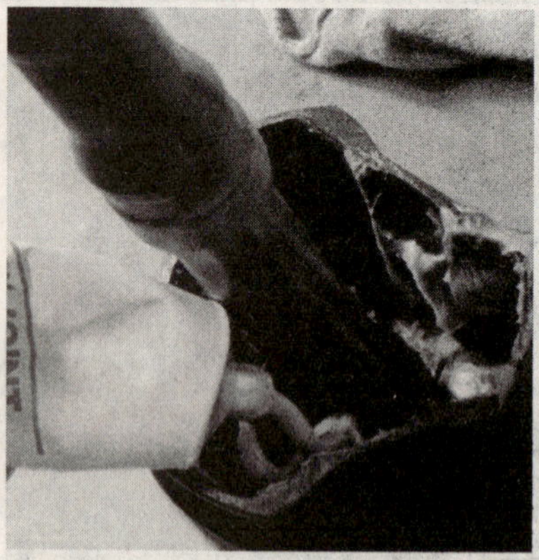

21.10d Install the boot clamps onto the axleshaft, then insert the tripod into the housing followed by the rest of the grease

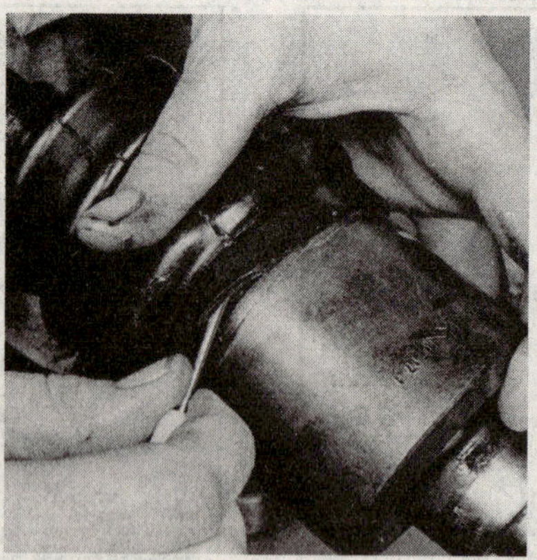

21.12a Equalize the pressure inside the boot by inserting a small, dull screwdriver between the boot and the outer race

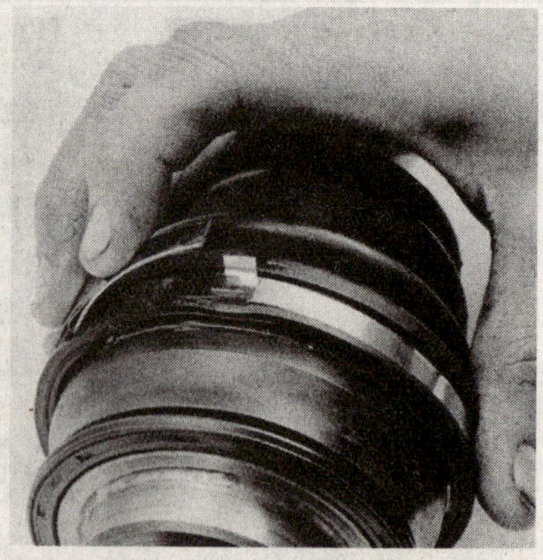

21.12b To install the new clamps, bend the tang down . . .

21.12c . . . then tap the tabs over to hold it in place

21.12d If your replacement boot came with crimp-type clamps, a special tool (available at most auto parts stores) will be required to tighten them properly

22 Automatic Disconnecting Differential (ADD) (4WD models) - description, removal and installation

DESCRIPTION

1 The Automatic Disconnecting Differential (ADD) connects the power flow through the right front axleshaft when 4WD mode is selected, and disconnects the power flow when 2WD mode is selected. If shifting into or out of 4WD becomes a problem, have the ADD system checked out by a dealer service department or other qualified repair shop that specializes in 4WD vehicles.

REMOVAL AND INSTALLATION

Differential carrier

2 Loosen the wheel lug nuts, raise the vehicle and support it securely on jackstands. Remove the wheels.
3 Remove the engine splash shields.
4 Drain the lubricant from the differential (see Chapter 1).
5 Remove the driveaxles (see Section 20).
6 Disconnect the driveshaft from the front differential (see Section 13) and support the front end of the driveshaft with a piece of wire.
7 Remove the front stabilizer bar and brackets (see Chapter 10).
8 Remove the ABS wheel speed sensors (see Chapter 9).
9 Disconnect the breather tube bracket, detach the fasteners that retain the vacuum tubing bracket to the differential, unplug the actuator electrical connector and detach the actuator vacuum hoses. Remove the tube and wire harness assembly from the differential.
10 Support the differential with a transmission jack or a floor jack.
11 Remove the differential rear mounting nut and the two differential front mounting bolts/nuts and lower the differential.
12 Installation is the reverse of removal. Refill the differential with the proper lubricant (see Chapter 1) and tighten the lug nuts to the torque listed in the Chapter 1 Specifications. Tighten the suspension fasteners to the torque values listed in the Chapter 10 Specifications.

ADD actuator

13 Raise the vehicle and support it securely on jackstands.
14 Remove the splash shield.
15 Remove the four retaining bolts.
16 Remove the actuator.
17 Before installing the actuator, remove the old RTV sealant from the mating surfaces of the differential and the actuator and apply a thin bead of new RTV sealant to those surfaces.
18 Tighten the bolts to the Specifications listed in this Chapter.
19 The remainder of installation is the reverse of removal.

23 Driveaxle oil seals (4WD models) - removal and installation

1 Raise the front of the vehicle and support it securely on jackstands. Block the rear wheels to prevent the vehicle from rolling. Place the transmission in Neutral with the parking brake off.
2 Remove the driveaxles (see Section 20).
3 Carefully pry out the side gear shaft oil seal with a seal removal tool or a large screwdriver; make sure you don't scratch the seal bore.
4 Using a seal installer or a large deep socket as a drift, install the new oil seal. Drive it into the bore squarely and make sure it's completely seated.
5 Lubricate the lip of the new seal with multi-purpose grease, then install the driveaxles (see Section 20). Be careful not to damage the lip of the new seal.
6 Refill the front differential with the proper lubricant (See Chapter 1).

24 Differential Lock System - description and component replacement

DESCRIPTION

1 The Differential Lock System (Auto LSD System) allows the rear differential to operate in two different modes. With the switch OFF, the differential operates normally, but when the switch is turned ON, the differential locks, allowing the axles to rotate equally with no slip. This mode is selected in mud, deep snow or other conditions when one of the wheels starts to spin. When activated, the rear differential lock actuator slides the differential lock sleeve into place, locking the axles together.

2 The Differential Lock System consists of the Differential Lock Switch, the 4WD Control Switch (4WD models only), the differential lock actuator mounted on the rear axle, the transfer indicator switch mounted on the transfer case, the center differential lock switch mounted on the differential lock actuator and the 4WD control ECU (4WD models) or the rear differential lock ECU (2WD models), both of which are mounted behind the right end of the instrument panel.

COMPONENT REPLACEMENT

Differential Lock Switch

3 Refer to Chapter 12, Section 10 to remove and install the Differential Lock Switch from the instrument panel.

Differential Lock Actuator

Refer to illustrations 24.8 and 24.9

4 Raise the rear of the vehicle and support it securely on jackstands. Block the front wheels to prevent the vehicle from rolling. Place the transmission in Neutral with the parking brake off.

5 Disconnect the electrical connectors from the differential lock actuator and remove the mounting bolts.

6 Separate the Differential Lock Actuator from the differential.

7 Installation is the reverse of removal.

8 Check the position of the rack teeth on the shift fork. The outermost tooth of the rack should be positioned above the centerline of the installation hole (see illustration).

9 Check that the mark on the pinion of the actuator is in range between 0 and 5 degrees clockwise in relation to the centerline of the actuator (see illustration). If not correctly aligned, use a 1.5 volt battery to apply power to the connector to rotate the pinion.

CAUTION:

Don't apply voltage any longer than necessary.

10 Install a new O-ring onto the actuator and apply grease to the gears.

11 Align the outside rack tooth of the shift fork with the matchmarks on the pinion of the actuator.

12 Install the actuator with the long hole positioned over the knock pin on the carrier side.

13 Rotate the actuator counterclockwise once the knock pin is moved over to the right side of the long hole.

14 Tighten the differential lock actuator bolts to the torque listed in this Chapter's Specifications.

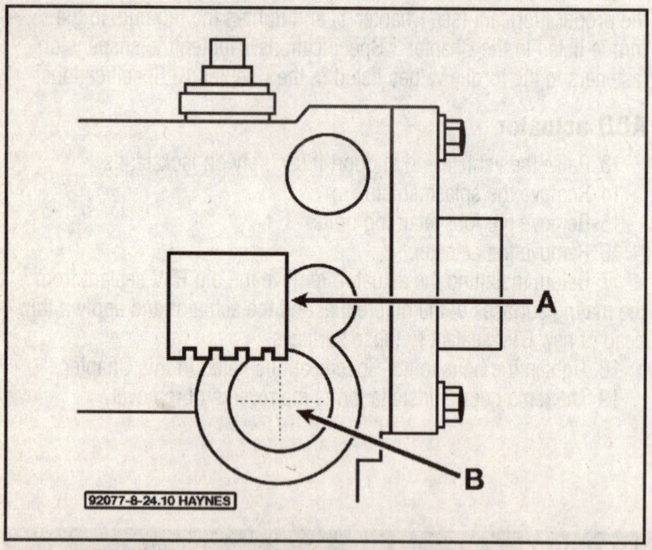

24.8 The outermost tooth of the rack should be positioned above the centerline of the installation hole

A Shift fork rack *B Actuator installation hole*

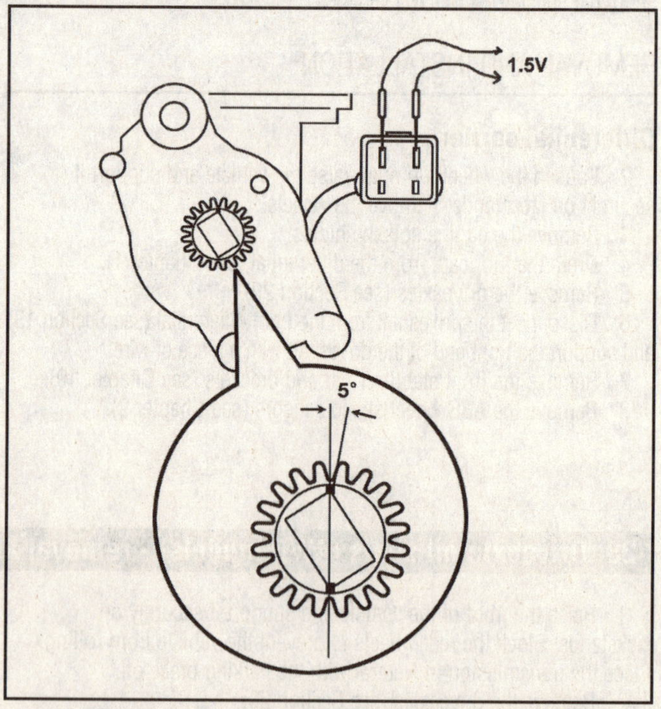

24.9 Use a 1.5 volt battery to apply power to the connector to rotate the pinion. Do not continue applying voltage or the pinion will become excessively rotated

Specifications

Clutch

Clutch fluid type	See Chapter 1
Clutch disc lining minimum rivet depth	1/32-inch (0.8 mm)
Clutch pedal height	6-5/8 to 7-1/32 inches (168.5 to 178.5 mm)
Clutch pedal freeplay	3/16 to 9/16 inch (5.0 to 15.0 mm)

Torque specifications

➡Note: One foot-pound (ft-lb) of torque is equivalent to 12 inch-pounds (in-lbs) of torque. Torque values below approximately 15 foot-pounds are expressed in inch-pounds, because most foot-pound torque wrenches are not accurate at these smaller values.

Torque specifications	Ft-lbs (unless otherwise indicated)	Nm
Clutch		
Accumulator bolts (6-speed models)	104 in-lbs	11.5
Master cylinder nuts	132 in-lbs	15
Pressure plate-to-flywheel bolts	168 in-lbs	19
Release cylinder bolts	104 in-lbs	11.5
Release cylinder bleeder plug	96 in-lbs	11
Driveshaft		
Flange bolts/nuts		
Front (4WD)	65	88
Rear	65	88
Center bearing bolts	27	37
Center bearing yoke nut		
Step 1	134	182
Step 2	Loosen one turn	
Step 3	51	69
Front driveaxle and differential (4WD models)		
Driveaxle/hub nut	173	234
ADD actuator mounting bolts	15	20
Rear axle		
Brake backing plate nuts	27	37
Differential lock actuator bolts	20	27

Notes

9 BRAKES

Section

Reference to other Chapters

Brake check - See Chapter 1
Brake fluid level check - See Chapter 1

1 General information

GENERAL

All models covered by this manual are equipped with hydraulically operated, power-assisted front disc and rear drum brakes.

The disc brakes are self-adjusting and automatically compensate for pad wear.

The hydraulic system has separate circuits for the front and rear brakes. If one circuit fails, the other circuit will remain functional and a warning indicator will light up on the dashboard when a substantial amount of brake fluid is lost, showing that a failure has occurred. However, in the event that the front brake circuit fails, braking effectiveness is greatly reduced, resulting in much longer stopping distances.

MODELS EQUIPPED WITH VEHICLE STABILITY AND TRACTION CONTROL (VSC TRAC)

Integrated hydraulic brake components

➧ Refer to illustration 1.4

Some models may be equipped with an optional VSC TRAC (vehicle stability and traction control) system. These models utilize an advanced integrated hydraulic control system (see illustration). The master cylinder, hydraulic brake booster and other ABS/VSC TRAC components are integrated and installed as an assembly. The assembly is located where a typical master cylinder and vacuum brake booster would be.

⁂ CAUTION:

This assembly requires special tools and expertise to diagnose and repair. Any concerns regarding this assembly should be addressed by a dealer service department or a qualified repair facility.

MODELS WITHOUT VEHICLE STABILITY AND TRACTION CONTROL (VSC TRAC)

Master cylinder

The master cylinder is located under the hood, mounted to the power brake booster, and can be identified by the large fluid reservoir on top. The master cylinder has separate primary and secondary piston assemblies for the front and rear circuits.

Power brake booster

The power brake booster uses engine manifold vacuum to provide assistance to the brakes. It is mounted on the firewall in the engine compartment, directly behind the master cylinder.

Parking brake

The parking brake holds the rear wheels only. A parking brake pedal or lever operates a series of cables attached to the parking brake system at each rear wheel. The cables pull on linkage that expand the rear brake shoes inside the brake drums to mechanically apply the rear brakes.

SERVICE

After completing any procedure involving disassembly of any part of the brake system, always test drive the vehicle to check for proper braking performance before resuming normal driving. When testing the brakes, perform the tests on a clean, dry and flat surface. Conditions other than these can lead to inaccurate test results.

Test the brakes at various speeds with both light and heavy pedal pressure. The vehicle should stop evenly without pulling to one side or the other. Under hard braking, the ABS system may engage resulting in brake pedal pulsation. This is considered normal operation.

Tires, vehicle load, and wheel alignment are factors which also affect braking performance.

PRECAUTIONS

There are some general cautions and warnings involving the brake system on this vehicle:

a) *Use only brake fluid conforming to DOT 3 specifications.*
b) *The brake pads and linings contain fibers which are hazardous to your health if inhaled. Whenever you work on brake system components, clean all parts with brake system cleaner. Do not allow the fine dust to become airborne. Also, wear an approved filtering mask.*
c) *Safety should be paramount whenever any servicing of the brake components is performed. Do not use parts or fasteners which are not in perfect condition, and be sure that all clearances and torque specifications are adhered to. If you are at all unsure about a certain procedure, seek professional advice. Upon completion of any brake system work, test the brakes carefully in a controlled area before putting the vehicle into normal service. If a problem is suspected in the brake system, don't drive the vehicle until it's fixed.*
d) *Used brake fluid is considered a hazardous waste and it must be disposed of in accordance with federal, state and local laws. DO NOT pour it down the sink, into septic tanks or storm drains, or on the ground.*
e) *Clean up any spilled brake fluid immediately and then wash the area with large amounts of water. This is especially true for any finished or painted surfaces.*

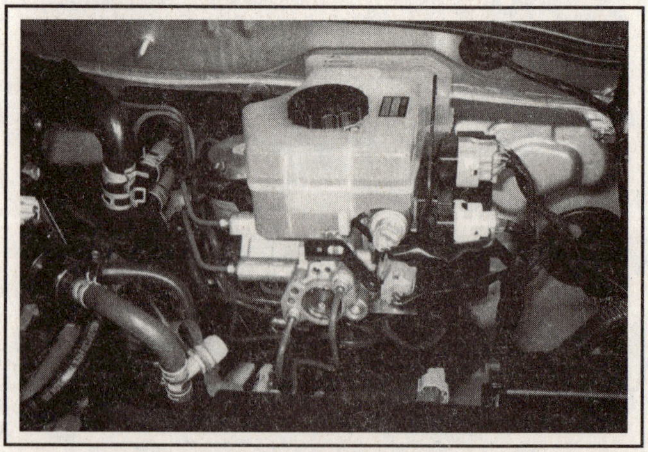

1.4 An integrated master cylinder assembly

2 Anti-lock Brake System (ABS), Vehicle Stability and Traction Control (VSC TRAC) - general information

1 All models are equipped with an Anti-lock Brake System (ABS). The Vehicle Stability Control and Traction Control (VSC TRAC) are optional features. These systems are designed to help maintain vehicle steerability, directional stability and optimum deceleration under severe braking conditions on most road surfaces. The ABS system is primarily designed to prevent wheel lockup during heavy or panic braking situations. It works by monitoring the rotational speed of each wheel and controlling the brake line pressure to each wheel when engaged. Data provided by the ABS wheel speed sensors is also shared with the Vehicle Stability Control feature. This system is designed to assist in correcting over/under steering. The Traction Control System is designed to keep wheels from spinning during vehicle acceleration when road conditions are slick. Overall, these very sophisticated systems help maintain vehicle control when conditions are less than ideal.

COMPONENTS

Modulator or hydraulic unit

➧ Refer to illustration 2.2

2 The modulator is part of the integrated master cylinder assembly on vehicles equipped with VSC TRAC (see Section 1). The modulator (along with a computer) (see illustration) controls hydraulic pressure to the brake calipers using two methods:

a) *An electric pump provides added hydraulic pressure to the braking system when needed.*
b) *Solenoid valves modulate brake line pressure during ABS and VSC TRAC operation.*

Wheel speed sensors

3 Generally, there is a wheel speed sensor designated for each wheel. Each sensor generates a signal in the form of a low-voltage electrical current or a frequency when the wheel is turning. A variable signal is generated as a result of a square-toothed ring (tone-ring, exciter-ring, reluctor, etc.) that rotates very close to the sensor. The signal is directly proportional to the wheel speed and is interpreted by an electronic module (computer).

2.2 An ABS hydraulic control unit

4 The front sensors are mounted in the steering knuckles. The tone-rings are integrated with the hub and wheel bearing assemblies.

5 The rear sensors are mounted in the rear axle hub assemblies. The tone-rings are integrated with the rear axle hub and bearing assemblies.

ABS/VSC TRAC computer

6 The ABS/VSC TRAC computer is the brain for these systems. The function of the computer is to accept and process information received from the various sensors to control the hydraulic line pressure, avoiding wheel lock up or wheel spin. The computer constantly monitors these systems for faults.

DIAGNOSIS AND REPAIR

7 If a dashboard warning light comes on and stays on while the vehicle is in operation, the ABS or VSC TRAC system requires attention. Although special diagnostic testing and tools are necessary to properly diagnose the system, you can perform a few preliminary checks before taking the vehicle to a dealer service department.

a) *Check the brake fluid level in the reservoir.*
b) *Verify that all electrical connectors at the master cylinder integrated assembly are securely connected (for VSC TRAC system equipped models).*
c) *Check the fuses.*
d) *Follow the wiring harness to each wheel and verify that all connections are secure and that the wiring is undamaged.*

8 If the above preliminary checks do not rectify the problem, the vehicle should be diagnosed by a dealer service department or other qualified repair shop. Due to the complexity of this system, all actual repair work must be done by a qualified automotive technician.

⁂ WARNING:

Do NOT try to repair an wheel speed sensor wiring harness. These systems are sensitive to even the smallest changes in resistance. Repairing the harness could alter resistance values and cause the system to malfunction. If the wiring harness is damaged in any way, it must be replaced.

⁂ CAUTION:

Make sure the ignition is turned off before unplugging or reattaching any electrical connections.

WHEEL SPEED SENSOR - REMOVAL AND INSTALLATION

➧ Refer to illustrations 2.11a and 2.11b

9 Loosen the wheel lug nuts, raise the vehicle and support it securely on jackstands. Remove the wheel.

10 Make sure the ignition key is turned to the Off position.

2.11a The front wheel speed sensor location

2.11b The rear wheel speed sensor location

11 Disconnect the electrical connector at the sensor (see illustrations).

12 Remove the mounting fastener and carefully pull the sensor out from the knuckle or rear axle assembly.

13 Installation is the reverse of the removal procedure. Tighten the mounting fastener to the torque listed in this Chapter's Specifications.

14 Install the wheel and lug nuts, lower the vehicle and tighten the lug nuts to the torque listed in the Chapter 1 Specifications.

3 Disc brake pads - replacement

➧ Refer to illustration 3.3

⁂ WARNING:

Disc brake pads must be replaced on both front wheels at the same time - never replace the pads on only one side. Also, the dust created by the brake system is harmful to your health. Never blow it out with compressed air and don't inhale any of it. An approved filtering mask should be worn when working on the brakes. Do not, under any circumstances, use petroleum-based solvents to clean brake parts. Use brake system cleaner only!

1 Remove the cap from the brake fluid reservoir and remove about two-thirds of the fluid. Discard the used brake fluid properly (see Section 1).

2 Loosen the wheel lug nuts, raise the front of the vehicle and support it securely on jackstands. Remove the front wheels.

3 Position a drain pan under the brake caliper assembly and thoroughly clean it with brake system cleaner (see illustration).

4 Inspect the brake disc carefully as outlined in Section 5. If machining is necessary, follow the information in that Section to remove the disc, at which time the calipers and pads can be removed as well.

FIXED CALIPERS (4WD AND PRERUNNER MODELS)

➧ Refer to illustrations 3.5a through 3.5u

5 Follow the accompanying illustrations beginning with illustration 3.5a, for the actual pad replacement procedure. Be sure to stay in order and read the caption under each illustration. Work on one brake assembly at a time using the assembled brake for reference if necessary. Once you have installed the new pads on both calipers, proceed to Step 8.

3.3 Wash the disc and caliper with brake system cleaner to remove the brake dust; DO NOT blow off the brake dust with compressed air

3.5a Remove the small retaining clips that hold the pad pins in place

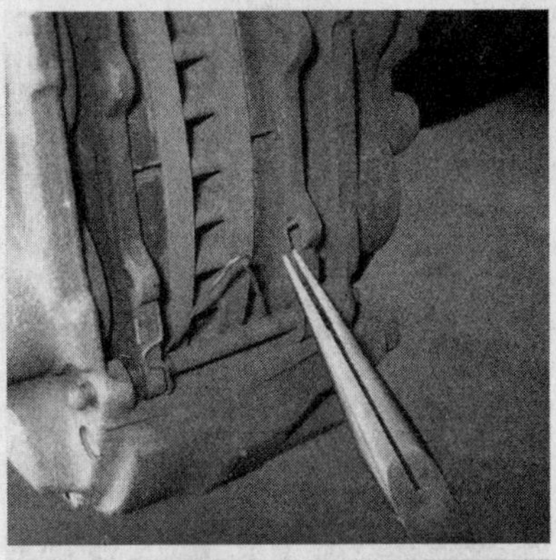

3.5b Release the anti-rattle spring from the hole in each brake pad plate

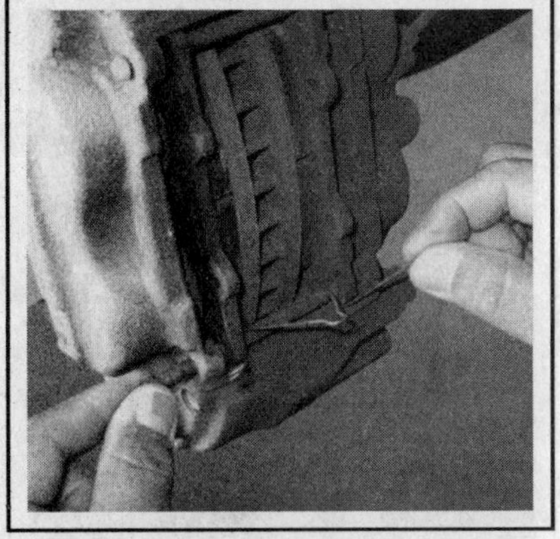

3.5c Withdraw the lower pad pin and remove the anti-rattle spring

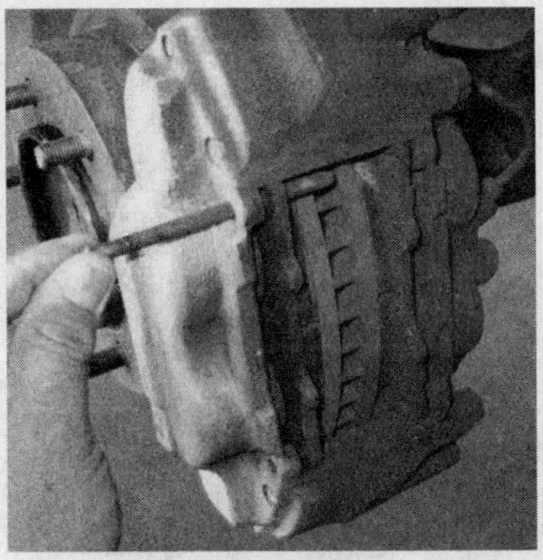

3.5d Withdraw the upper pad pin

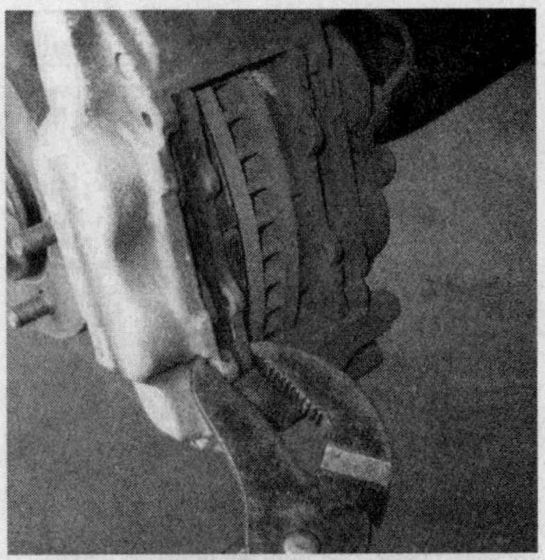

3.5e Squeeze the top and bottom of the outboard brake pad against the caliper to depress the pistons into their bores and to free up the brake pad. Note: *Do not squeeze the inboard brake pad yet*

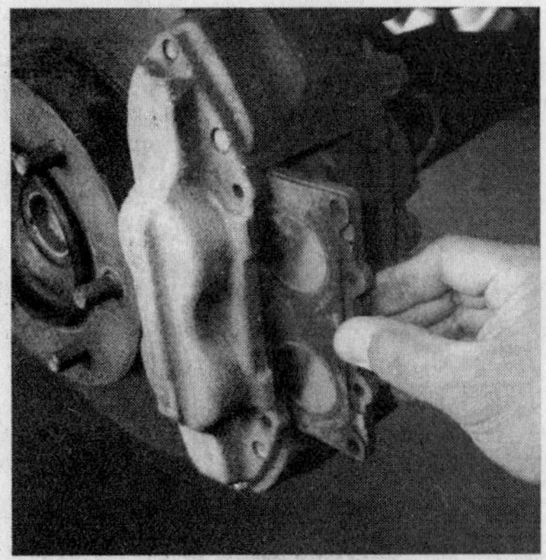

3.5f Remove the outboard brake pad

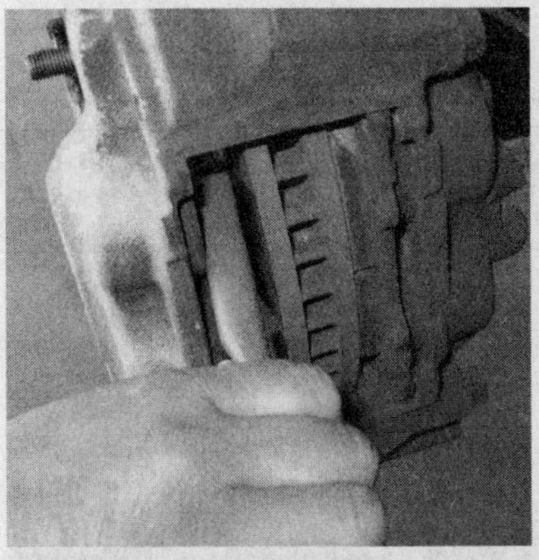

3.5g Push both pistons completely into their bores at the same time to provide room for the new pad. Be careful not to damage the pistons or dust boots

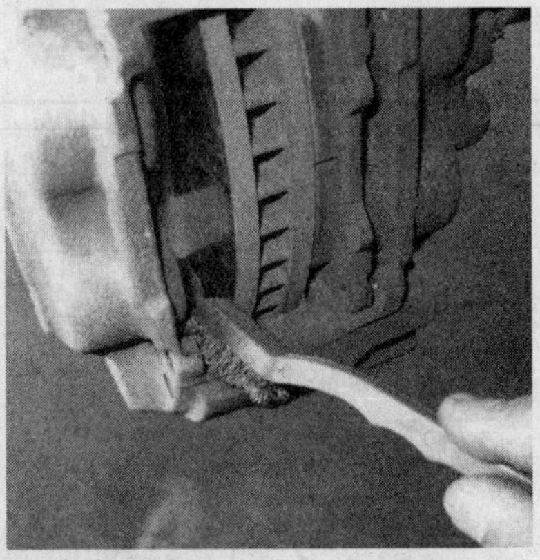

3.5h Use a small brush to clean the upper and lower brake pad plate contact surfaces

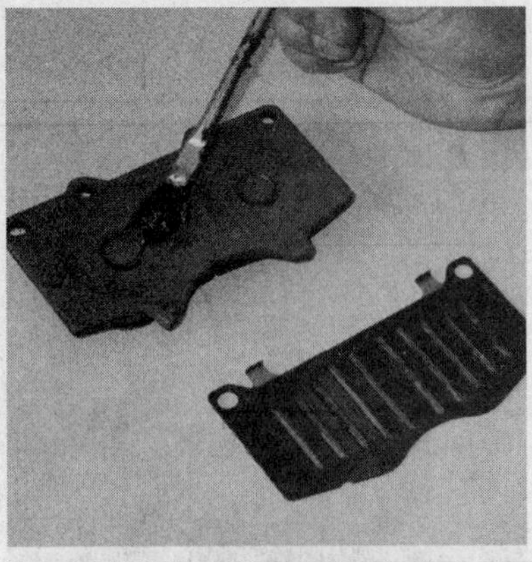

3.5i Lubricate the back of the new brake pad with a small amount of high-temperature brake grease, then install a clean anti-squeal shim to it. Always replace shims that are damaged or worn

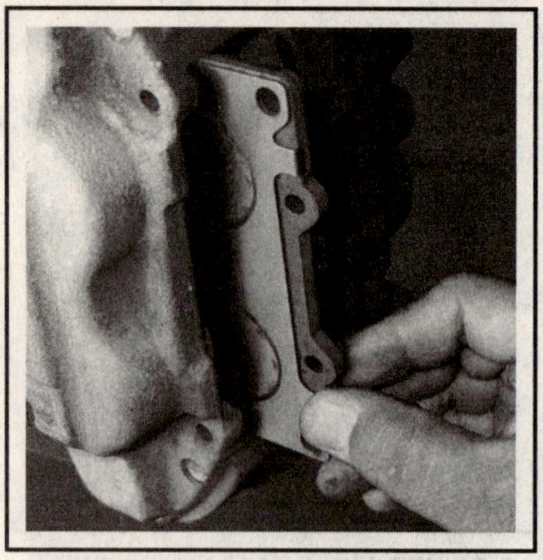

3.5j Install the outboard brake pad

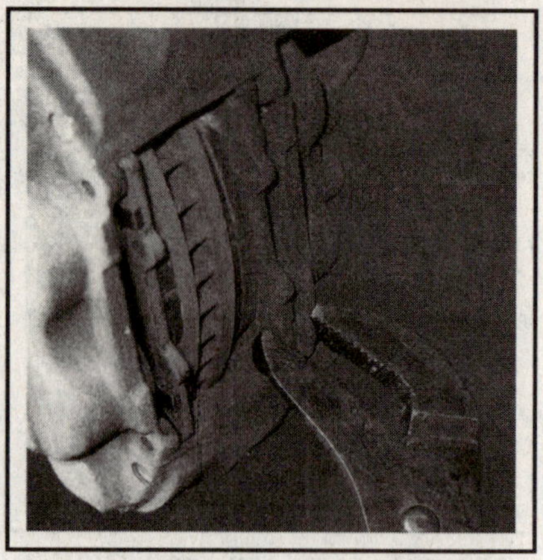

3.5k Squeeze the top and bottom of the inboard brake pad against the caliper to depress the pistons into their bores and to free up the brake pad

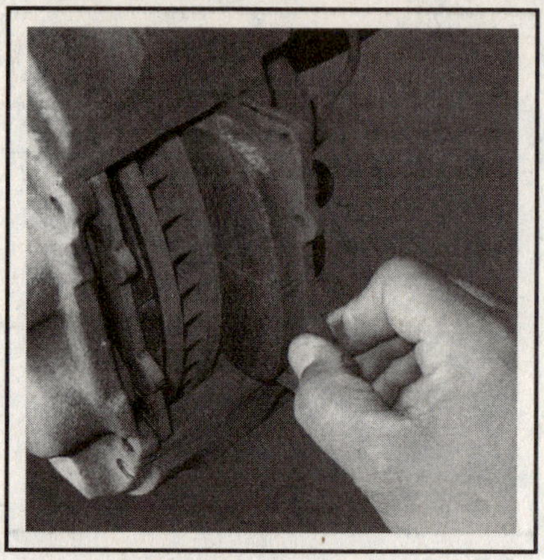

3.5l Remove the inboard brake pad

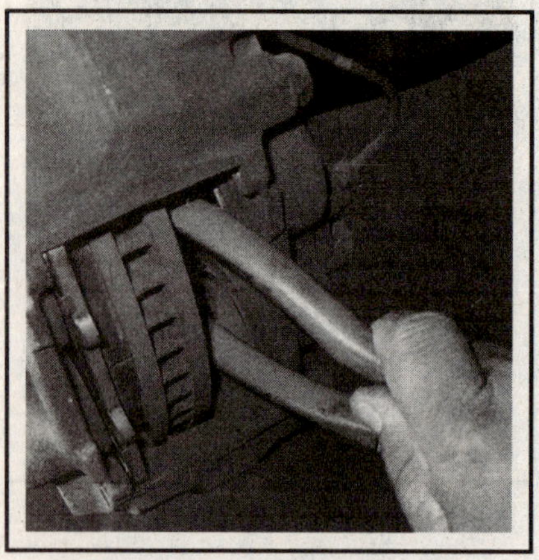

3.5m Push both pistons completely into their bores using the same method as the other brake pad

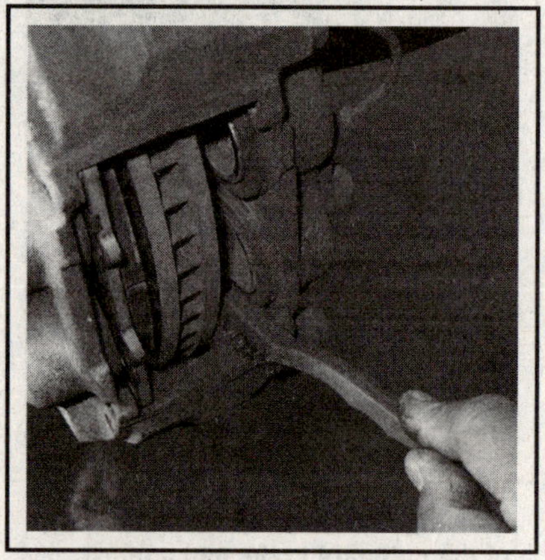

3.5n Clean the upper and lower brake pad plate contact surfaces

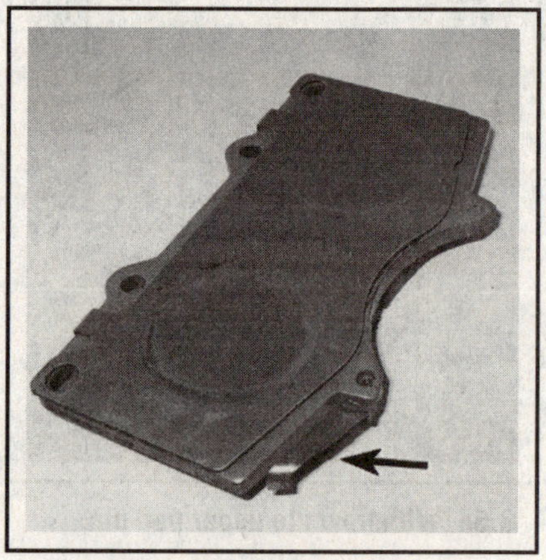

3.5o Lubricate the back of the new brake pad with a small amount of high-temperature brake grease, then install a clean anti-squeal shim to it. Always replace anti-squeal shims that are damaged or worn. Note the wear sensor mounted to this inboard brake pad plate

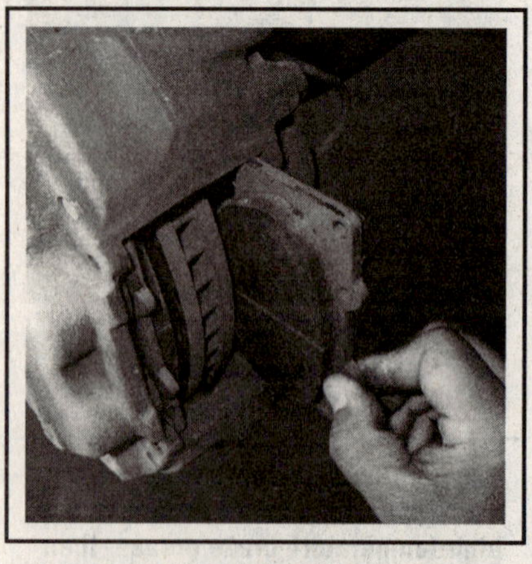

3.5p Install the inboard brake pad

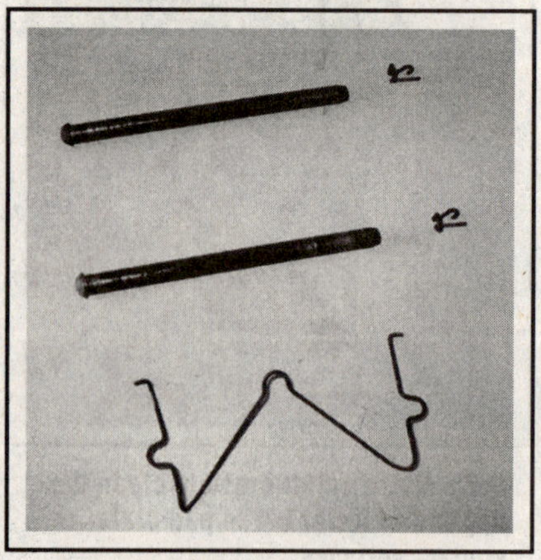

3.5q Inspect the pad pins, pin clips and anti-rattle spring for wear or damage. Replace them if necessary

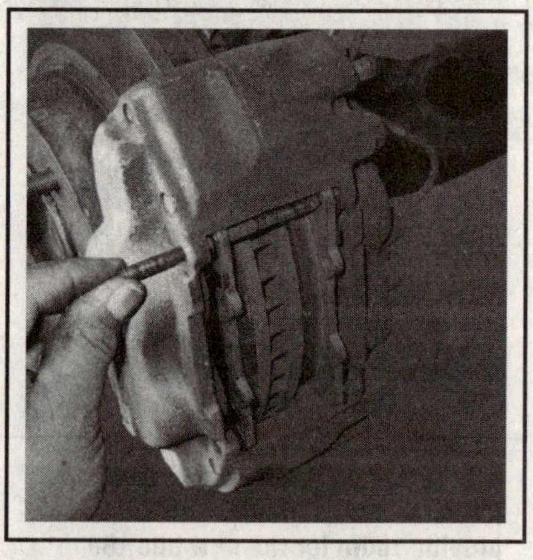

3.5r Install the upper pad pin

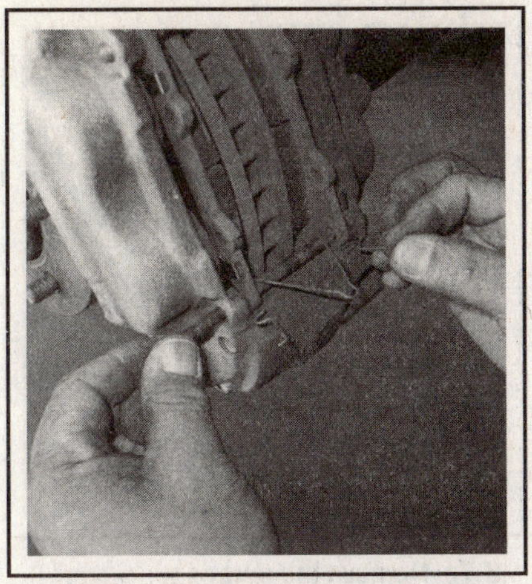

3.5s Place the anti-rattle spring into position and install the lower pad pin

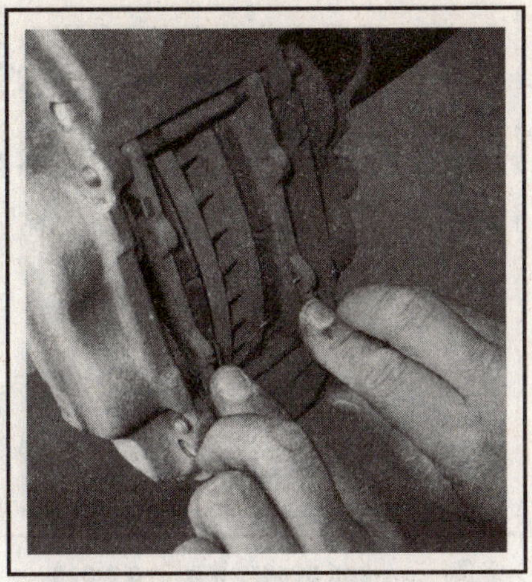

3.5t Place the ends of the anti-rattle spring into the brake pad plates

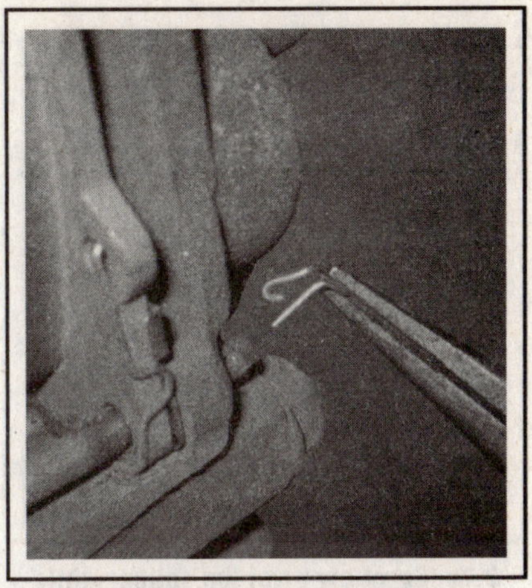

3.5u Install the pin clips to both pad pins with the handle end pointing away from the caliper

FLOATING CALIPERS (2WD MODELS)

➧ Refer to illustrations 3.6 and 3.7a through 3.7r

6 Push the piston back into the bore to provide room for the new brake pads. A C-clamp can be used to accomplish this (see illustration). As the piston is depressed to the bottom of the caliper bore, the fluid in the master cylinder will rise. Make sure it doesn't overflow. Remove more fluid if necessary.

7 Follow the accompanying illustrations beginning with illustration 3.7a, for the actual pad replacement procedure. Be sure to stay in order and read the caption under each illustration. Once you have installed the new pads, proceed to Step 8.

ALL CALIPERS

8 Install the wheel and lug nuts, lower the vehicle and tighten the lug nuts to the torque listed in the Chapter 1 Specifications.

9 Start the engine and apply and release the brake pedal several times (using short strokes) to bring the pads into contact with the brake discs. Turn the engine off.

10 Check the brake fluid level and add fluid, if necessary (see Chapter 1). Check the operation of the brakes in an isolated area before driving the vehicle in traffic.

➡Note: On VSC TRAC equipped models (see Section 1), pump the brake pedal 40 times with the engine off to remove the boost pressure; the brake pedal travel will increase and the resistance will decrease as result. Remove the brake reservoir cap and fill the reservoir to the MAX line.

3.6 Before removing the caliper, slowly depress the piston into the caliper bore by using a large C-clamp between the outer brake pad and the back of the caliper

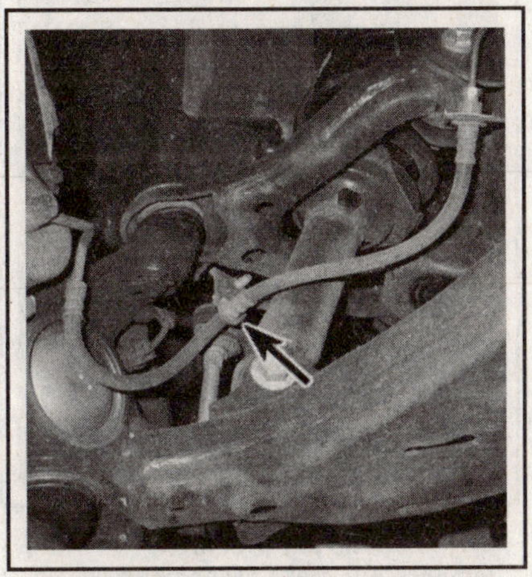

3.7a Detach the brake hose from the shock absorber bracket

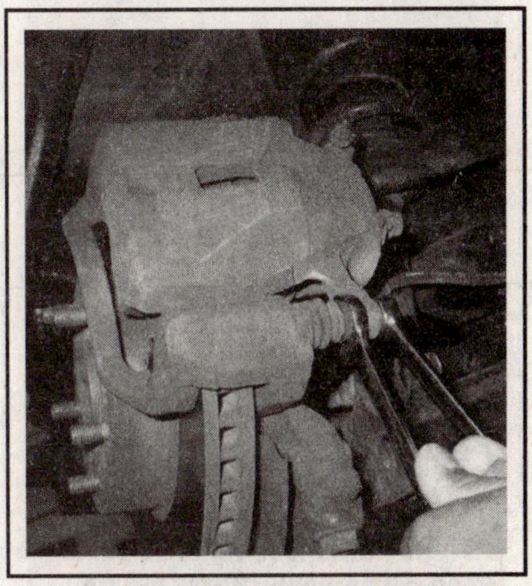

3.7b Using two wrenches, hold the slide-pin and remove the caliper lower mounting bolt

3.7c Carefully pivot the caliper while holding the upper slide-pin boot to keep it from twisting . . .

3.7d . . . and secure it with a piece of wire; do not allow it to hang by the flexible brake hose

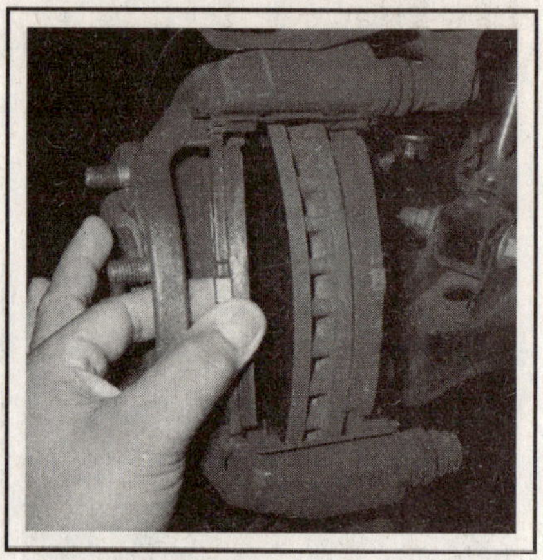

3.7e Remove the outer pad . . .

3.7f . . . and the inner pad

3.7g Remove the upper and lower pad support plates; make sure they fit tightly and aren't worn. Replace them if necessary

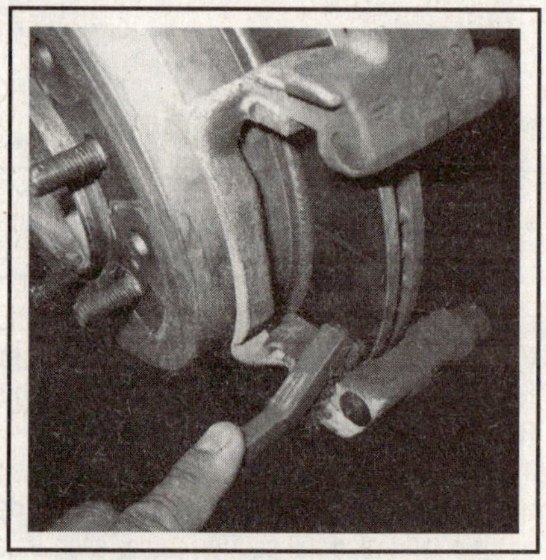

3.7h Clean the caliper mounting bracket where the pad support plates fit (typical shown)

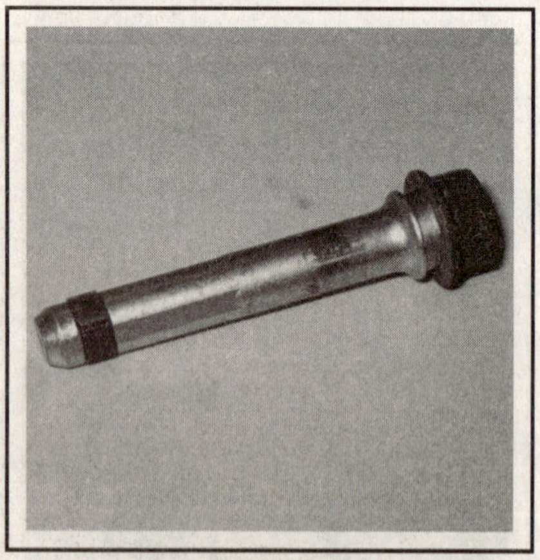

3.7i Remove the lower slide pin and clean it thoroughly. Check the pin and bushing for wear. Replace the bushing or the pin if necessary

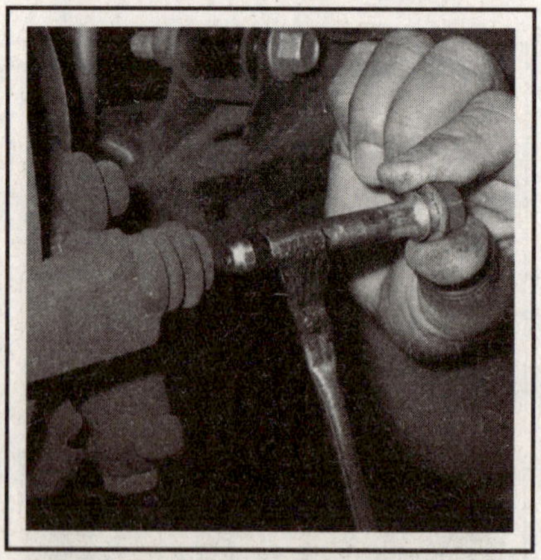

3.7j Coat the pin with high-temperature brake grease and reinstall it. Seat the slide-pin boot onto the pin

3.7k To clean and lubricate the upper sliding pin, pull the caliper away from the caliper mounting bracket while slowly withdrawing the pin. Pull the slide-pin boot off the pin and be careful not to damage it. Reinstall the slide-pin and caliper and be sure to seat the slide-pin boot onto the pin

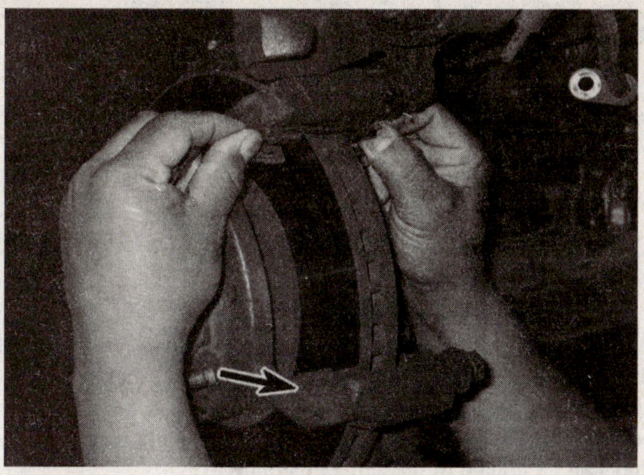

3.7l Install clean or new pad support plates

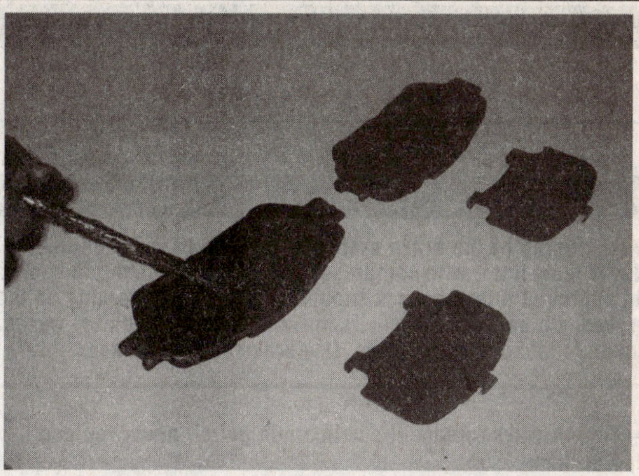

3.7m Lubricate the back of each pad with a small amount of high-temperature brake grease, then install clean anti-squeal shim(s) to them. Always replace shims that are damaged or worn

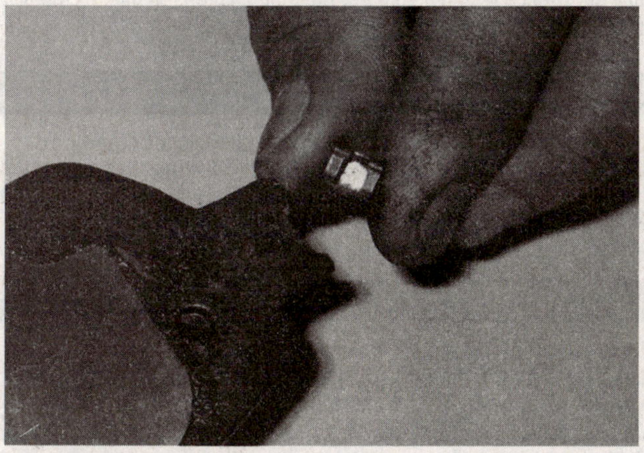

3.7n Install the pad wear indicator(s) to the bottom of the brake pad(s). Replace any wear indicators that are worn

3.7o Install the outer pad, making sure that the ends are seated correctly into the pad support plates . . .

3.7p . . . then install the inner pad in the same way

3.7q Carefully pivot the caliper back into position over the brake pads. Do not damage the upper slide-pin boot. Replace any boots that are damaged

3.7r Install the lower mounting bolt and tighten it to the torque listed in this Chapter's Specifications

4 Disc brake caliper - removal and installation

REMOVAL

⁂ WARNING:

Dust created by the brake system is harmful to your health. Never blow it out with compressed air and don't inhale any of it. An approved filtering mask should be worn when working on the brakes. Do not, under any circumstances, use petroleum-based solvents to clean brake parts. Use brake system cleaner only!

➡Note: Always replace the calipers in pairs - never replace just one of them.

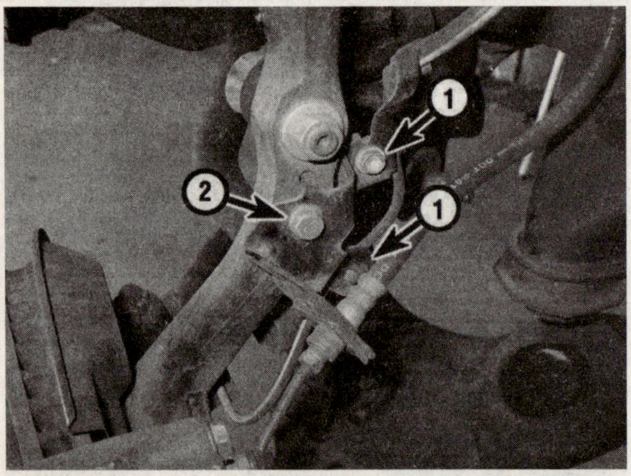

4.4a Brake hose/line bracket:

1 *Wheel speed sensor harness fasteners (remove first)*
2 *Bracket mounting bolt*

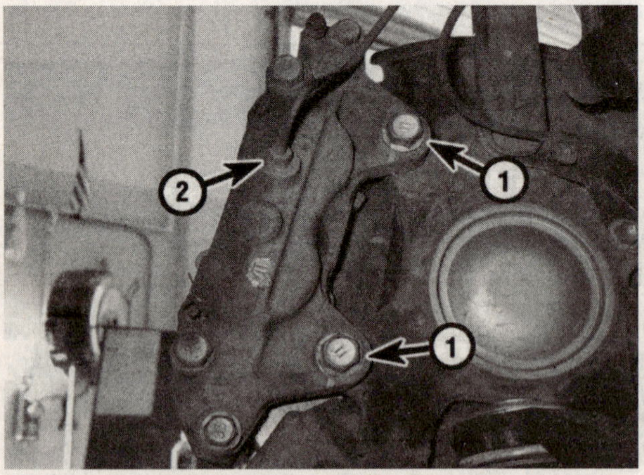

4.4b Fixed brake caliper mounting details:

1 *Caliper mounting bolts*
2 *Brake line fitting (tube nut)*

1 Loosen the front wheel lug nuts, raise the front of the vehicle and support it securely on jackstands. Apply the parking brake. Remove the wheels.

2 Position a drain pan under the brake assembly and clean the caliper and surrounding area with brake system cleaner.

Fixed type (4WD and PreRunner models)

➧ Refer to illustrations 4.4a, 4.4b and 4.4c

3 If the caliper is going to be replaced, remove the brake pads (see Section 3). Otherwise, squeeze the brake pads towards the caliper just a bit for clearance (see illustrations 3.5e and 3.5k).

➡Note 1: If the pads are going to be reused, mark them so that they can be placed in the same position.

➡Note 2: Make sure the brake fluid reservoir doesn't overflow when squeezing the brake pads; remove some brake fluid if necessary.

4 Detach the brake line and hose from the steering knuckle (see illustration), then detach the brake line fitting from the caliper using a flare-nut wrench (see illustration). Plug all openings to minimize brake fluid loss.

⁂ CAUTION 1:

Brake fluid will damage paint or finished surfaces. Cover all body parts and be careful not to spill fluid during this procedure. Clean up any spilled brake fluid immediately and wash the area with large amounts of water.

⁂ CAUTION 2:

If you're removing the caliper for access to other components, leave the line connected. Secure the caliper with rope or wire. DO NOT let it hang by the brake hose/line (see illustration).

5 Remove the caliper mounting bolts, then lift the caliper from the knuckle (see illustration 4.4b).

4.4c Hang the caliper with a piece of wire - don't let it hang by the brake line!

Floating type (2WD models)

Refer to illustrations 4.6a and 4.6b

6 Remove the brake hose banjo bolt and disconnect the hose from the caliper (see illustration). Plug the hose to keep contaminants out of the brake system and to prevent losing any more brake fluid than is necessary (see illustration).

➡Note: If you're just removing the caliper for access to other components, don't detach the hose.

7 Remove the caliper mounting bolts, lift the caliper from the bracket (see illustration 4.6a). If the hose is still attached, support the caliper with a length of wire - don't let it hang by the brake hose (see illustration 4.4c).

➡Note: If necessary, the caliper mounting bracket can be removed by removing the mounting bolts shown in illustration 4.6a.

INSTALLATION

8 Installation is the reverse of removal. Tighten the caliper mounting bolts to the torque listed in this Chapter's Specifications. For fixed type calipers, carefully install the brake line to the caliper using a flare-nut wrench and tighten it securely, if detached. For floating type calipers, install new sealing washers to each side of the brake hose fitting, then tighten the banjo bolt to the torque listed in this Chapter's Specifications.

9 Bleed the brake lines (see Section 9).

10 Install the wheels and lug nuts. Lower the vehicle and tighten the lug nuts to the torque listed in the Chapter 1 Specifications. Check the operation of the brakes carefully before placing the vehicle into normal service.

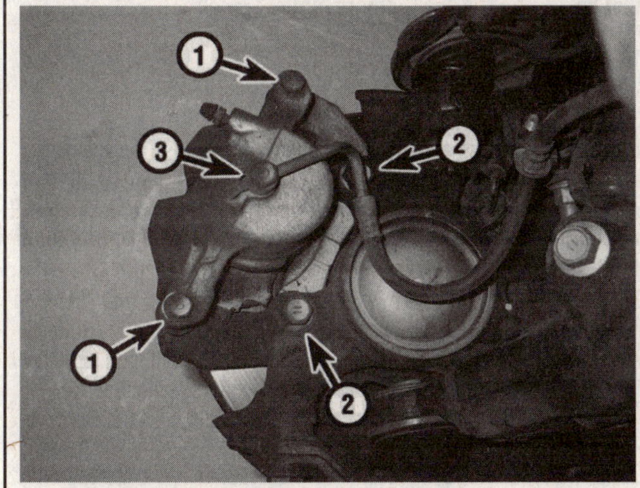

4.6a Floating brake caliper mounting details:

1 Caliper mounting bolts
2 Caliper mounting bracket bolts
3 Brake hose fitting and banjo bolt

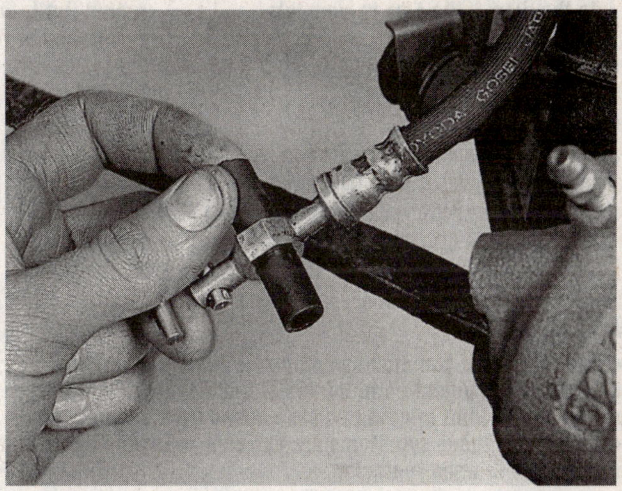

4.6b Using a piece of hose of the appropriate size, plug the banjo fitting to prevent brake fluid from dripping out of the hose and to prevent contaminants from entering the brake system

5 Brake disc - inspection, removal and installation

INSPECTION

➧ Refer to illustrations 5.4, 5.5a, 5.5b, 5.6a and 5.6b

1 Loosen the wheel lug nuts, raise the vehicle and support it securely on jackstands. Remove the wheel.

2 Remove the brake caliper as outlined in Section 4. It's not necessary to disconnect the brake line or hose for this procedure, but detach the hose/line bracket from the steering knuckle on front brakes. After removing the caliper bolts, suspend the caliper out of the way with a piece of wire. Don't let the caliper hang by the hose and don't stretch or twist the hose.

3 Reinstall two lug nuts with washers (for spacing) to hold the disc securely against the hub.

5.4 The brake pads on this vehicle were obviously neglected - they wore out completely and cut deep grooves into the disc; if the disc is worn this severely, replace it

5.5a Measure the brake disc runout with a dial indicator; if the reading exceeds the maximum allowable runout limit, the disc must be resurfaced or replaced

5.5b Using a swirling motion, remove the glaze from the disc surface with sandpaper or emery cloth

4 Visually check the disc surface for score marks, cracks and other damage. Light scratches and shallow grooves are normal after use and may not always be detrimental to brake operation. Deep score marks or cracks may require disc refinishing by an automotive machine shop or disc replacement (see illustration). Be sure to check both sides of the disc. If the brake pedal pulsates during brake application, suspect disc runout.

➡Note: The most common symptoms of damaged or worn brake discs are pulsation in the brake pedal when the brakes are applied or loud grinding noises caused from severely worn brake pads. If these symptoms are extreme, it is very likely that the disc(s) will need replacing.

5 To check disc runout, place a dial indicator at a point about 1/2-inch from the outer edge of the disc (see illustration). Set the indicator to zero and turn the disc. Indicator readings that exceed 0.003 of an inch could cause pulsation upon brake application and will require disc refinishing by an automotive machine shop or disc replacement.

➡Note: If disc refinishing or replacement is not necessary, you can deglaze the brake pad surface on the disc with emery cloth or sandpaper (use a swirling motion to ensure a non-directional finish) (see illustration).

6 The disc must not be machined to a thickness less than the specified minimum refinish thickness. The minimum (or discard) thickness is cast into the front or backside of the disc (see illustration). The disc thickness can be checked with a micrometer (see illustration).

REMOVAL AND INSTALLATION

➧ Refer to illustrations 5.8a, 5.8b, 5.9a and 5.9b

7 On floating type brake calipers, remove the caliper mounting bracket (see illustration 4.6a).

5.6a Measure the brake disc thickness at several points with a micrometer

5.6b The minimum allowable thickness dimension is usually cast into the disc (typical shown)

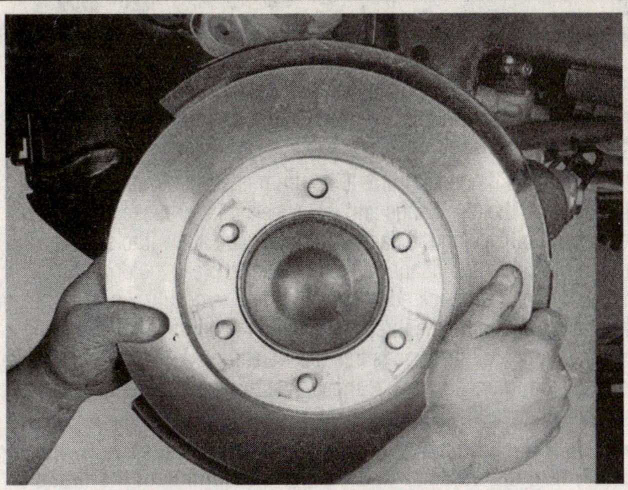

5.8a To detach the disc from the hub, simply pull it off (typical shown)

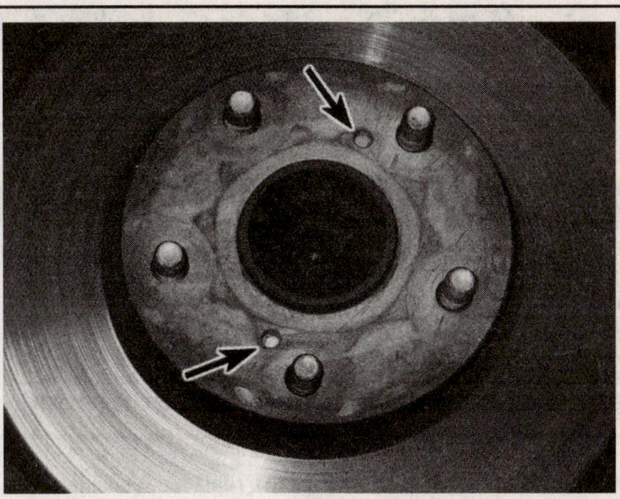

5.8b Use bolts in these threaded holes in the disc to push the disc off the hub

8 Mark the disc in relation to the hub so that it can be installed in its original position on the hub, then remove the disc (see illustration). If it's stuck, make sure you have removed any lug nuts installed during inspection. You can use a mallet to free a stuck disc from the hub. On 2WD (non PreRunner) models, install bolts into the threaded holes and alternately turn each of them a little until the disc is free (see illustration).

9 Clean the hub flange and the inside of the brake disc thoroughly, removing any rust or corrosion, then install the disc onto the hub assembly (see illustrations).

➡Note: A very thin coat of anti-seize compound between the brake disc and hub flange can prevent rust in this area and make subsequent service a little easier.

10 On floating type calipers, install the caliper mounting bracket, brake pads and caliper. Tighten all mounting bolts to the torque values listed in this Chapter's Specifications.

11 On fixed type calipers, install the caliper and brake pads. Tighten the caliper mounting bolts to the torque listed in this Chapter's Specifications.

12 Install the wheel, then lower the vehicle to the ground. Tighten the wheel lug nuts to the torque listed in the Chapter 1 Specifications. Depress the brake pedal a few times to bring the brake pads into contact with the disc. Bleeding of the brake lines will not be necessary unless the brake hose/line was disconnected from the caliper. Check the operation of the brakes carefully before placing the vehicle into normal service.

5.9a Clean any rust or corrosion from the areas of the hub flange that contact the disc. A wire brush or sanding tool, designed to be used with a power drill, can make the job a lot easier

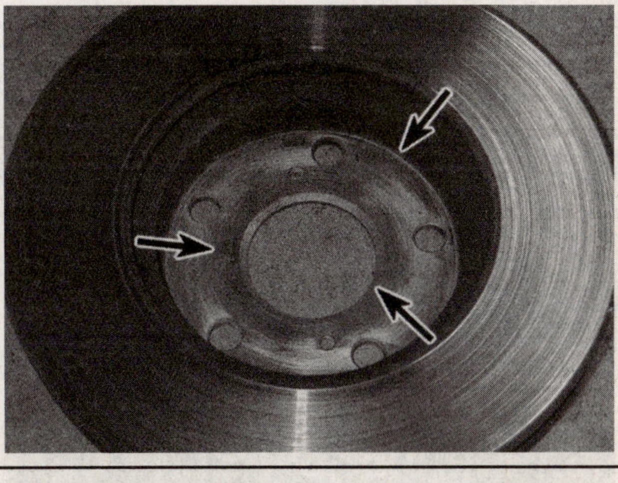

5.9b Clean any rust or corrosion from the inside of the disc that contacts the hub flange. Again, power tools are very useful for this job

6 Drum brake shoes - replacement

Refer to illustrations 6.2a, 6.2b, 6.2c, 6.2d, 6.3, 6.4a through 6.4gg, and 6.5

WARNING:

Drum brake shoes must be replaced on both wheels at the same time never replace the shoes on only one wheel. Also, the dust created by the brake system is harmful to your health. Never blow it out with compressed air and don't inhale any of it. An approved filtering mask should be worn when working on the brakes. Do not, under any circumstances, use petroleum-based solvents to clean brake parts. Use brake system cleaner only!

Note: Whenever the brake shoes are replaced, consider replacing the return and hold-down springs. Due to the continuous heating/cooling cycle that the springs are subjected to, they lose their tension over a period of time and may allow the shoes to drag on the drum and wear at a much faster rate than normal.

1 Loosen the wheel lug nuts, raise the rear of the vehicle and support it securely on jackstands. Block the front wheels to keep the vehicle from rolling. Release the parking brake. Remove the wheel.

Note: To avoid mixing up parts, work on only one brake assembly at a time.

2 Remove the brake drum. If the drum is stuck because of corrosion between the axle flange and the wheel studs or brake drum, spray penetrating oil around the flange and studs and allow it to soak in. Tap around the studs and flange with a hammer to break the drum loose, then tap around the back edge of the drum to remove it. Another method is to install bolts into the threaded holes on the drum and turn each one gradually until the disc is free (see illustration). If the drum still will not come off, retract the brake shoes by turning the brake shoe adjuster (star-wheel) (see illustrations).

Note: If the adjuster stops turning and the brake drum will not rotate, turn the adjuster in the other direction until the brake drum becomes loose and can be removed from the hub.

3 Clean the brake shoe assembly with brake system cleaner before beginning work (see illustration).

4 Follow the accompanying illustrations for the brake shoe replacement procedure (see illustrations 6.4a through 6.4gg). Be sure to stay in order and read the caption under each illustration.

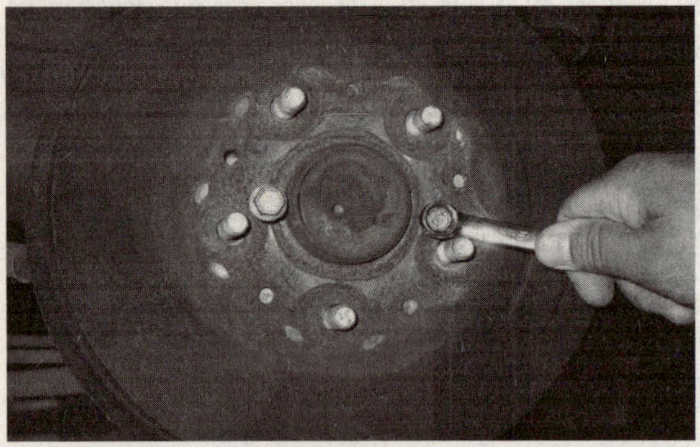

6.2a If the brake drum is rusted onto the wheel studs or axle flange, it can usually be loosened after spraying those areas with penetrant and letting it soak. If that doesn't work, thread a couple of bolts into the holes in the drum and tighten each one gradually until they push the drum off

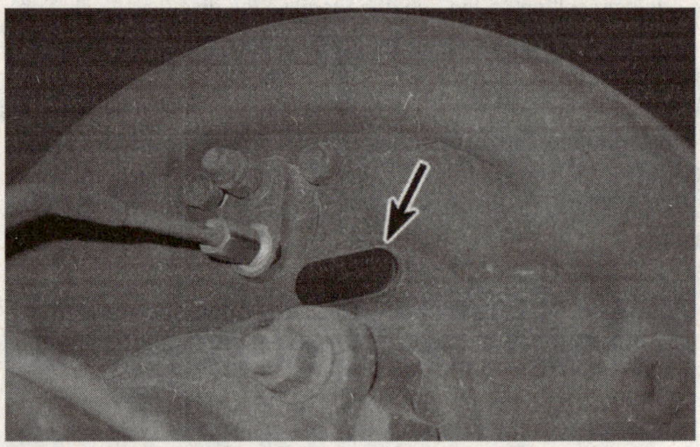

6.2b Remove the rubber plug from the port in the backing plate

6.2c The auto-adjust lever (A) and the adjuster (star-wheel) (B) are hidden from view when the drum is installed but can be moved by reaching through the port in the backing plate

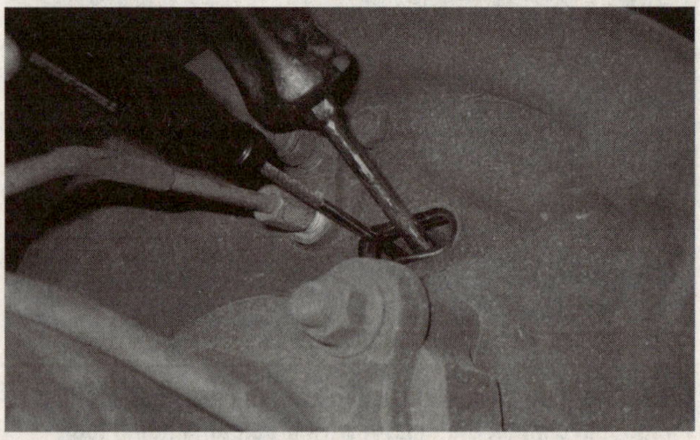

6.2d Use two screwdrivers; one to hold the auto-adjust lever away from the adjuster (star-wheel) and the other to turn the adjuster to retract the shoes

6.3 Clean the brake assembly with brake cleaner to remove any brake dust - position a drain pan under the brake assembly to catch the residue - DO NOT blow off the brake dust with compressed air

6.4a Remove the auto-adjust lever spring, . . .

6.4b . . . then remove the lever

6.4c Turn the adjuster (star-wheel) so that the strut assembly's length is minimized and the shoes contract

6.4d Firmly clamp the upper return spring with locking pliers and unhook it from the front shoe

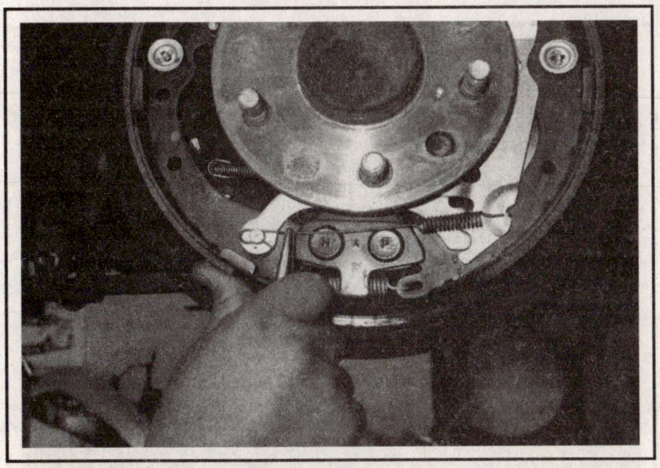

6.4e Remove the lower return spring

6.4f Compress the front shoe hold-down spring with a designated tool or needle-nose pliers. Turn the spring and retainer 90-degrees while holding the end of the pin at the rear of the backing plate. This will release the hold-down spring and shoe so be prepared for the parts to fall

6.4g Remove the lower brake shoe strut, detach the small tension spring from the bottom of both shoes, then remove the front shoe

6.4h Remove the rear shoe hold-down spring and . . .

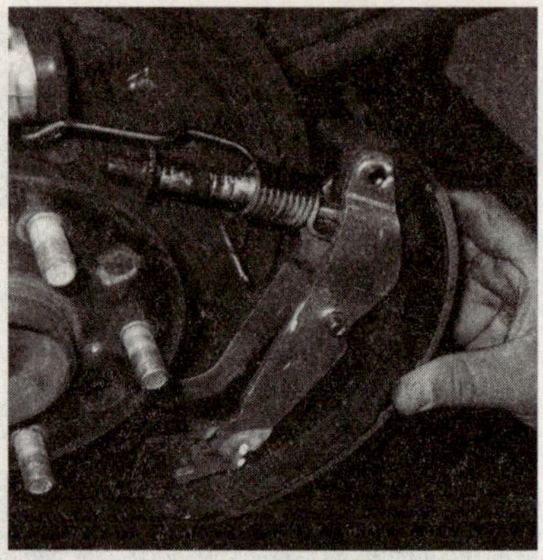

6.4i . . . separate the rear shoe with the upper return spring, the upper strut assembly and parking brake levers from the backing plate

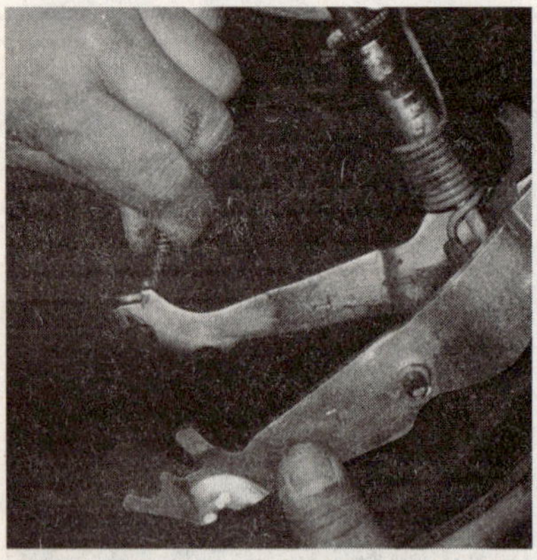

6.4j Detach the parking brake cable from the lever

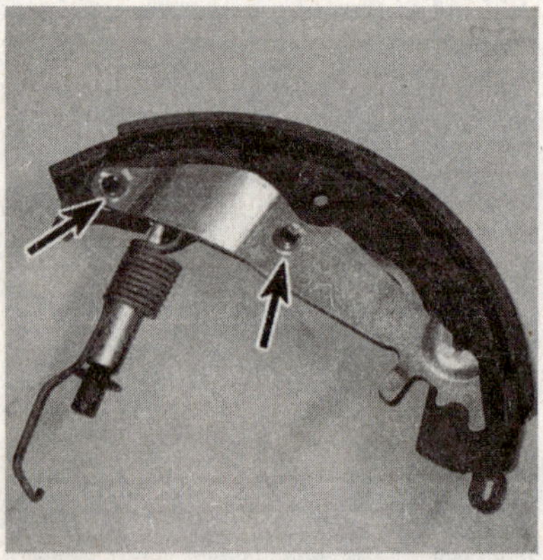

6.4k There are two C-clips holding the parking brake levers and the rear shoe together

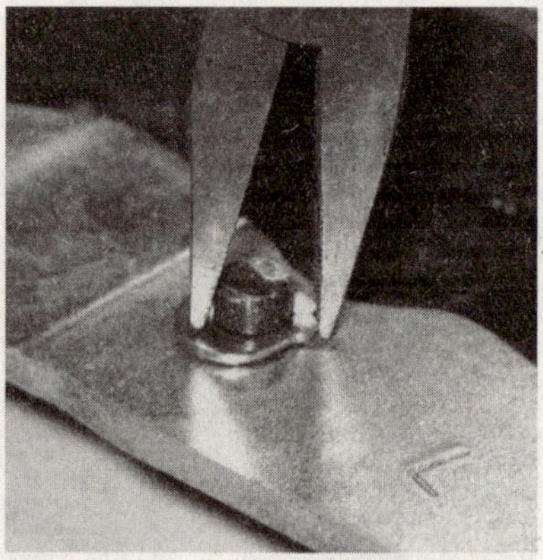

6.4l Remove the C-clips and discard them

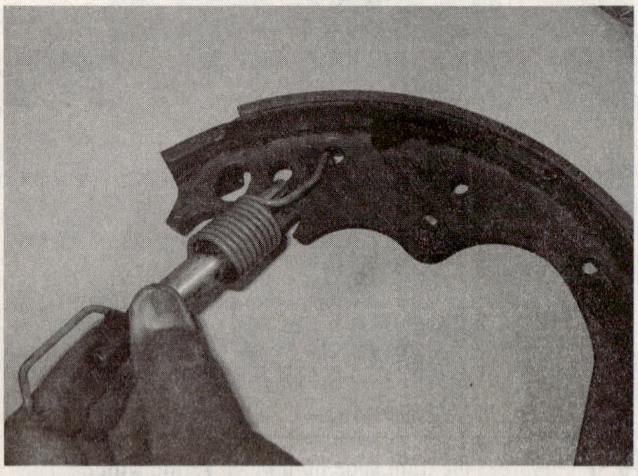

6.4m Unhook the upper return spring and upper strut assembly from the rear shoe

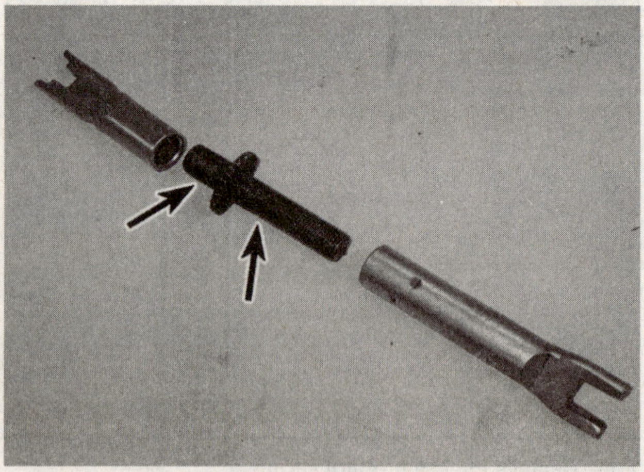

6.4n Separate the upper strut assembly from the upper return spring and lubricate these areas of the assembly (adjuster) with high-temperature brake grease

6.4o Reassemble the upper strut assembly with the upper return spring

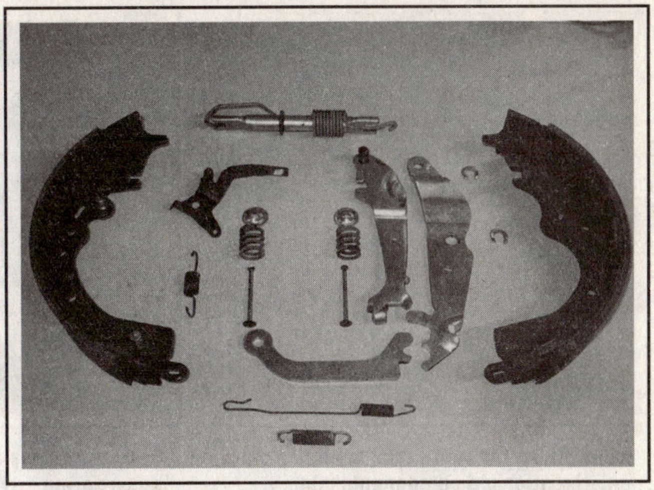

6.4p Brake shoes and related hardware (left side shown - right side similar)

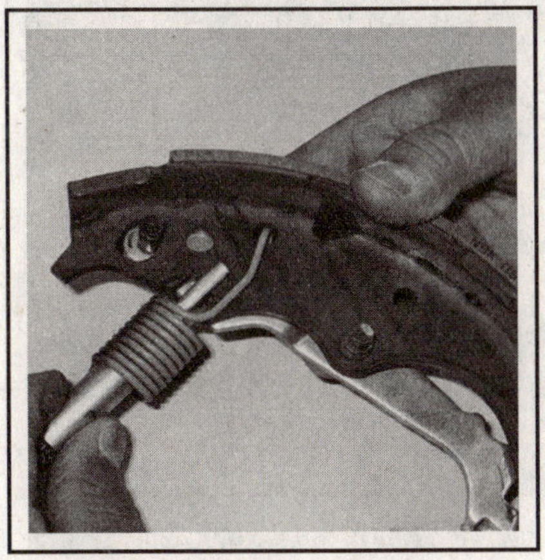

6.4q Reassemble the upper strut assembly and bottom parking lever to the rear brake

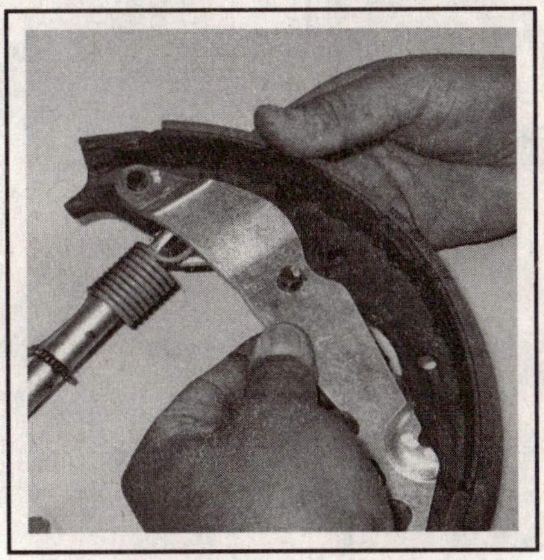

6.4r Place the parking brake lever in position . . .

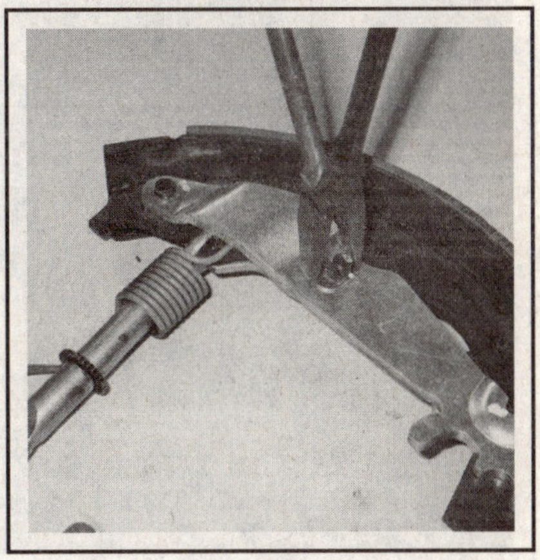

6.4s . . . install new C-clips onto the parking brake lever pivot pins . . .

6.4t . . . then squeeze the ends together

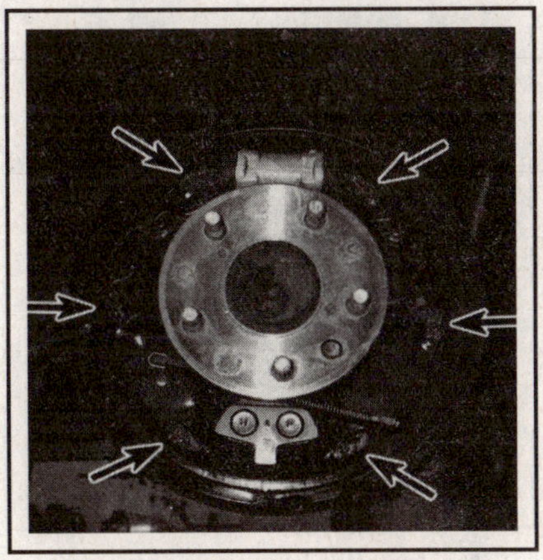

6.4u Lubricate the brake shoe contact points on the backing plate with high-temperature brake grease

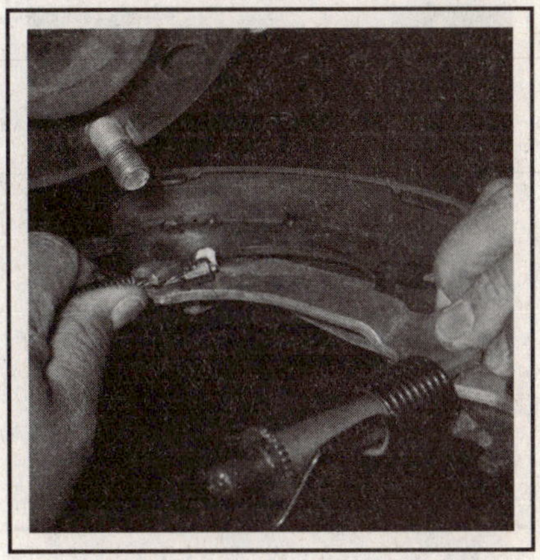

6.4v Reattach the parking brake cable to the parking brake lever

6.4w Place the rear brake shoe and the attached components into position with the hold-down spring pin held from the back

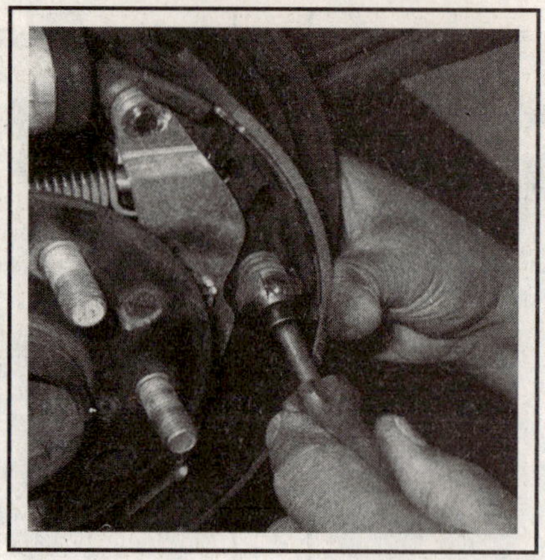

6.4x Install the rear brake shoe hold-down spring

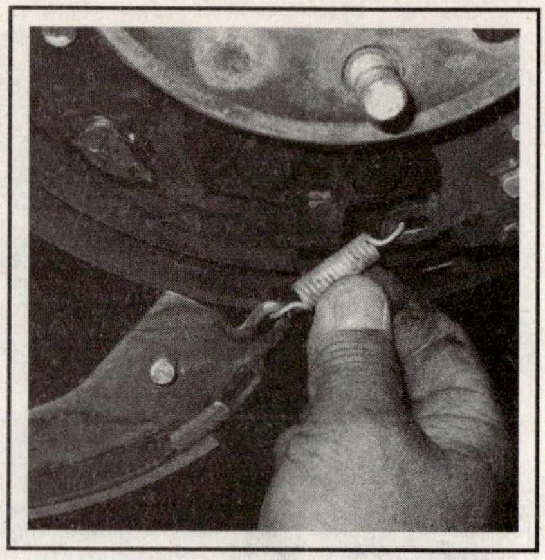

6.4y install the small tension spring to the bottom of both shoes

6.4z Place the front brake shoe into position with the hold-down spring pin held from the back

6.4aa Install the front brake shoe hold-down spring

6.4bb Install the lower brake shoe strut

6.4cc Install the lower return spring

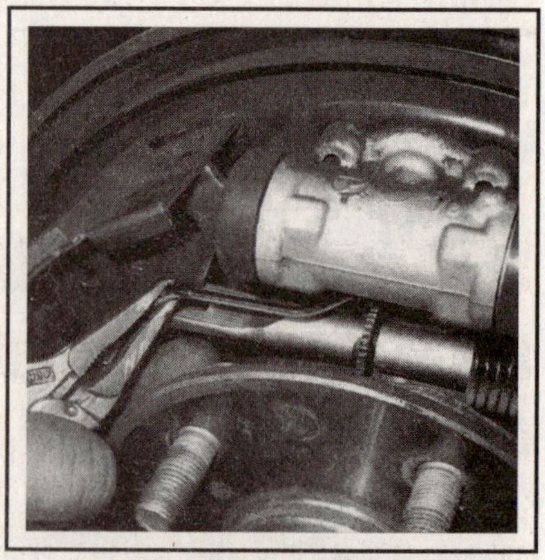

6.4dd Firmly clamp the upper return spring with locking pliers and stretch the end, then hook it into the hole in the front shoe

6.4ee Install the auto-adjust lever

6.4ff Install the auto-adjust lever spring

6.4gg This is how the assembled brake should look (left side shown - right side similar)

5 Before reinstalling the drum, remove the thin gasket from inside the drum where it mates with the axle flange and clean any rust from the drum and flange before installing a new one (see illustration). Also, carefully examine it for cracks, score marks, deep scratches and hard spots, which appear as small discolored areas around the center of the drum's friction surface. If any of the conditions listed above exist, the drum must be taken to an automotive machine shop for inspection, resurfacing or possibly scrapping.

➡Note: Professionals recommend resurfacing the drums whenever a brake job is performed. Resurfacing will eliminate the possibility of out-of-round drums. If the drums are worn so much that they can't be resurfaced without exceeding the maximum allowable diameter (stamped into the drum), then new ones will be required. At the very least, if you elect not to have the drums resurfaced, remove any glazing from the surface with emery cloth or sandpaper using a swirling motion.

6 Install the brake drum onto the hub flange, then remove the rubber plug covering the adjuster port in back (see illustration 6.2b). Using two screwdrivers, turn the adjuster (star-wheel) until the brake shoes lock the drum (see illustrations 6.2c and 6.2d). Back off the adjustment 15 notches and reinstall the rubber plug into the backing plate.

7 Mount the wheel and install the lug nuts. Lower the vehicle and tighten the lug nuts to the torque listed in the Chapter 1 Specifications.

8 Make a number of forward and reverse stops and operate the parking brake several times.

9 Confirm that the brakes are working properly before resuming normal driving.

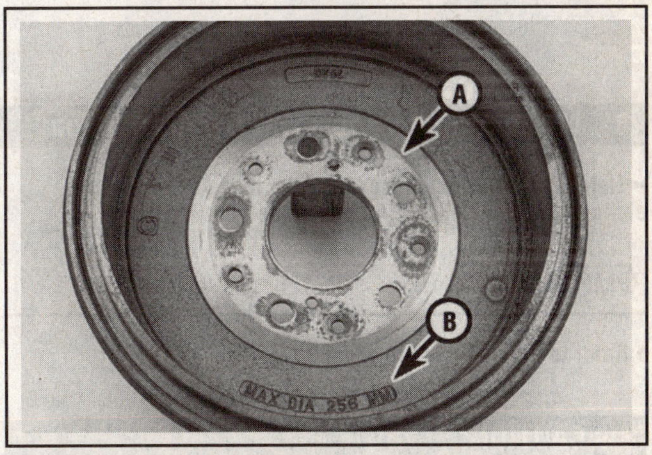

6.5 The gasket on the inside of the drum (A) and the maximum diameter for the brake drum is cast or stamped on the drum (B)

7 Wheel cylinder - removal and installation

⁂ WARNING:

The dust created by the brake system is harmful to your health. Never blow it out with compressed air and don't inhale any of it. An approved filtering mask should be worn when working on the brakes. Do not, under any circumstances, use petroleum-based solvents to clean brake parts. Use brake system cleaner only!

➡Note: If replacement is indicated (usually because of fluid leakage or sticky operation), it is recommended that the wheel cylinders be replaced, not overhauled. Always replace the wheel cylinders in pairs - never replace just one of them.

REMOVAL

➧ Refer to illustration 7.4

1 Raise the rear of the vehicle and support it securely on jackstands. Block the front wheels to keep the vehicle from rolling.

2 Remove the brake shoes (see Section 6).

3 Remove all dirt and foreign material from around the wheel cylinder.

4 Using a flare-nut wrench, disconnect the brake line (see illustration). Don't pull the brake line away from the wheel cylinder.

5 Remove the wheel cylinder mounting bolts.

6 Detach the wheel cylinder from the brake backing plate and immediately plug the brake line to prevent fluid loss and contamination.

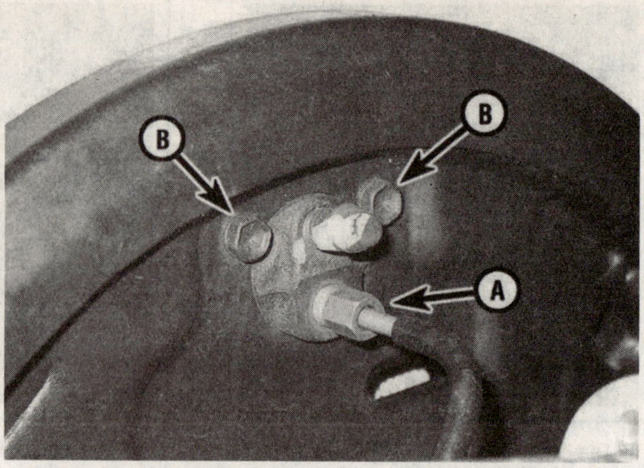

7.4 To remove the wheel cylinder, unscrew the brake line fitting (A) with a flare-nut wrench, then remove the wheel cylinder mounting bolts (B)

INSTALLATION

7 Place the wheel cylinder in position and install the mounting bolts finger tight. Connect the brake line to the cylinder, being careful not to cross-thread the fitting. Tighten the wheel cylinder mounting bolts to the torque listed in this Chapter's Specifications, then tighten the brake line fitting securely.

8 Install the brake shoes (see Section 6).

9 Bleed the rear brake lines (see Section 10).

10 Check the operation of the brakes carefully before placing the vehicle into normal service.

8 Master cylinder - removal and installation

➡**Note: The following procedure is for models that are not equipped with the VSC TRAC option (see Section 1).**

REMOVAL

➧ **Refer to illustrations 8.3 and 8.7**

⁂ CAUTION:

Brake fluid will damage paint. Cover all body parts and be careful not to spill fluid during the Steps in this procedure. Clean up any spilled brake fluid immediately and then wash the area with large amounts of water. This is especially true for any finished or painted surfaces.

1 Disconnect the cable from the negative terminal of the battery (see Chapter 5, Section 1).

2 Remove as much fluid as you can from the fluid reservoir with a syringe or an old turkey baster.

⁂ WARNING:

If a baster is used, never again use it for the preparation of food.

3 Disconnect the electrical connector for the fluid level sensor (see illustration).

4 On models equipped with a manual transmission, disconnect the fluid hose for the clutch master cylinder from the reservoir and plug it to minimize fluid loss or contamination.

5 Place rags under the fluid fittings and prepare caps or plastic bags to cover the ends of the lines once they are disconnected.

6 Use a flare-nut wrench to loosen the fittings at the ends of the brake lines where they enter the master cylinder (see illustration 8.3). Pull the brake lines slightly away from the master cylinder and plug the ends to prevent contamination.

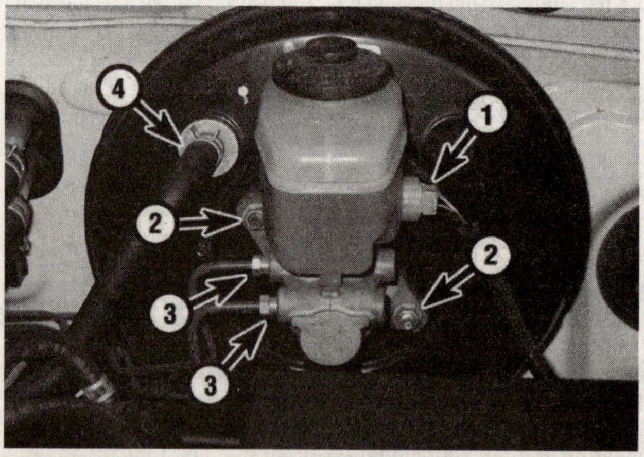

8.3 Master cylinder and other details:

1 *Fluid level sensor electrical connector*
2 *Mounting nuts*
3 *Brake line fittings (remove with a flare-nut wrench)*
4 *Vacuum hose at the power brake booster*

8.7 Replace the O-ring after removing the master cylinder

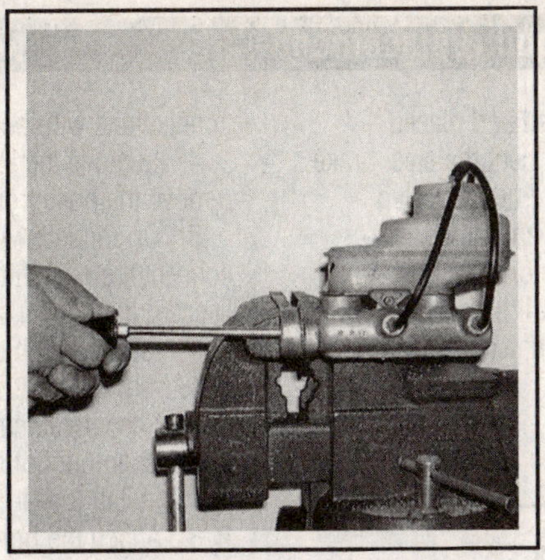

8.11 The best way to bleed air from the master cylinder before installing it on the vehicle is with a pair of bleeder tubes that direct brake fluid into the reservoir during bleeding

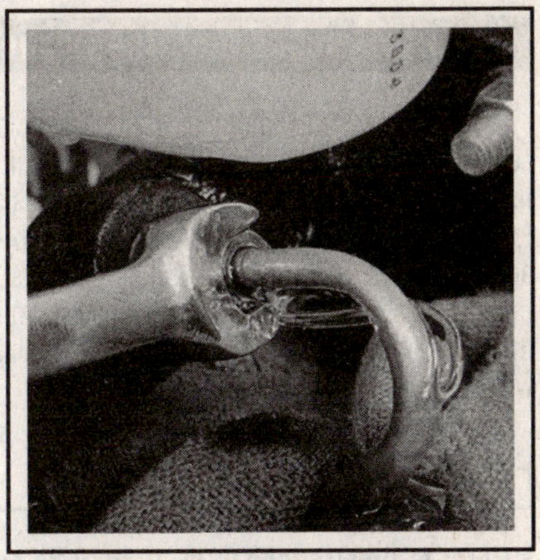

8.20 Have an assistant depress the brake pedal and hold it down. Secondly, loosen the fitting nut briefly and allow air and fluid to escape then close the fitting nut quickly. Repeat these steps on both fittings until the fluid is clear of air bubbles

7 Remove the master cylinder mounting nuts and pull the master cylinder off the studs and out of the engine compartment. Again, be careful not to spill the fluid as this is done. Discard the O-ring seal (see illustration).

8 If a new master cylinder is being installed, it may be necessary to remove the reservoir from the master cylinder and transfer it to the new master cylinder. Use a punch to remove the roll pin that secures the reservoir to the master cylinder and pull the reservoir off. Use a new grommet seal between the old reservoir and the new master cylinder during reassembly.

INSTALLATION

➧ Refer to illustrations 8.11 and 8.20

9 If a new master cylinder is being installed, measure the master cylinder to booster pushrod clearance (see Section 11).

10 Bench bleed the new master cylinder before installing it. Mount the master cylinder in a vise, with the jaws of the vise clamping on the mounting flange.

11 Attach a pair of master cylinder bleeder tubes to the outlet ports of the master cylinder (see illustration).

12 Fill the reservoir with brake fluid of the recommended type (see Chapter 1). On models with manual transmissions, plug or cap the fitting for the fluid hose that protrudes from the reservoir.

13 Slowly push the pistons into the master cylinder (a large Phillips screwdriver can be used for this) - air will be expelled from the pressure chambers and into the reservoir. Because the tubes are submerged in fluid, air can't be drawn back into the master cylinder when you release the pistons.

14 Repeat the procedure until no more air bubbles are present.

15 Remove the bleed tubes, one at a time, and install plugs in the open ports to prevent fluid leakage and air from entering. Install the reservoir cap.

16 Install a new O-ring seal onto the master cylinder (see illustration 8.7).

17 Make sure that the bore in the power brake booster is clean, then install the master cylinder over the studs on the power brake booster and tighten the mounting nuts finger tight.

18 Thread the brake line fittings into the master cylinder. Since the master cylinder is still a bit loose, it can be moved slightly in order for the fittings to thread in easily. Do not strip the threads as the fittings are tightened.

19 Fully tighten the mounting nuts, then the brake line fittings securely. Tighten the nuts to the torque listed in this Chapter's Specifications. On models with manual transmissions, connect the clutch master cylinder fluid hose to the reservoir.

20 Fill the master cylinder reservoir with fluid, then bleed the master cylinder and the brake system as described in Section 10. To bleed the cylinder on the vehicle, have an assistant depress the brake pedal and hold the pedal to the floor. Loosen the fitting to allow air and fluid to escape. Repeat this procedure on both fittings until the fluid is clear of air bubbles (see illustration).

⁂ CAUTION:

Have plenty of rags on hand to catch the fluid - brake fluid will ruin painted surfaces. After the bleeding procedure is completed, rinse the area under the master cylinder with clean water.

21 The remainder of installation is the reverse of removal. Test the operation of the brake system carefully before placing the vehicle into normal service.

⁂ WARNING:

Do not operate the vehicle if you are in doubt about the effectiveness of the brake system. If the pedal continues to feel spongy after repeated bleedings or the BRAKE or ANTI-LOCK (ABS) light stays on, have the vehicle towed to a dealer service department or other qualified shop to be bled.

9 Brake hoses and lines - check and replacement

1 About every six months, with the vehicle raised and placed securely on jackstands, the flexible hoses which connect the steel brake lines with the front and rear brake assemblies should be inspected for cracks, chafing of the outer cover, leaks, blisters and other damage. These are important and vulnerable parts of the brake system and inspection should be complete. A light and mirror will be needed for a thorough check. If a hose exhibits any of the above defects, replace it with a new one.

FLEXIBLE HOSES

Refer to illustrations 9.2a, 9.2b, 9.2c and 9.3

2 Clean all dirt away from the ends of the hose (see illustrations).
3 Disconnect the brake line from the hose fitting (see illustration). Be careful not to bend the frame bracket or line. If necessary, soak the connections with penetrating oil.
4 Remove the U-clip from the female fitting at the bracket and remove the hose from the bracket.
5 On models with the hose connected directly to the caliper, disconnect the hose fitting from the caliper by removing the banjo bolt. Discard the copper washers on both sides of the fitting. Use new copper washers and attach the new brake hose to the caliper. Also, detach the hose from the front shock absorber bracket.
6 Pass the female fitting through the frame or frame bracket. With the least amount of twist in the hose, install the fitting in this position.
7 Install the U-clip in the female fitting at the frame bracket.
8 Attach the brake line to the hose fitting using a back-up wrench on the fitting. Tighten the tube nut securely.
9 Carefully check to make sure the suspension or steering components don't make contact with the hose. Have an assistant push down on the vehicle and also turn the steering wheel lock-to-lock during inspection.

9.2a A front brake hose on 4WD and Pre-Runner models

9.2b A front brake hose on 2WD models

9.2c The location of the rear brake hoses above the rear axle housing

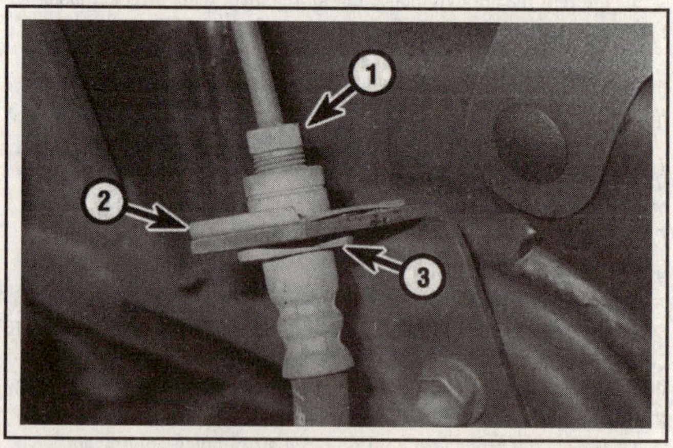

9.3 Brake hose fitting details:

1 *Metal tube nut (remove first - use a flare-nut wrench here)*
2 *Retaining clip (pull straight out with pliers to release the hose from the bracket)*
3 *Brake hose fitting (use a back-up wrench here while loosening the tube nut)*

10 Bleed the brake lines as described in Section 10.

METAL BRAKE LINES

11 When replacing brake lines, be sure to use the correct parts. Don't use copper tubing for any brake system components. Purchase steel brake lines from a dealer parts department or auto parts store.

12 Prefabricated brake lines, with the ends already flared and fittings installed, are available at auto parts stores and dealer service departments. If necessary, carefully bend the line to the proper shape. A tube bender is recommended for this.

⁂ CAUTION:

Do not crimp or damage the line.

13 When installing the new line, make sure it's well supported in the brackets and has plenty of clearance between moving or hot components.

14 After installation, check the brake fluid reservoir level and add fluid as necessary (see Chapter 1). Bleed the brake system as outlined in Section 10 and test the brakes carefully before placing the vehicle into normal operation.

10 Brake hydraulic system - bleeding

⁂ WARNING:

Wear eye protection when bleeding the brake system. If the fluid comes in contact with your eyes, immediately rinse them with water and seek medical attention.

⁂ CAUTION:

On models equipped with VCS TRAC (see Section 1), if air has found its way into the integrated master cylinder assembly, the vehicle will have to be towed to a dealership service department or qualified repair facility where the system can be bled with the use of a scan tool. Bleeding the brake system on these vehicles is limited to calipers and wheel cylinders, and brake lines and hoses that are near the front calipers or above the rear axle housing.

1 Remove the master cylinder fluid reservoir cap and fill the reservoir with brake fluid. Reinstall the cap.

2 Have an assistant on hand, as well as a supply of new brake fluid, an empty clear plastic container, a length of tubing to fit over the bleeder valve and a wrench to open and close the bleeder valve. Clear tubing is best so you can see if the fluid has bubbles in it as it leaves the bleeder valve.

VSC TRAC EQUIPPED MODELS

Front brake lines

➧ Refer to illustration 10.4

3 Loosen the bleeder valve (screw) on the front caliper slightly, then tighten it to a point where it's snug but can still be loosened quickly and easily.

➡Note: Either front wheel caliper can be bled first.

4 Place one end of the tubing over the bleeder valve fitting and submerge the other end in brake fluid in the container (see illustration).

5 Turn the ignition switch to the ON position and wait for the pump motor for the hydraulic brake booster to stop.

6 Have your assistant slowly pump the brake pedal several times, then hold it in a depressed position.

7 With the pedal depressed, open the bleeder valve just enough to allow the fluid to flow. Watch for air bubbles to exit the submerged end of the tube. When the fluid flow slows and stops after a couple of seconds, tighten the valve and have your assistant release the pedal.

8 Repeat Steps 6 and 7 until no more air is seen leaving the tube, then tighten the bleeder valve and proceed to the other wheel and perform the same procedure as necessary.

⁂ CAUTION:

Check the fluid level often during the bleeding operation and add fluid as necessary to prevent the fluid level from falling low enough to allow air into the master cylinder assembly.

Rear brake lines

⁂ CAUTION:

When the brake pedal is held down, the hydraulic brake booster will continually pressurize the rear brake circuit. If the reservoir runs out of fluid, air will get into the master cylinder assembly and require professional service to be bled.

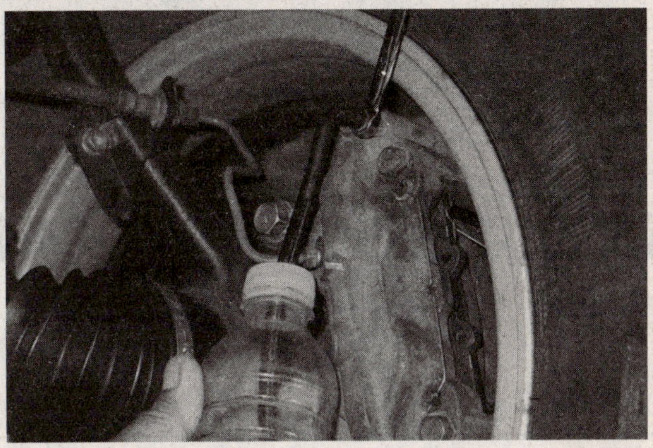

10.4 When bleeding the brakes, a hose is connected to the bleeder valve at the caliper or wheel cylinder and the other end is submerged in brake fluid. Air will be seen as bubbles in the tube and container. All air must be expelled before moving to the next wheel

9 Loosen the bleeder valve (screw) on the rear wheel cylinder slightly, then tighten it to a point where it's snug but can still be loosened quickly and easily.

➡**Note: Either rear wheel cylinder can be bled first.**

10 Place one end of the tubing over the bleeder valve fitting and submerge the other end in brake fluid in the container (see illustration 10.4).

11 Turn the ignition switch to the ON position, then depress brake pedal and hold it down.

12 With the pedal depressed, open the bleeder valve just enough to allow the fluid to flow. Watch for air bubbles to exit the submerged end of the tube. Allow the fluid to flow few a few seconds, then close the bleeder valve. The fluid will continually flow until the bleeder valve is closed and the brake pedal is released. Check the reservoir fluid level if more bleeding is necessary, but make sure that the bleeder valve is closed and the brake pedal is released.

13 When all of the air is bled from the line, tighten the bleeder valve and have your assistant release the pedal.

14 Proceed to the other wheel and perform the same procedure as necessary.

CAUTION:

Check the fluid level often during the bleeding operation and add fluid as necessary to prevent the fluid level from falling low enough to allow air into the master cylinder assembly.

NON-VSC TRAC EQUIPPED MODELS

15 Remove any residual vacuum from the brake power booster by applying the brake several times with the engine off.

16 Remove the master cylinder reservoir cover and fill the reservoir with brake fluid. Reinstall the cover.

➡**Note: Check the fluid level often during the bleeding operation and add fluid as necessary to prevent the fluid level from falling low enough to allow air bubbles into the master cylinder.**

17 Have an assistant on hand, as well as a supply of new brake fluid, a clear container partially filled with clean brake fluid, a length of tubing to fit over the bleeder valve and a wrench to open and close the bleeder valve.

18 Beginning at the right rear wheel, loosen the bleeder valve slightly, then tighten it to a point where it's snug but can still be loosened quickly and easily.

19 Place one end of the tubing over the bleeder valve and submerge the other end in brake fluid in the container (see illustration 10.4).

20 Have your assistant depress the brake pedal slowly, then hold the pedal down firmly.

21 While the pedal is held down, open the bleeder valve just enough to allow a flow of fluid to leave the valve. Watch for air bubbles to exit the submerged end of the tube. When the fluid flow slows after a couple of seconds, close the valve and have your assistant release the pedal.

22 Repeat Steps 20 and 21 until no more air is seen leaving the tube, then tighten the bleeder valve and proceed to the left front wheel, the left rear wheel and the right front wheel, in that order, and perform the same procedure. Be sure to check the fluid in the master cylinder reservoir frequently.

ALL MODELS

23 Never use old brake fluid. It can damage and disable your brake system. Place the old brake fluid in a suitable container and discard it properly (see Section 1).

24 On VSC TRAC equipped models, pump the brake pedal 40 times with the engine OFF to remove the boost pressure; the brake pedal travel will increase and the resistance will decrease as result.

25 Remove the brake reservoir cap and fill the reservoir to the MAX line.

26 Check the operation of the brakes. The pedal should feel solid when depressed, with no sponginess and the brakes should be responsive. If necessary, repeat the entire process.

WARNING:

Do not operate the vehicle if you are in doubt about the effectiveness of the brake system. If the pedal continues to feel spongy after repeated bleedings or the BRAKE or ANTI-LOCK (ABS) light stays on, have the vehicle towed to a dealer service department or other qualified shop to be bled.

11 Power brake booster - check, removal and installation

➡**Note: The following procedure is for non-VSC TRAC equipped models (see Section 1).**

OPERATING CHECK

1 Depress the brake pedal several times with the engine off and make sure there's no change in the pedal reserve distance.

2 Depress the pedal and start the engine. If the pedal goes down slightly, operation is normal.

AIRTIGHTNESS CHECK

3 Start the engine and turn it off after one or two minutes. Depress the brake pedal slowly several times. If the pedal depresses less each time, the booster is airtight.

4 Depress the brake pedal while the engine is running, then stop the engine with the pedal depressed. If there's no change in the pedal reserve travel after holding the pedal for 30 seconds, the booster is airtight.

11.9 Rod operating adapter (VSC TRAC) or pushrod (non-VSC TRAC) details:

1. *Clevis pin and retaining clip*
2. *Clevis*
3. *Rod operating adapter (or pushrod)*
4. *Clevis locknut (loosen only for adjustment)*
5. *Brake pedal lever*

REMOVAL

➧ Refer to illustrations 11.9 and 11.10

5 Power brake booster units shouldn't be disassembled. They require special tools not normally found in most automotive repair stations or shops. Because of its critical relationship to brake performance, the booster should be replaced with a new or rebuilt one.

6 Disconnect the vacuum hose connected to the booster. Be careful not to damage the vacuum check valve or the hose when disconnecting it (see illustration 8.3).

7 Remove the brake master cylinder (see Section 8).

8 Remove the left side lower finish panel (knee bolster) from the instrument panel (see Chapter 11).

9 Detach the pushrod from the brake pedal lever by detaching the clevis (see illustration). Remove the clevis pin retaining clip with pliers and pull out the pin.

10 Remove the four nuts holding the brake booster to the firewall (see illustration); you may need a light to see them. Slide the booster straight out from the firewall until the studs clear the holes.

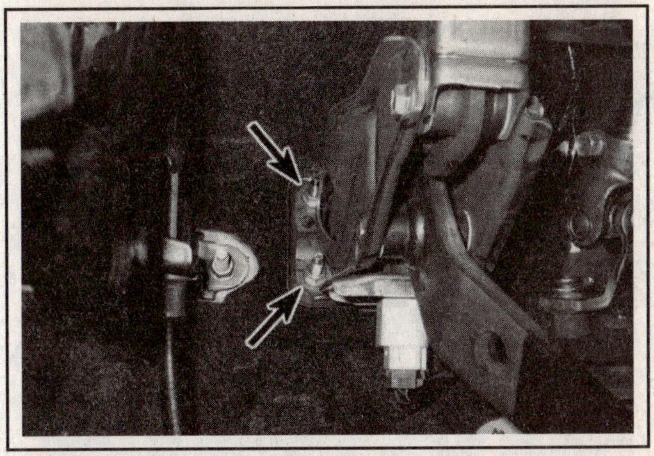

11.10 The location of the brake booster mounting nuts (two hidden from view in this photo - the passenger compartment fuse and relay box is removed for clarity)

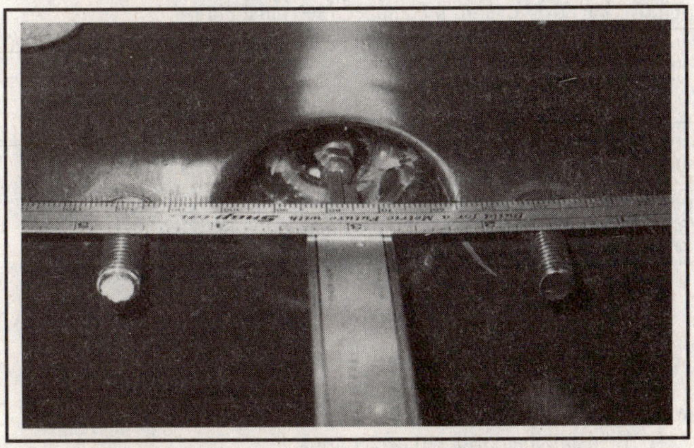

11.12a Measure the distance from the master cylinder mounting surface to the top of the booster pushrod (dimension A)

INSTALLATION

➧ Refer to illustrations 11.12a, 11.12b, 11.12c and 11.12d

11 Installation procedures are basically the reverse of removal. Install a new gasket and tighten the clevis locknut securely (if loosened or removed). Tighten the booster mounting nuts to the torque listed in this Chapter's Specifications.

➡Note: Apply some all-purpose grease to the clevis pin and to the end of the pushrod that meets the master cylinder piston during installation.

12 If a new power brake booster unit is being installed, check the pushrod clearance using the following Steps (see illustration).

➡Note: If a new genuine Toyota power brake booster is being installed, it is likely that the pushrod is pre-adjusted to achieve the correct clearance.

a) *Measure the distance from the face of the booster to the tip of the pushrod. Write down this measurement (see illustration). This is dimension A.*
b) *Measure the distance from the mounting flange to the end of the master cylinder (see illustration). Write down this measurement. This is dimension B.*
c) *Measure the distance from the piston pocket (inside the master cylinder) to the end of the master cylinder (see illustration). Write down this measurement. This is dimension C.*
d) *Subtract measurement C from measurement B, then subtract measurement A from the difference between C and B. This the pushrod clearance.*
e) *Compare your calculated pushrod clearance to the pushrod clearance listed in this Chapter's Specifications. If necessary, adjust the pushrod length to achieve the correct clearance (see illustration).*

13 After the final installation of the master cylinder and brake lines, the system must be bled (see Section 10) and the brake pedal height and freeplay must be checked (see Section 13).

14 Carefully test the operation of the brakes before placing the vehicle in normal operation.

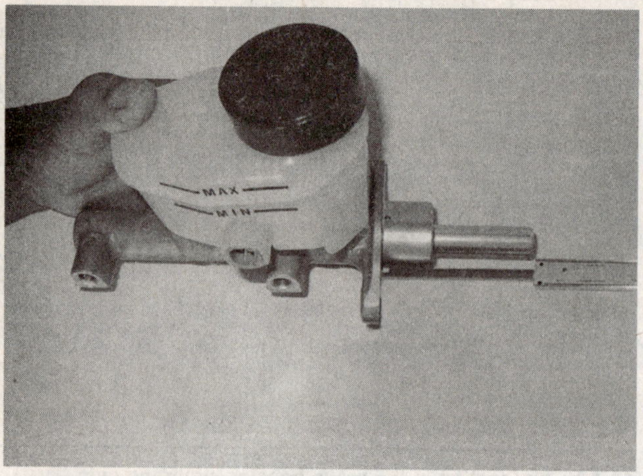

11.12b Measure the distance from the mounting flange to the end of the master cylinder (dimension B)

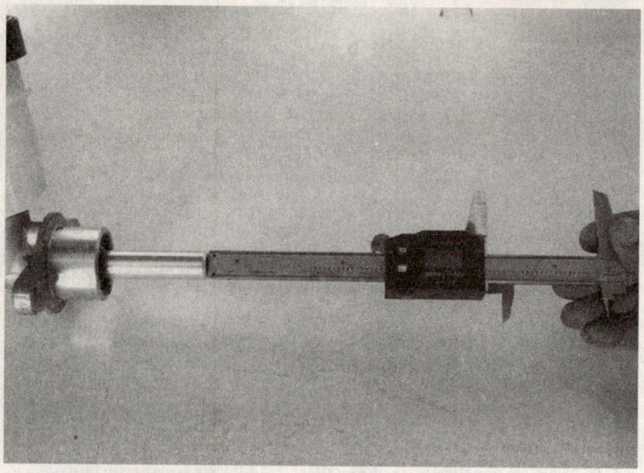

11.12c Measure the distance from the piston pocket to the end of the master cylinder (dimension C)

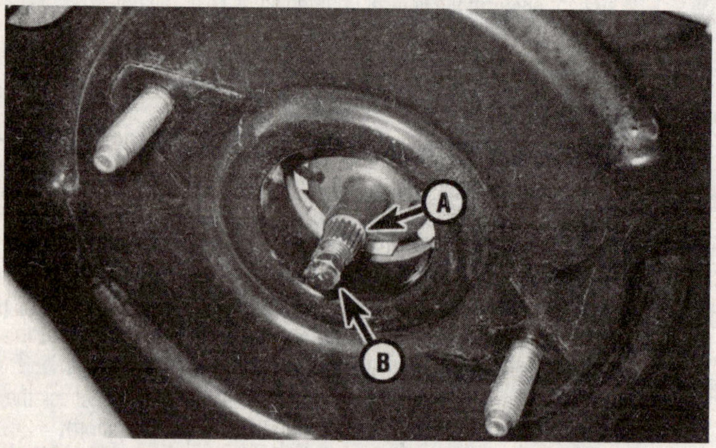

11.12d To adjust the length of the booster pushrod, have an assistant press the brake pedal down and hold it. This moves the front booster pushrod out of the booster for access. Hold the stationary portion of the pushrod (A) with pliers and turn the adjusting screw (B) in or out. Note: *Remember to release the brake pedal fully while checking the clearance after each adjustment and don't apply any grease to the pushrod until after the clearance has been achieved*

12 Parking brake - adjustment

Refer to illustrations 12.4 and 12.5

1 The parking brake is hand-operated by a handle under the dash or by a pedal under the left end of the instrument panel. All parking brake systems are self-adjusting; automatic adjusters compensate for brake shoe wear. However, additional adjustment may be needed in the event of cable stretch, which can occur as the vehicle ages. The parking brake pedal or lever, when properly adjusted, should travel a specified number clicks when set. If it travels less than specified, there's a chance the parking brake might not be releasing completely and the parking brake shoes might be dragging on the drum. If the pedal or lever travels more than specified, the parking brake may not hold the vehicle adequately on an incline; allowing the car to roll.

2 Make sure that the rear brake shoes are in good condition (see Chapter 1) and properly adjusted (see Section 6).

3 Set the parking brake and count the number of clicks that it travels. Compare the number of clicks to the value in this Chapter's Specifications. If adjustment is needed, remove the center console (see Chapter 11).

4 Pull the carpet up from above the cable adjustment assembly and remove the cover over it, if equipped (see illustration).

5 Locate the adjusting nut at the end of the parking brake cable (see illustration).

6 With the parking brake released, loosen the locknut by holding the adjusting nut. Tighten or loosen the adjusting nut to achieve the proper number of clicks when the parking brake is set. Hold the flat portion of the threaded stud at the end of the cable if necessary. Tightening the nut (towards the cable) decreases the number of clicks, while the opposite is achieved by loosening the nut (turning it toward the threaded stud at the cable's end).

7 Confirm that the parking brake is fully engaged within the specified number of clicks, then tighten the locknut.

8 Release the parking brake and confirm that the brakes don't drag when the rear wheels are turned.

9 Reinstall the cover and center console.

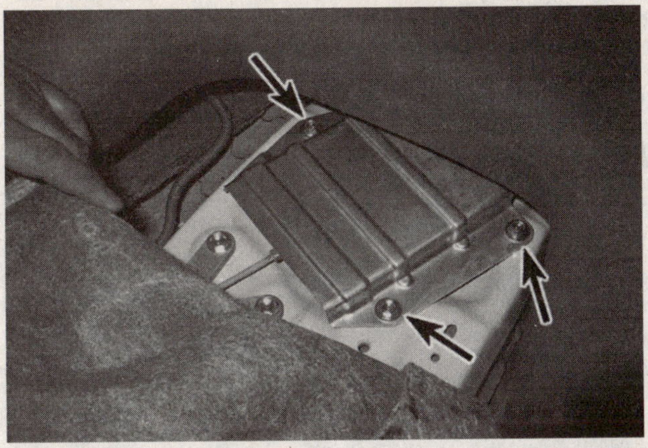

12.4 Remove the cover for access to the parking brake cable adjuster, if equipped

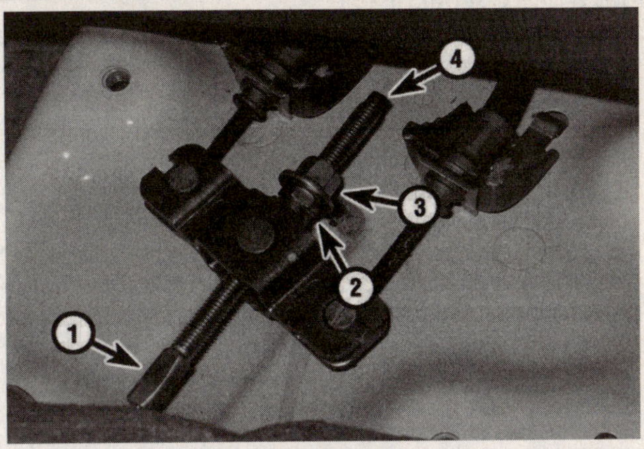

12.5 Parking brake cable adjuster details:

1. *Flat area used to hold the threaded stud (hold while turning adjusting nut)*
2. *Cable adjusting nut (hold while loosening the locknut, then turn to adjust)*
3. *Adjuster locknut (tighten after adjustment)*
4. *Threaded stud on the end of the cable*

13 Brake pedal - adjustment

BRAKE PEDAL HEIGHT

➧ Refer to illustrations 13.2 and 13.5

1 Remove and install the brake light switch (see Section 14).

2 With the brake pedal fully released, measure the distance from the top of the pad to the floor (see illustration).

3 Compare the height measurement to the Specifications listed in this Chapter. Continue to the next Step for the pedal height adjustment.

4 Loosen the clevis locknut on the rod operating adapter (VSC TRAC) or pushrod (non-VSC TRAC), then separate the clevis from the brake pedal lever (see illustration 11.9).

5 Rotate the clevis to adjust the rod's length to the value in this Chapter's Specifications (see illustration), then reattach the clevis to the brake pedal and confirm that the pedal height is correct. Repeat as necessary.

6 With the pedal height adjusted, lubricate and reinstall the clevis pin and install the locking clip, tighten the locknut securely, then check the pedal freeplay.

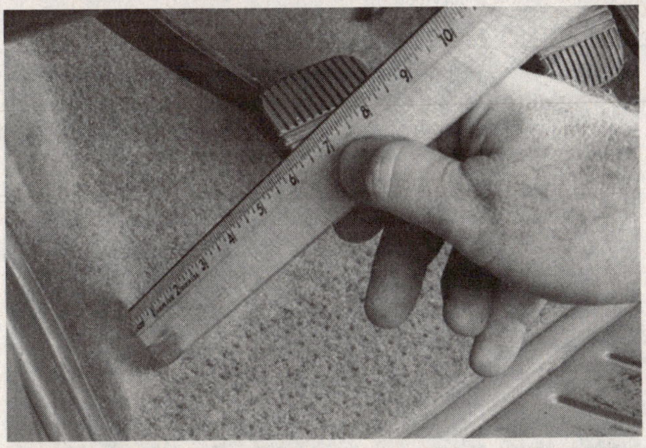

13.2 Measure the brake pedal height and compare your measurement to the pedal height listed in this Chapter's Specifications

13.5 Measure the length of the rod operating adapter (VSC TRAC) or pushrod (non-VSC TRAC) from the center of the hole in the clevis to the mounting face of the brake pedal bracket. Compare your measurement to the values listed in this Chapter's Specifications

BRAKE PEDAL FREEPLAY

➧ **Refer to illustration 13.8**

7 On non-VSC TRAC equipped models (see Section 1), press the brake pedal several times with the engine OFF.

8 Press down lightly on the brake pedal and measure the distance that it moves freely before resistance is felt (see illustration). The freeplay should be within the values listed in this Chapter's Specifications. If it isn't, check the clevis, clevis pin and the hole in the brake pedal arm for excessive wear.

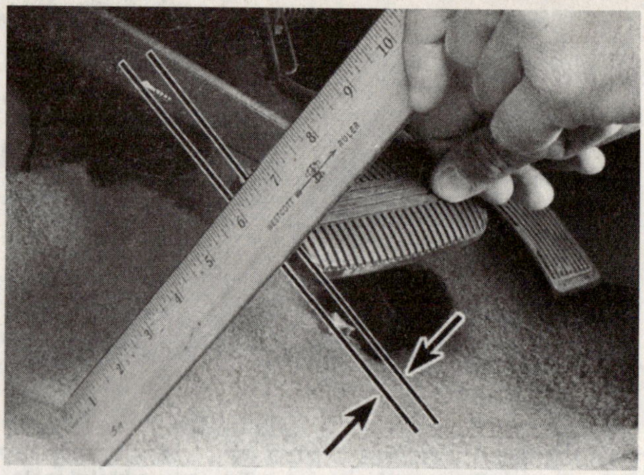

13.8 Push the brake pedal down with your hand to the point at which you feel initial resistance, measure the distance between the released pedal and this point and compare your measurement to the values listed in this Chapter's Specifications

14 Brake light switch - check and replacement

CHECK

➧ **Refer to illustration 14.1**

1 The brake light switch is located on the brake pedal bracket (see illustration). You'll need to remove the trim panel beneath the steering column to get to the switch and connector (see Chapter 11).

2 With the brake pedal in the fully released position, the switch opens the brake light circuit. When the brake pedal is depressed, the switch closes the circuit and sends current to the brake lights.

3 If the brake lights are inoperative, check the fuse and the bulbs (see Chapter 12).

4 If the fuse and bulbs are okay, verify that voltage is available at the switch.

5 If there's no voltage to the switch, search for an open circuit condition between the fuse block and the switch. If there is voltage to the switch, close the switch (depress the brake pedal) and verify that there's voltage on the other side of the switch.

6 If there's no voltage on the other side of the switch, replace the switch. If there is voltage but the brake lights still don't work, look for an open circuit condition between the switch and the brake lights.

➡ **Note: There is always the remote possibility that all of the brake light bulbs are burned out, but this is not very likely.**

REPLACEMENT

7 Disconnect the electrical connector from the brake light switch (see illustration 14.1).

8 Rotate the switch counterclockwise 90-degrees, so that it unlocks from its holder, then pull it directly out of the holder.

9 To install the switch, insert it into its holder and push it in until the switch body contacts the rubber stop on the brake pedal bracket. Hold the brake pedal up while placing the switch in position. Rotate the switch 90-degrees clockwise to lock it into place. It will achieve the proper clearance (0.020 to 0.098 inch) to the rubber stop automatically.

⁂ CAUTION:

Do not press in on the switch while turning it.

10 Plug the electrical connector into the switch.

11 Confirm that the brake lights work properly before placing the vehicle into normal service.

14.1 The electrical connector (A) for the brake light switch. Turn the switch (B) 90-degrees counterclockwise to release it, then pull it straight out to remove it

Specifications

General

Brake fluid type	See Chapter 1
Brake booster pushrod-to-master cylinder piston clearance	0.0 inch (0.0 mm)
Brake pedal	
Height	6.22 to 6.62 inches (158.1 to 168.1 mm)
Freeplay	0.04 to 0.24 inch (1 to 6 mm)
Parking brake travel (pedal or lever)	7 to 10 clicks
Booster rod operating adapter (pushrod)	
With VSC TRAC	8.43 to 8.82 inches (214 to 224 mm)
Without VSC TRAC	8.35 to 8.74 inches (212 to 222 mm)

Disc brakes

Brake pad minimum lining thickness	See Chapter 1
Disc lateral runout limit	0.0020 inch (0.05 mm)
Disc minimum (discard) thickness	Cast into disc

Drum brakes

Brake shoe minimum lining thickness	See Chapter 1
Maximum drum diameter	Cast into drum

Torque specifications

➡Note: One foot-pound (ft-lb) of torque is equivalent to 12 inch-pounds (in-lbs) of torque. Torque values below approximately 15 ft-lbs are expressed in inch-pounds, since most foot-pound torque wrenches are not accurate at these smaller values.

Torque specifications	Ft-lbs (unless otherwise indicated)	Nm
Brake hose-to-caliper banjo bolt	22	30
Caliper mounting bolts		
Fixed type	91	123
Floating type	27	37
Caliper mounting bracket bolts (floating type)	80	108
Master cylinder mounting nuts	108 in-lbs	12
Power brake booster mounting nuts	120 in-lbs	13.5
Wheel cylinder mounting bolts	84 in-lbs	9.5
Wheel lug nuts	See Chapter 1	
Wheel speed sensor mounting bolt/nut (front and rear)	71 in-lbs	8

Notes

10

SUSPENSION AND STEERING SYSTEMS

Section

Reference to other Chapters

Power steering fluid level check - See Chapter 1
Tire and tire pressure checks - See Chapter 1

1 General information and precautions

FRONT SUSPENSION

Refer to illustrations 1.2a, 1.2b, 1.2c and 1.2d

The suspension components for the 4WD and PreRunner models are similar to those used for the 2WD models, but are designed for off-road use and also increase the vehicle's ride height.

The front suspension system is fully independent. The steering knuckles are connected to upper and lower control arms by balljoints. The control arms are bolted to the frame. The shock absorbers and coil springs are integral assemblies; the upper ends are bolted to brackets on the frame and the lower ends are bolted to the lower control arms. All models use a front stabilizer bar to reduce vehicle roll during cornering (see illustrations).

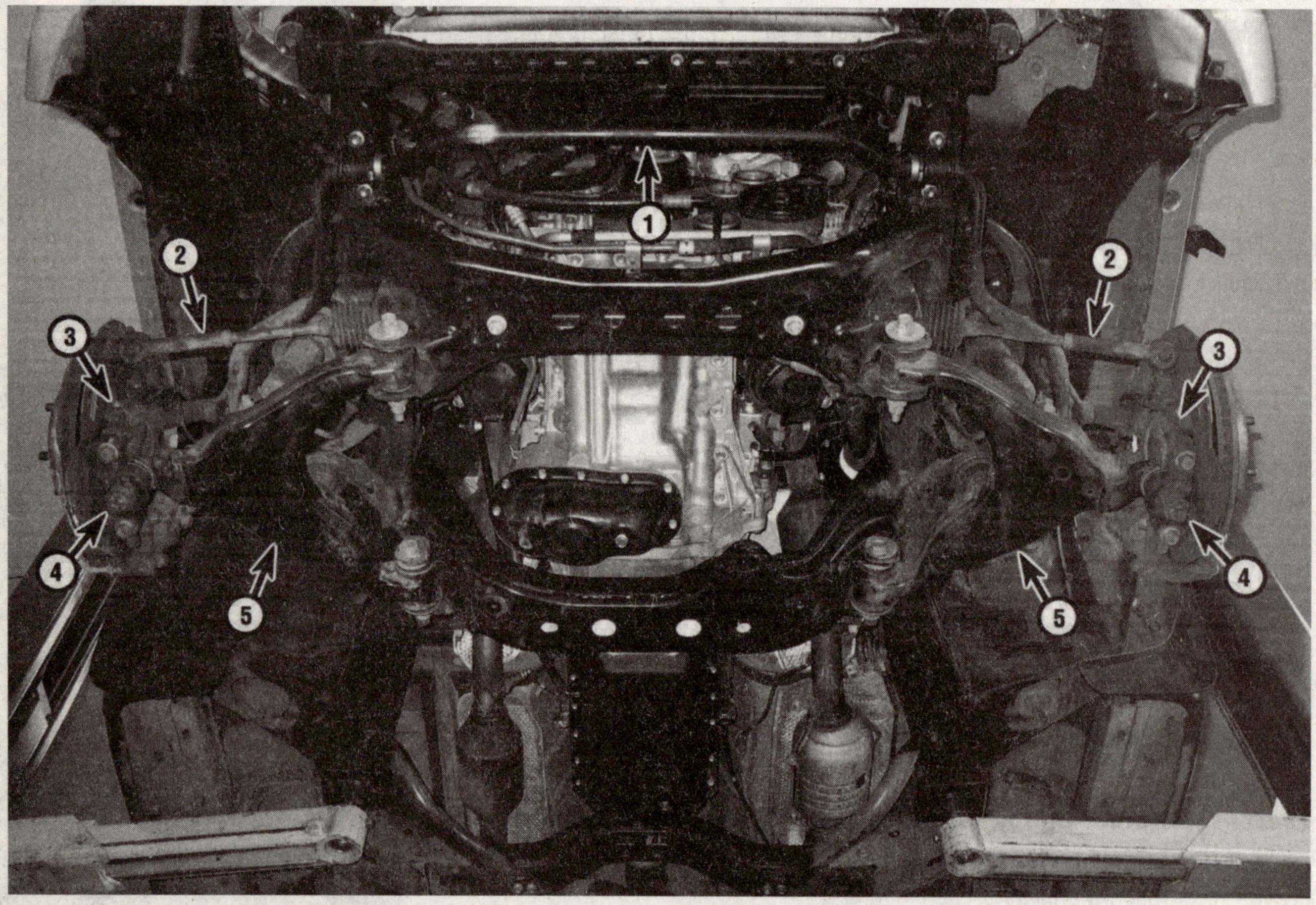

1.2a Front suspension and steering components - bottom view (typical shown for 4WD and PreRunner models):

1 Stabilizer bar
2 Tie-rod end
3 Steering knuckle
4 Lower balljoint and bracket
5 Lower control arm

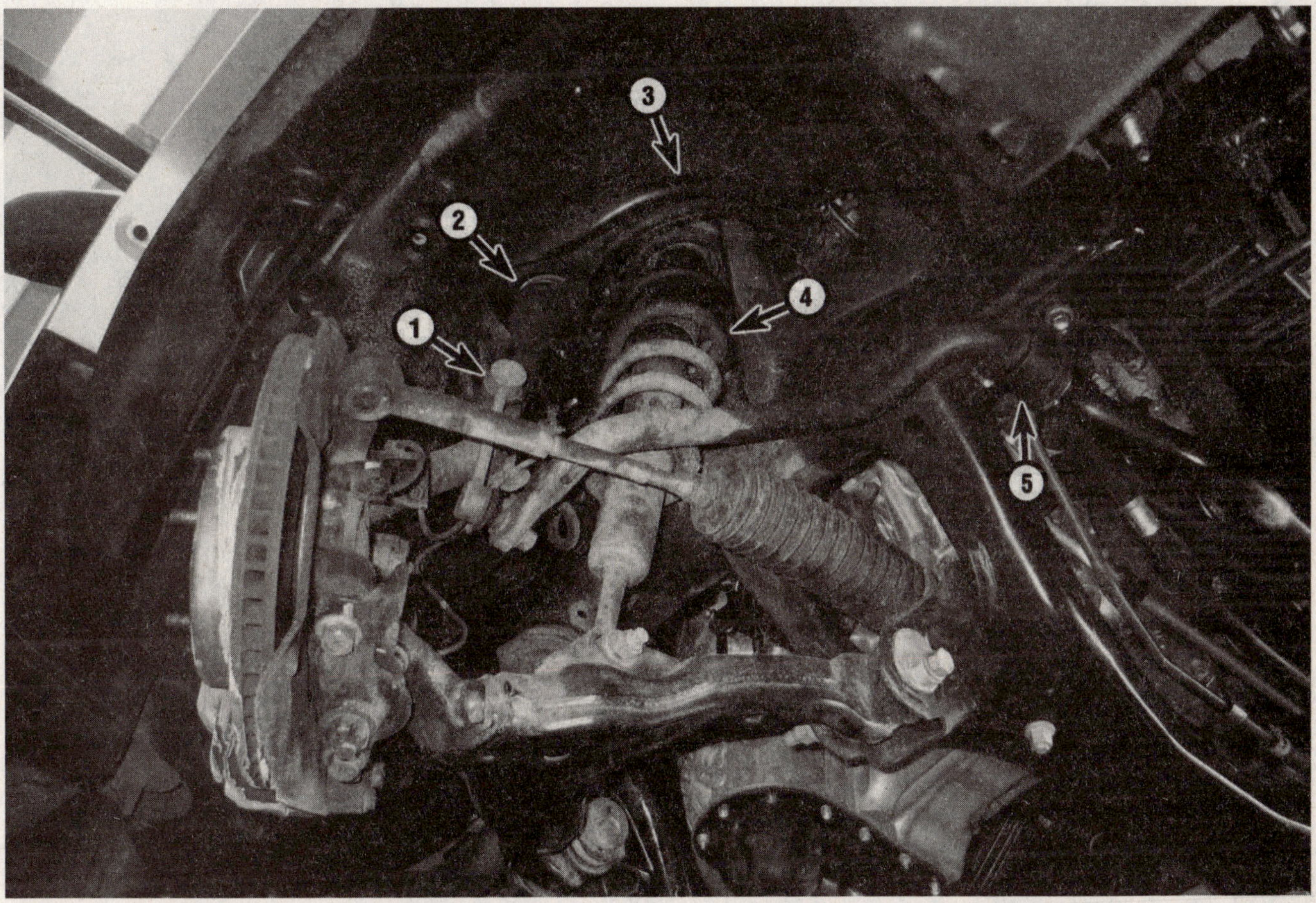

1.2b Front suspension and steering components - side view (typical shown for 4WD and PreRunner models):

1 *Stabilizer bar link*
2 *Upper balljoint*
3 *Upper control arm*
4 *Shock absorber/coil spring assembly*
5 *Stabilizer bar bushing and bracket*

1.2c Front suspension and steering components - bottom view (2WD models):

1. *Stabilizer bar*
2. *Steering gear*
3. *Tie-rod end*
4. *Lower balljoint and bracket*
5. *Lower control arm*

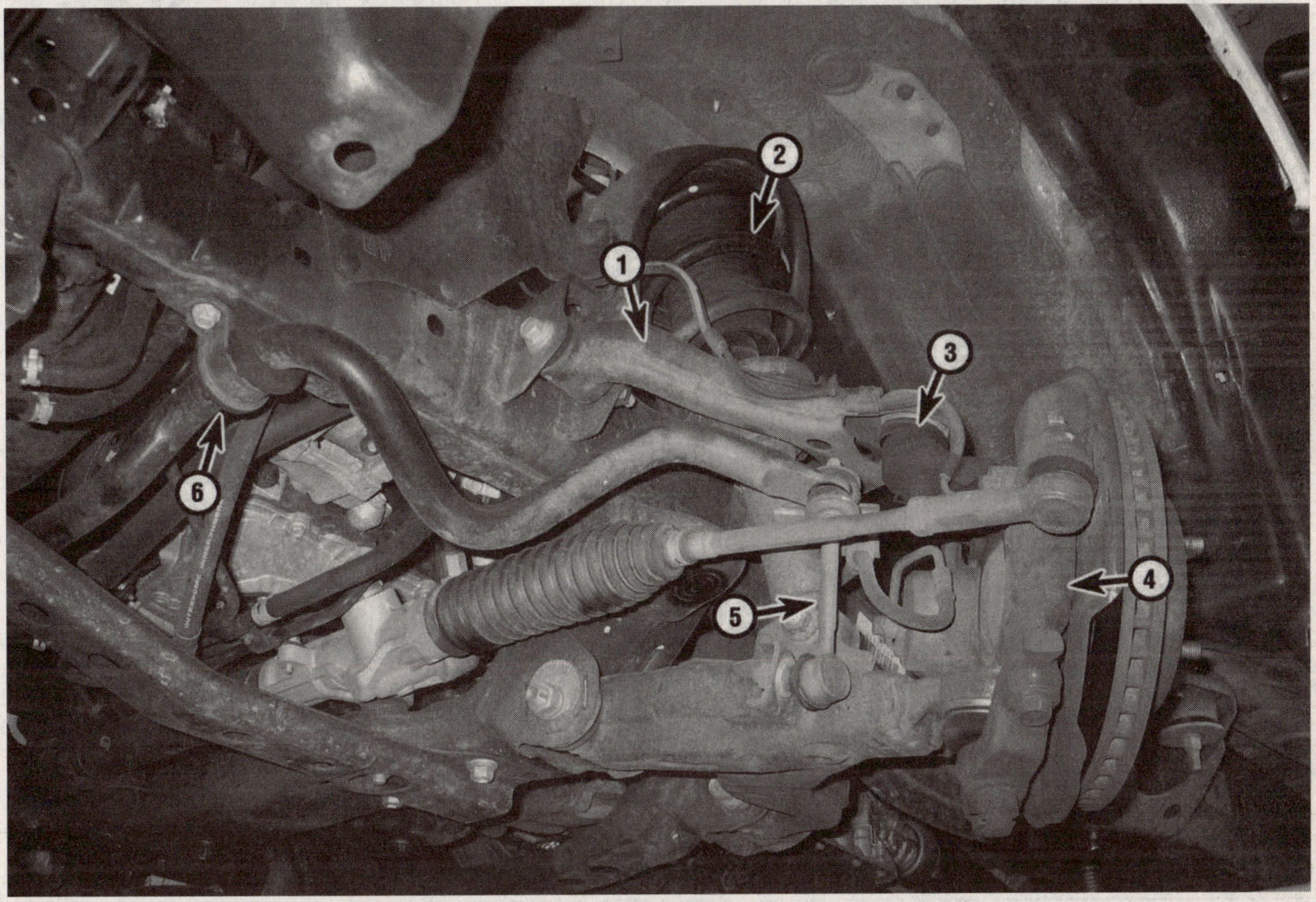

1.2d Front suspension and steering components - side view (2WD models):

1 *Upper control arm*
2 *Shock absorber/coil spring assembly*
3 *Upper balljoint*
4 *Steering knuckle*
5 *Stabilizer bar link*
6 *Stabilizer bar bushing and bracket*

REAR SUSPENSION

➧ **Refer to illustration 1.3**

The rear suspension consists of a pair of multi-leaf springs and two shock absorbers. The rear axle assembly is attached to the leaf springs by U-bolts. The front ends of the springs are attached to the frame at the front hangers, through rubber bushings. The rear ends of the springs are attached to the frame by shackles which allow the springs to alter their length when the vehicle is in operation. Some models use a rear stabilizer bar to reduce vehicle roll during cornering (see illustration).

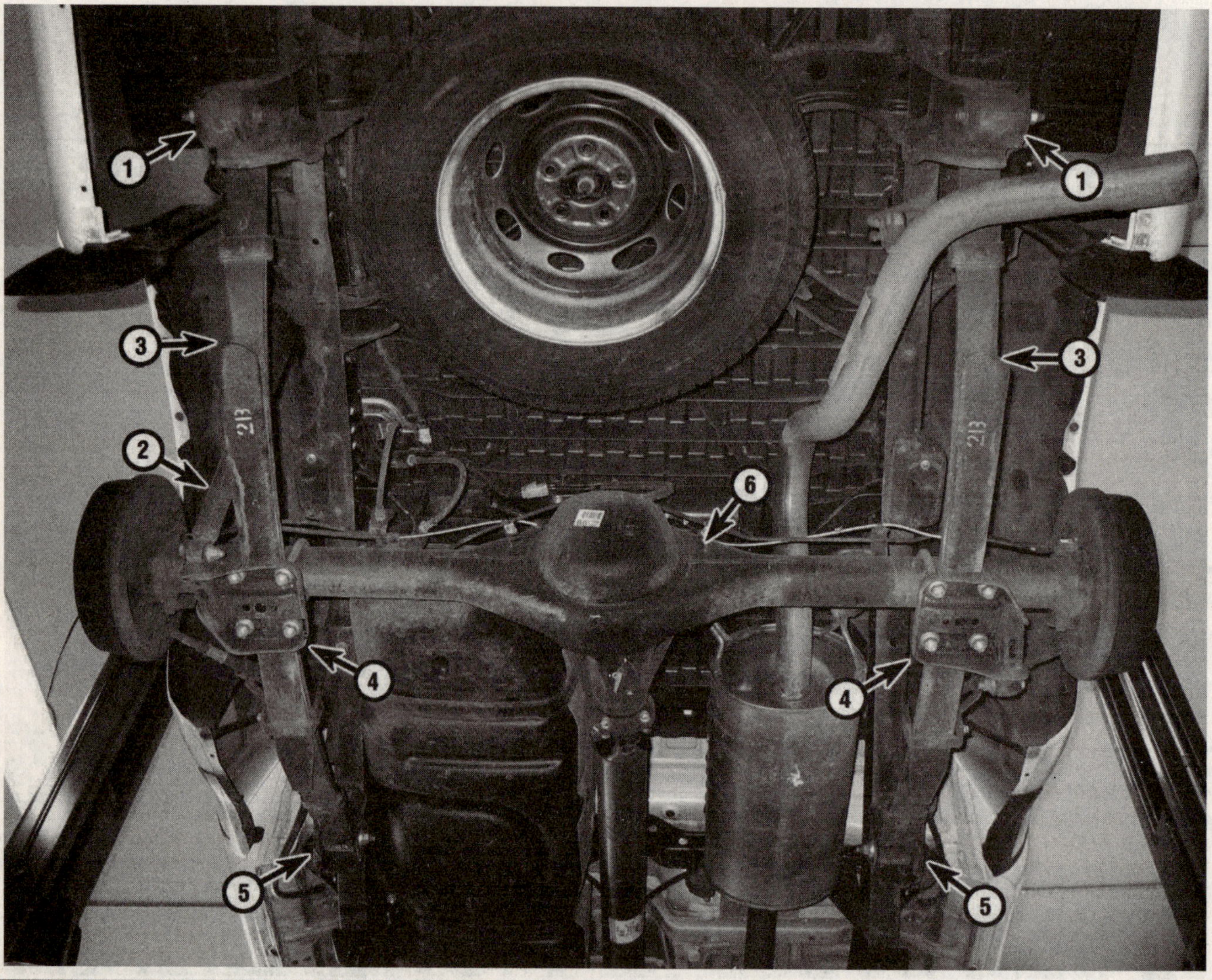

1.3 Rear suspension components:

1 *Leaf spring shackle*
2 *Shock absorber (right side hidden from view in this photo)*
3 *Leaf spring*
4 *Spring seat*
5 *Spring hanger*
6 *Rear axle housing*

STEERING

All models are equipped with power-assisted rack-and-pinion steering systems. The steering gear is bolted to the back of the crossmember and is connected to the steering knuckles by a pair of tie-rods.

PRECAUTIONS

Frequently, when working on the suspension or steering system components, you may come across fasteners which seem impossible to loosen. These fasteners on the underside of the vehicle are continually subjected to water, road grime, mud, etc., and can become rusted or frozen, making them extremely difficult to remove. In order to unscrew these stubborn fasteners without damaging them (or other components), be sure to use lots of penetrating oil and allow it to soak in for a while. Using a wire brush to clean exposed threads will also ease removal of the nut or bolt and prevent damage to the threads. Sometimes a sharp blow with a hammer and punch is effective in breaking the bond between a nut and bolt threads, but care must be taken to prevent the punch from slipping off the fastener and ruining the threads. Heating the stuck fastener and surrounding area with a torch sometimes helps too, but isn't recommended because of the obvious dangers associated with fire. Long breaker bars and extension, or cheater, pipes will increase leverage, but never use an extension pipe on a ratchet - the ratcheting mechanism could be damaged. Sometimes, turning the nut or bolt in the tightening (clockwise) direction first will help to break it loose. Fasteners that require drastic measures to unscrew should always be replaced with new ones.

Since most of the procedures that are dealt with in this Chapter involve jacking up the vehicle and working underneath it, a good pair of jackstands will be needed. A hydraulic floor jack is the preferred type of jack to lift the vehicle, and it can also be used to support certain components during various operations.

⁂ WARNING:

Never, under any circumstances, rely on a jack to support the vehicle while working on it. Also, whenever any of the suspension or steering fasteners are loosened or removed they must be inspected and, if necessary, replaced with new ones of the same part number or of original equipment quality and design. Torque specifications must be followed for proper reassembly and component retention. Never attempt to heat or straighten suspension or steering components. Instead, replace bent or damaged parts with new ones.

2 Shock absorber/coil spring (front) - removal, component replacement and installation

⁂ WARNING:

Always replace shock absorbers or shock absorber/coil spring assemblies in pairs - never replace just one of them.

➡Note: If the shocks or coil springs exhibit the telltale signs of wear (leaking fluid, loss of damping capability, chipped, sagging or cracked coil springs) explore all options before beginning any work. The shock absorbers or coil springs are not serviceable individually and must be replaced if a problem develops. However, complete assemblies may be available on an exchange basis, which eliminates much time and work. Whichever route you choose to take, check on the cost and availability of parts before disassembling your vehicle.

REMOVAL

1 Loosen the front wheel lug nuts. Raise the vehicle and support it securely on jackstands. Remove the front wheels.

4WD and PreRunner models

2 Remove the lower engine splash shield (front).
3 Remove the stabilizer bar and links (see Section 3).

➡Note: Separate the stabilizer bar links from the steering knuckle for removal.

4 Separate the tie-rod end from the steering knuckle (see Section 14).

2WD models

➧ Refer to illustration 2.5

5 Detach the brake hose from the shock absorber (see illustration).
6 Remove the upper control arm (see Section 4).

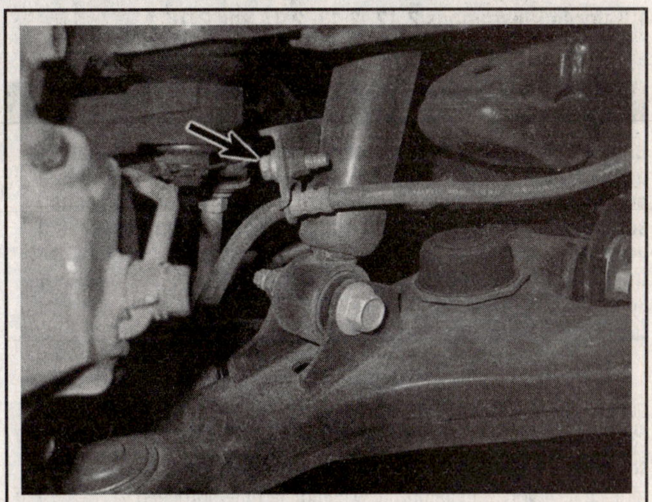

2.5 Detach the brake hose from the bracket on the shock absorber

2.7 Shock absorber/coil spring assembly lower fastener location on the lower control arm

2.8 To detach the upper end of the shock absorber/coil spring assembly from its mounting bracket, remove the mounting nuts (one mounting nut not visible in this photo) (NOT the nut in the middle, which is the damper rod nut; it must never be removed unless the spring is compressed with a spring compressor)

All models

➧ Refer to illustrations 2.7 and 2.8

7 Remove the shock absorber assembly's lower mounting fasteners to the lower control arm (see illustration). On 4WD and PreRunner models, push the bolt through the shock absorber and bracket as far as you can (but be careful not to nick the driveaxle), then pry the lower control arm down and remove the bolt.

8 While supporting the shock absorber assembly, remove the upper mounting nuts (see illustration).

9 Remove the shock absorber assembly.

➡Note: On 4WD and PreRunner models, it may be necessary to pry down on the lower control arm for clearance to remove the assembly.

COMPONENT REPLACEMENT

➧ Refer to illustrations 2.12, 2.13, 2.14 and 2.18

⁂ WARNING:

Before undertaking the following procedure, be aware that disassembling the shock absorber/coil spring assemblies is a potentially dangerous job. Careless or unsafe work can cause serious injury. Use only a high-quality spring compressor and be sure to follow the spring compressor manufacturer's instructions. After removing the compressed spring, set it aside in a safe, isolated place.

10 Inspect the shock absorber for leaking fluid, dents, cracks and other damage. Inspect the coil spring for chips and cracks which could cause premature failure. Inspect the spring seats for hardness and general deterioration. If either the shock or the spring is worn or damaged, replace it. If you're installing new complete units, proceed to Step 19; if you're going to install new shocks or coil springs, proceed to the next Step.

11 Secure the shock/coil assembly in a bench vise.

⁂ CAUTION:

Avoid placing the body of the shock in the jaws of the vice because this can easily damage the shock.

12 Install a spring compressor(s) in accordance with the tool manufacturer's instructions (see illustration).

➡Note: You can buy a spring compressor at most auto parts stores or rent one from most equipment yards on a daily basis.

Compress the spring far enough to relieve all pressure from the spring seat; when you can wiggle the spring, it's compressed enough to disassemble the shock/spring assembly.

⁂ WARNING:

Don't compress the spring any more than necessary.

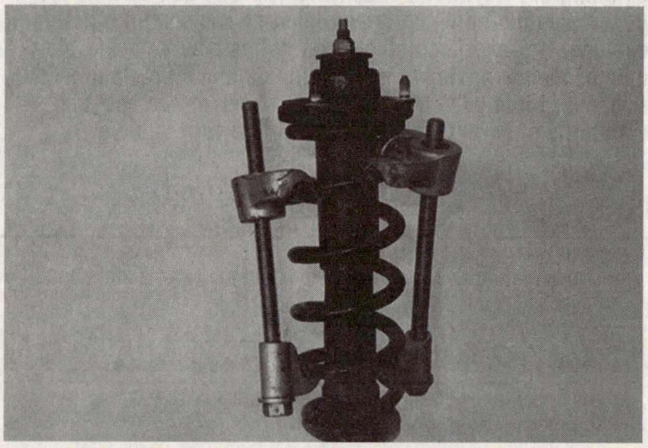

2.12 Install the spring compressor(s) in accordance with the manufacturer's instructions; compress the spring until you can wiggle it before removing the damper rod nut

2.13 Hold the damper rod (A) while removing the damper rod nut (B)

13 Hold the damper rod with a wrench, remove the damper rod nut and discard it (see illustration).

14 Remove the components that are above the spring while noting the order that they are installed (see illustration).

15 Remove the compressed spring assembly and set it safely aside.

16 Inspect the rubber parts for cracks and tears and general deterioration; replacing any that are damaged.

17 Inspect the washers and the suspension support (the top part with the three mounting studs) for damage and distortion; replace any parts that are damaged.

18 Reassembly is the reverse of disassembly. Make sure the lower end of the coil spring is correctly seated in the low spot in the lower spring seat (see illustration). Tighten a new damper rod nut to the torque listed in this Chapter's Specifications before releasing tension on the spring.

INSTALLATION

19 Installation is the reverse of removal. Be sure to tighten the upper and lower fasteners to the torque values listed in this Chapter's Specifications. Tighten the brake hose bracket mounting bolt securely.

20 Tighten the lug nuts to the torque listed in the Chapter 1 Specifications.

2.14 Components of the shock absorber/coil spring assembly (4WD and PreRunner model shown - 2WD models similar)

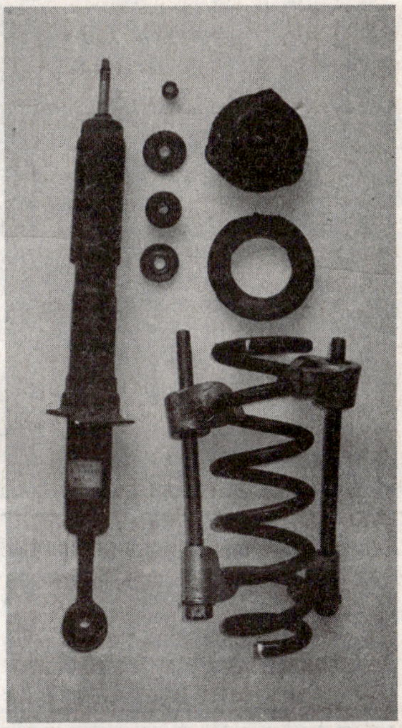

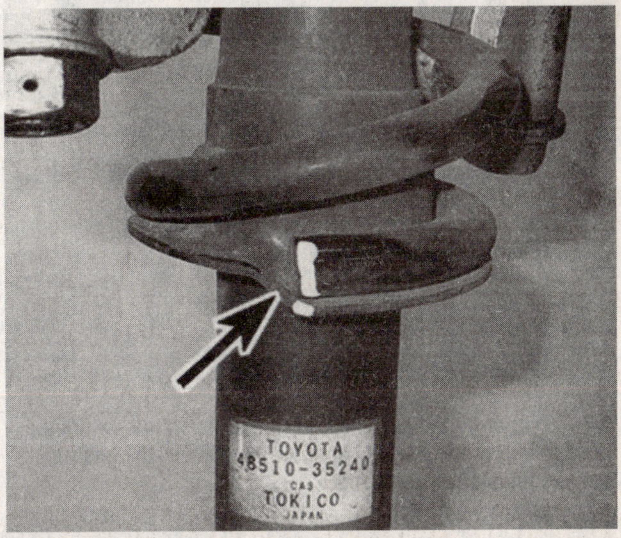

2.18 The coil spring seated correctly

3 Stabilizer bar, bushings and links (front) - removal and installation

➧ Refer to illustrations 3.2a, 3.2b and 3.3

1 Raise the front of the vehicle and support it securely on jackstands.

2 Remove the nuts from the stabilizer bar links and detach the links from the bar. Also, remove the nuts attaching the links to the steering knuckles (4WD or PreRunner models) or the lower control arms (2WD models), then remove the links (see illustrations). On 2WD models remove the rubber bushings from the top of the link.

➡Note: Use an Allen wrench to prevent the ballstud from turning when removing the link nuts, if necessary.

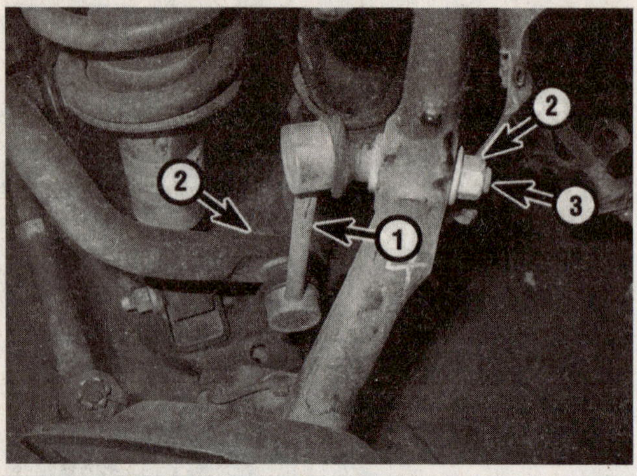

3.2a Stabilizer bar link details (4WD and PreRunner models)

1 Link
2 Link mounting nuts
3 Upper ballstud

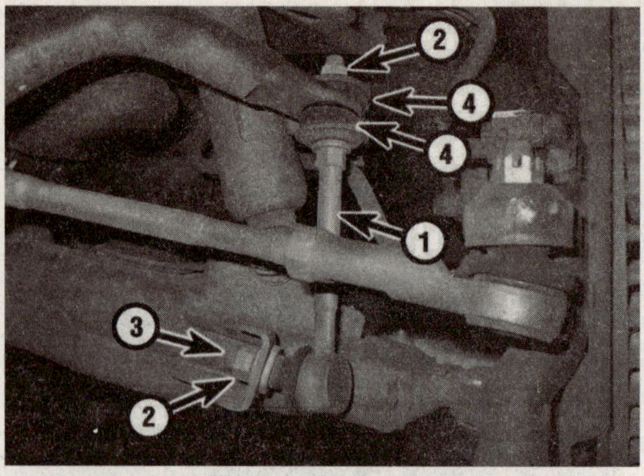

3.2b Stabilizer bar link details (2WD models)

1 Link
2 Link mounting nuts
3 Lower ballstud
4 Bushings

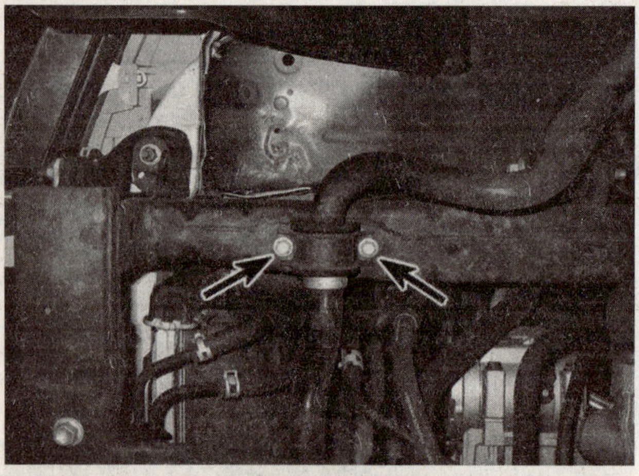

3.3 To separate the stabilizer bar from the frame, remove the bushing bracket bolts. Note the position of the bushing against the sleeve on the bar

3 Remove the stabilizer bar bushing bracket bolts and brackets (see illustration).

4 Remove the stabilizer bar.

5 Remove the rubber bushings from the stabilizer bar, noting the position of the slit openings.

6 Inspect the rubber bushings for cracks, tears and deterioration. If they're worn or damaged, replace them.

7 Check the balljoints on the ends of each link for looseness or other signs of excessive wear. You can check the rotational torque required to turn the balljoint by threading the nut onto the ballstud and turning it with an inch-pound torque wrench and comparing your reading to the value listed in this Chapter's Specifications.

8 Install the rubber bushings against the bushing stoppers on the stabilizer bar with the slit openings facing towards the front.

9 Installation is otherwise the reverse of removal. Be sure to tighten all fasteners to the torque values listed in this Chapter's Specifications.

4 Upper control arm - removal and installation

Refer to illustrations 4.3, 4.4, 4.7 and 4.8

1 Disconnect the cable from the negative terminal of the battery (see Chapter 5, Section 1).

2 Loosen the wheel lug nuts, raise the front of the vehicle and support it securely on jackstands. Apply the parking brake. Remove the wheel.

3 Disconnect the ABS wheel speed sensor and separate the wiring harness bracket from the upper control arm and steering knuckle (see illustration).

4 On 4WD and PreRunner models, remove the wiring harness bracket for clearance to remove the upper arm pivot bolt (see illustration).

5 On 2WD models, detach the brake hose from the shock absorber (see illustration 2.5).

6 Support the lower control arm with a floor jack.

7 Remove the cotter pin, then loosen the upper balljoint nut about 1/4 inch. Use a two-jaw puller or a balljoint separator to separate the balljoint from the steering knuckle (see illustration).

CAUTION:

Don't allow the steering knuckle to fall outward as the brake hose may be damaged. It's a good idea to tie the steering knuckle to the coil spring so this doesn't happen.

4.3 The wheel speed sensor wiring harness bracket (A) and the upper control arm mounting fasteners (B) (2WD model shown)

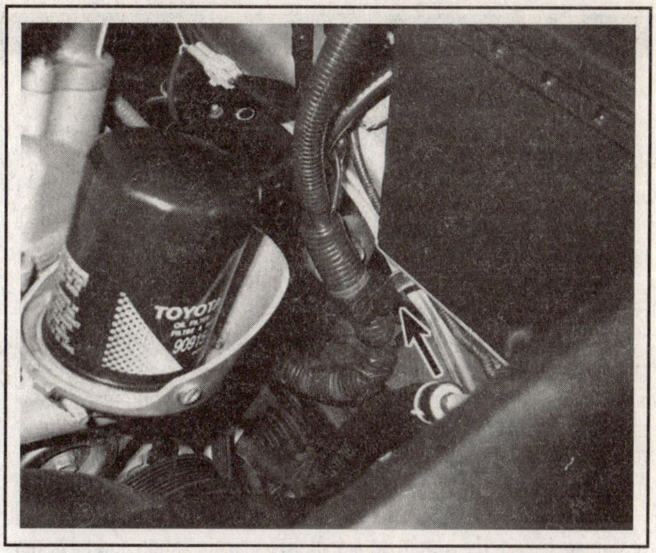

4.4 Remove the wiring harness bracket for clearance (the mounting bolt is located where the bracket meets the body)

4.7 Balljoint separator tools or two-jaw pullers (shown) are available at most auto parts stores and will not damage the balljoint boot when used correctly

4.8 Upper control arm fasteners (4WD and PreRunner model shown)

8 Remove the fasteners and detach the upper control arm from the frame, then remove the arm (see illustration).

➡Note: On 4WD and PreRunner models, the control arm is attached with one long pivot bolt. On 2WD models, it's attached using two shorter pivot bolts (see illustration 4.3).

9 Inspect the bushings for wear and deterioration. If they're cracked or damaged, take the arm to an automotive machine shop and have new bushings installed.

10 Installation is the reverse of removal. Tighten the wiring harness bracket fasteners securely. Be sure to tighten all suspension fasteners to the torque values listed in this Chapter's Specifications, and use a new cotter pin and castle nut on the upper control arm balljoint. If necessary, tighten the castle nut a little more to align the hole in the ballstud with the slots in the nut - don't loosen the nut to achieve this alignment.

➡Note: Tighten the pivot bolt(s) with the vehicle at normal ride height. This can be done after the vehicle has been assembled, lowered to the ground and bounced few times as if inspecting the shock absorbers. Alternatively, the normal ride height can be simulated by raising the lower control arm with a floor jack.

11 Tighten the lug nuts to the torque listed in the Chapter 1 Specifications.

12 It's a good idea to have the wheel alignment checked and, if necessary, adjusted.

5 Lower control arm - removal and installation

Refer to illustrations 5.4, 5.5a and 5.5b

1 Loosen the wheel lug nuts, raise the front of the vehicle and support it securely on jackstands. Apply the parking brake. Remove the wheel.

2 On 2WD models (except PreRunner), remove the stabilizer bar link from the lower control arm (see Section 3).

3 Unbolt the shock absorber/coil spring assembly from the lower control arm (see Section 2).

4 Remove the bracket mounting bolts holding the control arm to the steering knuckle at the balljoint (see illustration).

5 Make alignment marks to both sides of the front and rear pivot fasteners where the cam adjusters meet the frame (see illustration). Remove the pivot bolts and detach the control arm from the frame (see illustration).

6 Inspect the bushings for wear and deterioration. If they're cracked or damaged, take the arm to an automotive machine shop and have new bushings installed.

7 If you are replacing the lower control arm, you'll need to separate the bracket from the balljoint after the control arm is removed:

a) Place the arm securely in a vise.
b) Remove the cotter pin from the ballstud, then loosen the lower balljoint nut about 1/4 inch.
c) Using a two-jaw puller or a balljoint separator, separate the balljoint from the bracket (see illustration 4.7).

8 Installation is the reverse of removal. Make sure that the alignment marks you made prior to disassembly are lined up. Be sure to tighten all suspension fasteners to the torque values listed in this Chapter's Specifications, and use a new cotter pin and castle nut on the lower control arm balljoint. If necessary, tighten the castle nut a little more to align the hole in the ballstud with the slots in the nut - don't loosen the nut to achieve this alignment.

Note: The pivot bolts should be tightened with the vehicle at normal ride height. This can be done after the vehicle has been assembled, lowered to the ground and bounced few times as if inspecting the shock absorbers. Alternatively, the normal ride height can be simulated by raising the lower control arm with a floor jack.

9 Tighten the lug nuts to the torque listed in the Chapter 1 Specifications.

10 Have the wheel alignment checked and, if necessary, adjusted.

5.4 Lower balljoint bracket mounting fasteners

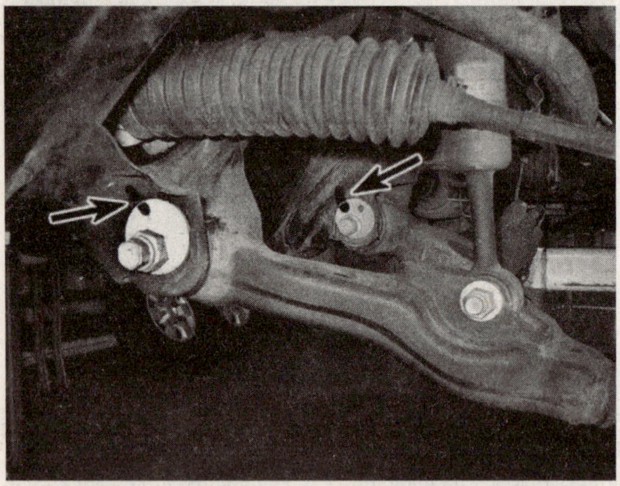

5.5a Mark the relationship of all four adjusting cams to the frame - consider using different color markers or paint or label them so that they do not get mixed up. They must go back in the exact same position to maintain wheel alignment (only two adjusting cams are visible in this photo)

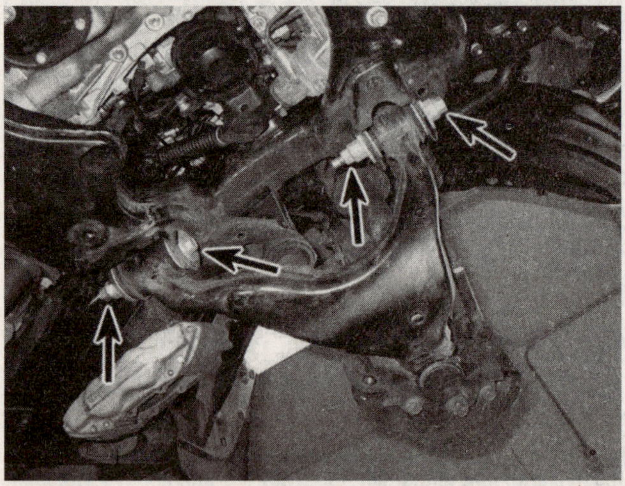

5.5b Lower control arm pivot fasteners

6 Steering knuckle - removal and installation

➧ Refer to illustrations 6.3a and 6.3b

1 Disconnect the cable from the negative terminal of the battery (see Chapter 5, Section 1).

2 Loosen the wheel lug nuts. Raise the front of the vehicle and support it securely on jackstands. Apply the parking brake. Remove the wheel.

3 Remove the brake hose/line bracket (4WD and PreRunner only) (see illustration), the wheel speed sensor and the wiring harness bracket from the steering knuckle (see illustration) (see Chapter 9).

4 Remove the brake caliper and brake disc. Hang the caliper with a length of wire - don't let it hang by the brake line/hose (see Chapter 9).

5 Disconnect the tie-rod end from the steering knuckle (see Section 14).

6 On 4WD and PreRunner models, disconnect the stabilizer bar link from the steering knuckle (see Section 3)

7 On 4WD models, remove the driveaxle/hub nut (see Chapter 8).

8 Disconnect the upper and lower control arms from the steering knuckle (see Sections 4 and 5), then remove the steering knuckle. On 4WD models, guide the driveaxle out of the hub, being careful to not overextend the inner CV joint. Support the driveaxle with a length of wire - don't let it hang by the inner CV joint.

9 Installation is the reverse of removal. Tighten all suspension fasteners to the torque values listed in this Chapter's Specifications. Tighten all other fasteners securely.

➡Note: The upper and lower control arm pivot bolts should be tightened with the vehicle at normal ride height. This can be done after the vehicle has been assembled, lowered to the ground and bounced few times as if inspecting the shock absorbers. Alternatively, the normal ride height can be simulated by raising the lower control arm with a floor jack.

10 Tighten the lug nuts to the torque listed in the Chapter 1 Specifications.

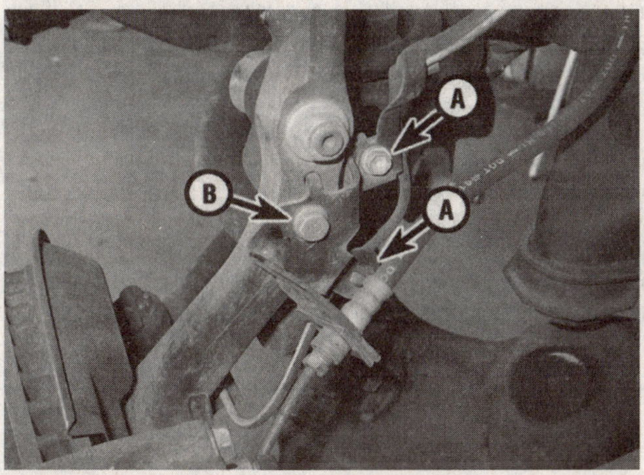

6.3a Brake hose/line bracket (4WD and PreRunner models):

- A *ABS wheel speed sensor wiring harness bracket and fasteners (remove first)*
- B *Bracket mounting bolt*

6.3b Wheel speed sensor wiring harness bracket (2WD models)

7 Hub and wheel bearing assembly (front) - removal and installation

Due to the special tools and expertise required to remove the hub and wheel bearing from the steering knuckle and disassemble them, this task is best left to experienced mechanics. However, on 2WD models (except PreRunner), the steering knuckle and hub may be removed and the assembly taken to a repair facility equipped with the necessary tools. On PreRunner and 4WD models, the hub and bearing assembly can be unbolted from the steering knuckle and taken to a repair facility equipped with the necessary tools. See Section 6 for the steering knuckle and hub removal procedure. On Pre Runner and 4WD models, be sure to tighten the hub assembly to steering knuckle bolts to the torque listed in this Chapter's Specifications.

8 Balljoints - check

1 Inspect the upper and lower balljoints for looseness whenever the vehicle is raised for any reason. You can check the balljoints with the suspension assembled as follows:

2 Raise the front of the vehicle and support it securely on jackstands.

3 Wipe the balljoints clean and inspect the seals for cuts and tears. If a balljoint seal is damaged, it can be replaced by simply removing the wire clip at the bottom and pulling it off the joint. Replace any lost grease, place the seal into position, then carefully work the retaining clip into the bottom groove.

LOWER BALLJOINT

Refer to illustration 8.4

4 With the vehicle raised and supported on jackstands and the lower control arm supported by a floor jack, attach a dial indicator to the lower control arm, with the plunger of the dial indicator touching the steering knuckle. Push up and down on the brake disc with about 65 pounds of force and check the dial indicator (see illustration). If there is more than 0.020-inch (0.05 mm) of play, replace the suspension arm. If you do not have a dial indicator, perform the test anyway. If you can feel play when applying pressure to the brake disc, replace the suspension arm.

Note: It may be possible to have the balljoint pressed out of the arm by an automotive machine shop or a repair shop that specializes in suspension work. A replacement part may be available as an aftermarket (non-OEM) part, but this could not be confirmed at the time of this manual's writing.

UPPER BALLJOINT

5 With the vehicle raised and supported on jackstands by the frame, and the front suspension components hanging freely, pry up and down on the upper control arm while feeling for play in the balljoint. It's possible that you can just use just your hands to move the upper control arm. If there is significant play, replace the suspension arm.

Note: It may be possible to have the balljoint pressed out of the arm by an automotive machine shop or a repair shop that specializes in suspension work. A replacement part may be available as an aftermarket (non-OEM) part, but this could not be confirmed at the time of this manual's writing.

ALL BALLJOINTS

Refer to illustration 8.7

6 Balljoints should also be checked whenever they're separated from the steering knuckle or the upper or lower control arm. See if you can turn the ballstud in its socket with your fingers. If the balljoint is loose, or if the ballstud can be turned easily, replace the balljoint or the suspension arm.

7 Toyota specifies a more accurate version of this bench test. Flip the balljoint stud back and forth five times, then install the nut. Using an inch-pound torque wrench, measure the turning torque as follows: turn the nut continuously at a rate of one turn every two to four seconds (see illustration). On the fifth turn, read the indicated torque and compare it to the acceptable torque range listed in this Chapter's Specifications. If the indicated turning torque is not within the range, the balljoint is worn out.

8.4 Check for play in the lower balljoint with a dial indicator mounted on the lower control arm and touching the part of the balljoint mounted on the steering knuckle; push up and down on the steering knuckle and read the dial indicator

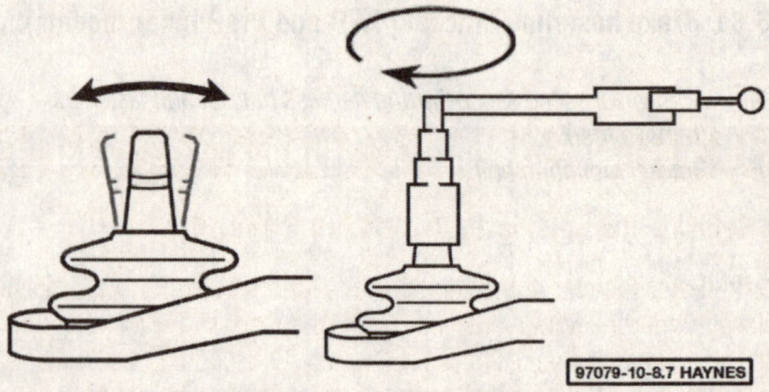

8.7 To bench test a balljoint, flip the balljoint stud back and forth five times, then measure the turning torque with an inch-pound torque wrench. Do this by installing the nut onto the ballstud and turning the balljoint continuously with the torque wrench and noting the torque reading on the fifth turn

9 Rear shock absorber - removal and installation

Refer to illustrations 9.2 and 9.3

WARNING:

Always replace shock absorbers in pairs - never replace just one of them.

1 Loosen the rear wheel lug nuts. Raise the rear of the vehicle and support securely on jackstands. Block the front wheels so the vehicle doesn't roll off the stands. Remove the rear wheels.

2 Support the rear axle with a floor jack. Remove the shock absorber lower mounting fasteners (see illustration).

Note: The shock absorber is mounted to the spring seat on 2WD models or to the rear axle housing on 4WD and PreRunner models.

3 Remove the shock absorber upper mounting nut (see illustration). Remove the shock absorber. Note the arrangement of the mounting bushings and fitted washers.

4 Installation is the reverse of removal. Tighten the mounting fasteners to the torque values listed in this Chapter's Specifications.

Note: Tighten the lower mounting fasteners with the vehicle at normal ride height (on vehicles with adequate clearance). This can be done after the vehicle has been assembled, lowered to the ground and bounced a few times as if inspecting the shock absorbers. Alternatively, the normal ride height can be simulated by raising the rear axle with a floor jack.

5 Tighten the lug nuts to the torque listed in the Chapter 1 Specifications.

9.2 Rear shock absorber lower mounting fasteners

9.3 To detach the upper end of the shock absorber from the frame, remove this nut

10 Leaf spring/shackle - removal and installation

Refer to illustrations 10.3, 10.4, 10.6 and 10.7

1 Raise the rear of the vehicle and support it securely on jackstands placed under the frame rails (***not*** under the rear axle). Block the front wheels to keep the vehicle from rolling off the stands. Support the rear axle with a floor jack so that there is no tension on the leaf springs.

2 Detach the shock absorber from the rear axle housing (on 4WD or PreRunner models) or from the spring seat (2WD models) (see Section 9).

3 Detach the parking brake cable bracket from the spring (see illustration).

4 Remove the four U-bolt fasteners (see illustration).

5 Remove the spring seat and U-bolts.

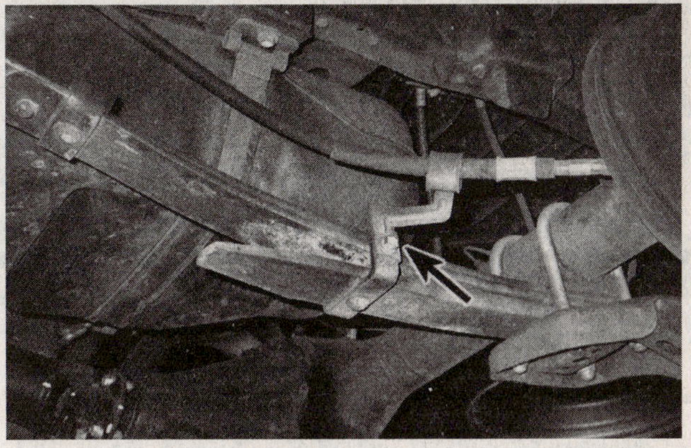

10.3 Detach the parking brake cable bracket from the leaf spring

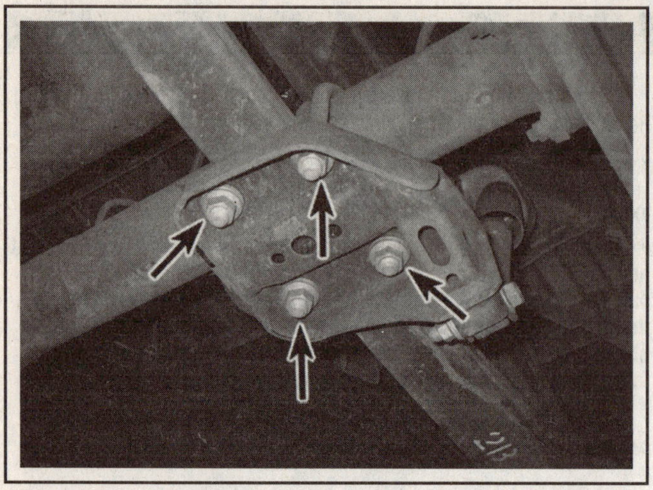

10.4 Leaf spring U-bolt fasteners (2WD model shown, 4WD and PreRunner models similar)

10.6 Leaf spring front fasteners

6 Remove the front leaf spring nut and through-bolt (see illustration).

CAUTION:

On 2WD models (excluding PreRunners), support the leaf spring so that it does not fall when the through-bolt is removed.

7 Remove the rear leaf spring shackle-to-frame nut, support the leaf spring and withdraw the through-bolt (see illustration). Carefully remove the spring.

Note: The shackle can be removed from the leaf spring after the spring is removed, if necessary.

8 If the bushings at the ends of the spring are worn or deteriorated, an automotive machine shop can press the old ones out and press new ones in. Be sure to inspect the spring bumpers and replace them as necessary.

9 Installation is the reverse of removal. Gradually tighten the U-bolt nuts equally in a criss-cross pattern so that the same amount of threads are showing on each U-bolt and to the torque listed in this Chapter's Specifications. Tighten the mounting fasteners at the end of the leaf spring to the torque listed in this Chapter's Specifications. Tighten the parking brake cable bracket securely.

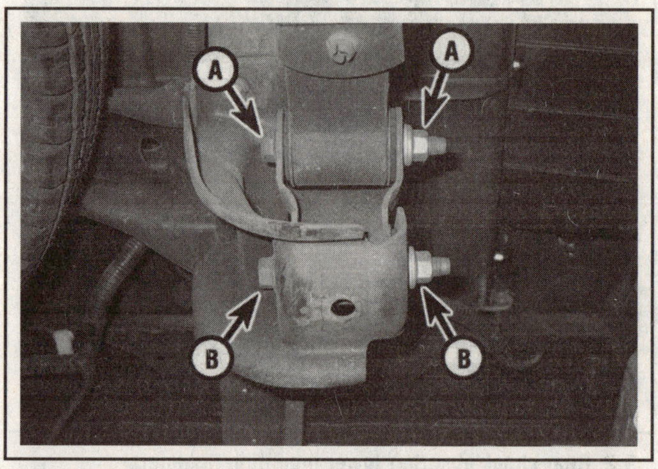

10.7 Leaf spring rear fasteners (A) and the shackle-to-frame fasteners (B)

Note: Tighten the front and rear leaf spring fasteners (including the shackle) with the vehicle at normal ride height. This can be done after the vehicle has been assembled, lowered to the ground and bounced a few times.

11 Stabilizer bar, bushings and links (rear) - removal and installation

Refer to illustration 11.3

Note: Not all models are equipped with a rear stabilizer bar.

1 Loosen the rear wheel lug nuts. Raise the rear of the vehicle and support securely on jackstands. Block the front wheels so the vehicle doesn't roll off the stands. Remove the rear wheels.

2 Locate the rear stabilizer bar that is attached to the rear axle housing with links connecting the bar to the frame. Remove the stabilizer bar link nuts and remove both links.

Note: Use an Allen wrench to prevent the ballstud from turning when removing the link nuts, if necessary.

3 Remove the bushing bracket bolts and brackets (see illustration).
4 Remove the stabilizer bar.
5 Inspect the rubber bushings for cracks, tears and deterioration. If they're worn or damaged, replace them.
6 Check the balljoint on the lower end of each link for looseness or other signs of excessive wear. You can check the rotational torque required to turn the balljoint by threading the nut onto the ballstud and turning it with an inch-pound torque wrench and comparing your reading to the value listed in this Chapter's Specifications.
7 Install the rubber bushings on the stabilizer bar with the slit opening facing down.
8 Installation is otherwise the reverse of removal. Be sure to tighten all fasteners to the torque values listed in this Chapter's Specifications.

11.3 To separate the stabilizer bar from the rear axle, remove the bushing bracket bolts

12 Steering wheel - removal and installation

➧ Refer to illustrations 12.2, 12.3a, 12.3b, 12.5, 12.6 and 12.7

⁂ WARNING 1:

These models are equipped with a Supplemental Restraint System (SRS), more commonly known as airbags. Always disable the airbag system before working in the vicinity of any airbag system component to avoid the possibility of accidental deployment of the airbag(s), which could cause personal injury (see Chapter 12).

⁂ WARNING 2:

Do not use a memory saving device to preserve the PCM or radio memory when working on or near airbag system components.

1 Make sure the front wheels are pointed straight ahead, then disconnect the cable from the negative terminal of the battery (see Chapter 5, Section 1). Wait at least three minutes before proceeding.
2 Pry the screw covers off each side of the steering wheel, and loosen the two Torx screws that retain the airbag module (see illustration).

➡Note: Each screw has a groove around its head that will catch a ridge around the inside of each hole. When the screw is loosened enough to release the airbag module, the ridge simply holds the screw in the hole.

3 Lift the airbag module off the steering wheel and disconnect the airbag and horn electrical connectors (see illustrations).

⁂ WARNING:

Carry the airbag module with the trim side facing away from you, and set the airbag module down with the trim side facing up. Don't place anything on top of the airbag module.

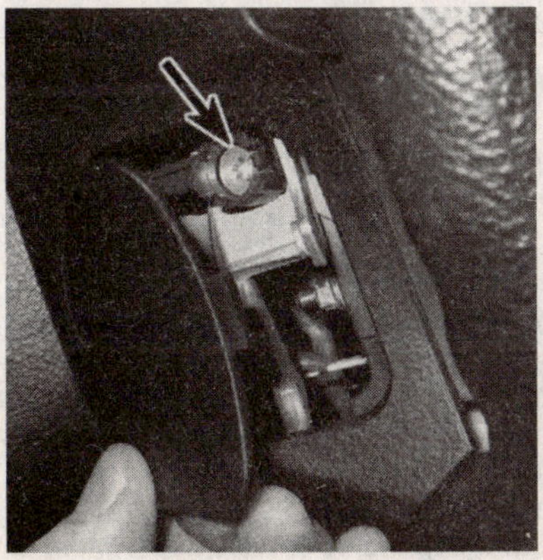

12.2 Pry the covers off each side of the steering wheel and loosen the Torx screws to release the airbag module

12.3a To unplug the airbag module connectors, release the locking tabs first . . .

12.3b . . . then remove all of the connectors from the module

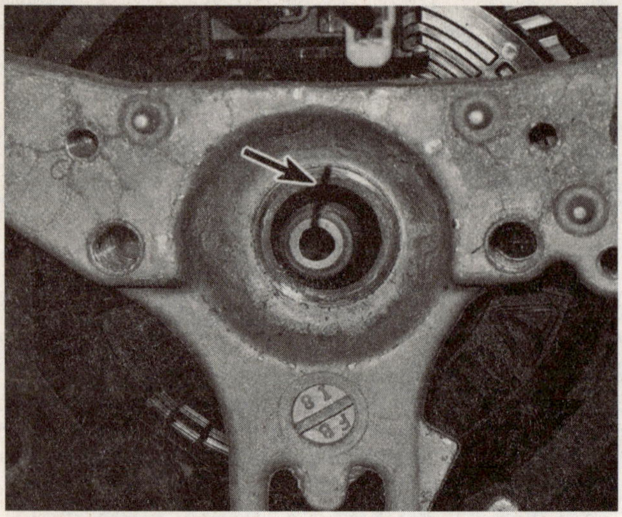

12.5 Remove the steering wheel nut, then mark the relationship of the steering wheel to the steering shaft

12.6 Use a steering wheel puller to remove the steering wheel and DO NOT screw the puller bolts into the steering wheel hub more than five turns each

4 Unplug any other electrical connectors attached to the steering wheel that may exist for accessories (cruise control, audio remote control, etc.).

5 Remove the steering wheel retaining nut and mark the position of the steering wheel to the shaft, if marks don't already exist or don't line up (see illustration).

6 Use a puller to detach the steering wheel from the shaft (see illustration). Don't hammer on the shaft to dislodge the wheel.

⁑ WARNING 1:

When installing the puller tool, DO NOT screw the puller bolts into the steering wheel hub more than five turns each or the spiral cable could be damaged.

⁑ WARNING 2:

While the steering wheel is removed, do NOT turn the steering shaft. If the steering shaft is turned, the spiral cable for the airbag system could be damaged when the steering wheel is put back on and used.

7 If necessary, re-center the spiral cable as follows: Verify that the front wheels are pointing straight ahead. Turn the spiral cable housing counterclockwise by hand until it becomes hard to turn (don't apply too much force). Turn the spiral cable clockwise about 2-1/2 turns and align the marks (see illustration). The spiral cable can be removed from the combination switch if necessary; disconnect the electrical connectors, then release the retaining clips and remove the spiral cable.

8 To install the wheel, thread the spiral cable wires through the steering wheel hub and slide the wheel onto the shaft while aligning the marks made in Step 5. Install the nut and tighten it to the torque listed in this Chapter's Specifications.

9 Installation is the reverse of removal.

12.7 Spiral cable details:

1 *Alignment marks and centering window (color visible inside when centered)*
2 *Mounting clips (one hidden from view - vicinity given)*
3 *Electrical connectors*

➡Note: On models equipped with VSC TRAC (vehicle stability and traction control), any repairs to the steering components may require calibration of the system's various sensors. A warning light at the instrument cluster for the VSC TRAC system will confirm this. If no warning light is present, the system should be normal. Calibration requires special tools and expertise. The vehicle can be taken to a dealership service department or qualified repair for calibration, if necessary. Keep in mind that the VSC TRAC, ABS and DAC features may not be operational until the calibration procedure has been performed.

13 Steering column - removal and installation

⁂ WARNING 1:

These models are equipped with a Supplemental Restraint System (SRS), more commonly known as airbags. Always disable the airbag system before working in the vicinity of any airbag system component to avoid the possibility of accidental deployment of the airbag(s), which could cause personal injury (see Chapter 12).

⁂ WARNING 2:

Do not use a memory saving device to preserve the PCM or radio memory when working on or near airbag system components.

REMOVAL

➧ Refer to illustrations 13.5, 13.6, 13.7a, 13.7b and 13.8

1 Make sure the front wheels are pointed straight ahead, then disconnect the cable from the negative terminal of the battery (see Chapter 5, Section 1). Wait at least three minutes before proceeding.

2 Remove the steering wheel (see Section 12), then turn the ignition key to the LOCK position to prevent the steering shaft from turning.

⁂ CAUTION:

Damage to the airbag spiral cable could occur if the shaft is turned and the steering wheel is installed out-of-sync with the shaft.

3 Remove the instrument cluster finish panel and knee bolster located below the column (see Chapter 11).

4 Remove the steering column covers (see Chapter 11).

5 Disconnect the electrical connectors and wiring harness fasteners from the steering column (see illustration).

6 Remove the steering column hole cover at the firewall (see illustration).

7 Mark the relationship of the steering shaft to the intermediate shaft U-joint, then remove the U-joint pinch bolt (see illustration). Remove the lower heating and air conditioning duct, if necessary (see illustration).

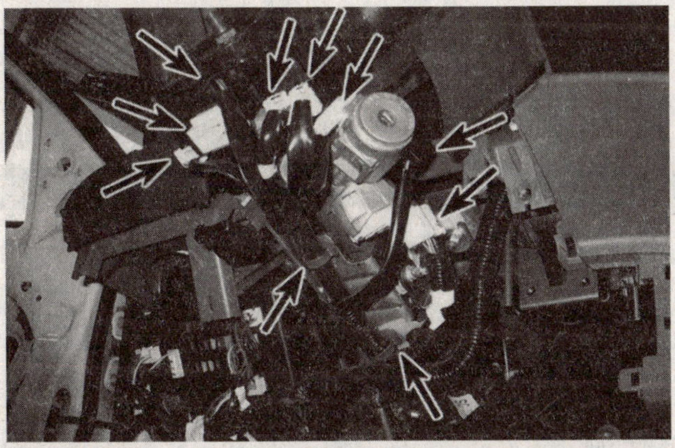

13.5 Electrical connectors and harness fasteners for the steering column (most connectors shown)

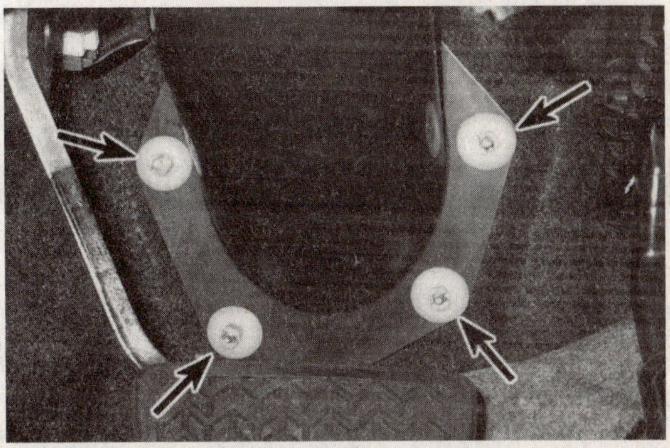

13.6 Steering column hole cover fasteners

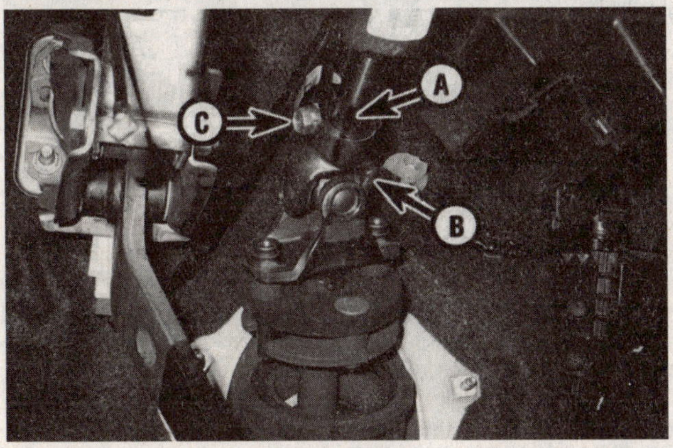

13.7a Mark the steering shaft (A) in relation to the intermediate shaft U-joint (B), then remove the pinch bolt (C)

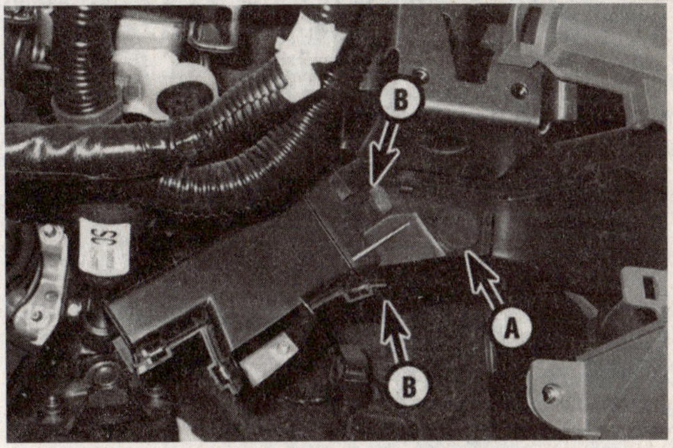

13.7b Remove the heater duct by removing the plastic fastener (A), then separating the retaining tabs (B) (not all tabs are visible in this photo)

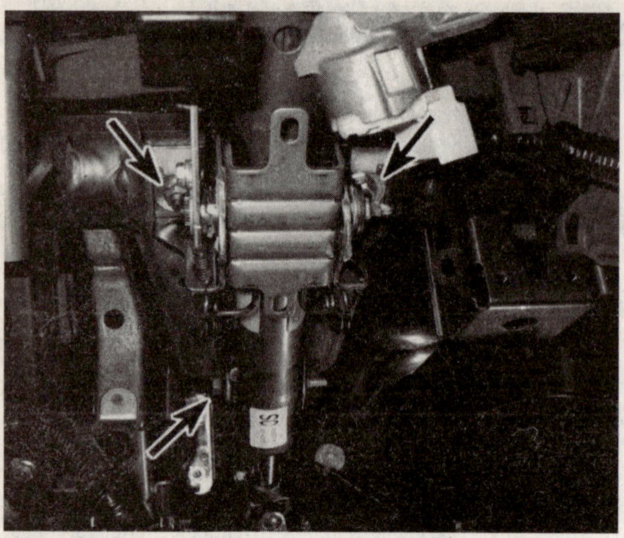

13.8 Steering column mounting fasteners

8 Remove the steering column mounting fasteners (see illustration), then carefully lower the column. While making sure nothing is still connected, pull the column towards the rear of the vehicle to separate the steering shaft from the U-joint.

INSTALLATION

9 Guide the steering column into position while aligning the marks made in Step 7 and connecting the intermediate shaft.

10 Install and tighten the column mounting fasteners to the torque listed in this Chapter's Specifications.

11 Install the pinch bolt, tightening it to the torque listed in this Chapter's Specifications.

12 The remainder of installation is the reverse of removal.

➡Note: On models equipped with VSC TRAC (vehicle stability and traction control), any repairs to the steering components may require calibration of the system's various sensors. A warning light at the instrument cluster for the VSC TRAC system will confirm this. If no warning light is present, the system should be normal. Calibration requires special tools and expertise. The vehicle can be driven to a dealership service department or qualified repair for calibration, if necessary. Keep in mind that the VSC TRAC, ABS and DAC features may not be operational until the calibration procedure has been performed.

14 Tie-rod ends - removal and installation

➧ Refer to illustrations 14.2a, 14.2b and 14.3

1 Loosen the wheel lug nuts, raise the vehicle and place it securely on jackstands. Remove the wheel.

2 Loosen the tie-rod end locknut and mark the position of the tie-rod end on the threaded portion of the tie-rod (see illustrations).

3 Remove the cotter pin and discard it. Loosen the castle nut about 1/4 inch from the tie-rod end balljoint stud, then install a balljoint separator or puller and separate the tie-rod end from the steering knuckle (see illustration). Remove the nut and detach the tie-rod end from the steering knuckle arm.

4 Unscrew the old tie-rod end and install the new one. Make sure the new tie-rod end is aligned with the mark you made on the threads of the tie-rod.

14.2a Loosen the tie-rod end locknut using two wrenches . . .

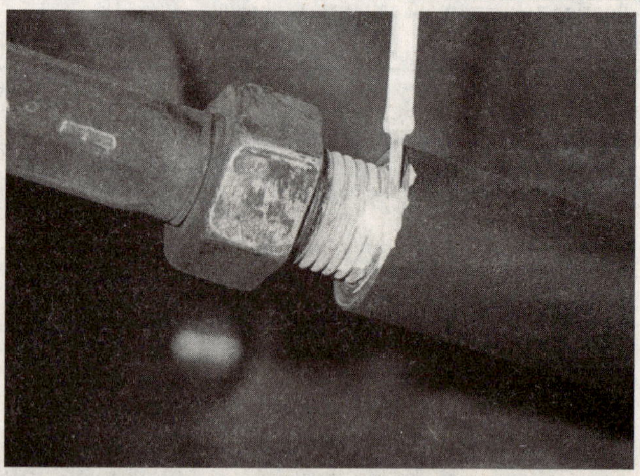

14.2b . . . then mark the position of the tie-rod end on the threaded portion of the tie-rod

5 Installation is the reverse of removal. Be sure to tighten the tie-rod end castle nut to the torque listed in this Chapter's Specifications and use a new cotter pin. If necessary, tighten the castle nut a little more to align the hole in the ballstud with the slots in the nut - don't loosen the nut to achieve this alignment. Tighten the locknut securely.

6 Tighten the lug nuts to the torque listed in the Chapter 1 Specifications.

7 Have the front end alignment checked and, if necessary, adjusted.

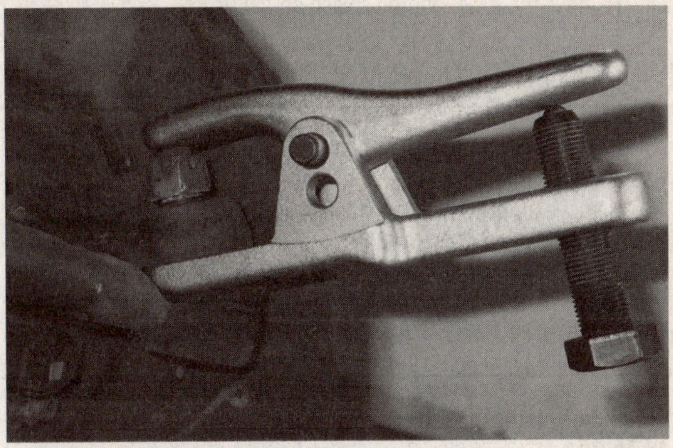

14.3 Loosen (but don't remove) the castle nut from the tie-rod end ballstud, then install a balljoint separator or puller and separate the tie-rod end from the steering knuckle

15 Steering gear boots - replacement

Refer to illustration 15.4

1 If a steering gear boot is torn, dirt and moisture can damage the steering gear. Replace it.

2 Loosen the wheel lug nuts, raise the vehicle and place it securely on jackstands. Remove the front wheels.

3 Remove the tie-rod end and locknut (see Section 14).

4 Remove the boot clamps (see illustration) and slide the boot off the tie-rod.

5 Installation is the reverse of removal. Be careful not to damage the boot while sliding it into place. Refer to Chapter 8 regarding procedures for installing the inner clamp. Be sure to tighten the tie-rod end castle nut to the torque listed in this Chapter's Specifications and use a new cotter pin. Tighten the locknut securely. Tighten the lug nuts to the torque listed in the Chapter 1 Specifications.

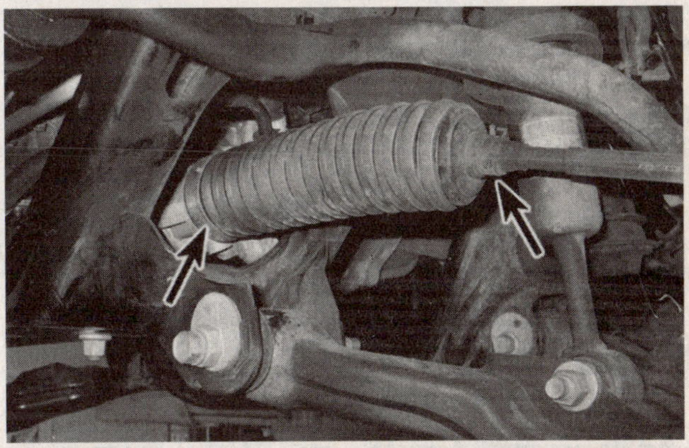

15.4 The outer clamp (small) on the steering gear boot can be removed with a pair of pliers - the inner clamp (large) must be cut off

16 Steering gear - removal and installation

WARNING 1:

These models are equipped with a Supplemental Restraint System (SRS), more commonly known as airbags. Always disable the airbag system before working in the vicinity of any airbag system component to avoid the possibility of accidental deployment of the airbag(s), which could cause personal injury (see Chapter 12).

WARNING 2:

Do not use a memory saving device to preserve the PCM or radio memory when working on or near airbag system components.

WARNING 3:

Make sure the steering wheel is not turned while the steering gear is removed. This could result in damage to the airbag system spiral cable. To prevent the steering wheel from turning, place the ignition key in the LOCK position or thread the seat belt through the steering wheel and clip it into place.

1 Make sure the front wheels are pointed straight ahead and apply the parking brake.

2 Disconnect the cable from the negative terminal of the battery (see Chapter 5, Section 1).

3 Loosen the wheel lug nuts. Raise the front of the vehicle and support it securely on jackstands. Remove the front wheels.

4 Remove the lower engine splash shield (front).

4WD AND PRERUNNER MODELS

5 Remove the front exhaust pipe assemblies that include the catalytic converters and oxygen sensors (see Chapter 4).

6 Remove all driveshaft assemblies including the front driveshaft for 4WD models (see Chapter 8).

7 Remove the stabilizer bar (see Section 3).

ALL MODELS

➧ Refer to illustrations 16.9, 16.10, 16.11, 16.12, 16.13, 16.15 and 16.16

8 Disconnect the tie-rod ends from the steering knuckles (see Section 14).

9 Mark the upper and lower intermediate shaft in relation to the shaft coupler, then remove both pinch bolts (see illustration).

10 Remove the U-joint coupler pinch bolt at the bottom of the lower intermediate shaft where it connects to the steering gear input shaft (see illustration).

11 Slide the U-joint up from the steering gear and mark the steering gear input shaft in relation to the U-joint, then slide it back down to its normal position (see illustration).

12 Slide the shaft coupler up onto the upper intermediate shaft allowing the two shafts to separate, then pull the lower intermediate shaft U-joint off the steering gear input shaft and remove it (see illustration).

➡Note: The shaft coupler may fall from the upper intermediate steering shaft. It can be removed until the shafts are ready to be connected again.

13 Detach the pressure and return line bracket(s) from the steering gear (see illustration).

14 Use a flare wrench to disconnect the pressure line from the steering gear. Squeeze the hose clamp on the return hose and move it up the hose a bit, then pull the hose from the fitting (see illustration 16.10).

15 Remove the steering gear mounting fasteners (see illustration).

⁂ CAUTION:

Hold the mounting nuts and turn the mounting bolts when removing or installing the fasteners.

16 On 4WD and PreRunner models, support the rear crossmember (beneath the rear part of the transmission) with a floor jack. Remove the mounting fasteners and lower it with the jack so that it tilts the front of the engine up (see illustration). This will provide the necessary clearance to remove the steering gear.

17 Remove the steering gear assembly by carefully angling it down and moving it from side to side until it's clear of the crossmember.

18 Installation is the reverse of removal. Be sure to align all matchmarks made during the removal procedure and tighten all suspension and steering gear fasteners to the torque values listed in this Chapter's Specifications. Tighten the wheel lug nuts to the torque listed in the Chapter 1 Specifications.

➡Note: On models equipped with VSC TRAC (vehicle stability and traction control), any repairs to the steering components may require calibration of the system's various sensors. A warning light at the instrument cluster for the VSC TRAC system will confirm this. If no warning light is present, the system should be normal. Calibration requires special tools and expertise. The vehicle can be driven to a dealership service department or qualified repair for calibration, if necessary. Keep in mind that the VSC TRAC, ABS and DAC features may not be operational until the calibration procedure has been performed.

19 Bleed the power steering system (see Section 18).

20 Have the front end alignment checked and, if necessary, adjusted.

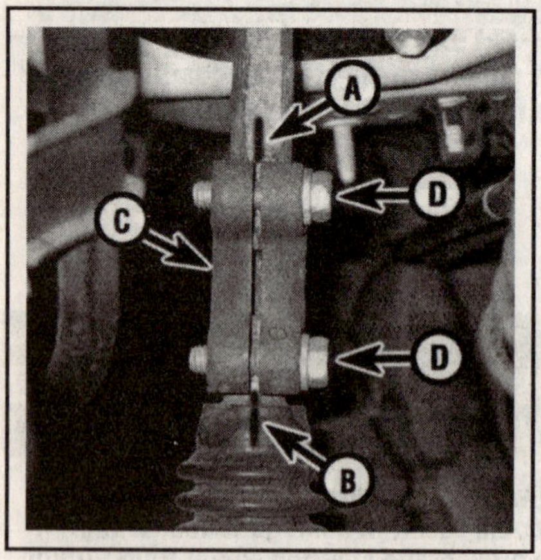

16.9 Mark the upper (A) and lower (B) intermediate shafts in relation to the shaft coupler (C), then remove the pinch bolts (D) (2WD model shown, 4WD and PreRunner models similar)

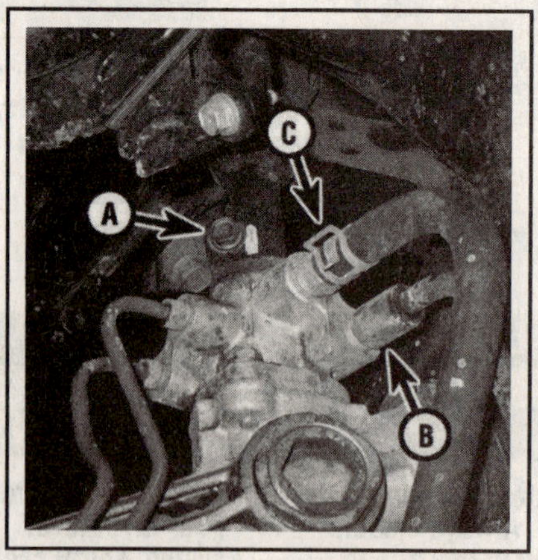

16.10 Location of the U-joint pinch bolt (A) on the lower intermediate shaft and the pressure line (B) and return hose (C) connections at the steering gear (4WD and PreRunner model shown, 2WD models similar)

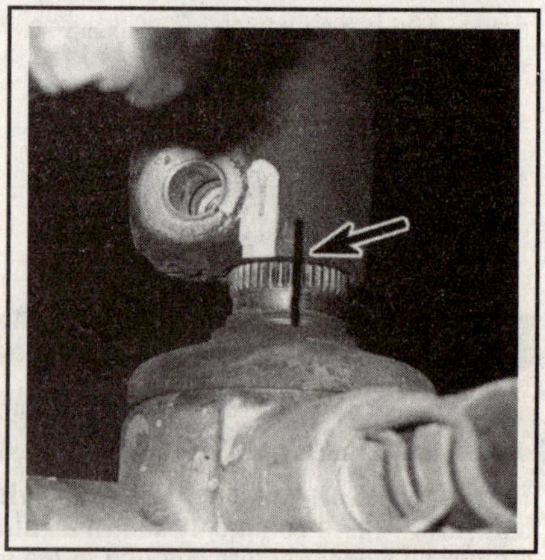

16.11 Mark the relationship of the steering gear input shaft to the U-joint on the lower intermediate shaft

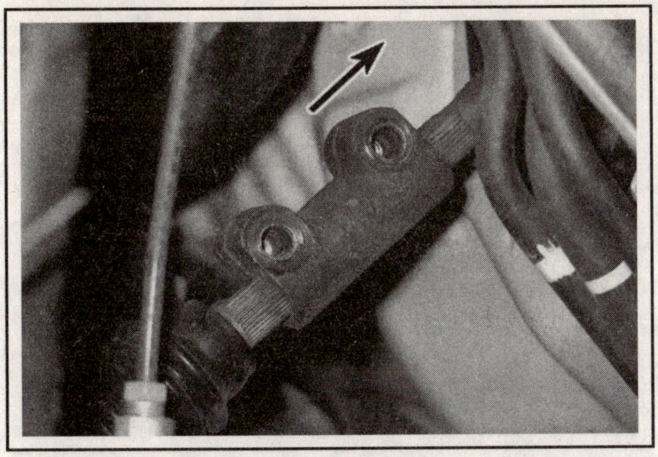

16.12 Slide the shaft coupler up on the upper intermediate shaft enough to separate it from the lower intermediate shaft

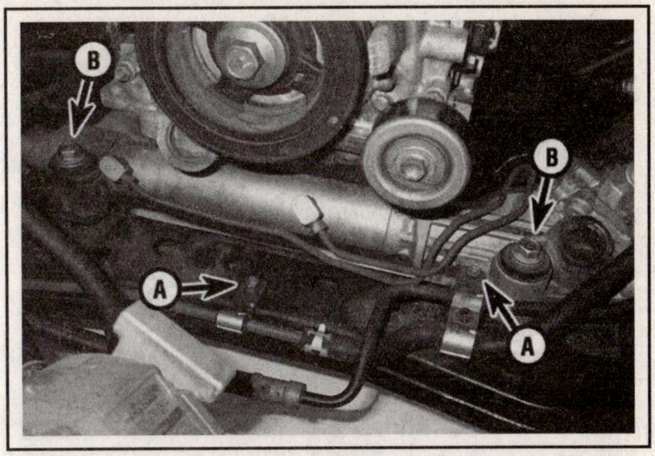

16.13 Location of the pressure and return line brackets (A) mounted to the crossmember and the steering gear mounting bolts (B) (4WD and PreRunner model shown, 2WD models similar)

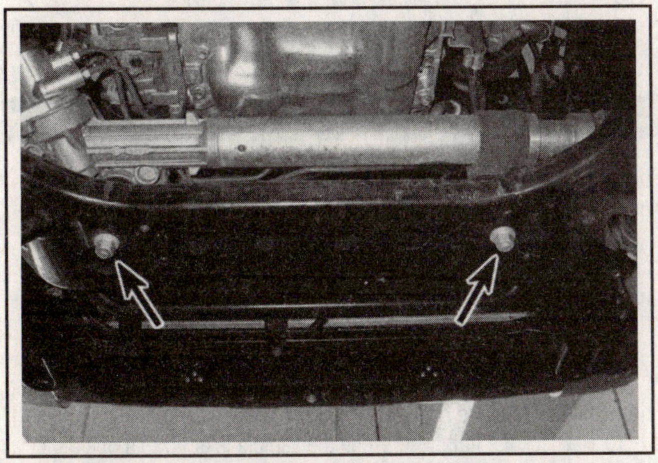

16.15 The location of the steering gear mounting nuts (4WD and PreRunner model shown, 2WD models similar)

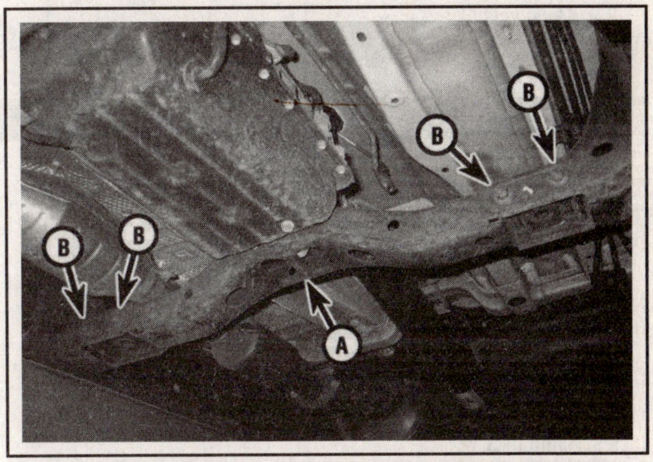

16.16 On 4WD and PreRunner models, support the rear crossmember (A) with a floor jack, remove the mounting fasteners (B), then slowly lower it. This will tilt the front of the engine up and provide the clearance needed to remove the steering gear

17 Power steering pump - removal and installation

Refer to illustrations 17.5a and 17.5b

1 Disconnect the cable from the negative terminal of the battery (see Chapter 5, Section 1).

2 On 4WD and PreRunner models, remove the lower engine splash shield (front).

3 Remove the drivebelt (see Chapter 1).

4 Remove the power steering fluid from the reservoir. This can be accomplished with a suction tool or large syringe, or by disconnecting the fluid hose and draining the fluid into a container.

5 Disconnect the electrical connector from the pressure switch (see illustrations). On V6 engine models, detach the wiring harness clip from the bracket on the pump.

6 Position a drain pan under the power steering pump. Disconnect the pressure line and fluid hose from the pump (see illustrations 17.5a and 17.5b). Plug all openings to prevent contaminants from entering any part of the system. Discard the sealing washer(s).

7 Remove the pump mounting fasteners and lift the pump from the engine compartment, taking care not to spill fluid on the painted surfaces (see illustrations 17.5a and 17.5b).

8 Installation is the reverse of removal. Use new sealing washers on both sides of the pressure line fitting. Tighten the pressure line banjo bolt and the pump mounting fasteners to the torque values listed in this Chapter's Specifications.

9 Fill the power steering reservoir with the recommended fluid (see Chapter 1) and bleed the system (see Section 18).

17.5a Power steering pump mounting details on four-cylinder models:

1 *Power steering fluid pressure sensor electrical connector (vicinity shown)*
2 *Mounting bolts*
3 *Return hose and clamp*
4 *Pressure line fitting and banjo bolt*

17.5b Power steering pump mounting details on V6 models (typical shown):

1 *Power steering fluid pressure sensor electrical connector*
2 *Mounting bolts*
3 *Feed hose and clamp*
4 *Pressure line fitting and banjo bolt*

18 Power steering system - bleeding

1 Following any operation in which the power steering fluid lines have been disconnected, the power steering system must be bled to remove all air and obtain proper steering performance.

2 With the front wheels in the straight ahead position, check the power steering fluid level and, if low, add fluid until it reaches the Cold range on the reservoir (see Chapter 1).

3 Raise the front of the vehicle and support it securely on jackstands.

4 With the engine off, turn the steering wheel slowly from lock-to-lock several times.

5 Lower the vehicle.

6 Start the engine and recheck the fluid level. Add more fluid if necessary, but do not exceed the Cold range on the reservoir. Allow the engine to warm up.

7 Bleed the system by turning the wheels from lock-to-lock (as far as the wheel will go in each direction). Hold the steering wheel at each lock position for two to three seconds, but no more than that at any one time. Repeat this Step several times.

8 Stop the engine and check the fluid level.

9 Road test the vehicle to be sure the steering system is functioning normally and is noise free.

10 Recheck the fluid level while the vehicle is at normal operating temperature.

19 Wheels and tires - general information

Refer to illustration 19.1

All vehicles covered by this manual are equipped with metric-size fiberglass or steel belted radial tires (see illustration). Use of other size or type of tires may affect the ride and handling of the vehicle. Don't mix different types of tires, such as radials and bias belted, on the same vehicle as handling may be seriously affected. It's recommended that tires be replaced in pairs on the same axle, but if only one tire is being replaced, be sure it's the same size, structure and tread design as the other.

Because tire pressure has a substantial effect on handling and wear, the pressure on all tires should be checked at least once a month or before any extended trips (see Chapter 1).

Wheels must be replaced if they're bent, dented, leak air, have elongated bolt holes, are heavily rusted, out of vertical symmetry or if the lug nuts won't stay tight. Wheel repairs that use welding or peening are not recommended.

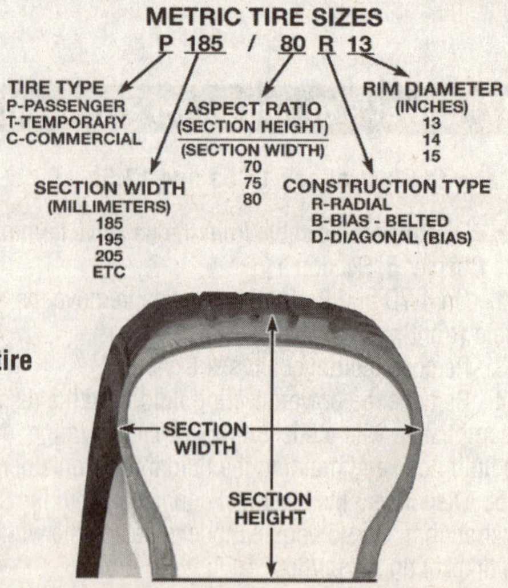

19.1 Metric tire size code

Tire and wheel balance is important to the overall handling, braking and performance of the vehicle. Unbalanced wheels can adversely affect handling and ride characteristics as well as tire life. Whenever a tire is installed on a wheel, the tire and wheel should be balanced by a shop with the proper equipment.

20 Front end alignment - general information

➧ Refer to illustration 20.1

A front end alignment refers to the adjustments made to the front wheels so they're in proper angular relationship to the suspension and the ground (see illustration). Front wheels that are out of proper alignment not only affect steering control, but also increase tire wear.

Getting the proper front wheel alignment is a very exacting process, one in which complicated and expensive machines are necessary to perform the job properly. Because of this, you should have a technician with the proper equipment perform these tasks. We will, however, use this space to give you a basic idea of what is involved with front end alignment so you can better understand the process and deal intelligently with the shop that does the work.

Toe-in is the turning in of the front wheels. The purpose of a toe specification is to ensure parallel rolling of the front wheels. In a vehicle with zero toe-in, the distance between the front edges of the wheels will be the same as the distance between the rear edges of the wheels. The actual amount of toe-in is normally only a fraction of an inch. Toe-in adjustment is controlled by the tie-rod length. Incorrect toe-in will cause the tires to wear improperly by making them scrub against the road surface.

Camber is the tilting of the front wheels from vertical when viewed from the front of the vehicle. When the wheels tilt out at the top, the camber is said to be positive (+). When the wheels tilt in at the top, the camber is negative (-). The amount of tilt is measured in degrees from vertical and this measurement is called the camber angle. This angle affects the amount of tire tread which contacts the road and compensates for changes in the suspension geometry when the vehicle is cornering or traveling over an undulating surface. Camber is adjusted by rotating cam-shaped adjusters at the front and rear of the lower control arm.

Caster is the tilting of the top of the front steering axis from the vertical. A tilt toward the rear is positive caster and a tilt toward the front is negative caster. Caster is also adjusted by rotating cam-shaped adjusters at the front and rear of the lower control arm.

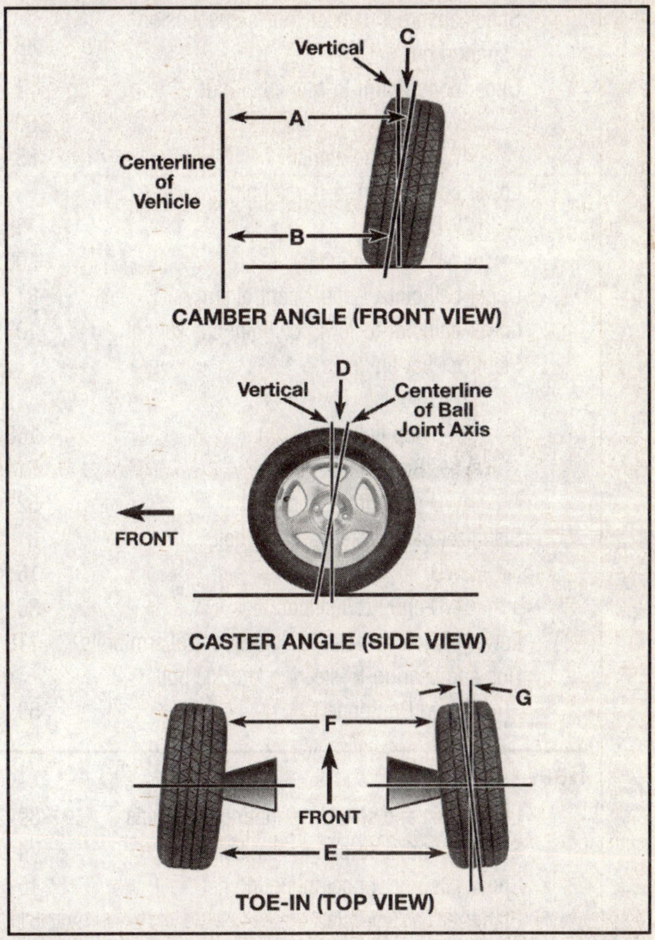

20.1 Front end alignment details:

A minus B = C (degrees camber)
D = degrees caster
E minus F = toe-in (measured in inches)
G = toe-in (expressed in degrees)

Specifications

General

Power steering fluid type	See Chapter 1
Balljoint stud turning torque	
Stabilizer bar links	
Front	18 in-lbs (2 Nm) or less
Rear	18 in-lbs (2 Nm) or less
Lower control arm balljoint	27 in-lbs (3 Nm) or less
Upper control arm balljoint	40 in-lbs (4.5 Nm) or less

Torque specifications	Ft-lbs (unless otherwise indicated)	Nm

➡**Note: One foot-pound (ft-lb) of torque is equivalent to 12 inch-pounds (in-lbs) of torque. Torque values below approximately 15 ft-lbs are expressed in inch-pounds, since most foot-pound torque wrenches are not accurate at these smaller values.**

Torque specifications	Ft-lbs (unless otherwise indicated)	Nm
Front suspension		
Shock absorber-to-lower control arm bolt/nut	61	83
Shock absorber upper mounting nuts	47	64
Shock absorber damper rod-to-suspension support nut	20	27
Upper control arm-to-frame bolt/nut		
2WD	60	81
4WD and PreRunner	85	115
Lower control arm-to-frame bolt		
2WD	155	210
4WD and PreRunner	100	136
Upper balljoint-to-upper control arm nut	81	110
Lower balljoint-to-lower balljoint bracket nut	103	140
Stabilizer bar link nuts		
2WD		
Top link nut	168 in-lbs	19
Bottom link nut	51	69
4WD and PreRunner	52	71
Stabilizer bar bracket-to-frame bolts		
2WD	16	22
4WD and PreRunner	30	40
Lower balljoint bracket-to-lower control arm bolts	118	160
Hub and bearing-to-steering knuckle bolts (4WD and PreRunner)	59	80
Rear suspension		
Leaf spring and shackle mounting bolts/nuts	89	121
Shock absorber-to-spring seat bolt/nut	74	100
Shock absorber upper mounting nut	15	20
Stabilizer bar link nuts	51	69
Stabilizer bar bracket-to-axle bolts	20	27
U-bolt nuts	37	50
Steering		
Airbag module mounting screws	78 in-lbs	9
Power steering pressure line fitting banjo bolt	38	52
Power steering pump mounting bolts	15	20
Steering wheel nut	37	50
Steering gear mounting bolts/nuts	68	92
Steering column mounting bolts/nuts	16	22
U-joint pinch bolt	26	35
Upper and lower intermediate shaft coupler pinch bolts	26	35
Tie-rod end-to-steering knuckle nut		
2WD	36	49
4WD and PreRunner	67	91

11

BODY

Section

1 General information

GENERAL INFORMATION

These models feature a body-on-frame construction. The frame is a ladder-type, consisting of two box sectioned steel side rails joined by crossmembers. The crossmembers are welded to the side rails, with the exception of the transmission crossmember, which is bolted in place for easy removal.

Certain components are particularly vulnerable to accident damage and can be unbolted and repaired or replaced. Among these parts are the body moldings, bumpers, hood, fenders, doors, tailgate and all glass.

Only general body maintenance practices and body panel repair procedures within the scope of the do-it-yourselfer are included in this Chapter.

FASTENERS AND TRIM REMOVAL

➧ Refer to illustrations 1.6a through 1.6i, 1.7 and 1.8

There is a variety of plastic fasteners used to hold trim panels, splash shields and other parts in place, in addition to typical screws, nuts and bolts. Once you are familiar with them, they can usually be removed without too much difficulty.

The proper tools and approach can prevent added time and expense to a project by minimizing the number of broken fasteners and/or parts.

The following illustrations show various types of fasteners that are typically used on most vehicles and how to remove and install them (see illustrations).

Trim panels are typically made of plastic and their flexibility can help

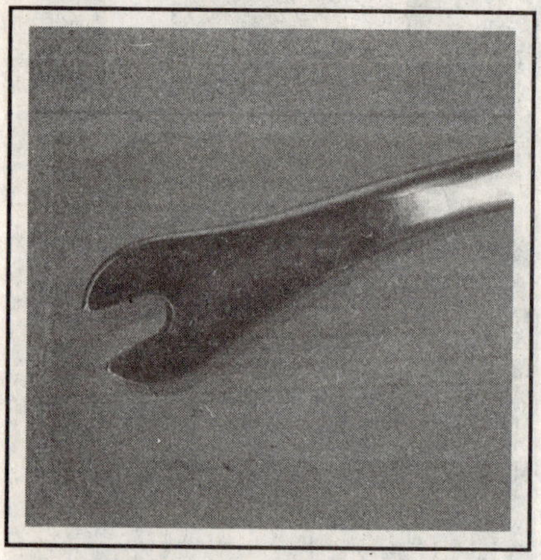

1.6a This tool is designed to remove special fasteners. A small pry tool used for removing nails will also work well in place of this tool

1.6b This fastener is used for interior panels. A Phillips head screwdriver can be used to release the center portion, but light pressure must be used because the plastic is easily damaged. Once the center is up, the fastener can easily be pried from its hole

1.6c Here is a view with the center portion fully released. Install the fastener as shown, then press the center in to set it

1.6d This fastener is used for interior panels. The center portion is pressed in to release the fastener

1.6e Here is a view of the fastener in the released position

1.6f Here is a view of the fastener in a reset position for installation

during removal. The key to their removal is to use a tool to pry the panel near its retainers (or fasteners) to release it without damaging surrounding areas or breaking off any retainers. The retainers or fasteners will usually snap out of their designated slot or hole after force is applied to them. Stiff plastic tools designed for prying on trim panels can be found at specialty automotive or tool supply stores (see illustration). Tools that are tapered and wrapped in protective tape, such as a screwdriver or small pry tool, are also very effective when used with care.

There are also special bolts used to align the hinges of the hood, doors and other related components to a centered position. If those components need repositioning that is off-center for alignment purposes, then the factory centering bolts will need to be replaced with standard bolts and washers of the same grade (strength) (see illustration).

1.6g This fastener is used for exterior panels and shields. The center portion must be pried up fully before the entire fastener can be removed or installed. Also, the center portion remains out for installation

1.6h This fastener is used for exterior and interior panels. The center portion must be pried up fully before the entire fastener can be removed or installed. Use a pointed tool to raise the center because it is sunk into the fastener

1.6i This fastener is used for exterior and interior panels. Simply pry the fastener from its hole

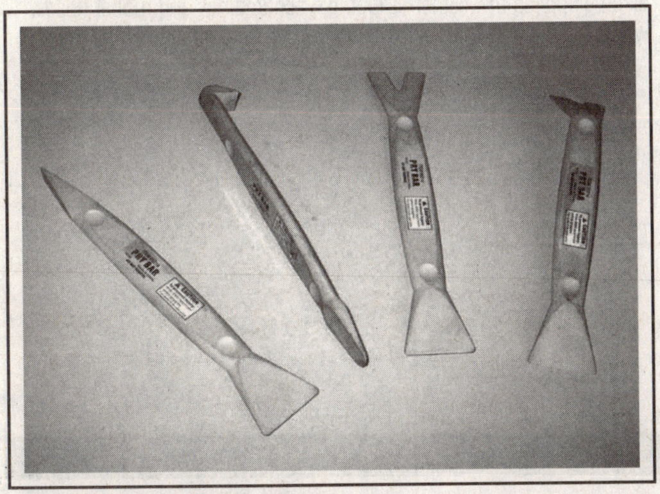

1.7 These small plastic pry tools are the best choice of tool to use when prying off trim panels

1.8 This special mounting bolt is designed to center the holes of body components. Hoods, doors, and some latches are mounted with these types of bolts from the factory. Replace them with standard bolts and washers if those components require adjustment to an off-center position

2 Body - maintenance

1 The condition of your vehicle's body is very important, because the resale value depends a great deal on it. It's much more difficult to repair a neglected or damaged body than it is to repair mechanical components. The hidden areas of the body, such as the wheel wells, the frame and the engine compartment, are equally important, although they don't require as frequent attention as the rest of the body.

2 Once a year, or every 12,000 miles, it's a good idea to have the underside of the body steam-cleaned. All traces of dirt and oil will be removed and the area can then be inspected carefully for rust, damaged brake lines, frayed electrical wires, damaged cables and other problems.

3 At the same time, clean the engine and the engine compartment with a steam cleaner or water-soluble degreaser.

4 The wheel wells should be given close attention, since undercoating can peel away and stones and dirt thrown up by the tires can cause the paint to chip and flake, allowing rust to set in. If rust is found, clean down to the bare metal and apply an anti-rust paint.

5 The body should be washed about once a week. Wet the vehicle thoroughly to soften the dirt, and then wash it down with a soft sponge and plenty of clean soapy water. If the surplus dirt is not washed off very carefully, it can wear down the paint.

6 Spots of tar or asphalt thrown up from the road should be removed with a cloth soaked in kerosene. Scented lamp oil is available in most hardware stores and the smell is easier to work with than straight kerosene.

7 Once every six months, wax the body and chrome trim. If chrome cleaner is used to remove rust from any of the vehicle's plated parts, remember that the cleaner also removes part of the chrome, so use it sparingly. On any plated parts where chrome cleaner is used, use a good paste wax over the plating for extra protection.

3 Vinyl trim - maintenance

Don't clean vinyl trim with detergents, caustic soap or petroleum-based cleaners. Plain soap and water works just fine, with a soft brush to clean dirt that may be ingrained. Wash the vinyl as frequently as the rest of the vehicle.

After cleaning, application of a high quality rubber and vinyl protectant will help prevent oxidation and cracks. The protectant can also be applied to weather stripping, vacuum lines and rubber hoses, which often fail as a result of chemical degradation, and to the tires.

4 Upholstery and carpets - maintenance

1 Every three months, remove the floormats and clean the interior of the vehicle (more frequently if necessary). Use a stiff whisk broom to brush the carpeting and loosen dirt and dust, and then vacuum the upholstery and carpets thoroughly, especially along seams and crevices.

2 Dirt and stains can be removed from carpeting with basic household or automotive carpet shampoos available in spray cans. Follow the directions and vacuum again, then use a stiff brush to bring back the nap of the carpet.

3 Most interiors have cloth or vinyl upholstery, either of which can be cleaned and maintained with a number of material-specific cleaners or shampoos available in auto supply stores. Follow the directions on the product for usage, and always spot-test any upholstery cleaner on an inconspicuous area (bottom edge of a backseat cushion) to ensure

that it doesn't cause a color shift in the material.

4 After cleaning, vinyl upholstery should be treated with a protectant.

➡**Note: Make sure the protectant container indicates the product can be used on seats - some products may make a seat too slippery.**

⁂ CAUTION:

Do not use protectant on steering wheels.

5 Leather upholstery requires special care. It should be cleaned regularly with saddlesoap or leather cleaner. Never use alcohol, gasoline, nail polish remover or thinner to clean leather upholstery.

6 After cleaning, regularly treat leather upholstery with a leather conditioner, rubbed in with a soft cotton cloth. Never use car wax on leather upholstery.

7 In areas where the interior of the vehicle is subject to bright sunlight, cover leather seating areas of the seats with a sheet if the vehicle is to be left out for any length of time.

5 Body repair - minor damage

FLEXIBLE PLASTIC BODY PANELS (FRONT AND REAR BUMPER FASCIA)

The following repair procedures are for minor scratches and gouges. Repair of more serious damage should be left to a dealer service department or qualified auto body shop. Below is a list of the equipment and materials necessary to perform the following repair procedures on plastic body panels. Although a specific brand of material may be mentioned, it should be noted that equivalent products from other manufacturers may be used instead.

Wax, grease and silicone removing solvent
Cloth-backed body tape
Sanding discs
Drill motor with three-inch disc holder
Hand sanding block
Rubber squeegees
Sandpaper
Non-porous mixing palette
Wood paddle or putty knife
Curved-tooth body file
Flexible parts repair material

1 Remove the damaged panel, if necessary or desirable. In most cases, repairs can be carried out with the panel installed.

2 Clean the area(s) to be repaired with a wax, grease and silicone removing solvent applied with a water-dampened cloth.

3 If the damage is structural, that is, if it extends through the panel, clean the backside of the panel area to be repaired as well. Wipe dry.

4 Sand the rear surface about 1-1/2 inches beyond the break.

5 Cut two pieces of fiberglass cloth large enough to overlap the break by about 1-1/2 inches. Cut only to the required length.

6 Mix the adhesive from the repair kit according to the instructions included with the kit, and apply a layer of the mixture approximately 1/8-inch thick on the backside of the panel. Overlap the break by at least 1-1/2 inches.

7 Apply one piece of fiberglass cloth to the adhesive and cover the cloth with additional adhesive. Apply a second piece of fiberglass cloth to the adhesive and immediately cover the cloth with additional adhesive in sufficient quantity to fill the weave.

8 Allow the repair to cure for 20 to 30 minutes at 60-degrees to 80-degrees F.

9 If necessary, trim the excess repair material at the edge.

10 Remove all of the paint film over and around the area(s) to be repaired. The repair material should not overlap the painted surface.

11 With a drill motor and a sanding disc (or a rotary file), cut a V along the break line approximately 1/2-inch wide. Remove all dust and loose particles from the repair area.

12 Mix and apply the repair material. Apply a light coat first over the damaged area; then continue applying material until it reaches a level slightly higher than the surrounding finish.

13 Cure the mixture for 20 to 30 minutes at 60-degrees to 80-degrees F.

14 Roughly establish the contour of the area being repaired with a body file. If low areas or pits remain, mix and apply additional adhesive.

15 Block sand the damaged area with sandpaper to establish the actual contour of the surrounding surface.

16 If desired, the repaired area can be temporarily protected with several light coats of primer. Because of the special paints and techniques required for flexible body panels, it is recommended that the vehicle be taken to a paint shop for completion of the body repair.

STEEL BODY PANELS

➧ **See photo sequence**

Repair of minor scratches

17 If the scratch is superficial and does not penetrate to the metal of the body, repair is very simple. Lightly rub the scratched area with a fine rubbing compound to remove loose paint and built up wax. Rinse the area with clean water.

18 Apply touch-up paint to the scratch, using a small brush. Continue to apply thin layers of paint until the surface of the paint in the scratch is level with the surrounding paint. Allow the new paint at least two weeks to harden, and then blend it into the surrounding paint by rubbing with a very fine rubbing compound. Finally, apply a coat of wax to the scratch area.

19 If the scratch has penetrated the paint and exposed the metal of the body, causing the metal to rust, a different repair technique is required. Remove all loose rust from the bottom of the scratch with a pocket knife, and then apply rust inhibiting paint to prevent the formation of rust in the future. Using a rubber or nylon applicator, coat the scratched area with glaze-type filler. If required, the filler can be mixed with thinner to provide a very thin paste, which is ideal for filling narrow scratches. Before the glaze filler in the scratch hardens, wrap a piece of smooth cotton cloth around the tip of a finger. Dip the cloth in thinner and then quickly wipe it along the surface of the scratch. This will ensure that the surface of the filler is slightly hollow. The scratch can now be painted over as described earlier in this Section.

These photos illustrate a method of repairing simple dents. They are intended to supplement Body repair - minor damage in this Chapter and should not be used as the sole instructions for body repair on these vehicles.

1 If you can't access the backside of the body panel to hammer out the dent, pull it out with a slide-hammer-type dent puller. In the deepest portion of the dent or along the crease line, drill or punch hole(s) at least one inch apart . . .

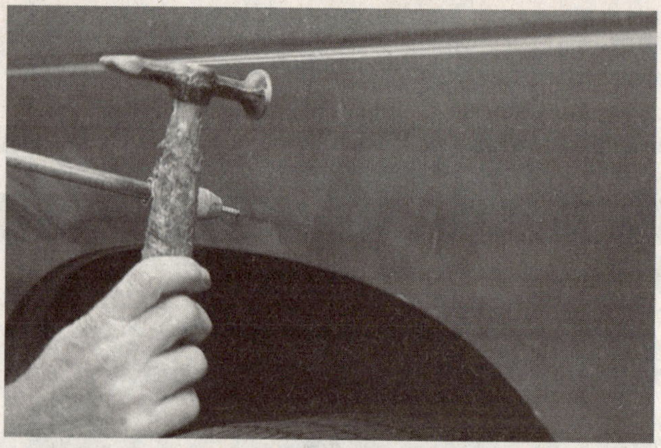

2 . . . then screw the slide-hammer into the hole and operate it. Tap with a hammer near the edge of the dent to help 'pop' the metal back to its original shape. When you're finished, the dent area should be close to its original contour and about 1/8-inch below the surface of the surrounding metal

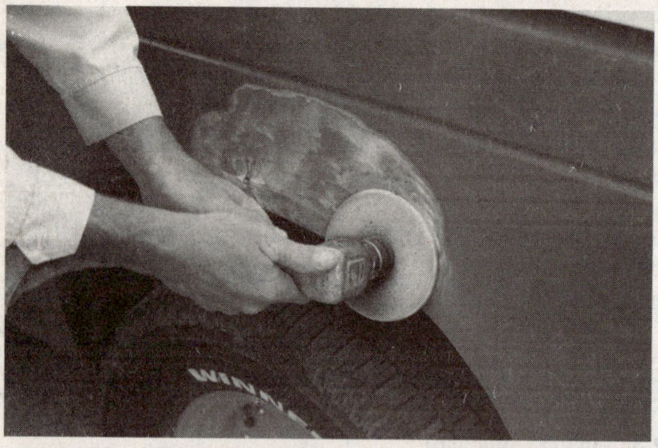

3 Using coarse-grit sandpaper, remove the paint down to the bare metal. Hand sanding works fine, but the disc sander shown here makes the job faster. Use finer (about 320-grit) sandpaper to feather-edge the paint at least one inch around the dent area

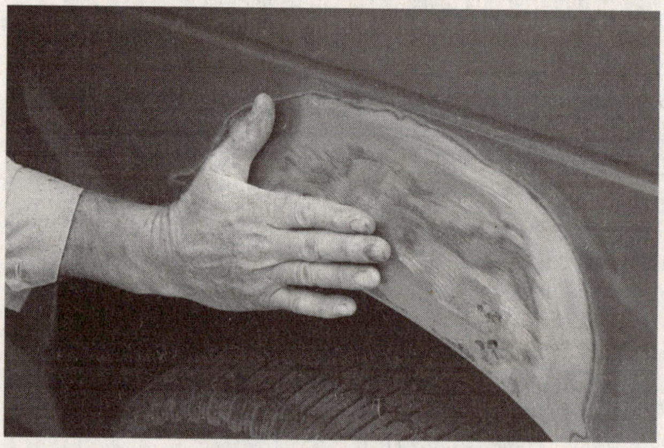

4 When the paint is removed, touch will probably be more helpful than sight for telling if the metal is straight. Hammer down the high spots or raise the low spots as necessary. Clean the repair area with wax/silicone remover

5 Following label instructions, mix up a batch of plastic filler and hardener. The ratio of filler to hardener is critical, and, if you mix it incorrectly, it will either not cure properly or cure too quickly (you won't have time to file and sand it into shape)

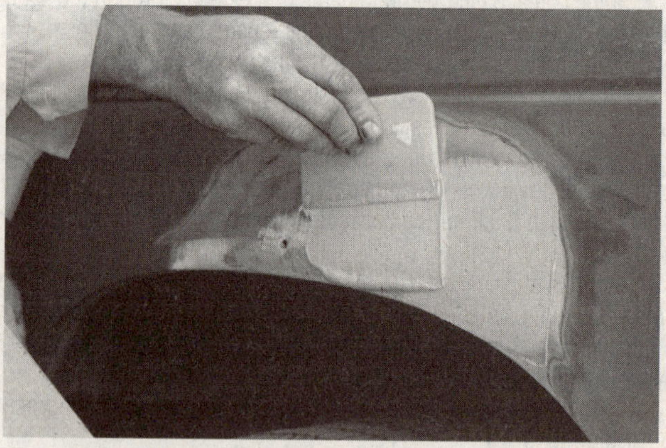

6 Working quickly so the filler doesn't harden, use a plastic applicator to press the body filler firmly into the metal, assuring it bonds completely. Work the filler until it matches the original contour and is slightly above the surrounding metal

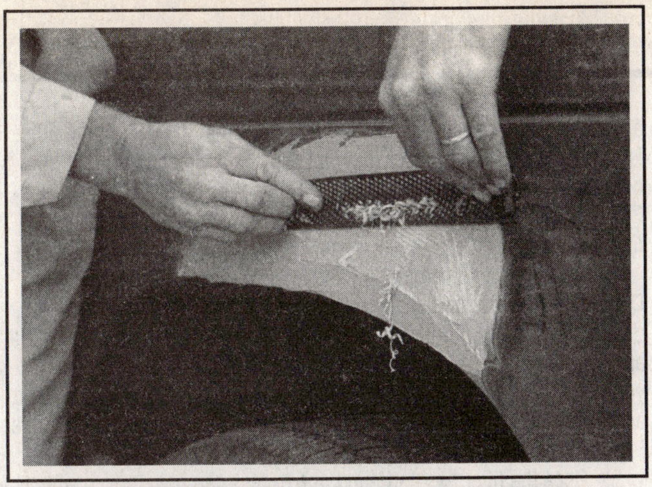

7 Let the filler harden until you can just dent it with your fingernail. Use a body file or Surform tool (shown here) to rough-shape the filler

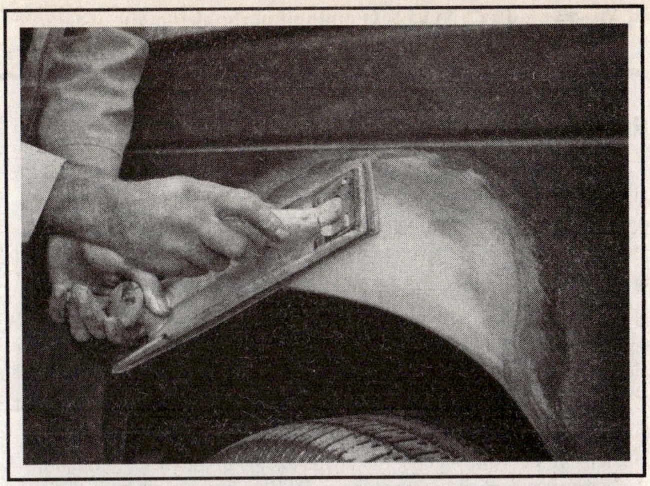

8 Use coarse-grit sandpaper and a sanding board or block to work the filler down until it's smooth and even. Work down to finer grits of sandpaper - always using a board or block - ending up with 360 or 400 grit

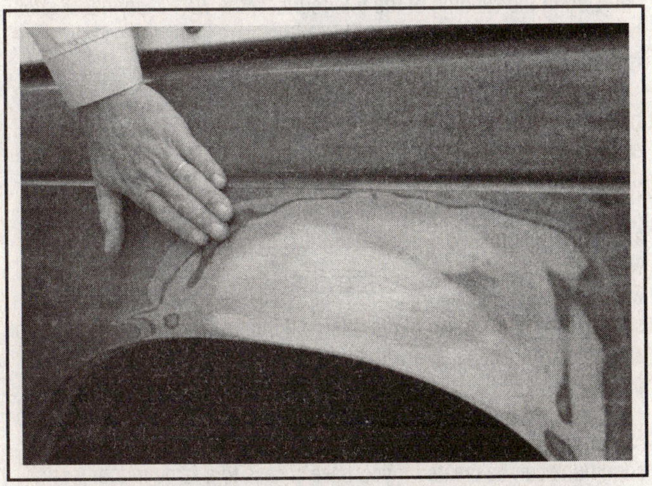

9 You shouldn't be able to feel any ridge at the transition from the filler to the bare metal or from the bare metal to the old paint. As soon as the repair is flat and uniform, remove the dust and mask off the adjacent panels or trim pieces

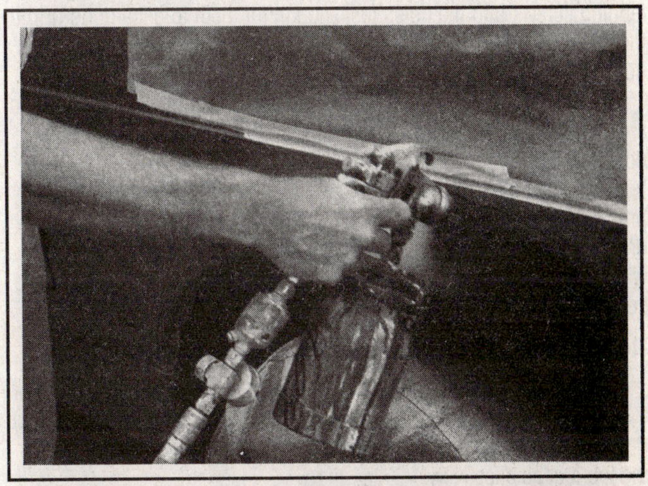

10 Apply several layers of primer to the area. Don't spray the primer on too heavy, so it sags or runs, and make sure each coat is dry before you spray on the next one. A professional-type spray gun is being used here, but aerosol spray primer is available inexpensively from auto parts stores

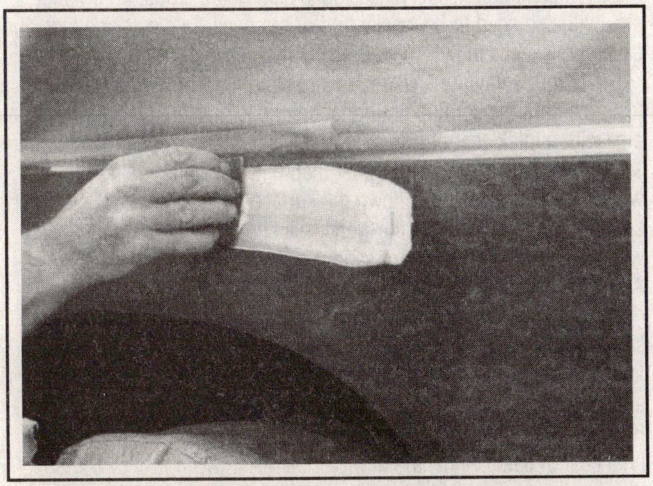

11 The primer will help reveal imperfections or scratches. Fill these with glazing compound. Follow the label instructions and sand it with 360 or 400-grit sandpaper until it's smooth. Repeat the glazing, sanding and respraying until the primer reveals a perfectly smooth surface

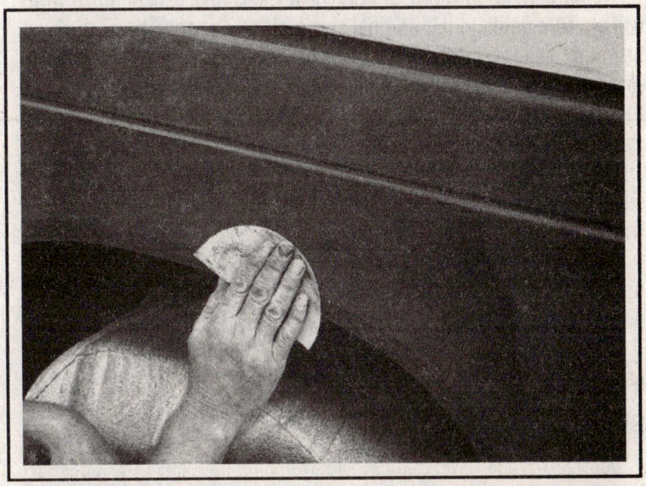

12 Finish sand the primer with very fine sandpaper (400 or 600-grit) to remove the primer overspray. Clean the area with water and allow it to dry. Use a tack rag to remove any dust, then apply the finish coat. Don't attempt to rub out or wax the repair area until the paint has dried completely (at least two weeks)

Repair of dents

20 When repairing dents, the first job is to pull the dent out until the affected area is as close as possible to its original shape. There is no point in trying to restore the original shape completely as the metal in the damaged area will have stretched on impact and cannot be restored to its original contours. It is better to bring the level of the dent up to a point that is about 1/8-inch below the level of the surrounding metal. In cases where the dent is very shallow, it is not worth trying to pull it out at all.

21 If the back side of the dent is accessible, it can be hammered out gently from behind using a soft-face hammer. While doing this, hold a block of wood firmly against the opposite side of the metal to absorb the hammer blows and prevent the metal from being stretched.

22 If the dent is in a section of the body that has double layers, or some other factor makes it inaccessible from behind, a different technique is required. Drill several small holes through the metal inside the damaged area, particularly in the deeper sections. Screw long, self-tapping screws into the holes just enough for them to get a good grip in the metal. Now the dent can be pulled out by pulling on the protruding heads of the screws with locking pliers.

23 The next stage of repair is the removal of paint from the damaged area and from an inch or so of the surrounding metal. This is easily done with a wire brush or sanding disk in a drill motor, although it can be done just as effectively by hand with sandpaper. To complete the preparation for filling, score the surface of the bare metal with a screwdriver or the tang of a file or drill small holes in the affected area. This will provide a good grip for the filler material. To complete the repair, see the Section on filling and painting.

Repair of rust holes or gashes

24 Remove all paint from the affected area and from an inch or so of the surrounding metal using a sanding disk or wire brush mounted in a drill motor. If these are not available, a few sheets of sandpaper will do the job just as effectively.

25 With the paint removed, you will be able to determine the severity of the corrosion and decide whether to replace the whole panel, if possible, or repair the affected area. New body panels are not as expensive as most people think and it is often quicker to install a new panel than to repair large areas of rust.

26 Remove all trim pieces from the affected area except those which will act as a guide to the original shape of the damaged body, such as headlight shells, etc. Using metal snips or a hacksaw blade, remove all loose metal and any other metal that is badly affected by rust. Hammer the edges of the hole in to create a slight depression for the filler material.

27 Wire brush the affected area to remove the powdery rust from the surface of the metal. If the back of the rusted area is accessible, treat it with rust inhibiting paint.

28 Before filling is done, block the hole in some way. This can be done with sheet metal riveted or screwed into place, or by stuffing the hole with wire mesh.

29 Once the hole is blocked off, the affected area can be filled and painted. See the following subsection on filling and painting.

Filling and painting

30 Many types of body fillers are available, but generally speaking, body repair kits which contain filler paste and a tube of resin hardener are best for this type of repair work. A wide, flexible plastic or nylon applicator will be necessary for imparting a smooth and contoured finish to the surface of the filler material. Mix up a small amount of filler on a clean piece of wood or cardboard (use the hardener sparingly). Follow the manufacturer's instructions on the package, otherwise the filler will set incorrectly.

31 Using the applicator, apply the filler paste to the prepared area. Draw the applicator across the surface of the filler to achieve the desired contour and to level the filler surface. As soon as a contour that approximates the original one is achieved, stop working the paste. If you continue, the paste will begin to stick to the applicator. Continue to add thin layers of paste at 20-minute intervals until the level of the filler is just above the surrounding metal.

32 Once the filler has hardened, the excess can be removed with a body file. From then on, progressively finer grades of sandpaper should be used, starting with a 180-grit paper and finishing with 600-grit wet-or-dry paper. Always wrap the sandpaper around a flat rubber or wooden block, otherwise the surface of the filler will not be completely flat. During the sanding of the filler surface, the wet-or-dry paper should be periodically rinsed in water. This will ensure that a very smooth finish is produced in the final stage.

33 At this point, the repair area should be surrounded by a ring of bare metal, which in turn should be encircled by the finely feathered edge of good paint. Rinse the repair area with clean water until all of the dust produced by the sanding operation is gone.

34 Spray the entire area with a light coat of primer. This will reveal any imperfections in the surface of the filler. Repair the imperfections with fresh filler paste or glaze filler and once more smooth the surface with sandpaper. Repeat this spray-and-repair procedure until you are satisfied that the surface of the filler and the feathered edge of the paint are perfect. Rinse the area with clean water and allow it to dry completely.

35 The repair area is now ready for painting. Spray painting must be carried out in a warm, dry, windless and dust free atmosphere. These conditions can be created if you have access to a large indoor work area, but if you are forced to work in the open, you will have to pick the day very carefully. If you are working indoors, dousing the floor in the work area with water will help settle the dust that would otherwise be in the air. If the repair area is confined to one body panel, mask off the surrounding panels. This will help minimize the effects of a slight mismatch in paint color. Trim pieces such as chrome strips, door handles, etc., will also need to be masked off or removed. Use masking tape and several thickness of newspaper for the masking operations.

36 Before spraying, shake the paint can thoroughly, and then spray a test area until the spray-painting technique is mastered. Cover the repair area with a thick coat of primer. The thickness should be built up using several thin layers of primer rather than one thick one. Using 600-grit wet-or-dry sandpaper, rub down the surface of the primer until it is very smooth. While doing this, the work area should be thoroughly rinsed with water and the wet-or-dry sandpaper periodically rinsed as well. Allow the primer to dry before spraying additional coats.

37 Spray on the top coat, again building up the thickness by using several thin layers of paint. Begin spraying in the center of the repair area and then, using a circular motion, work out until the whole repair area and about two inches of the surrounding original paint is covered. Remove all masking material 10 to 15 minutes after spraying on the final coat of paint. Allow the new paint at least two weeks to harden, then use a very fine rubbing compound to blend the edges of the new paint into the existing paint. Finally, apply a coat of wax.

6 Body repair - major damage

1 Major damage must be repaired by an auto body shop specifically equipped to perform unibody repairs. These shops have the specialized equipment required to do the job properly.

2 If the damage is extensive, the body must be checked for proper alignment or the vehicle's handling characteristics may be adversely affected and other components may wear at an accelerated rate.

3 Due to the fact that some of the major body components (hood, fenders, doors, etc.) are separate and replaceable units, any seriously damaged components should be replaced rather than repaired. Sometimes the components can be found in a wrecking yard that specializes in used vehicle components, often at considerable savings over the cost of new parts.

7 Hinges and locks - maintenance

Once every 3000 miles, or every three months, the hinges and latch assemblies on the doors, hood and trunk should be given a few drops of light oil or lock lubricant. The door latch strikers should also be lubricated with a thin coat of grease to reduce wear and ensure free movement. Lubricate the door and trunk locks with spray-on graphite lubricant.

8 Windshield and fixed glass - replacement

Replacement of the windshield and fixed glass requires the use of special fast-setting adhesive/caulk materials and some specialized tools and techniques. These operations should be left to a dealer service department or a shop specializing in glass work.

9 Hood - removal, installation and adjustment

➡Note: The hood is somewhat awkward to remove and install; at least two people should perform this procedure.

REMOVAL AND INSTALLATION

➧ Refer to illustrations 9.3 and 9.4

1 Open the hood, then place blankets or pads over the fenders and cowl area of the body. This will protect the body and paint as the hood is lifted off.

2 Disconnect any cables or wires that will interfere with removal. Disconnect the windshield washer tubing near the right-side hinge.

3 Mark the hood in relation to the hinges to ensure proper alignment during installation (see illustration).

4 Have an assistant support one side of the hood. Take turns removing the hinge-to-hood bolts and lift off the hood (see illustration).

5 Installation is the reverse of removal. Align the hood using the marks made in Step 3.

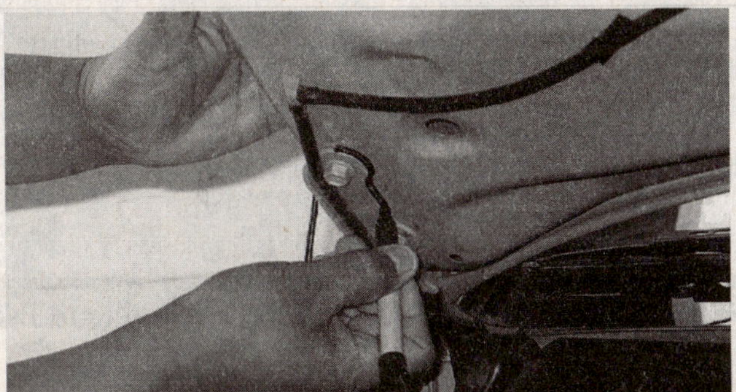

9.3 Draw alignment marks on the hood around the hinge plates to align it when it's reinstalled

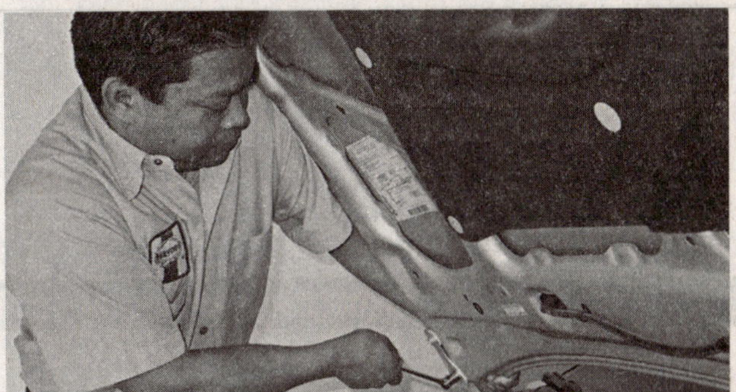

9.4 Support the hood with your shoulder while removing the hood mounting bolts

ADJUSTMENT

➧ **Refer to illustrations 9.11a, 9.11b and 9.12**

➡**Note: Replace the hinge and latch factory mounting bolts with regular bolts (of the same grade) and washers (see Section 1).**

6 Fore-and-aft and side-to-side adjustment of the hood is done by moving the hood in relation to the hinge plate slots while the bolts are loose.

7 Mark the hood around the hinge plates so you can determine the amount of movement (see illustration 9.3).

8 Loosen the bolts and move the hood into correct alignment. Move it only a little at a time. Tighten the hinge bolts and carefully lower the hood to check its position. Look along the edge of each side of the hood observing the small gap between each fender. The gaps should be uniform on both sides.

9 If necessary, the entire hood latch assembly (at the radiator support) can be moved up-and-down and side-to-side to adjust the hood so that it closes securely and is relatively flush with the fenders.

10 Remove the radiator grill (see Section 13).

11 Remove the hood latch cover (see illustration) and mark the relationship of the hood latch to the radiator support, then loosen the mounting bolts and reposition the latch assembly as necessary (see illustration). Following adjustment, retighten the mounting bolts.

12 Finally, adjust the hood cushions so that when the hood is closed, it's flush with the fenders (see illustration).

13 The hood latch assembly, as well as the hinges, should be periodically lubricated with white, lithium-base grease to prevent binding and wear.

14 The remainder of the installation is the reverse of removal.

9.11a Remove the mounting bolts for the latch cover

9.11b With loose mounting bolts, adjust the hood latch horizontally or vertically

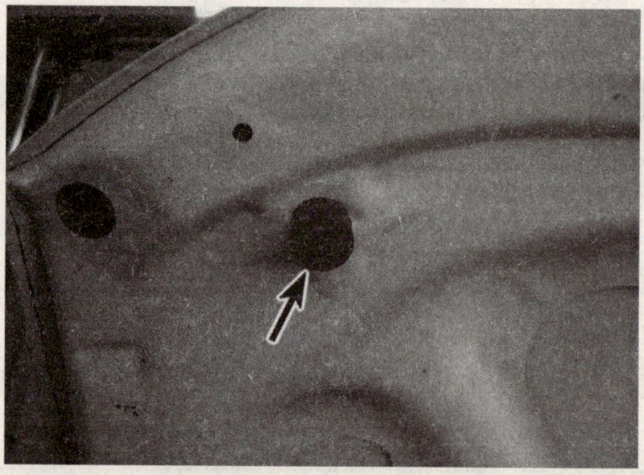

9.12 Rotate the hood cushions (one on each side) to adjust the height of the hood's leading edge so that it's flush with the fenders. Turn the cushions clockwise to lower the hood and counterclockwise to raise it

10 Hood latch and release cable - removal and installation

1 Remove the radiator grill (see Section 13).

LATCH

➧ **Refer to illustration 10.3**

2 Remove the hood latch cover (see illustration 9.11a) and mark the relationship of the hood latch to the radiator support to aid reassembly. Remove the mounting bolts and the latch assembly (see illustration 9.11b).

3 Disconnect the hood release cable from the latch (see illustration).

4 Installation is the reverse of removal. Use the index marks made in Step 1 to align the latch to its original position.

10.3 **Carefully pry or pull the cable retainer (A) from the small bracket on the hood latch assembly, then disengage the cable from the lever (B)**

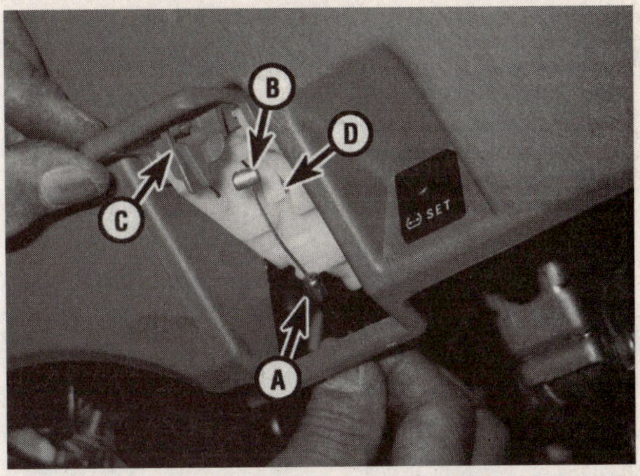

10.5 **Pull the cable housing (A) out of its retainer, then guide the cable out of the slot in the side. Detach the cable end (B) from the lever (C). If necessary, pull outward on the retaining tab (D) while pushing the assembly forward to release it from the panel, then remove it**

CABLE

➧ Refer to illustration 10.5

5 Working in the passenger compartment, disengage the cable from the hood release lever (see illustration). If the lever assembly needs to be replaced, pull outward on the assembly retaining tab while pushing the assembly towards the firewall to release it. Once it's loose, guide it out of the lower instrument panel.

6 Attach a piece of thin wire or string to the end of the cable.

7 Working in the engine compartment, disconnect the hood release cable from the latch as described in Steps 1 and 2. Follow the path of the cable from the latch towards the firewall while unclipping the cable retaining clips.

8 Pull the cable forward into the engine compartment until you can see the wire or string, then remove the wire or string from the old cable and fasten it to the new cable.

9 With the new cable attached to the wire or string, pull the wire or string back through the firewall grommet and inside to the passenger compartment until the new cable reaches the hood release lever.

➡Note: Pull on the cable housing with your fingers from the passenger compartment until the cable stop seats in the grommet on the firewall.

10 Install the new cable to the hood release lever, making sure the cable housing fits snugly into its retainer on the assembly.

11 The remainder of the installation is the reverse of removal.

11 Bumpers - removal and installation

FRONT BUMPER COVER

➧ Refer to illustrations 11.4, 11.5a, 11.5b and 11.5c

1 Remove the radiator grille (see Section 13).

2 Apply the parking brake, raise the vehicle and support it securely on jackstands.

3 Disconnect the electrical connectors for the fog lights, if equipped.

4 Remove the bottom fasteners for the bumper cover (see illustration). Refer to Section 1 for fastener types and removal techniques.

5 Remove the side and top fasteners for the bumper cover (see illustrations). Disengage the bumper cover from the retainer on the bottom of each fender (see illustration), then remove the cover from the vehicle.

➡Note: Some models may be equipped with additional bumper extensions, valances, spoilers, etc. Remove any fasteners from these items as necessary to remove the bumper cover. Otherwise, they can be detached after the bumper cover is removed.

11.4 **Location of the bumper cover's lower fasteners**

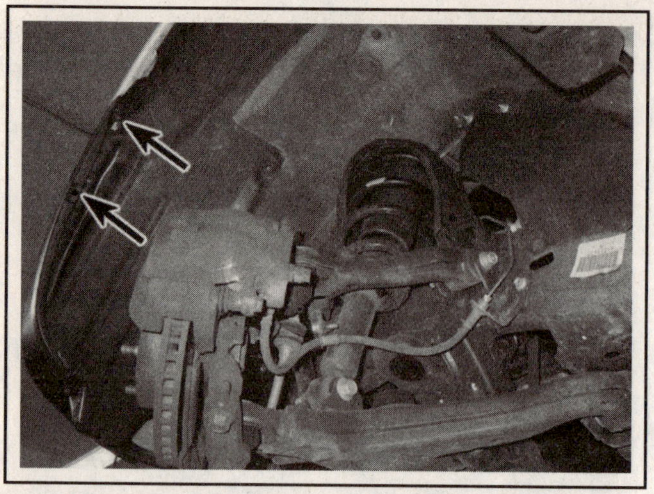

11.5a Location of the bumper cover's side fasteners

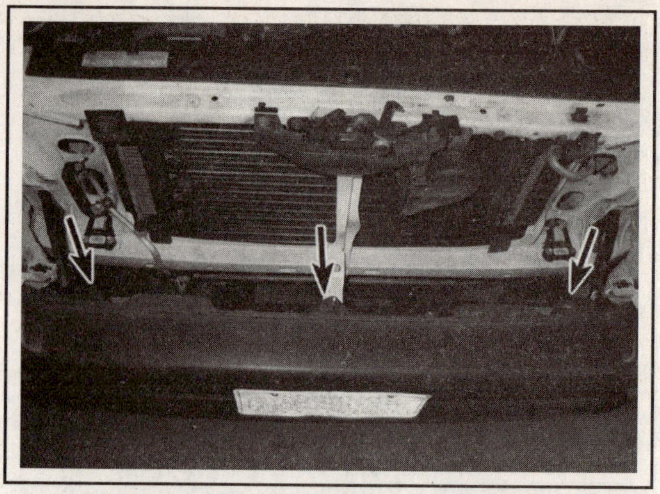

11.5b Location of the bumper cover's top fasteners

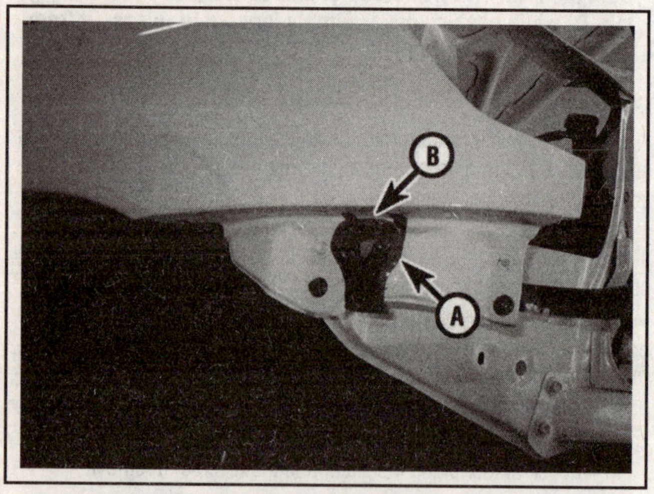

11.5c Each large retainer (A) has a tab (B) that fits into a slot on the bumper cover (bumper cover removed for clarity)

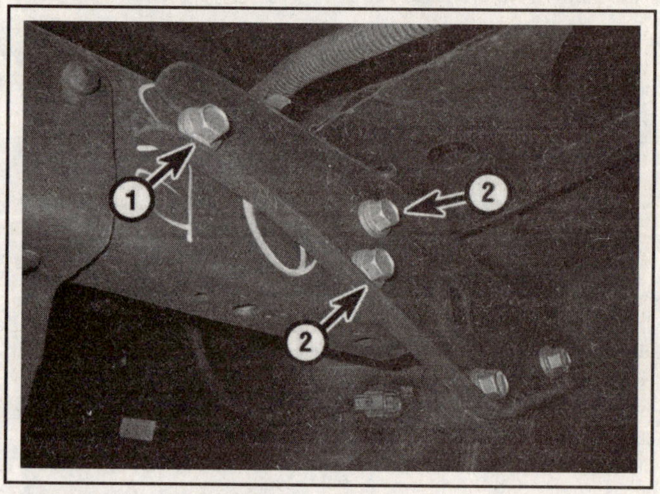

11.8 Bumper mounting bracket details:

1 *Mounting bolt (loosen and leave in place)*
2 *Mounting bolts (remove)*

6 Installation is the reverse of removal. Make sure any tabs for the bumper cover fit into the corresponding slots on the body before attaching any fasteners. An assistant would be helpful for installation.

REAR BUMPER

Refer to illustrations 11.8 and 11.9

7 Locate and unplug any electrical connectors that would interfere with bumper removal.

8 With an assistant holding the bumper to the frame rails, remove the bumper mounting bracket-to-frame rail bolts (see illustration)

Note: On models equipped with a factory installed trailer hitch, remove the hitch along with the bumper as an assembly.

9 Carefully move the bumper while disengaging the mounting bracket pins from the slots in the frame rails (see illustration).

10 Installation is the reverse of removal.

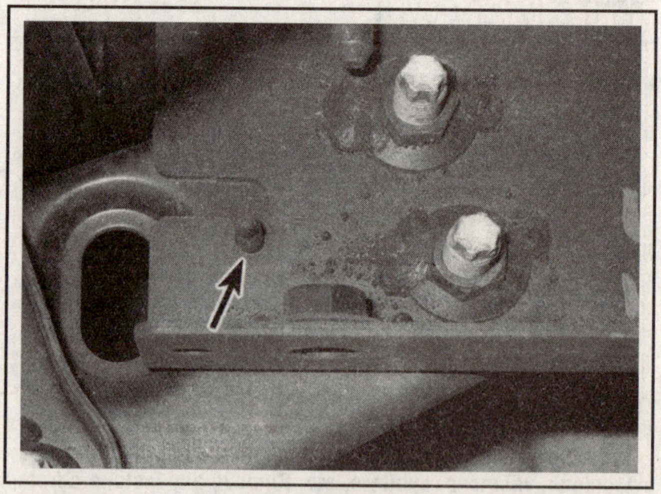

11.9 Pins on the mounting brackets engage in slots in the frame rails to ease removal and installation

12 Front fender - removal and installation

Refer to illustrations 12.5, 12.6, 12.7, 12.9, 12.10 and 12.11

1 Loosen the front wheel lug nuts. Raise the vehicle, support it securely on jackstands and remove the front wheel.
2 Open the hood.
3 Remove the front bumper cover (see Section 11).
4 Remove the headlight housing (see Chapter 12).
5 Remove the inner fender splash shield fasteners, then remove the shield (see illustration). Refer to Section 1 for fastener types and removal techniques.
6 On models equipped with trim molding around the wheel opening on the fender and the rocker panels, the trim will have to be removed first. Look for screws or bolts along the bottom edge of each type of trim and remove them. Use a flat tool wrapped in tape to pry the trim away from the rocker panel and fender (see illustration).
7 Remove the mounting fasteners at the front of the fender (see illustration).
8 If you're removing the passenger side fender, remove the radio antenna (see Chapter 12).
9 Working inside the inner fenderwell, remove the fasteners securing the fender to the body (see illustration).

12.5 The inner fender splash shield fasteners (some fasteners are not visible in this photograph)

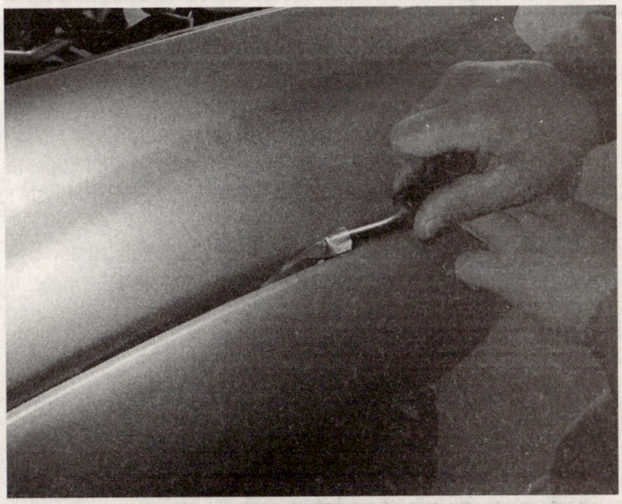

12.6 Carefully remove the trim using a tool wrapped with tape

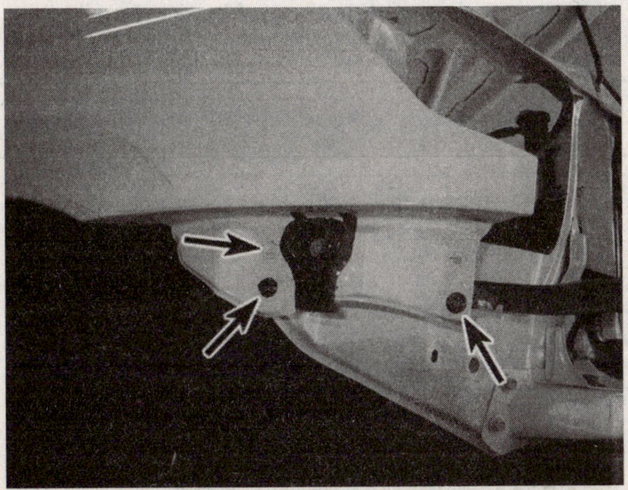

12.7 Remove the fender front mounting fasteners

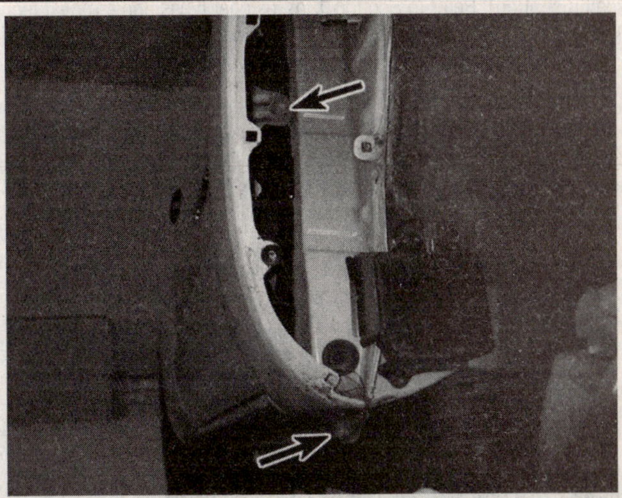

12.9 Remove the lower rear mounting fasteners from inside the fenderwell

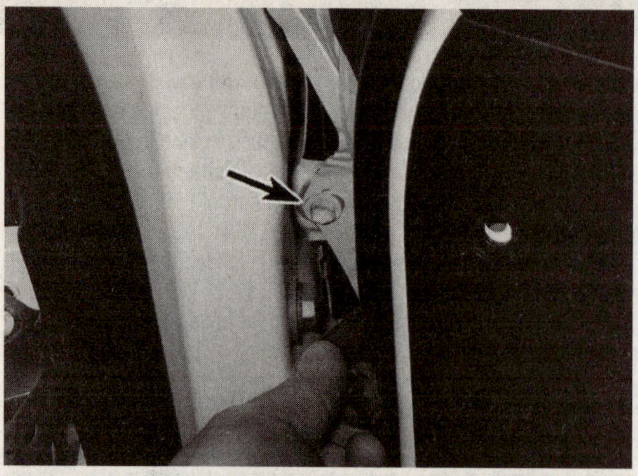

12.10 Remove the upper fender mounting fastener at the door pillar

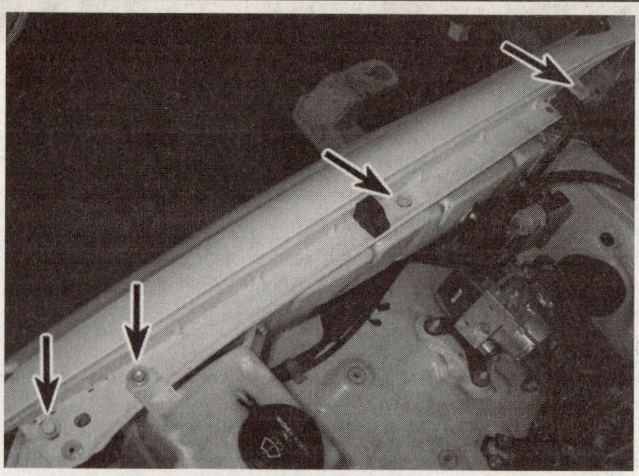

12.11 Remove the mounting fasteners along the top of the fender

10 Open the front door. Remove the lining fastener near the top and pull the lining back enough to remove the upper fender-to-body fastener (see illustration).

11 Remove the fender upper mounting fasteners (see illustration).

12 Lift off the fender. It's a good idea to have an assistant support the fender while its being moved away from the vehicle to prevent damage to the surrounding body panels.

13 Installation is the reverse of removal. Check the alignment of the fender to the hood and front edge of the door before final tightening of the fender fasteners.

13 Radiator grille - removal and installation

Refer to illustrations 13.1 and 13.2

1 Open the hood. Remove the fasteners along the top of the grille (see illustration). Refer to Section 1 for fastener types and removal techniques.

2 Working from the back of the grille, release the grille from the small clips on the bottom corner of each side of the grille while pulling the grille up (see illustration).

3 Once the retaining clips are disengaged, remove the grille.

4 Installation is the reverse of removal.

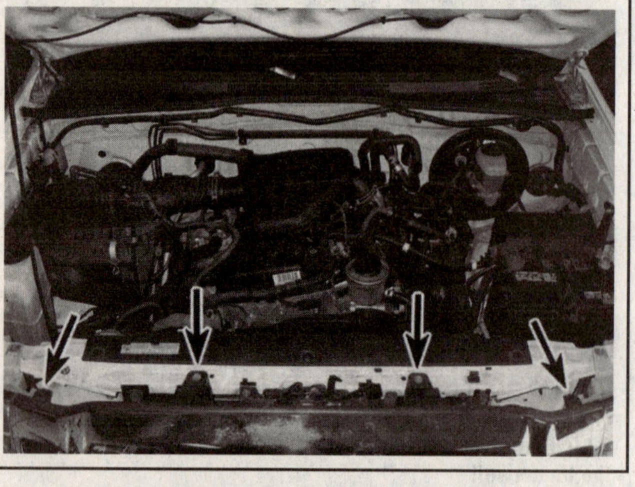

13.1 Radiator grille upper fasteners

13.2 The location of the two clips on the back side of the radiator grille (grille removed for clarity)

14 Cowl cover - removal and installation

➧ Refer to illustration 14.2

1 Remove the wiper arms (see Chapter 12).
2 Remove the push pin fastener on each end of the cowl cover (see illustration). Refer to Section 1 for fastener types and removal techniques.
3 Use a flat tool to pry the lower edge of the cowl cover up from the tray in the body. Pry at the points where the retainers are located. Be careful not to damage the weather seal.
4 With the bottom of the cowl cover loosened, pull the cover away from the windshield. The cowl seals at each end of the cover may break away from the cowl during removal or can be detached to ease removal (see illustration 14.2).
5 Installation is the reverse of removal.

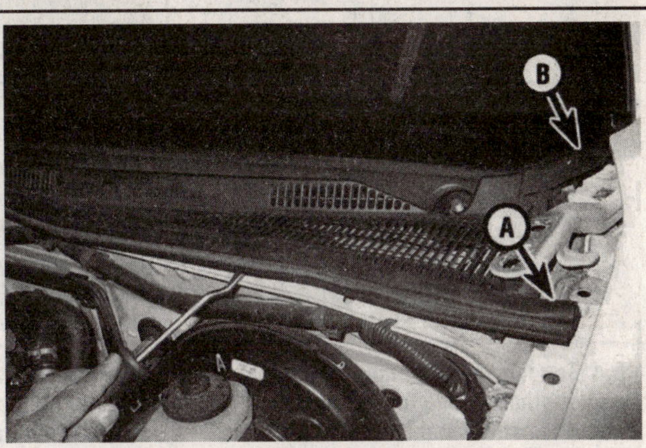

14.2 The vicinity of the cowl fastener (A) and the cowl seal (B) (left side shown - right side similar)

15 Door trim panels - removal and installation

⁂ WARNING:

The edges around the door frame openings are very sharp. Wear gloves for protection against cuts when working inside the door.

➡Note: Refer to Section 1 for general information on removing trim panels.

1 Disconnect the cable from the negative battery terminal (see Chapter 5, Section 1).

FRONT AND REAR DOORS

➧ Refer to illustrations 15.2, 15.3, 15.4, 15.5, 15.7a, 15.7b, 15.8, 15.9 and 15.10

2 Remove the small trim cover from the upper part of the door (see illustration).
3 On models with power windows and door locks, pry the window switch or switch plate off the door panel, disconnect the electrical connectors beneath and remove it (see illustration).

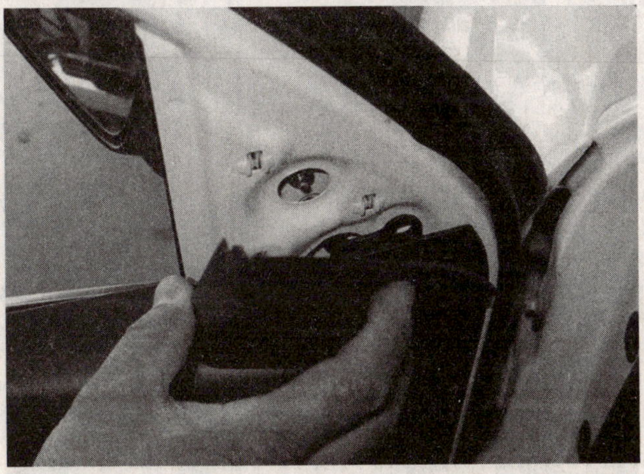

15.2 Carefully pry the small trim cover off the door

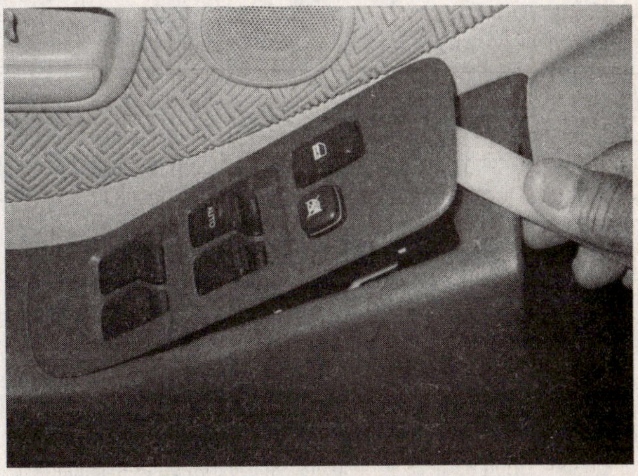

15.3 Pry the power window switch plate up, then disconnect the electrical connectors and remove the switch plate

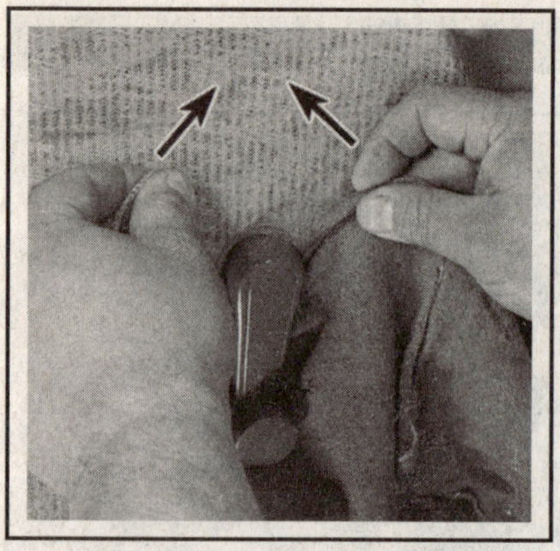

15.4 To remove the window regulator handle, place a shop towel directly under the handle and in front of the round plastic washer. Move the towel back and forth while also pulling it up. This will release the small clip that holds the handle to the regulator

15.5 The location of the small fastener on the door panel (front door shown - rear door similar)

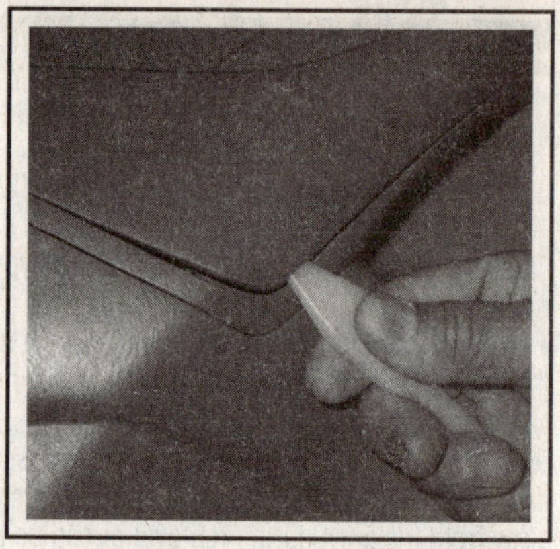

15.7a Pry the trim cover off the door grip . . .

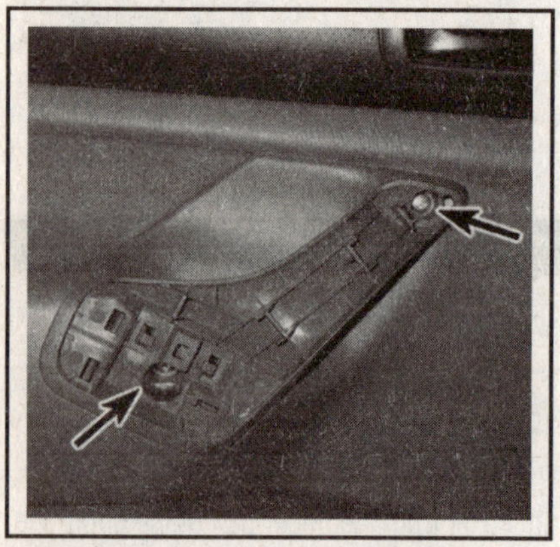

15.7b . . . then remove the door panel mounting screws

15.8 At the door release lever, pry the small screw cover aside, then remove the panel mounting screw

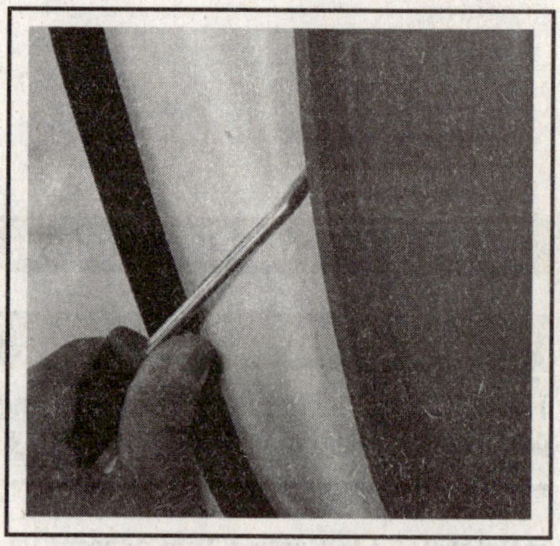

15.9 Carefully pry the door panel away from the door

4 On models without power windows, remove the window regulator handle (see illustration).

5 Remove the fastener in the upper corner of the door panel (see illustration). Refer to Section 1 for fastener types and removal techniques.

6 On models with power windows, remove the panel mounting screw that is behind the door grip that is just in front of the armrest portion of the panel on front doors. On rear doors, the mounting screw is at the bottom of the door grip trim.

7 On models without power windows, pry the cover off the door grip, then remove the mounting screws (see illustrations).

8 Remove the panel mounting screw that is located behind the door release lever (see illustration).

9 Pry the panel away from the door until the fasteners disengage (see illustration). Work slowly and carefully around the bottom and side edges of the trim panel. Pull the bottom of the trim away from the door, unplug any wiring harness connectors, lift the trim up to disengage it from the window opening, then remove it.

10 For access to the outside door handle or the door window regulator, raise the window fully, then carefully peel back the plastic watershield (see illustration).

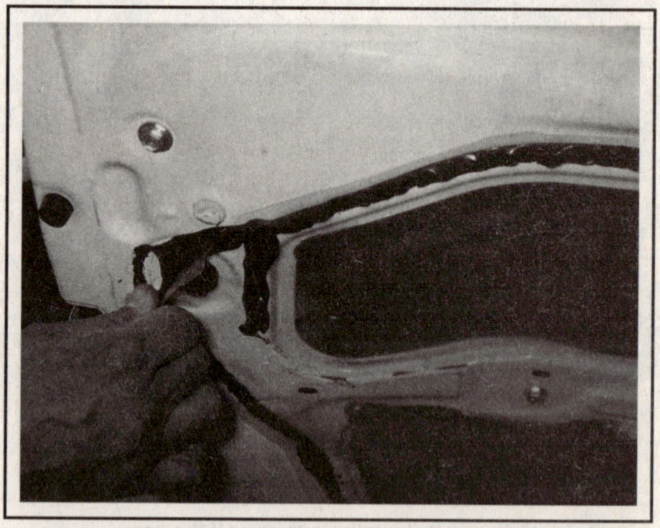

15.10 Starting in the upper corner, carefully peel back the plastic watershield for access to the inner door

11 Installation is the reverse of removal.

REAR DOORS ON ACCESS-CAB MODELS

12 Remove the weather-stripping from the door by carefully removing the top mounting fasteners and pulling it away from the edge around the door.

13 Remove the plastic trim covering the seatbelt anchor near the bottom of the door, then remove the anchor mounting bolt and anchor.

14 Locate the panel mounting screw at the back of the door release lever's trim, then remove it and the trim.

➡**Note: The mounting screw has a small cover over it; simply pry it aside to reveal the screw.**

15 Locate the mounting screw at the bottom of the door-pull handle just below the window, then remove it and the trim.

16 Pry the panel away from the door until the fasteners disengage (see illustration 15.9). Work slowly and carefully around the bottom and side edges of the trim panel. Pull the bottom of the trim away from the door, lift the trim up to disengage it, then remove it.

17 Installation is the reverse of removal.

16 Door - removal, installation and adjustment

➡**Note: The door is heavy and somewhat awkward to remove and install - at least two people should perform this procedure.**

REMOVAL AND INSTALLATION

➧ Refer to illustrations 16.6 and 16.8

1 Raise the window completely in the door, then disconnect the cable from the negative battery terminal (see Chapter 5, Section 1).

2 Open the door all the way and support it from the ground on jacks or blocks covered with rags to prevent damaging the paint.

3 Remove the door trim panel and watershield as described in Section 15.

4 Disconnect all electrical connections, ground wires and harness retaining clips from the door.

➡**Note: It is a good idea to label all connections to aid the reassembly process.**

5 From the door side, detach the rubber conduit between the body and the door. Then pull the wiring harness through the conduit hole and remove it from the door.

6 Remove the door stop strut bolt (see illustration).

7 Mark around the door hinges with a pen or equivalent to use as a reference for reassembly.

8 With an assistant holding the door, remove the hinge-to-door bolts (see illustration) and lift off the door.

9 Installation is the reverse of removal.

ADJUSTMENT

➧ Refer to illustration 16.12

➡**Note: Replace the hinge factory mounting bolts with regular bolts (of the same grade) and washers (see Section 1).**

10 Having proper door-to-body alignment is a critical part of a well-functioning door assembly. First check the door hinge pins for excessive play. Fully open the door and lift up and down on the door without lifting the body. If a door has 1/16-inch or more excessive play, the hinges should be replaced.

11 Door-to-body alignment adjustments are made by loosening the hinge-to-body or hinge-to-door (non-factory) mounting bolts and moving the door. Proper body alignment is achieved when the top of the doors are parallel with the roof section, the front door is flush with the fender, the rear door is flush with the rear quarter panel and the bottom of the doors are aligned with the lower rocker panel. If these goals can't be reached by adjusting the hinge-to-body or hinge-to-door bolts, body alignment shims may have to be purchased and inserted behind the hinges to achieve correct alignment.

12 To adjust the door-closed position, scribe a line or mark around the striker plate to provide a reference point, then check that the door latch is contacting the center of the latch striker. If not, adjust the up and down position first (see illustration).

13 Finally adjust the latch striker sideways position, so that the door panel is flush with the center pillar or rear quarter panel and provides positive engagement with the latch mechanism.

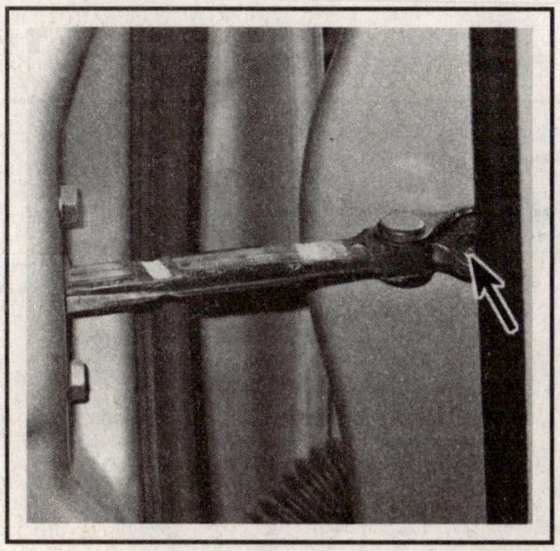

16.6 Remove the bolt retaining the door stop strut

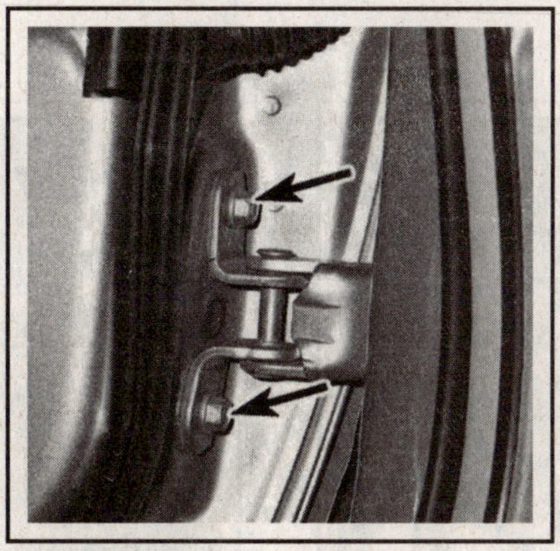

16.8 Remove the door hinge bolts with the door supported

16.12 Adjust the door latch striker by loosening the mounting screws to the point of being snug and gently tapping the striker to the desired position

17 Door latch, lock cylinder and handles - removal and installation

WARNING:

Wear gloves when working inside the door openings to protect against cuts from sharp metal edges.

1 Raise the window, then remove the door trim panel and watershield (see Section 15).

DOOR LATCH

Refer to illustrations 17.2 and 17.5

2 Working through the large access hole, disengage the rod from the handle (see illustration). All door lock rods are attached by plastic clips. The plastic clips can be removed by unsnapping the portion engaging the connecting rod, then pulling the rod out of its locating hole.
3 Remove the door lock cylinder (see Steps 7 and 8).
4 Disconnect the electrical connectors at the latch.
5 Remove the screws securing the latch to the door (see illustration). Remove the latch assembly through the door opening.
6 Installation is the reverse of removal.

OUTSIDE HANDLE AND DOOR LOCK CYLINDER

Refer to illustrations 17.7 and 17.8

7 Remove the plug from the end of the door and remove the lock cylinder retaining screw (see illustration).
8 Withdraw the lock cylinder from the door (see illustration).
9 Remove the mounting fastener for the handle from the inside of the door and remove the handle frame (see illustration 17.2).
10 Installation is the reverse of removal.

INSIDE DOOR HANDLE

Refer to illustrations 17.11 and 17.12

11 Release the lever retainers to disengage it from the door (see illustration).
12 Disengage the lever-to-latch cables and remove the lever (see illustration).
13 Installation is the reverse of removal.

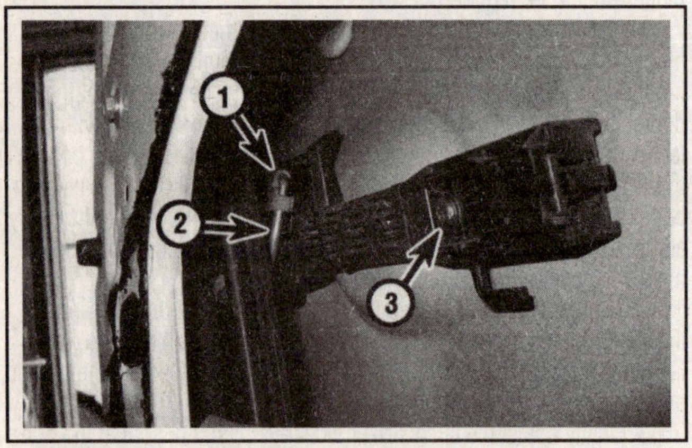

17.2 Door handle details:

1 Rod retaining clip (rotate the clip off the rod to release it)
2 Door handle-to-latch rod
3 Handle frame mounting screw

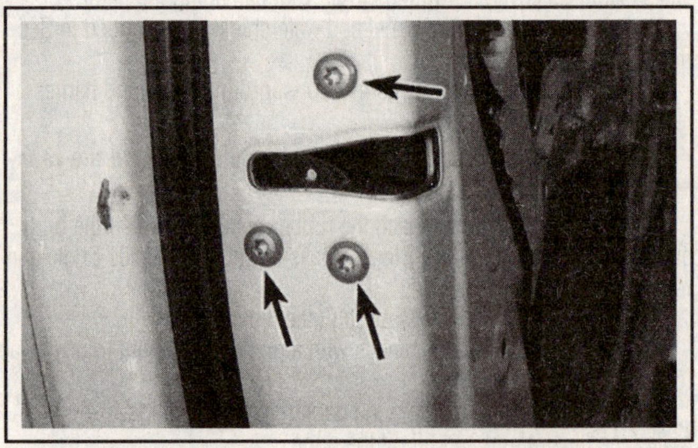

17.5 Remove the door latch mounting fasteners

17.7 Remove the plug from the end of the door to access the door lock cylinder retaining bolt . . .

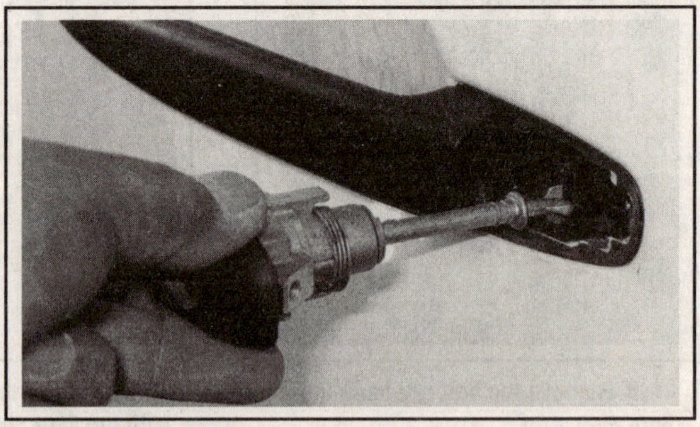

17.8 . . . then slide the lock cylinder out

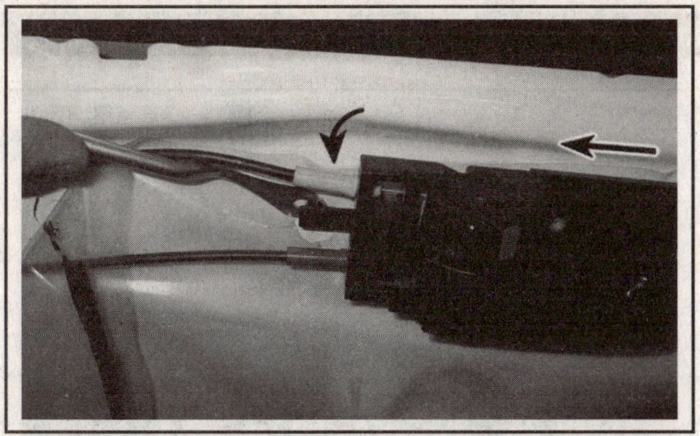

17.11 Release the rear retaining tab and pull the rear of the assembly away from the door and towards the rear to disengage the front retaining tabs

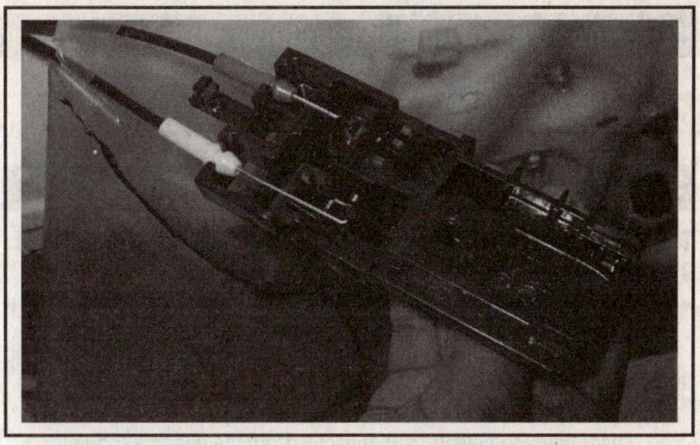

17.12 Detach the cables from the inside of the door handle assembly

18 Door window glass - removal and installation

WARNING:

The edges around the door frame openings are very sharp. Wear gloves for protection against cuts.

FRONT AND REAR DOORS ON REGULAR AND DOUBLE-CAB MODELS, FRONT DOORS ON ACCESS-CAB MODELS

Refer to illustrations 18.2 and 18.3

1 Remove the door trim panel and the plastic watershield (see Section 15).

Note: Temporarily reconnect the window switch or switch plate electrical connectors and battery or attach the window regulator handle (without the retaining clip) to move the window down.

2 Remove the inner metal panel from the door frame (see illustration).

WARNING:

The panel edges are very sharp.

3 Lower the window until the retaining bolts are visible through the holes in the door frame (see illustration).

4 Place a rag inside the door frame to protect the glass from scratches during removal, then remove the retaining bolts.

5 Raise the glass by tilting the rear of the glass upward at an angle (roughly 30-degrees). Pull it directly up and out at that angle.

6 Installation is the reverse of removal.

REAR DOORS ON ACCESS-CAB MODELS

7 Replacement of the rear door glass on Access-Cab models requires the use of special fast-setting adhesive/caulk materials and some specialized tools and techniques. These operations should be left to a dealer service department or a shop specializing in automotive glass work.

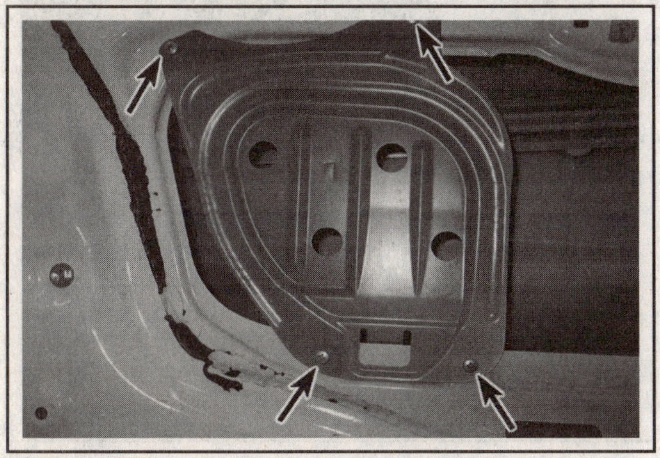

18.2 The inner metal panel and fastener locations

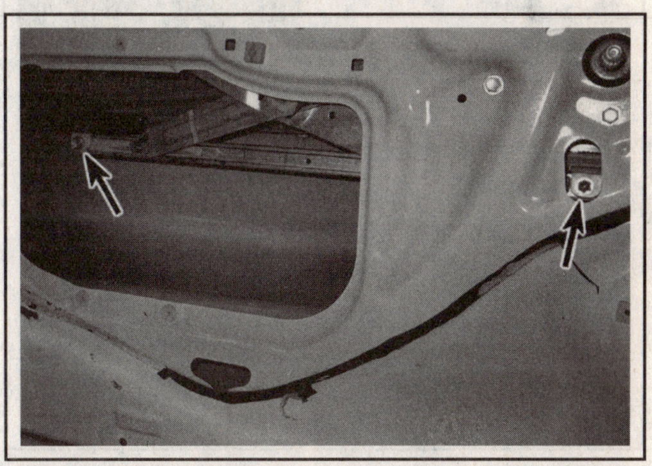

18.3 Lower the window to access the glass retaining bolts through the holes in the door frame

19 Door window glass regulator - removal and installation

➧ Refer to illustration 19.4

⁂ WARNING:

The edges around the door frame openings are very sharp. Wear gloves for protection against cuts.

1 Remove the door trim panel and the plastic watershield (see Section 15).
2 Remove the window glass (see Section 18).
3 Disconnect the electrical connector from the window regulator motor, if applicable.
4 Support the regulator, then remove the regulator/motor mounting bolts (see illustration).
5 Remove the regulator/motor assembly. Pull the equalizer arm and regulator assemblies through the service hole in the door frame to remove it.
6 Installation is the reverse of removal. Lubricate the rollers and wear points on the regulator with white grease before installation.

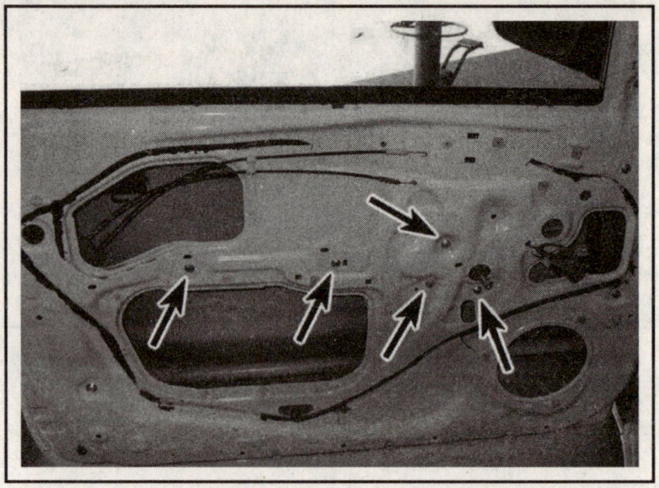

19.4 The location of the window regulator mounting bolts (front door/manual window shown - others similar)

20 Mirrors - removal and installation

OUTSIDE MIRRORS

➧ Refer to illustration 20.3

1 Remove the door trim panel as described in Section 15.
2 On models with power windows, follow the wiring harness from the mirror to the electrical connector and disconnect it.
3 Remove the three mirror mounting nuts and detach the mirror from the vehicle (see illustration).
4 Installation is the reverse of removal.

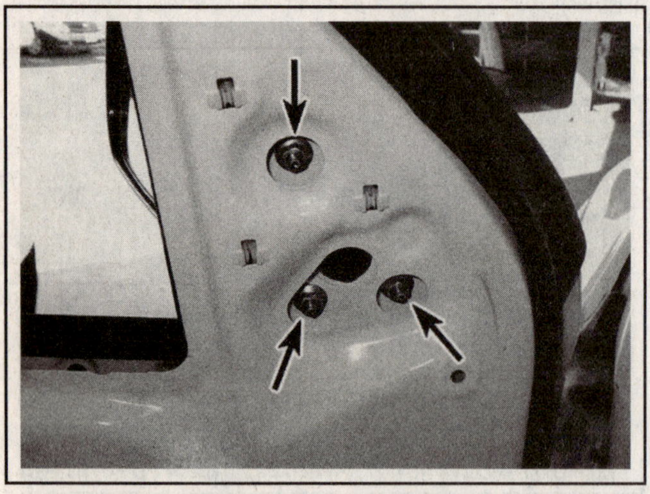

20.3 The location of the mirror mounting nuts

INSIDE MIRROR

➧ Refer to illustrations 20.6 and 20.7

5 Disconnect the electrical connector from the mirror, if equipped.
6 On models without an electrical connector, pry the plastic cover from the mirror base (see illustration). On models with an electrical connector, move the plastic cover that is over the mirror base and stalk towards the headliner.
7 Release the mirror base from the mounting plate on the wind-

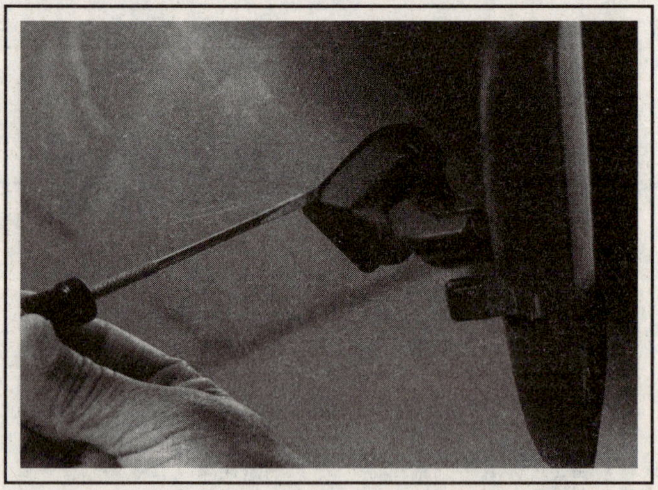

20.6 Pry the edge of the plastic cover to remove it from the mirror base

shield, then slide the mirror stalk upward and off the plate (see illustration).

8 Installation is the reverse of removal.

9 If the mounting plate itself has come off the windshield, adhesive kits are available at auto parts stores to reattach it. Follow the instructions included with the kit.

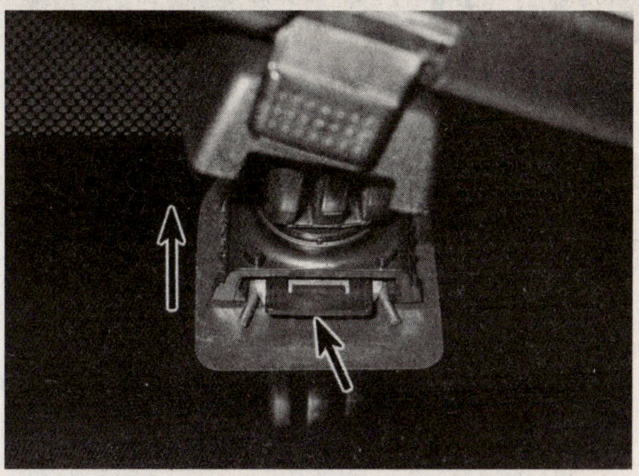

20.7 **Pull the center tab out slightly while moving the mirror upward off the mounting plate**

21 Tailgate - removal and installation

Refer to illustrations 21.1 and 21.2

1 Open the tailgate and detach the cables from the stays (see illustration).

2 Hold the tailgate at an angle on the right side and align the opening in the hinge-pocket with the elongated hinge-pin, then release the tailgate from the joint (see illustration). With the help of an assistant to support the weight, separate the joint from the other side, then remove the tailgate from the vehicle.

3 Installation is the reverse of removal.

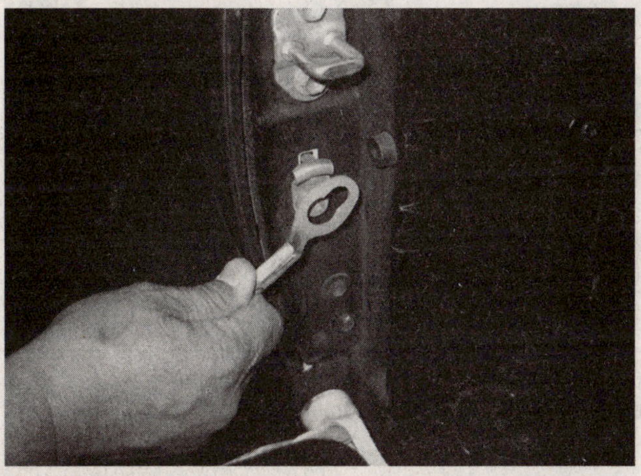

21.1 **Remove the cable ends from the stay**

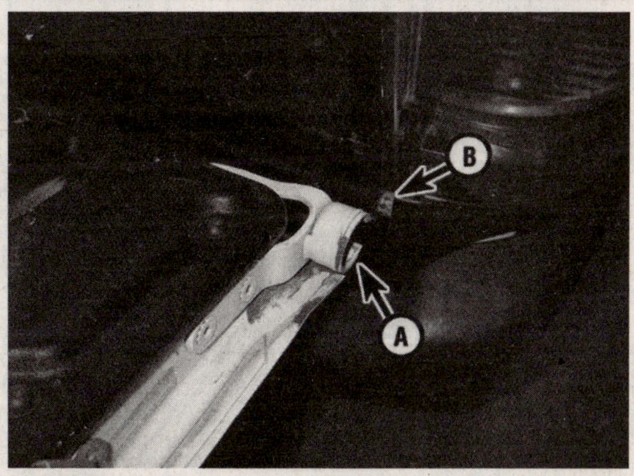

21.2 **Holding the tailgate at an angle, align the slot in the hinge-pocket (A) with the elongated hinge-pin (B) to separate the joint**

22 Tailgate latch and handle - removal and installation

➧ Refer to illustrations 22.1a and 22.1b

1 Lower the tailgate and remove the protective trim from the inside and top edge of the tailgate (see illustrations).

LATCH

➧ Refer to illustrations 22.2 and 22.3

2 Remove the latch mounting fasteners (see illustration).
3 Disconnect the control rod from the latch and remove the latch from the door (see illustration).
4 Installation is the reverse of removal.

RELEASE HANDLE

➧ Refer to illustration 22.5

5 Disconnect the control rods (see illustration).
6 Detach the mounting bolts and remove the handle assembly while separating its retaining tabs from the tailgate.
7 Installation is the reverse of removal.

22.1a Fastener locations for the tailgate protective trim (some fasteners are not visible in this photo)

22.1b Pull the trim from the top edge of the tailgate; it is retained by plastic tabs underneath

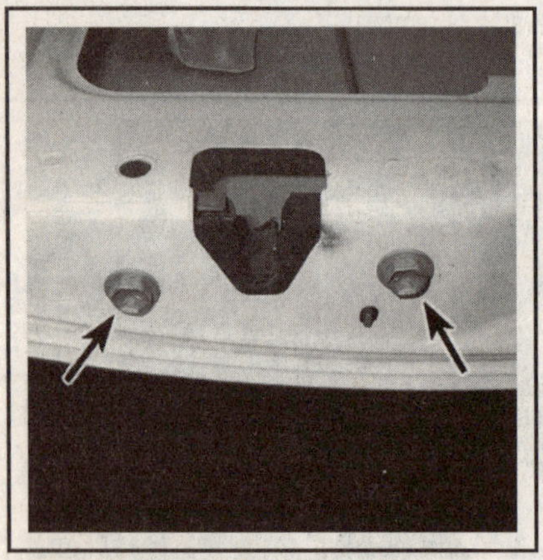

22.2 Tailgate latch mounting fasteners

22.3 Rotate the rod clip off the rod to release it, then remove the rod from the latch

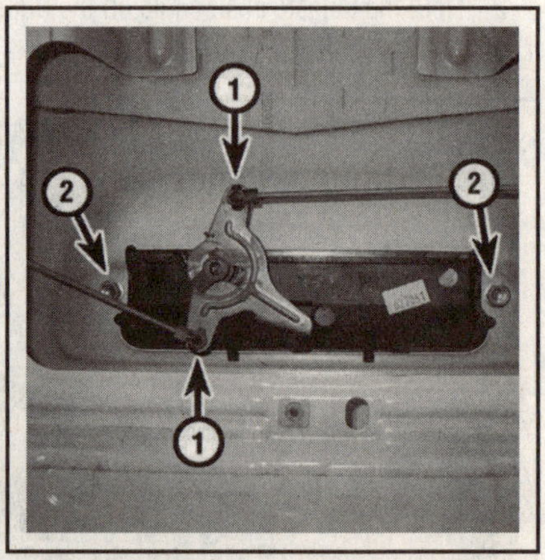

22.5 Release handle assembly details:

1 *Rod retaining clips (rotate the clip off the rod to release it)*
2 *Mounting bolts*

23 Center console - removal and installation

➡Note: Refer to Section 1 for general information on removing trim panels.

1 Remove the shift knob by turning it counterclockwise.

BENCH SEAT TYPE

➧ Refer to illustrations 23.2 and 23.3

2 Pull the cup holder and rubber mats from the console, then remove the mounting fasteners (see illustration). Also, remove the plastic fastener on the side.

3 Pull the console toward the rear of the vehicle to disengage the clips from the lower instrument panel trim, then remove the console (see illustration).

SEPARATE SEAT TYPE (WITH REAR BOX)

4 On models with an automatic transmission, pry the trim (bezel) from around the shifter and remove it.

5 On models with an automatic transmission, pry the cup holder panel from in front of the shifter with a trim stick or equivalent and remove it.

6 On models with a manual transmission, pry the drink tray panel from the console and remove it.

7 Remove the two exposed screws that attach the rear part of the console to the front.

8 Remove the carpet from the bottom of the console box, then remove the two exposed bolts.

9 Remove the rear section of the console.

10 Remove the two mounting screws at the rear part of the front console (behind the shifter).

11 Pull the front part of the console toward the rear of the vehicle to disengage the clips from the lower instrument panel trim, then remove the console.

ALL SEAT TYPES

12 Installation is the reverse of removal.

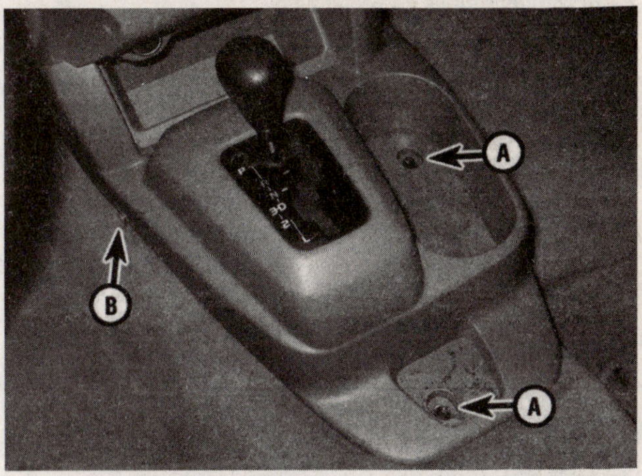

23.2 Remove the rubber mats and cup holder to expose the mounting fasteners (A) hidden beneath. There is also a plastic fastener (B) on the side

23.3 Pull the console directly toward the rear to disengage the retaining clips at the lower instrument panel trim

24 Dashboard trim panels - removal and installation

⁂ WARNING:

Models covered by this manual are equipped with a Supplemental Restraint System (SRS), more commonly known as airbags. Always disable the airbag system before working in the vicinity of any airbag system component to avoid the possibility of accidental deployment of the airbag, which could cause personal injury (see Chapter 12).

➡Note 1: Refer to Section 1 for general information on removing trim panels.

➡Note 2: Some of the trim panels may have electrical switches mounted in them. When removing these panels, do not pull the panel out too far before disconnecting all of the electrical connectors for the installed switches.

1 These trim panels cover many of the instrument panel (dashboard) mounting fasteners. Removable fasteners, tabs and slots or a combination of both, hold the trim panels in place. If you're going to remove the instrument panel (dashboard), first remove all of the tri panels and components listed in this section.

2 Disconnect the cable from the negative terminal of the ba (see Chapter 5, Section 1).

INSTRUMENT CLUSTER FINISH PANEL

➧ Refer to illustrations 24.5 and 24.6

3 Remove the knee bolster (see Steps 11 through 20).
4 On models with tilt steering, lower the steering column to its lowest position.
5 Remove the two plastic mounting fasteners (see illustration). Refer to Section 1 for fastener types and removal techniques.
6 Carefully pry the panel away from the instrument panel until the tabs are released (see illustration). Take care not to scratch the instrument panel. Disconnect any electrical connectors for switches mounted in the panel.
7 Installation is the reverse of the removal procedure. Make sure that all tabs are aligned properly and that all electrical connectors are properly connected before pushing the panel firmly into place.

CENTER TRIM PANEL

➧ Refer to illustration 24.9

8 Remove the heater and air conditioning control assembly (see Chapter 3) and the radio (see Chapter 12).
9 Carefully pry the panel away from the instrument panel until the tabs are released (see illustration). Take care not to scratch the instrument panel. Disconnect any electrical connectors for switches mounted in the panel.
10 Installation is the reverse of the removal procedure. Make sure that all tabs are aligned properly and that all electrical connectors are properly connected before pushing the panel firmly into place.

KNEE BOLSTER (AND REINFORCEMENT)

➧ Refer to illustrations 24.11, 24.12, 24.13, 24.14, 24.15, 24.18, 24.19 and 24.20

11 Remove the scuff plate by pulling it straight up (see illustration).
12 Remove the foot rest by pulling it away from the floor (see illustration).
13 Remove the kick panel mounting fastener, then pull it towards the rear of the vehicle (see illustration).
14 Open the coin tray on the left side of the bolster, then pull it back off its hinge and remove it (see illustration).
15 On models with a parking brake pedal, remove the accessory tray

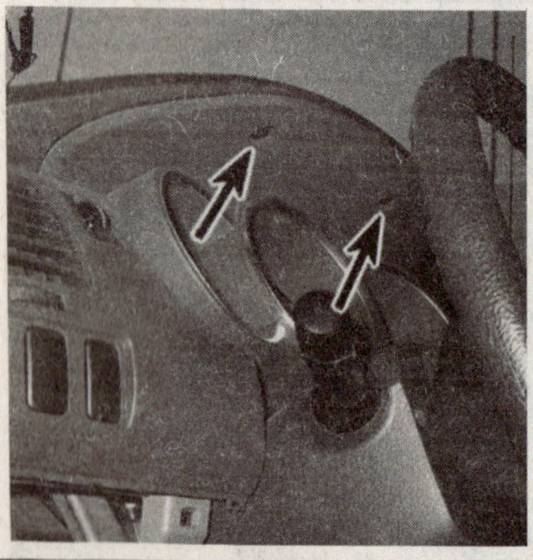

24.5 Location of the two upper mounting fasteners

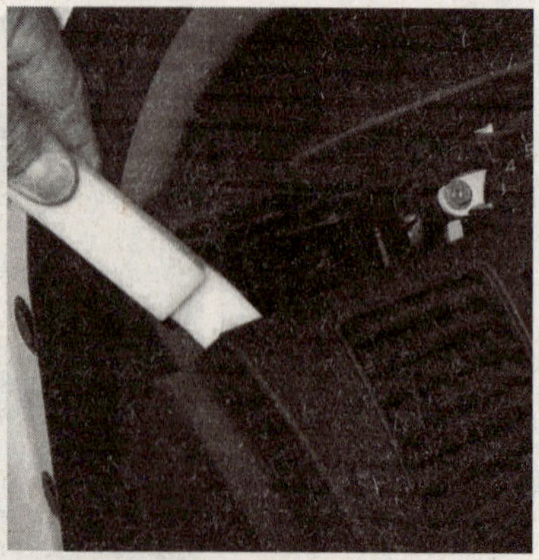

24.6 Carefully pry the finish panel away from the instrument panel until the tabs are released

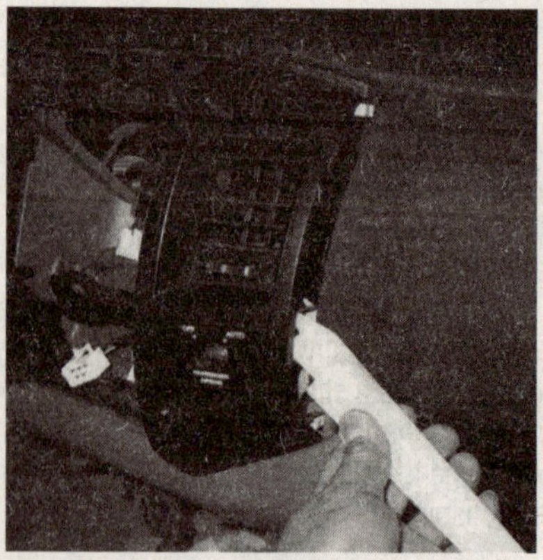

24.9 Carefully pry the center trim panel away from the instrument panel until the tabs are released

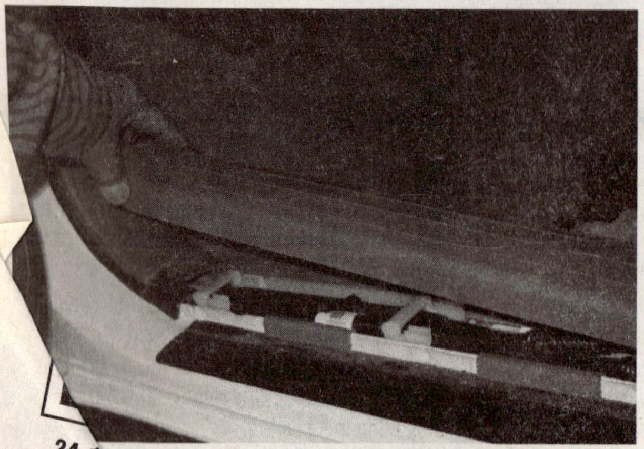

24.1[illegible]sp the scuff plate and pull it straight up to remove

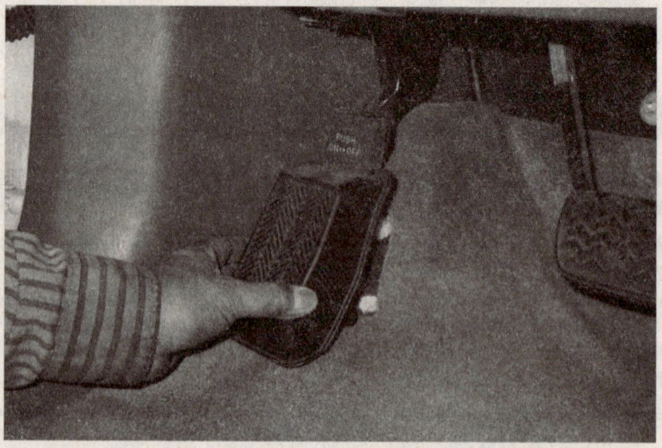

24.12 Grasp the foot rest and pull it away from the floor to remove it

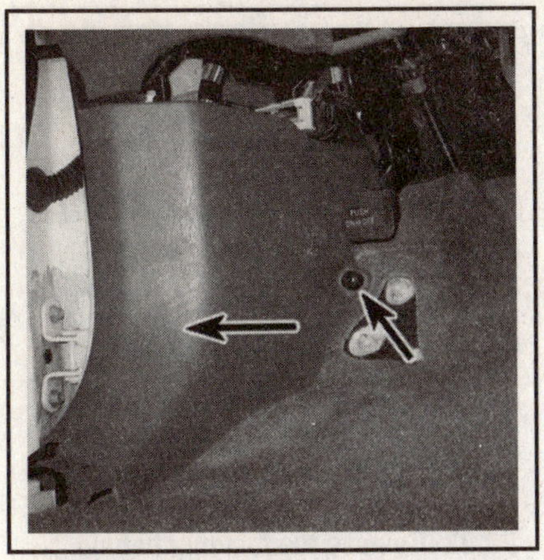

24.13 **Remove the plastic nut, then pull the kick panel towards the rear of the vehicle to remove it**

24.14 **Open the coin tray, then pull it off its hinge and out of the bolster**

24.15 **Pull the right accessory tray straight out from the bolster to remove it**

on the right side of the bolster by pulling it straight out from the bolster (see illustration).

16 On models with a parking brake lever, remove the trim around the lever by pulling it rearward.

17 Remove the hood release cable from the release lever mounted to the bolster (see Section 10).

18 Pry the small cover on the lower right side of the bolster loose, then remove the two knee bolster mounting fasteners (see illustration).

19 Carefully pry the panel away from the instrument panel until the tabs are released (see illustration). Take care not to scratch other panels. Disconnect any electrical connectors for switches mounted in the panel.

20 If necessary, remove the knee bolster reinforcement mounting fasteners, then remove the reinforcement (see illustration).

➡**Note: This Step is only necessary if access to components behind it is required.**

21 Installation is the reverse of the removal procedure. Make sure that all tabs are aligned properly and that all electrical connectors are properly connected before pushing the panel firmly into place.

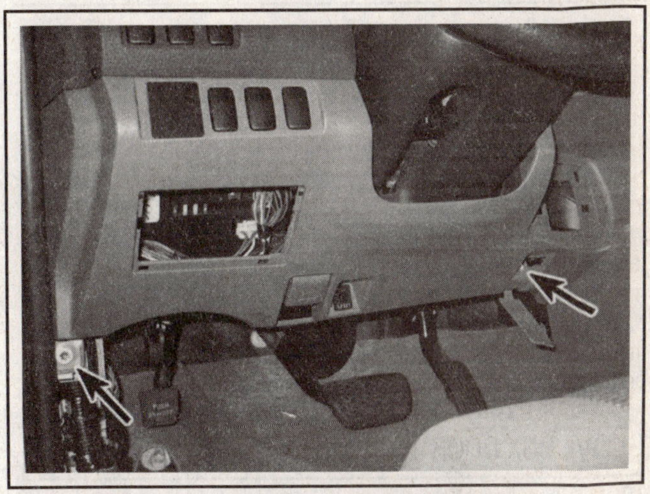

24.18 **Remove the knee bolster mounting fasteners**

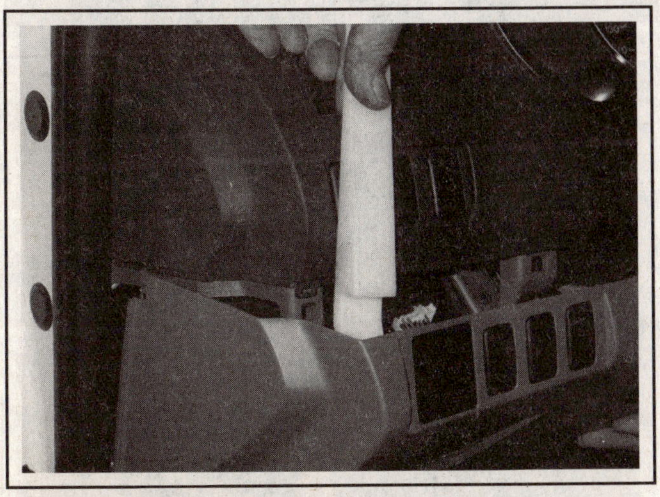

24.19 **Carefully pry the knee bolster away from the instrument panel until the tabs are released**

24.20 **The knee bolster reinforcement mounting fasten**

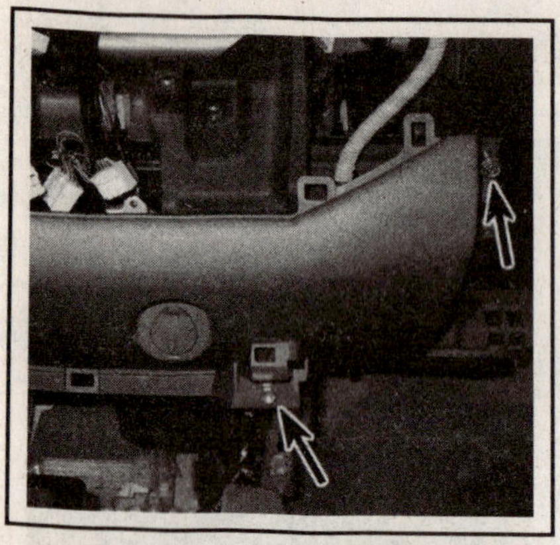

24.24 The lower center trim mounting fasteners

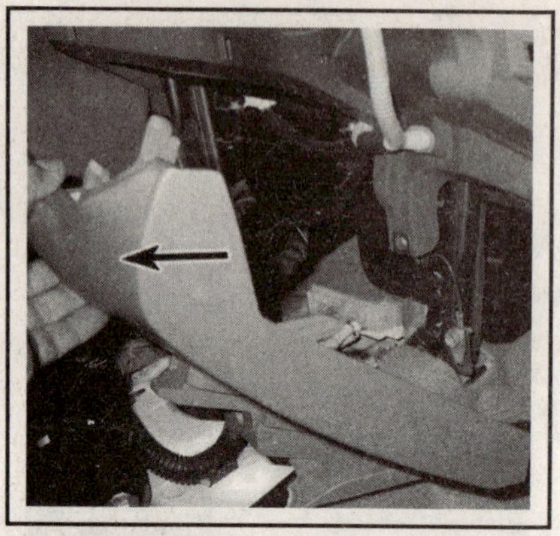

24.25 Grasp the lower center trim and pull it directly away from the instrument panel to remove it

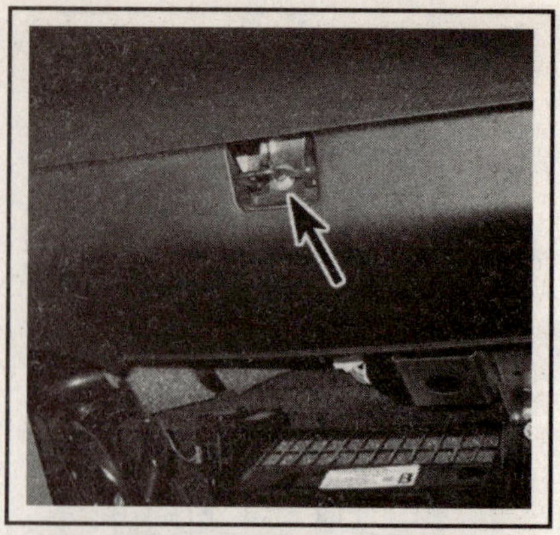

24.32 Remove the fastener behind the latch, then grasp the glove box trim panel and pull it directly away from the instrument panel to remove it

LOWER CENTER TRIM PANEL

➧ Refer to illustrations 24.24 and 24.25

22 Remove the center console (see Section 23).

23 Remove the instrument cluster finish and center trim panels and the knee bolster (see Steps 1 through 20).

24 Remove the lower center trim panel mounting fasteners (see illustration).

25 Pull the trim panel direct away from the instrument panel to remove it. Disconnect any electrical connectors for components mounted in the panel (see illustration).

26 Installation is the reverse of the removal procedure. Make sure that all tabs are aligned properly and that all electrical connectors are properly connected before pushing the panel firmly into place.

GLOVE BOX DOOR

27 Open the glove box door and squeeze the sides in to allow the door to open further (past the door stops).

28 Disengage the door stopper from the right side of the door by lightly prying it off the plastic retaining tab.

29 With the door fully open, pull it directly up to release it from its bottom hinge and remove it.

30 Installation is the reverse of the removal procedure.

GLOVE BOX TRIM PANEL

➧ Refer to illustration 24.32

31 Remove the glove box door (see Steps 27 through 29).

32 Remove the fastener under the door latch (see illustration).

33 Pull the trim panel directly away from the instrument panel to remove it.

➡Note: The retaining tabs are on each side of the trim panel near the rubber stops for the door.

34 Installation is the reverse of the removal procedure. Make sure that all tabs are aligned properly before pushing the panel firmly into place.

25 Steering column covers - removal and installation

➧ Refer to illustration 25.2

✲✲ WARNING:

lodels covered by this manual are equipped with a Supplemen-l Restraint System (SRS), more commonly known as airbags. vays disable the airbag system before working in the vicinity ny airbag system component to avoid the possibility of acci-al deployment of the airbag, which could cause personal y (see Chapter 12).

Disconnect the cable from the negative terminal of the battery Chapter 5, Section 1). Move the steering column or up down and steering wheel as needed for clearance.

unmove the mounting screws, separate the two halves of the col-

➡No at the seam, then remove them (see illustration).

the tabsssing on the covers near the seam will help release hold the two halves together.

3 Installa is the reverse of the removal procedure.

25.2 Remove the screws, then remove the upper and lower covers

26 Instrument panel - removal and installation

➧ **Refer to illustrations 26.3, 26.4a, 26.4b, 26.6, 26.7a, 26.7b, 26.7c and 26.8**

⁂ WARNING:

Models covered by this manual are equipped with a Supplemental Restraint System (SRS), more commonly known as airbags. Always disable the airbag system before working in the vicinity of any airbag system component to avoid the possibility of accidental deployment of the airbag, which could cause personal injury (see Chapter 12).

➡Note: We recommend that an experienced do-it-yourselfer perform the following procedure. Also, detailed notes of how various parts and electrical connectors are assembled or placed will make reassembly much easier.

1 Disconnect the cable from the negative battery terminal (see Chapter 5, Section 1).

2 Remove the center floor console (see Section 23) and all dashboard trim panels and components listed in Section 24.

3 Carefully pry the small side panel away from the instrument panel until the tabs are released (see illustration).

4 Remove the front pillar trim from both sides of the vehicle (see illustrations).

5 Remove the instrument cluster (see Chapter 12).

6 Disconnect the electrical connector for the passenger's side airbag and remove its bracket mounting bolts (see illustration).

7 Most components that are mounted to the instrument panel will be removed with it, such as the passenger's side airbag. Only disconnect the electrical connectors that must be disconnected in order to remove the instrument panel. Also, detach the wiring harnesses from the panel as well. Most are designed so that they will only fit on the matching connector (male or female), but if there is any doubt, label the connectors with masking tape and a pen before disconnecting them (see illustrations).

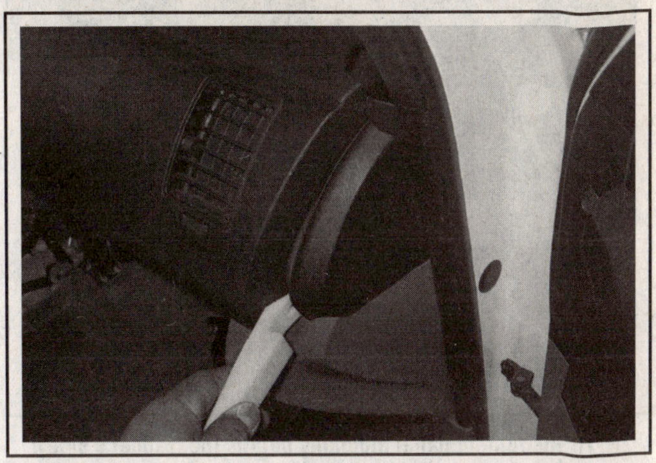

26.3 Carefully pry the side panel away from the instrument panel until the tabs are released

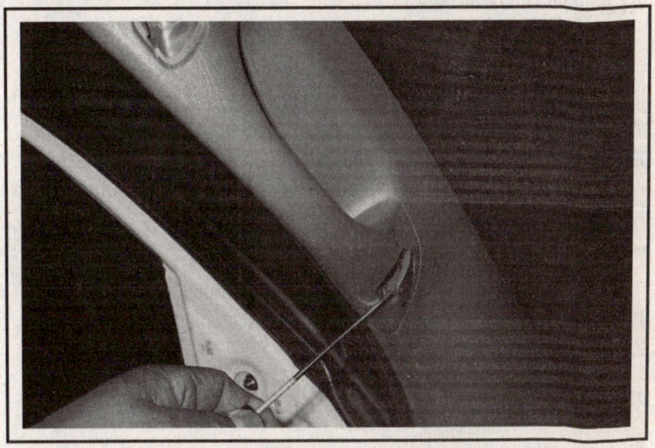

26.4a Carefully pry the bolt covers off of the pillar handle, then remove the exposed bolts

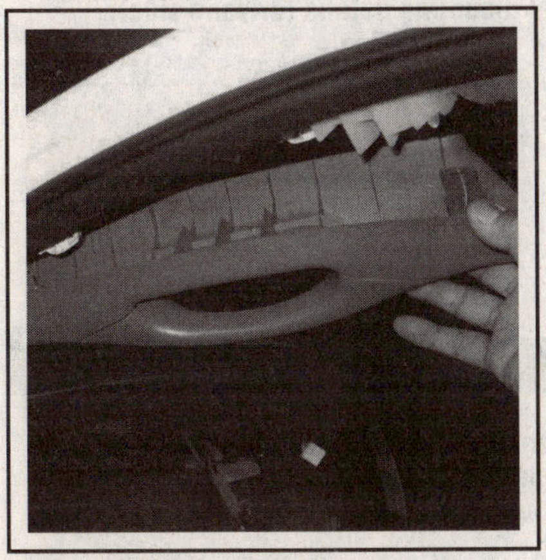

26.4b Pull the pillar trim (with the handle) away from the pillar to remove it

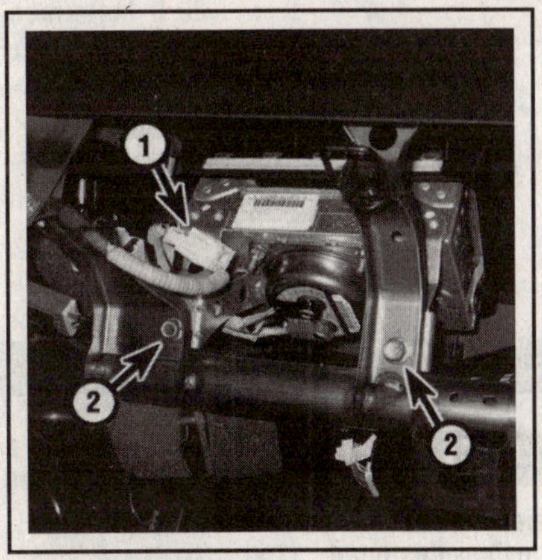

26.6 Passenger's side airbag mounting details:

1 Electrical connector
2 Bracket mounting bolts

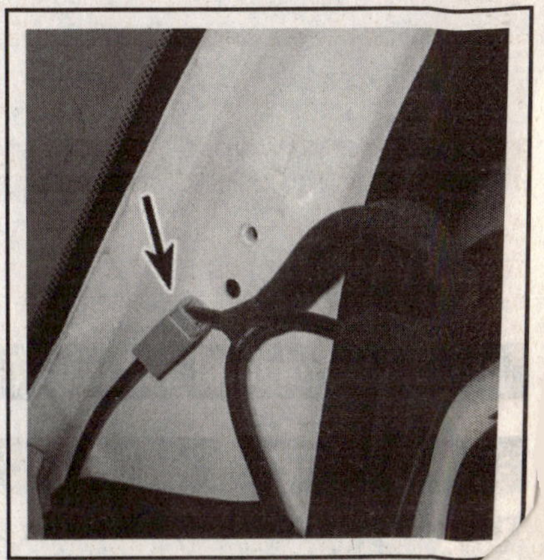

26.7a Disconnect the electrical connector at the passenger's side pillar

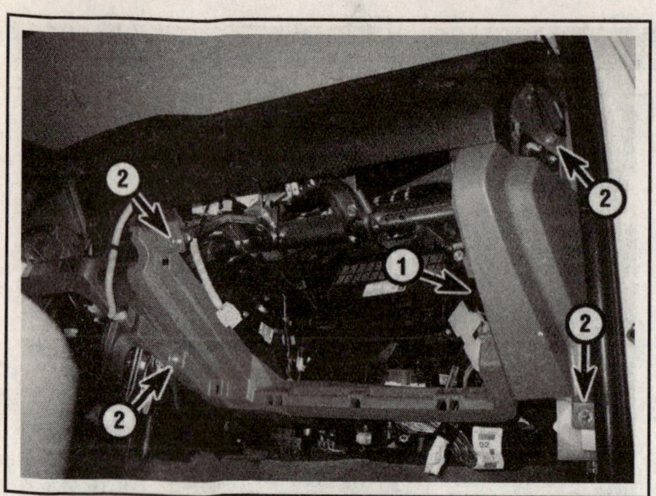

26.7b Instrument panel mounting details (passenger's side):

1 *Electrical connector or wiring harness (hidden from view in this photo - vicinity given)*
2 *Mounting fasteners*

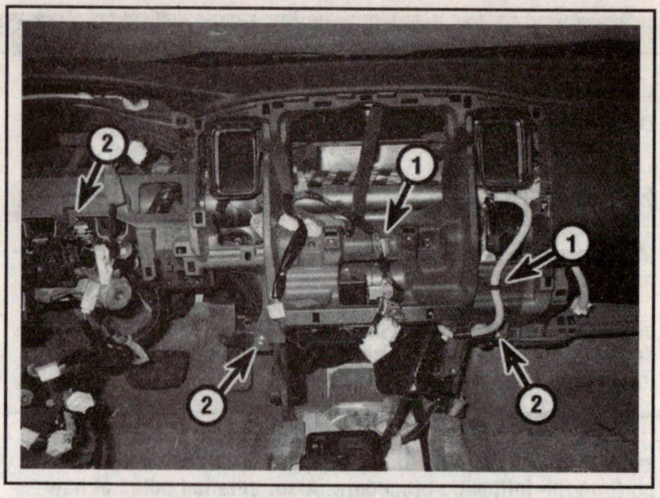

26.7c Instrument panel mounting details (center):

1 *Electrical connector or wiring harness*
2 *Mounting fasteners*

8 Remove the instrument panel mounting fasteners (see illustrations 26.7b, 26.7c and the accompanying illustration). Once all fasteners are removed, carefully pull the panel away from the reinforcement bar and the windshield. Make sure that there are no wiring harnesses still connected or attached, then guide the instrument panel out through the passenger's door opening.

➡**Note: It's best to have an assistant help with this Step.**

9 Installation is the reverse of the removal procedure. Make sure that that all wiring harnesses and electrical connectors are properly secured and connected.

26.8 Instrument panel mounting details (driver's side):

1 *Electrical connector or wiring harness*
2 *Mounting fasteners*

Seats - removal and installation

WARNING 1:

...ront seat belts on these models are equipped with pre-...ers, which are pyrotechnic (explosive) devices designed ...t the seat belts in the event of a collision. On models with pre-tensioners, do not remove the front seat belt ...ssemblies, and do not disconnect the electrical con-...ing to the assemblies. Problems with the pre-ten-...rn on the SRS (airbag) warning light on the dash. ...ioner problems are suspected, take the vehicle to a dealer service department. Also on these models, be sure to disable the airbag system (see Chapter 12).

WARNING 2:

On models with side-impact airbags in the outer sides of the front seat backs, be sure to disarm the airbag system before beginning this procedure (see Chapter 12).

FRONT SEAT

➧ **Refer to illustrations 27.1a and 27.1b**

1 Remove the seat mounting bolts (see illustrations).
2 On bench-type seats, push the center passenger seatbelt buckles through the seat towards the floor. Detach the seatbelt from the guide on top of the seat back.
3 Tilt the seat to access the underside, disconnect any electrical connectors from wire harnesses coming from the floor, then lift the seat out of the vehicle. It is best to use an assistant to help you if you are removing a bench-type seat.
4 Installation is the reverse of removal.

REAR SEAT

Double cab

5 Remove the mounting bolts in front of the rear seat cushion at the bottom, then remove the seat cushion.
6 If you are removing the right (passenger's side) seat, unbolt the seat belt anchor and buckle from the floor.
7 Fold both seat backs down and remove the plastic hinge cover that is at each bottom corner of the seat back.
8 Remove the seat back mounting bolts at each hinge, then remove the seat back.
9 Installation is the reverse of removal.

Access cab

10 Remove the mounting bolts in front of the rear seat cushion at the bottom.
11 Release the seat latch, then remove the seat cushion.
12 Installation is the reverse of removal.

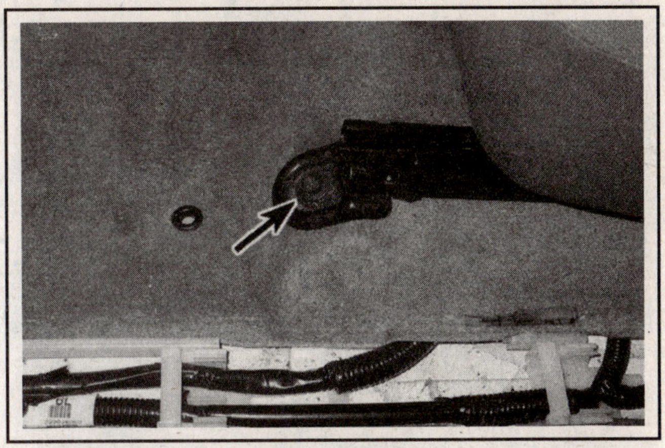

27.1a Remove the front mounting bolts (one side of bench seat shown - separate seat type similar)

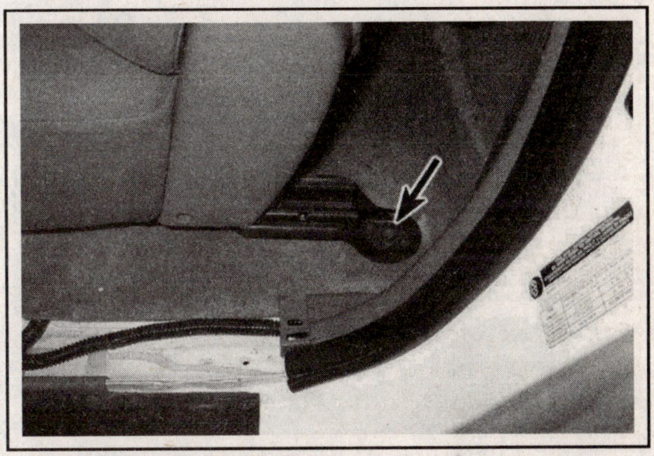

27.1b Remove the rear mounting bolts (one side of bench seat shown - separate seat type similar)

Notes

12 CHASSIS ELECTRICAL SYSTEM

1 General information

The electrical system is a 12-volt, negative ground type. Power for the lights and all electrical accessories is supplied by a lead/acid-type battery which is charged by the alternator.

This Chapter covers repair and service procedures for the various electrical components not associated with the engine. Information on the battery, alternator, ignition system and starter motor can be found in Chapter 5. It should be noted that when portions of the electrical system are serviced, the negative battery cable should be disconnected from the battery to prevent electrical shorts and/or fires.

2 Electrical troubleshooting - general information

➧ Refer to illustrations 2.5a, 2.5b, 2.6 and 2.9

A typical electrical circuit consists of an electrical component, any switches, relays, motors, fuses, fusible links or circuit breakers related to that component and the wiring and connectors that link the component to both the battery and the chassis. To help you pinpoint an electrical circuit problem, wiring diagrams are included at the end of this Chapter.

Before tackling any troublesome electrical circuit, first study the appropriate wiring diagrams to get a complete understanding of what makes up that individual circuit. Trouble spots, for instance, can often be narrowed down by noting if other components related to the circuit are operating properly. If several components or circuits fail at one time, chances are the problem is in a fuse or ground connection, because several circuits are often routed through the same fuse and ground connections.

Electrical problems usually stem from simple causes, such as loose or corroded connections, a blown fuse, a melted fusible link or a failed relay. Visually inspect the condition of all fuses, wires and connections in a problem circuit before troubleshooting the circuit.

If test equipment and instruments are going to be utilized, use the diagrams to plan ahead of time where you will make the necessary connections in order to accurately pinpoint the trouble spot.

The basic tools needed for electrical troubleshooting include a circuit tester or voltmeter (a 12-volt bulb with a set of test leads can also be used), a continuity tester, which includes a bulb, battery and set of test leads, and a jumper wire, preferably with a circuit breaker incorporated, which can be used to bypass electrical components (see illustrations). Before attempting to locate a problem with test instruments, use the wiring diagram(s) to decide where to make the connections.

VOLTAGE CHECKS

Voltage checks should be performed if a circuit is not functioning properly. Connect one lead of a circuit tester to either the negative battery terminal or a known good ground. Connect the other lead to a connector in the circuit being tested, preferably nearest to the battery or fuse (see illustration). If the bulb of the tester lights, voltage is present, which means that the part of the circuit between the connector and the battery is problem free. Continue checking the rest of the circuit in the same fashion. When you reach a point at which no voltage is present, the problem lies between that point and the last test point with voltage. Most of the time the problem can be traced to a loose connection.

➡Note: Keep in mind that some circuits receive voltage only when the ignition key is in the ACC or RUN position.

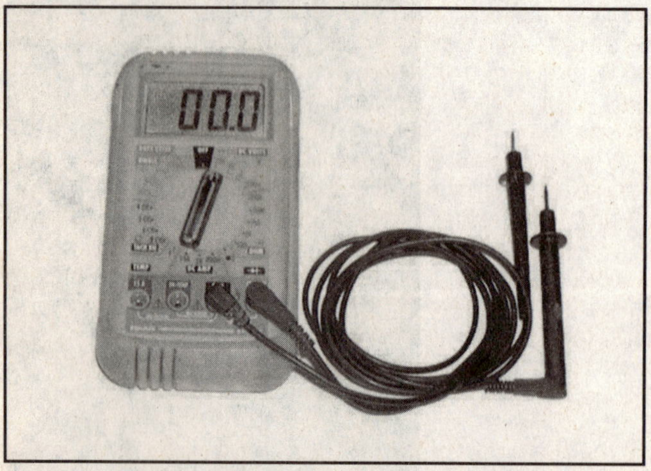

2.5a The most useful tool for electrical troubleshooting is a digital multimeter that can check volts, amps, and test continuity

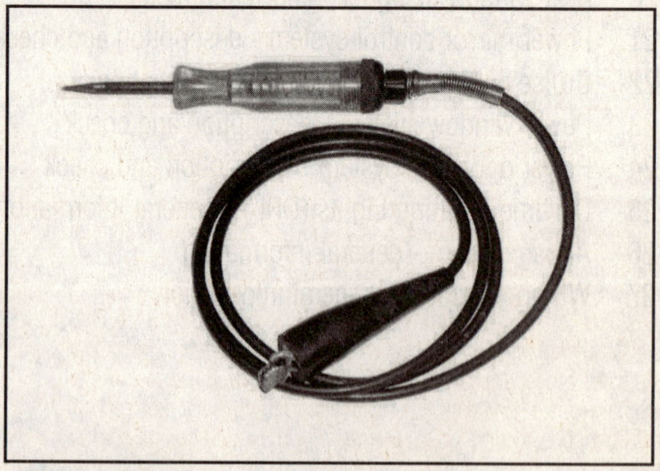

2.5b A simple test light is a very handy tool for testing voltage

2.6 In use, a basic test light's lead is clipped to a known good ground, then the pointed probe can test connectors, wires or electrical sockets - if the bulb lights, the circuit being tested has battery voltage

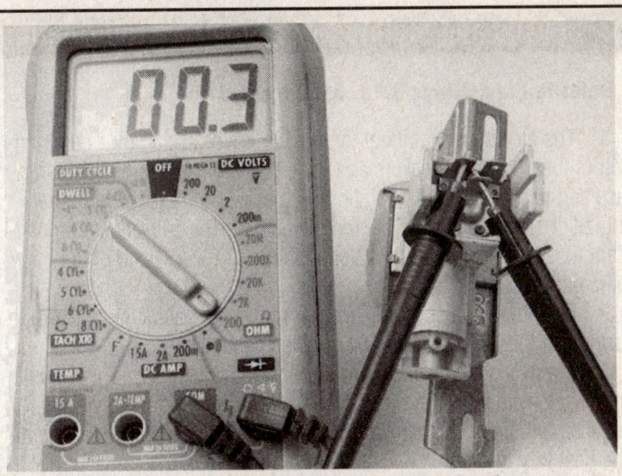

2.9 With a multimeter set to the ohms scale, resistance can be checked across two terminals - when checking for continuity, a low reading indicates continuity, a high reading or infinity indicates lack of continuity

FINDING A SHORT

One method of finding shorts in a live circuit is to remove the fuse and connect a test light in place of the fuse terminals (fabricate two jumper wires with small spade terminals, plug the jumper wires into the fuse box and connect the test light). There should be voltage present in the circuit. Move the suspected wiring harness from side-to-side while watching the test light. If the bulb goes off, there is a short to ground somewhere in that area, probably where the insulation has rubbed through.

GROUND CHECK

Perform a ground test to check whether a component is properly grounded. Disconnect the battery and connect one lead of a continuity tester or multimeter (set to the ohms scale), to a known good ground. Connect the other lead to the wire or ground connection being tested. If the resistance is low (less than 5 ohms), the ground is good. If the bulb on a self-powered test light does not go on, the ground is not good.

CONTINUITY CHECK

A continuity check is done to determine if there are any breaks in a circuit - if it is passing electricity properly. With the circuit off (no power in the circuit), a self-powered continuity tester or multimeter can be used to check the circuit. Connect the test leads to both ends of the circuit (or to the power end and a good ground), and if the test light comes on the circuit is passing current properly (see illustration). If the resistance is low (less than 5 ohms), there is continuity; if the reading is 10,000 ohms or higher, there is a break somewhere in the circuit. The same procedure can be used to test a switch, by connecting the continuity tester to the switch terminals. With the switch turned on, the test light should come on (or low resistance should be indicated on a meter).

FINDING AN OPEN CIRCUIT

When diagnosing for possible open circuits, it is often difficult to locate them by sight because the connectors hide oxidation or terminal misalignment. Merely wiggling a connector on a sensor or in the wiring harness may correct the open circuit condition. Remember this when an open circuit is indicated when troubleshooting a circuit. Intermittent problems may also be caused by oxidized or loose connections.

Electrical troubleshooting is simple if you keep in mind that all electrical circuits are basically electricity running from the battery, through the wires, switches, relays, fuses and fusible links to each electrical component (light bulb, motor, etc.) and to ground, from which it is passed back to the battery. Any electrical problem is an interruption in the flow of electricity to and from the battery.

CONNECTORS

Most electrical connections on these vehicles are made with multiwire plastic connectors. The mating halves of many connectors are secured with locking clips molded into the plastic connector shells. The mating halves of large connectors, such as some of those under the instrument panel, are held together by a bolt through the center of the connector.

To separate a connector with locking clips, use a small screwdriver to pry the clips apart carefully, then separate the connector halves. Pull only on the shell; never pull on the wiring harness as you may damage the individual wires and terminals inside the connectors. Look at the connector closely before trying to separate the halves. Often the locking clips are engaged in a way that is not immediately clear. Additionally, many connectors have more than one set of clips.

Each pair of connector terminals has a male half and a female half. When you look at the end view of a connector in a diagram, be sure to understand whether the view shows the harness side or the component side of the connector. Connector halves are mirror images of each other, and a terminal shown on the right side end-view of one half will be on the left side end view of the other half.

3 Fuses and fusible links - general information

➧ Refer to illustrations 3.1a, 3.1b and 3.3

1 The electrical circuits of the vehicle are protected by a combination of fuses, circuit breakers and fusible links. The engine compartment fuse and relay box (see illustration) is located in the left side of the engine compartment, right behind the battery. The passenger compartment fuse and relay box is located in the left (driver's) end of the instrument panel (see illustration).

2 Each of the fuses is designed to protect a specific circuit, and the various circuits are identified on the fuse panel itself.

3 Miniaturized fuses are employed in the fuse boxes. These compact fuses, with blade terminal design, allow fingertip removal and replacement. If an electrical component fails, always check the fuse first. The best way to check the fuses is with a test light. Check for power at the exposed terminal tips of each fuse. If power is present on one side of the fuse but not the other, the fuse is blown. A blown fuse can also be confirmed by visually inspecting it (see illustration).

4 Be sure to replace blown fuses with the correct type. Fuses of different ratings are physically interchangeable, but only fuses of the proper rating should be used. Replacing a fuse with one of a higher or lower value than specified is not recommended. Each electrical circuit needs a specific amount of protection. The amperage value of each fuse is molded into the fuse body.

5 If the replacement fuse fails immediately, don't replace it again until the cause of the problem is isolated and corrected. In most cases, the cause will be a short circuit in the wiring caused by a broken or deteriorated wire.

6 Some circuits are protected by fusible links. The links are used in circuits which are not ordinarily fused, such as the ignition circuit.

7 Most common fusible links are simply heavy-gauge wire, installed in high-current circuits, which burns through when the current exceeds the design threshold of the circuit. When one of these fusible links blows, you have to cut it out of the circuit and splice in a new fusible link. For some common applications, like the battery positive cable, you might be able to replace the damaged wiring with new wiring with the fusible link already installed.

8 Some fusible links are designed like an oversize version of a typical large fuse (see illustration 3.3). After disconnecting the negative battery cable, simply remove the damaged fusible link and install a new unit rated for the same amperage.

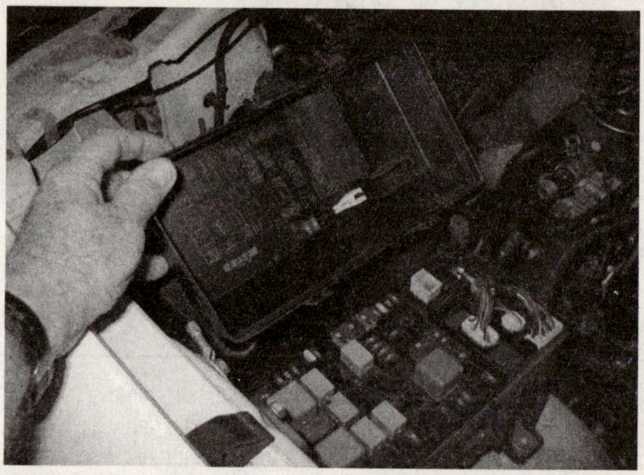

3.1a The engine compartment fuse/relay block is located on the driver's side of the engine compartment. There's a fuse and relay guide on the underside of the cover

3.1b The passenger compartment fuse/relay block is located in the left end of the instrument panel. To access it, remove the coin tray (note the fuse and relay guide on the panel)

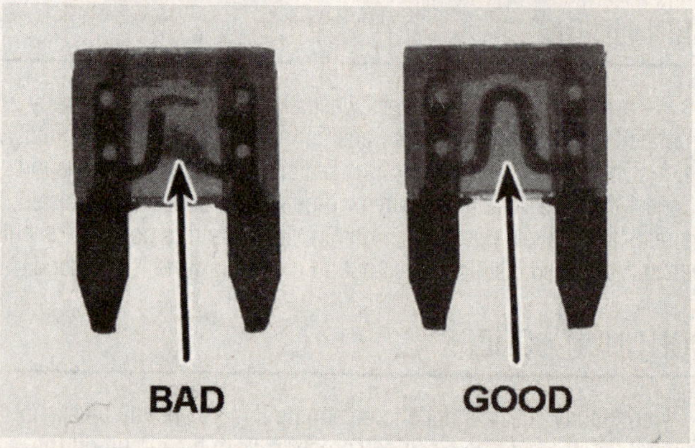

3.3 When a fuse blows, the element between the terminals melts

4 Circuit breakers - general information

Circuit breakers protect components such as power windows, power door locks and headlights.

On some models the circuit breaker resets itself automatically, so an electrical overload in a circuit breaker protected system will cause the circuit to fail momentarily, then come back on. If the circuit doesn't come back on, check it immediately. Once the condition is corrected, the circuit breaker will resume its normal function. Some circuit breakers must be reset manually.

5 Relays - general information and testing

GENERAL INFORMATION

1 Several electrical accessories in the vehicle, such as the fuel injection system, horns, starter, and fog lamps use relays to transmit the electrical signal to the component. Relays use a low-current circuit (the control circuit) to open and close a high-current circuit (the power circuit). If the relay is defective, that component will not operate properly. Most relays are mounted in the engine compartment fuse/relay boxes, with some specialized relays located above the interior fuse box under the dash. If a faulty relay is suspected, it can be removed and tested using the procedure below or by a dealer service department or a repair shop. Defective relays must be replaced as a unit.

TESTING

➧ Refer to illustrations 5.3a, 5.3b and 5.6

2 Refer to the wiring diagrams for the circuit to determine the proper connections for the relay you're testing. If you can't determine the correct connection from the wiring diagrams, however, you may be able to determine the test connections from the information that follows.

3 There are four basic types of relays used on these models (see illustrations). Some are normally open type and some normally closed, while others include a circuit of each type.

4 On most relays, two of the terminals are the relay control circuit (they connect to the relay coil which, when energized, closes the large contacts to complete the circuit). The other terminals are the power circuit (they are connected together within the relay when the control-circuit coil is energized).

5 Some relays may be marked as an aid to help you determine which terminals are the control circuit and which are the power circuit. If the relay is not marked, refer to the wiring diagrams at the end of this Chapter to determine the proper hook-ups for the relay you're testing.

6 To test a relay, connect an ohmmeter across the two terminals of the power circuit, continuity should not be indicated (see illustration). Now connect a fused jumper wire between one of the two control circuit terminals and the positive battery terminal. Connect another jumper wire between the other control circuit terminal and ground. When the connections are made, the relay should click and continuity should be indicated on the meter. On some relays, polarity may be critical, so, if the relay doesn't click, try swapping the jumper wires on the control circuit terminals.

7 If the relay fails the above test, replace it.

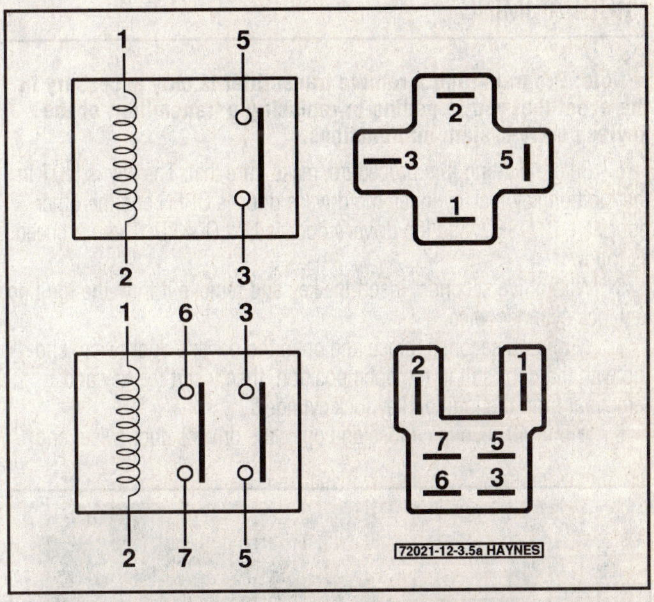

5.3a These two relays are typical normally open types; the one above completes a single circuit (terminal 5 to terminal 3) when energized - the lower relay type completes two circuits (6 and 7, and 3 and 5) when energized

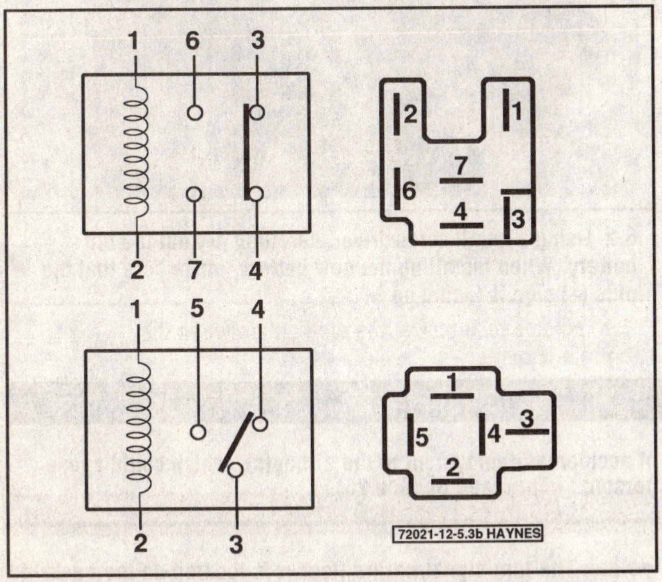

5.3b These relays are normally closed types, where current flows though one circuit until the relay is energized, which interrupts that circuit and completes the second circuit

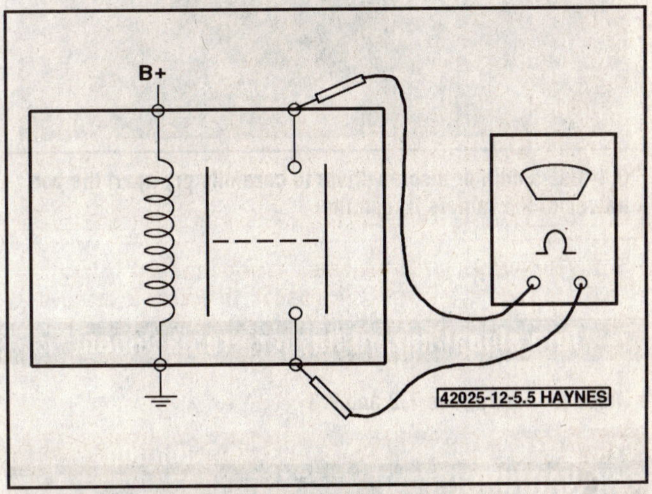

5.6 To test a typical four-terminal normally open relay, connect an ohmmeter to the two terminals of the power circuit - the meter should indicate continuity with the relay energized and no continuity with the relay not energized

6 Keyless entry remote transmitter - battery replacement and programming

BATTERY REPLACEMENT

➧ **Refer to illustrations 6.1 and 6.2**

1 Pry apart the two halves of the remote transmitter (see illustration).
2 Carefully pry out the old battery with a small screwdriver (see illustration).
3 Install the new battery with the "+" side facing up.
4 Snap the two halves of the remote transmitter back together.

PROGRAMMING

➡**Note: Programming a remote transmitter is only necessary in the event that you're adding or replacing a transmitter, or the keyless entry system malfunctions.**

5 Before starting this procedure, make sure that: The key is NOT in the ignition key lock cylinder, the driver's door is OPEN and the other doors are CLOSED, and the driver's door is UNLOCKED. Then, proceed as follows:
6 Within five seconds, insert the key and remove it from the ignition key lock cylinder twice.
7 Within 40 seconds, close and open the driver's door twice, ending with the door still in the open position, then insert the key and remove it from the ignition key lock cylinder.
8 Within 40 seconds, close and open the driver's door twice, ending with the door still in the open position.
9 Insert the key in the ignition key lock cylinder and close the driver's door.
10 Within 40 seconds, turn the key from the LOCK position to the ON position, then back to LOCK the following number of times, according to the following criteria:

One time, to program a remote transmitter code while retaining the original code

Two times, to program a remote transmitter code while erasing the original code

Five times, to erase all registered codes

11 Remove the key from the lock cylinder. The system should now lock and unlock the vehicle one, two or five times, depending upon which mode you selected in the previous step.
12 Within 45 seconds, press and hold the LOCK and UNLOCK remote transmitter buttons simultaneously for 1.5 seconds.
13 Within three seconds, press and release the LOCK or UNLOCK remote transmitter button.
14 The system should now lock and unlock the vehicle once to confirm that the registration procedure has been carried out successfully, or lock and unlock the vehicle twice to indicate that the procedure has not been correctly performed.
15 To exit the programming mode, open the driver's door. If you wish to program more remote transmitters repeat this entire procedure within 45 seconds.

➡**Note: You can program a maximum of four remote transmitters.**

6.1 Use a coin or a screwdriver to carefully pry apart the two halves of the remote transmitter

6.2 Using a small screwdriver, carefully pry out the old battery. When installing the new battery, make sure that the plus (+) sign is facing up

7 Turn signal/hazard flasher - replacement

➧ **Refer to illustrations 7.2 and 7.3**

⁂ WARNING:

The models covered by this manual are equipped with Supplemental Restraint Systems (SRS), more commonly known as airbags. Always disable the airbag system before working in the vicinity of any airbag system components to avoid the possibility of accidental deployment of the airbag(s), which could cause personal injury (see Section 26).

➡**Note: The turn signal/hazard flasher is located on the backside of the in-dash fuse and relay box.**

1 Disconnect the cable from the negative battery terminal (see Chapter 5, Section 1). Remove the knee bolster trim panel (see Chapter 11).

7.2 To detach the in-dash fuse and relay box, remove this bolt and these two nuts, then carefully pull down the box so that you can access the backside

2 Remove the in-dash fuse and relay box (see illustration).

3 Remove the turn signal/hazard flasher unit from the in-dash fuse and relay box (see illustration).

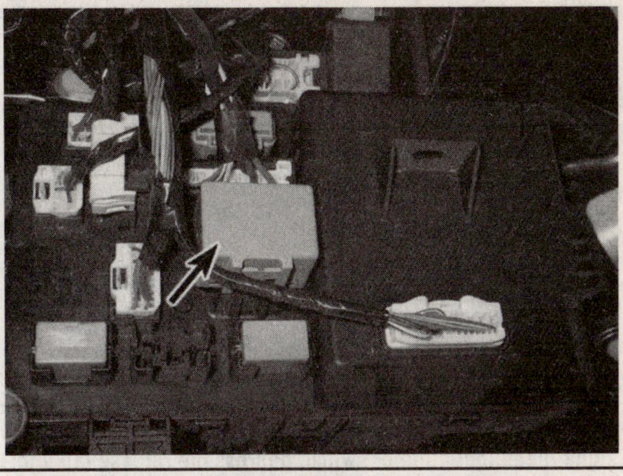

7.3 To remove the turn signal/hazard flasher relay from the in-dash fuse and relay box, grasp it firmly and pull it straight off

4 Make sure that the replacement unit is identical to the original. Compare the old one to the new one before installing it.

5 Installation is the reverse of removal.

8 Steering column switches - replacement

WARNING:

The models covered by this manual are equipped with Supplemental Restraint Systems (SRS), more commonly known as airbags. Always disable the airbag system before working in the vicinity of any airbag system components to avoid the possibility of accidental deployment of the airbag(s), which could cause personal injury (see Section 26).

1 Disconnect the cable from the negative battery terminal (see Chapter 5, Section 1).

2 Remove the steering column covers (see Chapter 11).

WIPER/WASHER SWITCH

Refer to illustrations 8.3 and 8.4

3 Disconnect the electrical connector from the wiper/washer switch (see illustration).

4 Using a small screwdriver, disengage the claw (see illustration) and remove the switch.

5 Installation is the reverse of removal.

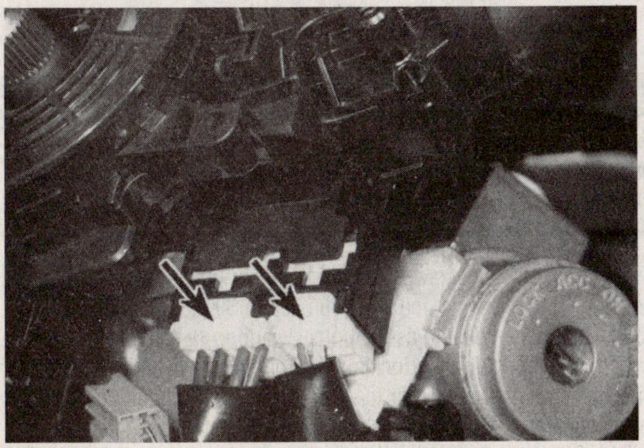

8.3 To disconnect the two electrical connectors from the wiper/washer switch, depress the release tabs and pull out the connectors

8.4 To detach the wiper/washer switch from the turn signal/headlight dimmer switch assembly, depress this locking tab with a small screwdriver and pull out the switch

8.7 To disconnect the electrical connector from the turn signal/headlight dimmer switch, depress the release tab and pull out the connector

8.8 To remove the turn signal/headlight dimmer switch from the steering column, squeeze together the two ends of this clamp and slide the switch off the column

TURN SIGNAL/HEADLIGHT DIMMER SWITCH

Refer to illustrations 8.7 and 8.8

6 Remove the steering wheel and spiral cable (see Chapter 10).

7 Disconnect the electrical connector from the turn signal switch/headlight switch (see illustration).

8 Release the switch retaining clamp (see illustration) and remove the switch from the steering column.

9 Installation is the reverse of removal.

9 Key lock cylinder and ignition switch - replacement

WARNING:

The models covered by this manual are equipped with Supplemental Restraint Systems (SRS), more commonly known as airbags. Always disable the airbag system before working in the vicinity of any airbag system components to avoid the possibility of accidental deployment of the airbag(s), which could cause personal injury (see Section 26).

1 Disconnect the cable from the negative battery terminal (see Chapter 5, Section 1).

2 Remove the steering column covers (see Chapter 11).

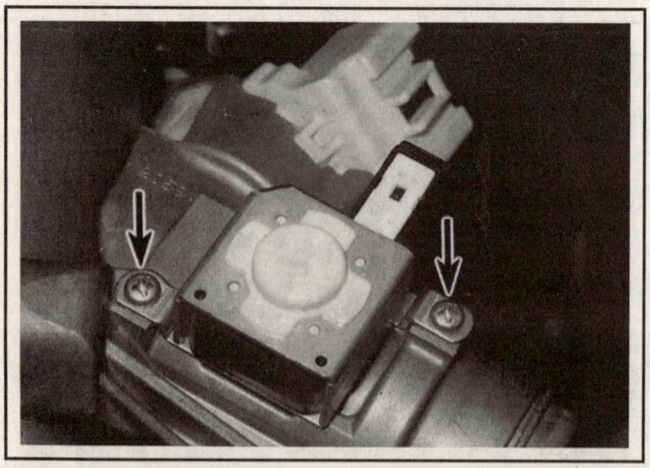

9.3 To detach the key interlock solenoid from the key lock cylinder housing, remove these two screws (vehicles with an automatic transmission)

KEY LOCK CYLINDER

Refer to illustrations 9.3, 9.5, 9.6 and 9.7

3 On vehicles with an automatic transmission, remove the key interlock solenoid from the key lock cylinder housing (see illustration).

4 Insert the ignition key into the key lock cylinder and turn it to the ACC position.

5 Insert an awl or a small punch into the hole in the lock cylinder housing, push down the stop pin and pull the lock cylinder straight out until it contacts the upper stopper (see illustration).

Note: The key lock cylinder does not come all the way out in this step; when it contacts the stopper, it comes out about 1/4-inch, then stops.

6 Remove the unlock warning switch (see illustration).

CAUTION:

Do NOT attempt to remove the key lock cylinder with the switch installed. Its spring-loaded plunger will break off if you attempt to force the key lock cylinder out without removing the switch.

7 Insert a small screwdriver into the other hole in the lock cylinder housing - not the same one that you used in Step 5; this one's on the same side of the housing as the unlock warning switch (see illustration) - until it contacts the locking claw of the lock cylinder, then push the handle of the screwdriver forward (toward the front of the vehicle) until you hear a click sound, which indicates that the lock cylinder is released. Pull out the lock cylinder.

8 Installation is the reverse of removal. Remember to put the key in the ACC position before inserting the key lock cylinder into the housing. You will hear a click when the lock cylinder locks into place.

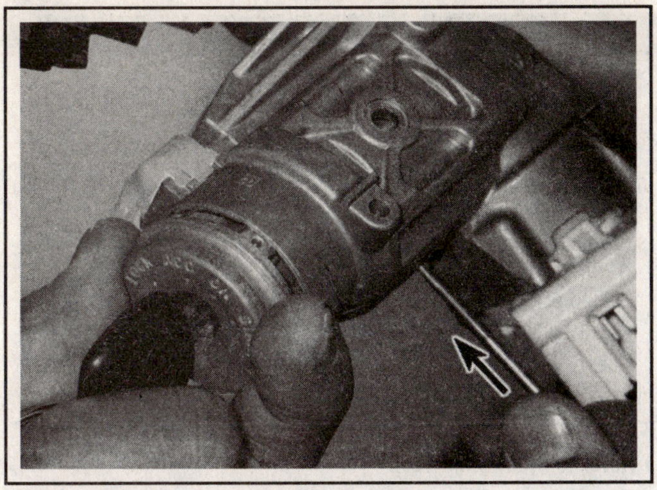

9.5 Locate the small hole in the lock cylinder housing near the ignition switch, insert a small screwdriver into the hole, then push down the stop pin and pull the lock cylinder out of the housing until it contacts the stopper pin (the lock cylinder comes out about 1/4-inch)

9.6 To detach and remove the unlock warning switch, pry the two locking lugs up (other lug not visible) and push the switch down. To install the switch, align it with its two mounting rails and push it up until the two locking lugs snap into place

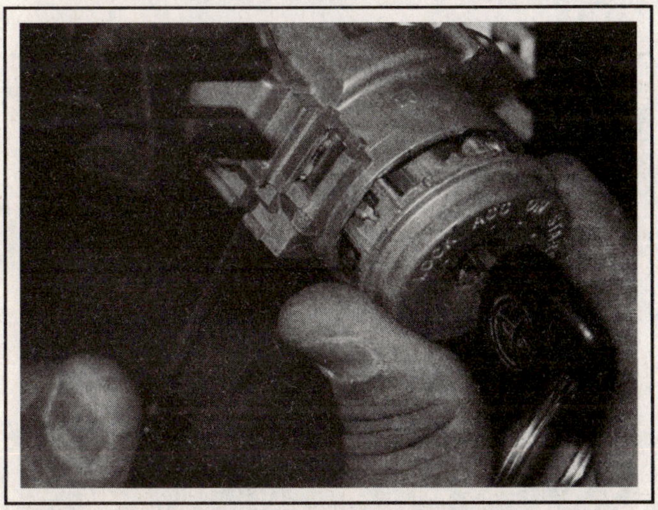

9.7 To remove the key lock cylinder, insert a small screwdriver into the hole in the lock cylinder housing until it engages the locking claw on the lock cylinder, then push the screwdriver handle forward (toward the front of the vehicle) until you hear a click sound, which indicates that the claw has disengaged

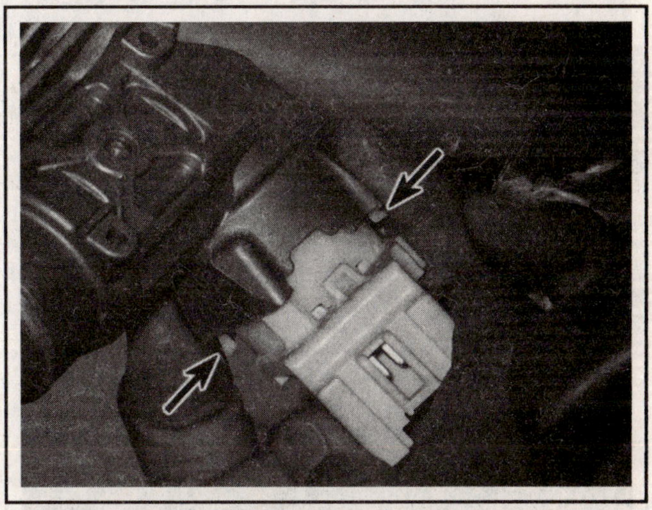

9.10 To remove the ignition switch, squeeze these two retaining tabs and pull out the switch

IGNITION SWITCH

➧ **Refer to illustration 9.10**

9 Disconnect the electrical connector from the ignition switch.

10 Depress the two switch retaining tabs (see illustration) and pull out the switch.

11 Installation is the reverse of removal. Make sure that the switch clicks into place.

10 Dashboard switches - replacement

⁂ WARNING:

The models covered by this manual are equipped with Supplemental Restraint Systems (SRS), more commonly known as airbags. Always disable the airbag system before working in the vicinity of any airbag system components to avoid the possibility of accidental deployment of the airbag(s), which could cause personal injury (see Section 26).

SWITCHES ON INSTRUMENT CLUSTER FINISH PANEL AND KNEE BOLSTER

➧ Refer to illustration 10.3

➡Note: All vehicles have an instrument cluster rheostat on the instrument cluster finish panel. Depending on the year and model, a vehicle might also have a variety of other switches on the left end of the finish panel and the knee bolster. All of these switches are removed the same way. The instrument cluster rheostat shown in the accompanying photograph is typical.

1 Disconnect the cable from the negative battery terminal (see Chapter 5, Section 1).
2 Depending on the switch that you need to replace, remove the instrument cluster finish panel and/or the knee bolster (see Chapter 11).
3 Remove the switch from the cluster finish panel or knee bolster (see illustration).
4 When installing a new switch, make sure that the retaining tabs snap into place.
5 Installation is otherwise the reverse of removal.

HAZARD FLASHER SWITCH AND CLOCK

➧ Refer to illustration 10.7

6 Remove the radio (see Section 12).
7 The hazard flasher/clock assembly is retained by lugs on each end. Using a couple of small screwdrivers, carefully lever these lugs loose and pry out the hazard flasher switch/clock assembly (see illustration).
8 When installing the unit, make sure that it snaps into place.
9 Installation is otherwise the reverse of removal.

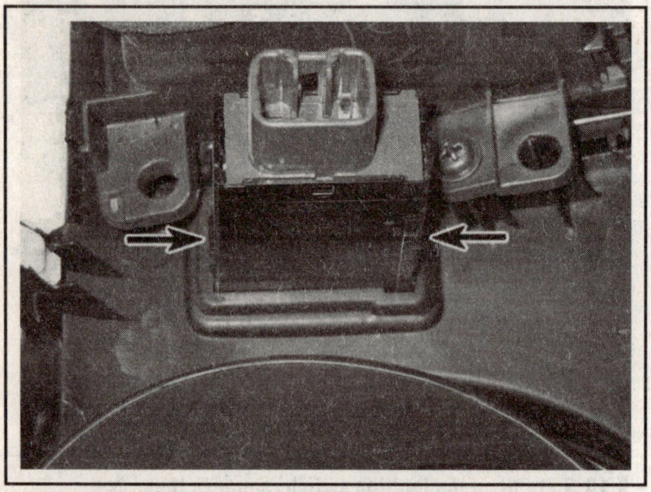

10.3 To remove any dashboard switch from the instrument cluster finish panel or the knee bolster, squeeze the two retaining tabs and push the switch through the front side

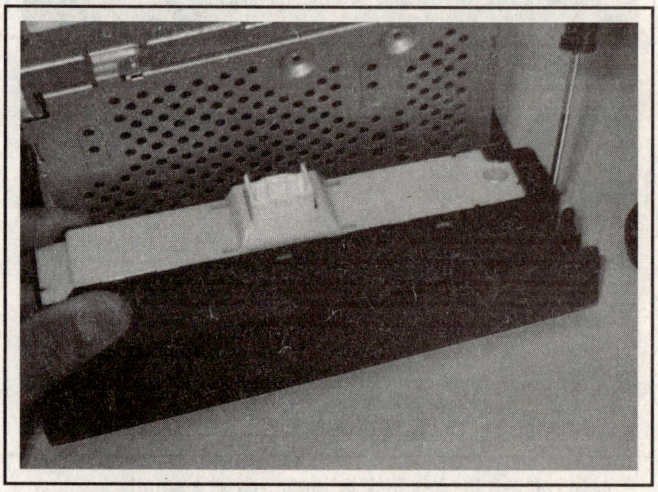

10.7 To remove the hazard flasher switch and clock unit from the radio assembly, carefully pry it loose with small screwdrivers

11 Instrument cluster - removal and installation

➧ Refer to illustrations 11.3 and 11.4

⁂ WARNING:

The models covered by this manual are equipped with Supplemental Restraint Systems (SRS), more commonly known as airbags. Always disable the airbag system before working in the vicinity of any airbag system components to avoid the possibility of accidental deployment of the airbag(s), which could cause personal injury (see Section 26).

1 Disconnect the cable from the negative terminal of the battery (see Chapter 5, Section 1).
2 Remove the instrument cluster finish panel (see Chapter 11).
3 Remove the four bolts that secure the cluster (see illustration).
4 Pull out the cluster and disconnect the electrical connectors (see illustration).
5 Installation is the reverse of removal.

11.3 To detach the instrument cluster, remove these four bolts

11.4 Pull out the cluster and disconnect the electrical connectors

12 Radio and speakers - removal and installation

⁂ WARNING:

The models covered by this manual are equipped with Supplemental Restraint Systems (SRS), more commonly known as airbags. Always disable the airbag system before working in the vicinity of any airbag system components to avoid the possibility of accidental deployment of the airbag(s), which could cause personal injury (see Section 26).

RADIO

➧ Refer to illustrations 12.3a, 12.3b, 12.3c and 12.5

1 Disconnect the cable from the negative battery terminal (see Chapter 5, Section 1).

2 Remove the heater and air conditioning control assembly (see Chapter 3).

3 Remove the radio mounting bracket bolts and pull out the radio and brackets as a single assembly, then disconnect the antenna lead and the electrical connector (see illustrations).

4 If you're replacing the radio, remove the left and right mounting brackets from the radio and install them on the new unit.

5 When installing the radio, make sure that the five locator pins (see illustration) on the radio assembly are aligned with their corresponding holes in the instrument panel. Installation is otherwise the reverse of removal.

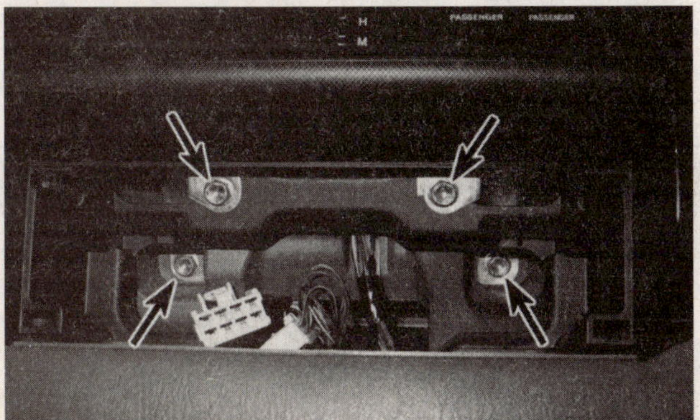

12.3a To detach the radio from the instrument panel, remove these four bolts

12.3b Pull out the radio and mounting brackets as a single assembly . . .

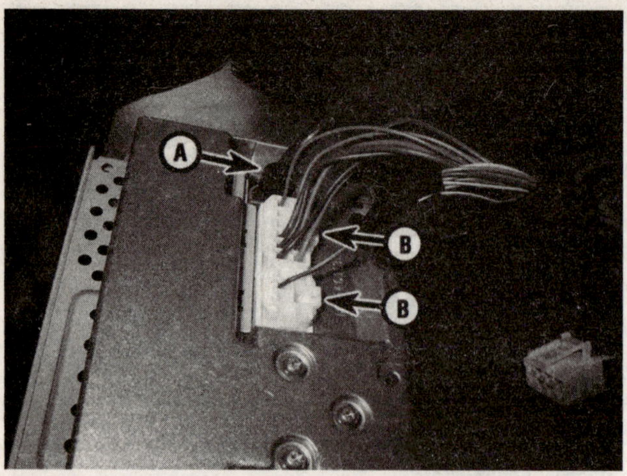

12.3c . . . and disconnect the antenna cable (A), then depress the release tabs and disconnect the electrical connectors (B)

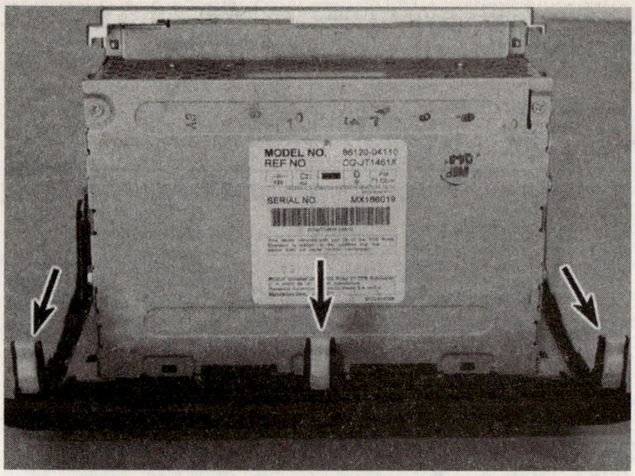

12.5 When installing the radio make sure that the five locator pins on the radio assembly are aligned with their corresponding holes in the instrument panel (these are the lower three pins; there are two more at the top)

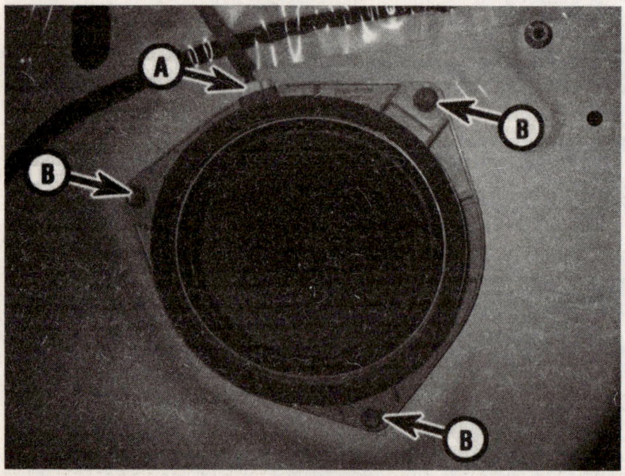

12.7 Depress the release tab and disconnect the electrical connector from the door speaker (A), then remove the three mounting bolts (B) and remove the speaker from the door

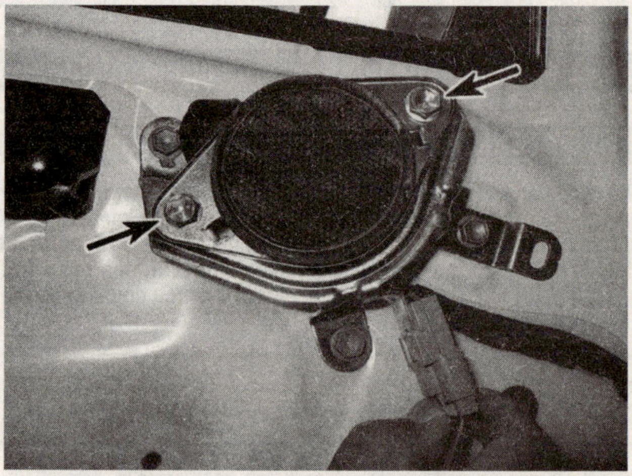

12.11 To detach the tweeter from the door, depress the release tab and disconnect the electrical connector, then remove the two tweeter mounting bolts

DOOR SPEAKERS

➧ Refer to illustration 12.7

6 Remove the door trim panel (see Chapter 11).
7 Disconnect the electrical connector (see illustration).
8 Remove the speaker mounting screws and remove the speaker.
9 Installation is the reverse of removal.

DOOR TWEETERS

➧ Refer to illustration 12.11

10 Remove the door trim panel (see Chapter 11).
11 Disconnect the tweeter electrical connector (see illustration).
12 Remove the tweeter mounting bolts and remove the tweeter.
13 Installation is the reverse of removal.

13 Antenna - removal and installation

ANTENNA MAST

➧ Refer to illustration 13.1

1 Unscrew the antenna mast (see illustration).
2 Installation is the reverse of removal.

ANTENNA MOUNTING BASE AND ANTENNA CABLE

➧ Refer to illustrations 13.4, 13.7 and 13.9

3 Remove the antenna mast (see illustration 13.1).
4 Unscrew the antenna mounting base retaining nut (see illustration).
5 Loosen the right front wheel lug nuts, raise the vehicle and place it securely on jackstands. Remove the right front wheel.
6 Remove the inner fender splash shield (see Chapter 11).
7 Remove the antenna base mounting bracket bolt (see illustration) and remove the weather grommet from its hole in the wheel well. From here the cable enters the right side of the cabin, where it connects to the main antenna cable, which is routed through the instrument panel to the radio. Trace the antenna mounting base cable to the connection with the main antenna cable and disconnect the two cables.
8 Remove the instrument panel (see Chapter 11).
9 Note the routing of the antenna cable through the instrument panel (see illustration), then detach all clips and reattach a new cable using exactly the same routing as shown.
10 Installation is the reverse of removal.

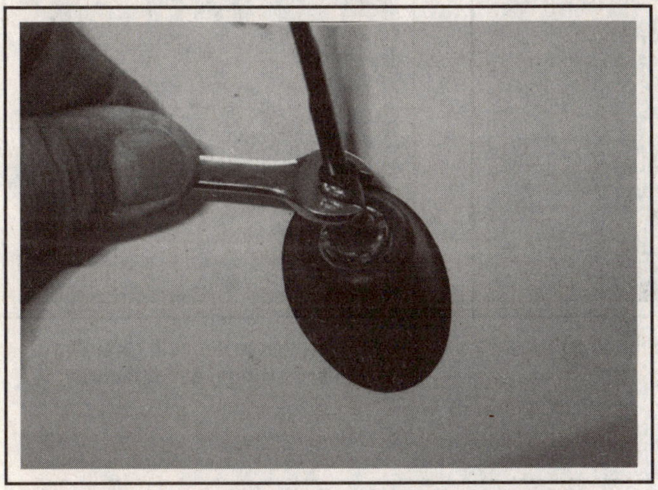

13.1 To remove the antenna mast, simply unscrew it with a wrench

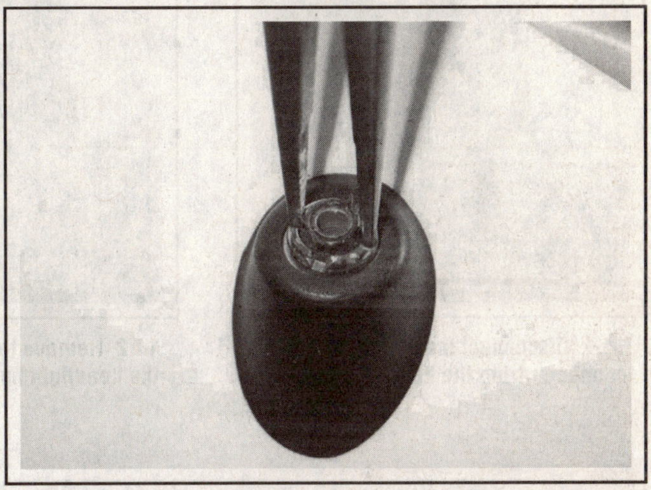

13.4 To detach the antenna mounting base, unscrew this retaining nut

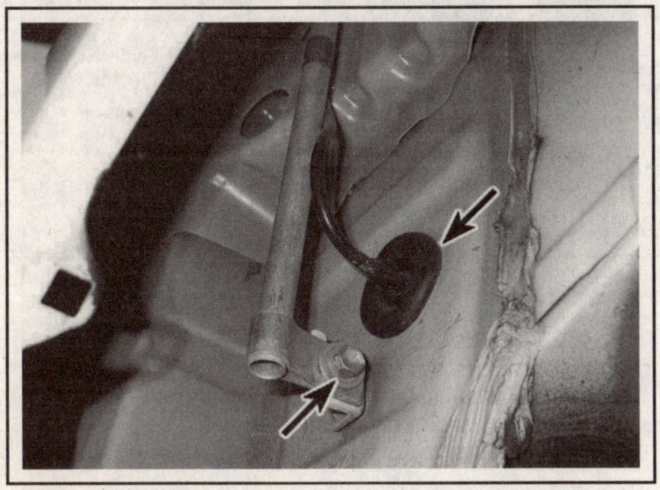

13.7 To detach the lower end of the antenna mounting base assembly, remove this bolt, then remove the weather grommet where the cable enters the cabin

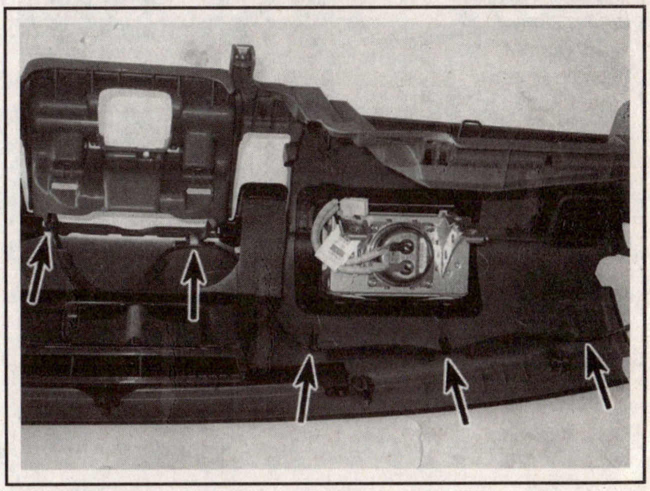

13.9 With the instrument panel removed, detach the cable from the clips

14 Headlight bulb - replacement

Refer to illustrations 14.1, 14.2, 14.3, 14.4 and 14.5

1 Depress the release tab and disconnect the electrical connector from the headlight bulb (see illustration).
2 Remove the rubber cover from the back of the headlight housing (see illustration).
3 Disengage the bulb retaining spring (see illustration) and swing it out of the way.
4 Remove the bulb from the headlight housing (see illustration).
5 When installing the bulb, make sure that the three lugs on the mounting base of the bulb mounting flange are aligned with their corresponding cutouts in the headlight housing (see illustration).

CAUTION:

Don't touch the bulb with your fingers. If you do, clean it with rubbing alcohol (the oil from your skin can cause the bulb to overheat and fail).

6 Installation is otherwise the reverse of removal.

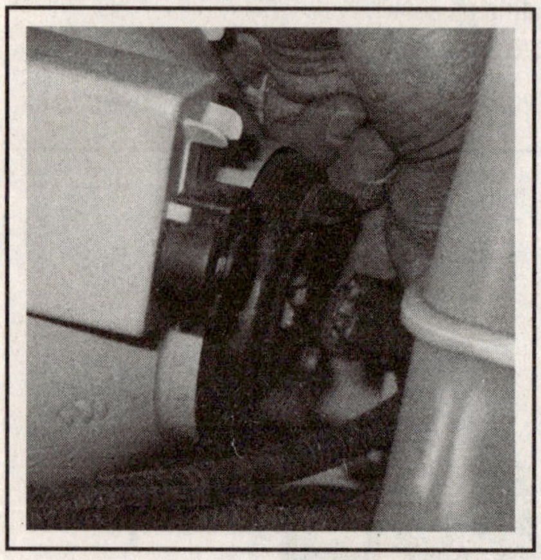

14.1 Disconnect the electrical connector from the headlight bulb

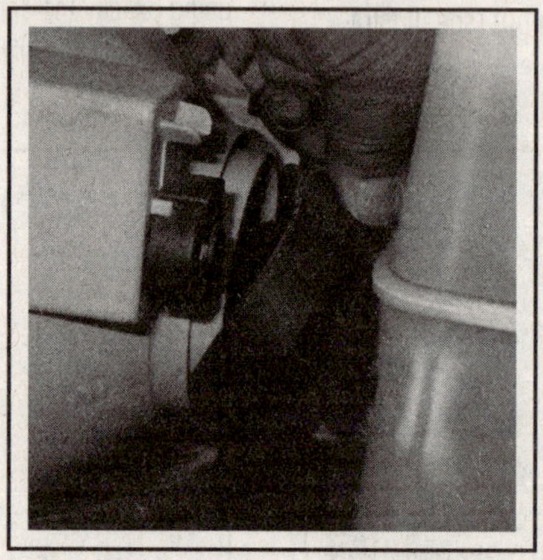

14.2 Remove the rubber cover from the headlight housing

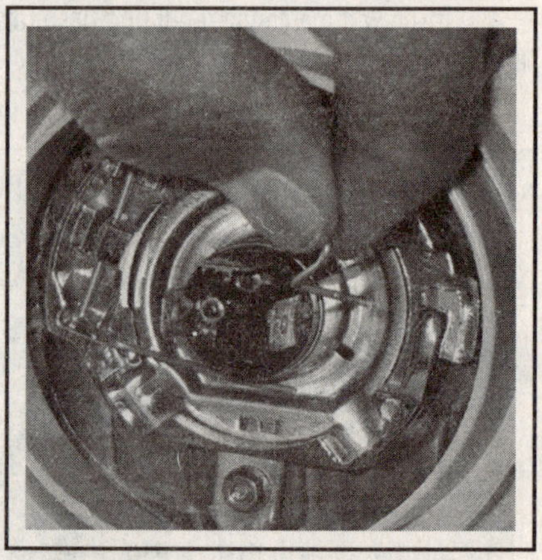

14.3 Disengage the bulb retaining spring and swing it out of the way

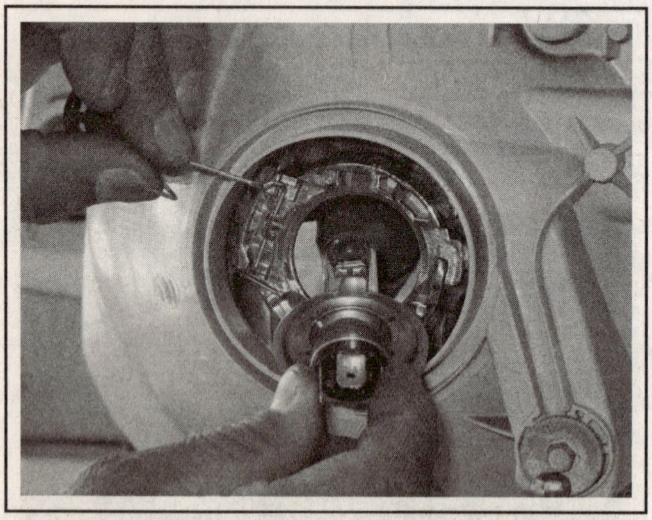

14.4 Remove the bulb from the headlight housing

14.5 When installing the new bulb, align the lugs on the bulb mounting flange with the cutouts in the headlight housing

15 Headlight housing - removal and installation

Refer to illustrations 15.2 and 15.4

1 Remove the radiator grille and the front bumper cover (see Chapter 11).

2 To remove the small trim panel below the headlight housing, remove the Phillips head cap screw and washer at the inner end of the trim piece (see illustration). Note how the tab in the center of the trim piece fits into a slot on the underside of the headlight housing and a slot on the outer end of the trim engages a tab on the front fender. Inspect the condition of screw and clip. If they're damaged, replace them.

3 Remove the headlight housing mounting bolts.

4 Pull out the headlight housing and disconnect the electrical connectors from the bulbs (see illustration).

5 Installation is the reverse of removal. When you're done, check the headlight adjustment (see Section 16).

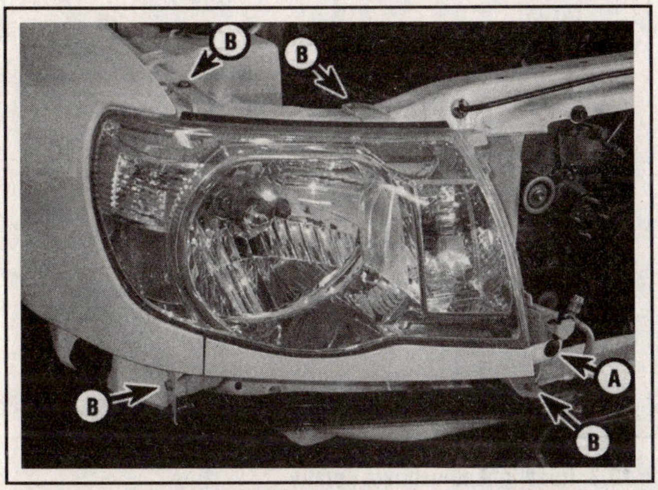

15.2 To detach the small trim panel below the headlight housing, remove the Phillips head cap screw and washer (A), then pull off the trim panel. To remove the headlight housing, remove the mounting bolts (B)

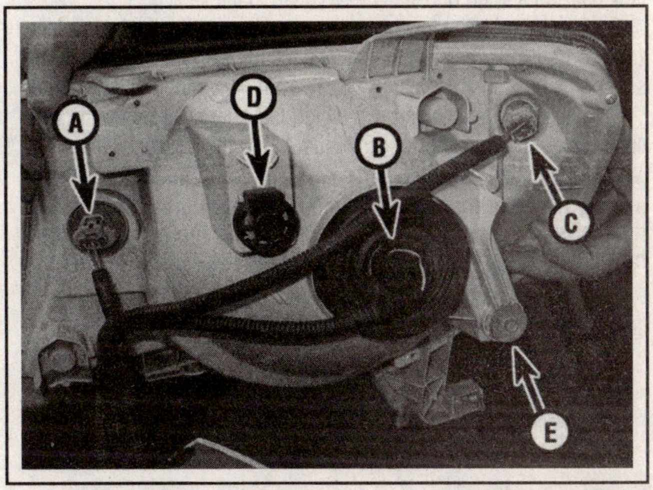

15.4 Headlight housing details:

A Front turn signal bulb
B Headlight bulb
C Parking and front sidemarker bulb
D Horizontal headlight housing adjuster
E Vertical headlight housing adjuster

16 Headlights - adjustment

Refer to illustrations 16.1 and 16.3

Note: The headlights must be aimed correctly. If adjusted incorrectly, they could blind the driver of an oncoming vehicle and cause a serious accident or seriously reduce your ability to see the road. The headlights should be checked for proper aim every 12 months and any time a new headlight is installed or front end body work is performed. It should be emphasized that the following procedure is only an interim step that will provide temporary adjustment until a properly equipped shop can adjust the headlights.

1 Each headlight housing has a vertical adjusting screw (see illustration). The vertical adjuster controls the up-and-down movement of the housing. You can access the vertical adjuster with a long Phillips screwdriver. Each headlight housing also has a horizontal adjuster screw. Note that the horizontal adjuster has a plastic retainer on it that prevents it from being adjusted unless you remove the retainer. Each headlight is adjusted horizontally at the factory, then sealed with the retainer.

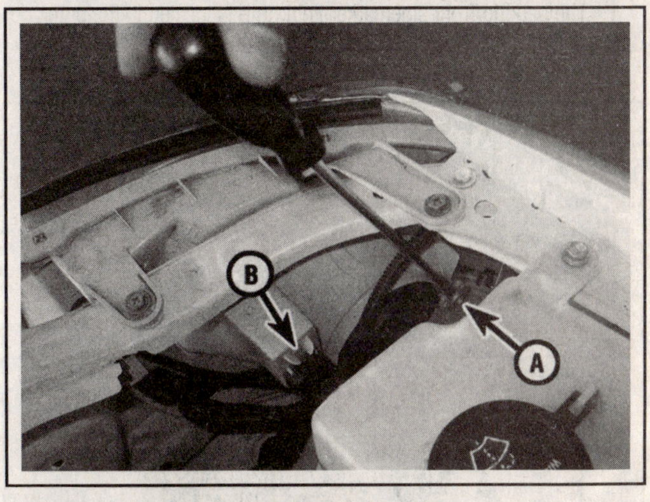

16.1 Use a Phillips screwdriver to turn the headlight adjusters

A Vertical adjuster
B Horizontal adjuster

2 There are several ways to adjust the headlights. The simplest method requires a blank wall 25-feet in front of the vehicle and a level floor.

3 Position masking tape vertically on the wall in reference to the vehicle centerline and the centerlines of both headlights (see illustration).

4 Position a horizontal tape line in reference to the centerline of all the headlights.

➡**Note: It may be easier to position the tape on the wall with the vehicle parked only a few inches away.**

5 Adjustment should be made with the vehicle sitting level, the gas tank half-full and no unusually heavy load in the vehicle.

6 Starting with the low beam adjustment, position the high intensity zone so it's two inches below the horizontal line and two inches to the side of the headlight vertical line away from oncoming traffic.

7 With the high beams on, the high intensity zone should be vertically centered with the exact center just below the horizontal line.

➡**Note: It may not be possible to position the headlight aim exactly for both high and low beams. If a compromise must be made, keep in mind that the low beams are the most used and have the greatest effect on driver safety.**

8 Have the headlights adjusted by a dealer service department or service station at the earliest opportunity.

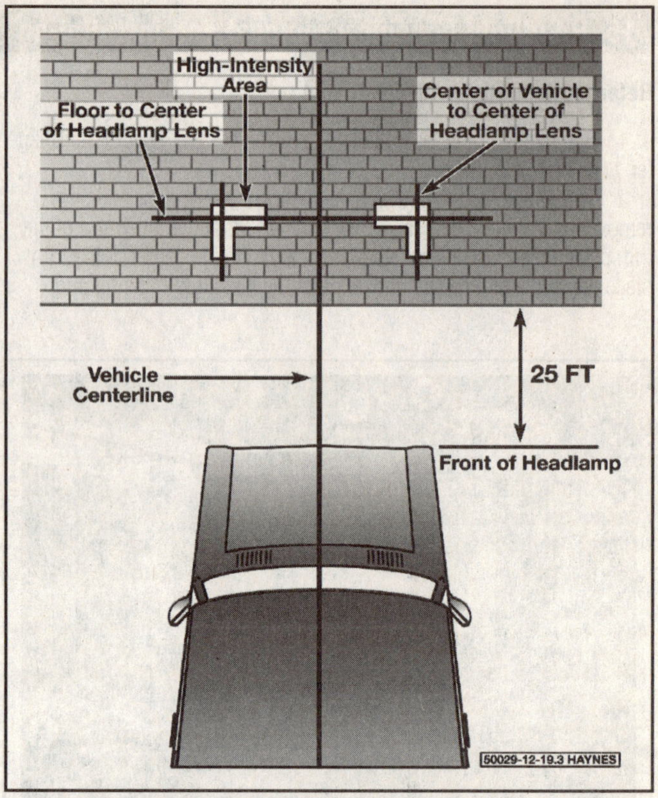

16.3 Headlight adjustment details

17 Bulb replacement

EXTERIOR LIGHT BULBS

Parking/front sidemarker light bulbs

➧ **Refer to illustrations 17.2 and 17.3**

1 The parking/front sidemarker light bulb is located in the outer, upper corner of the headlight housing (see illustration 15.4).

➡**Note: These bulbs are difficult to access with the headlight housing installed, but they ARE accessible.**

2 Disconnect the electrical connector from the bulb socket (see illustration).

3 Turn the bulb socket counterclockwise and pull it out of the headlight housing (see illustration). Note that there are four lugs on the socket and four cutouts in the headlight housing, and one of the lugs is slightly bigger than the other three.

17.2 The parking/front sidemarker light bulb sockets are located in the outer, upper corner of each headlight housing. Depress the release tab to disconnect the electrical connector

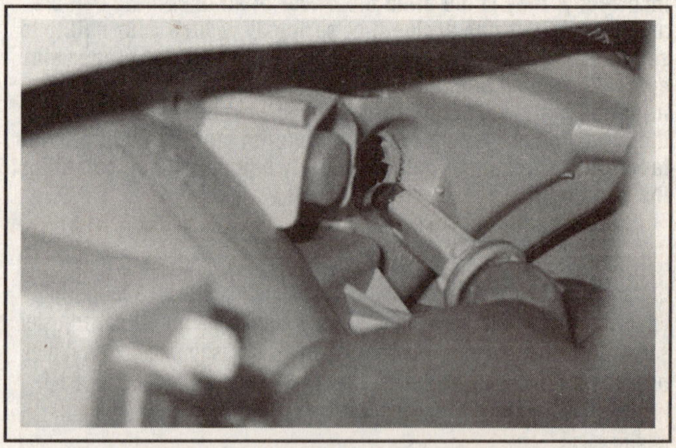

17.3 To remove the parking/front sidemarker light bulb socket from the headlight housing, turn it counterclockwise and pull it out

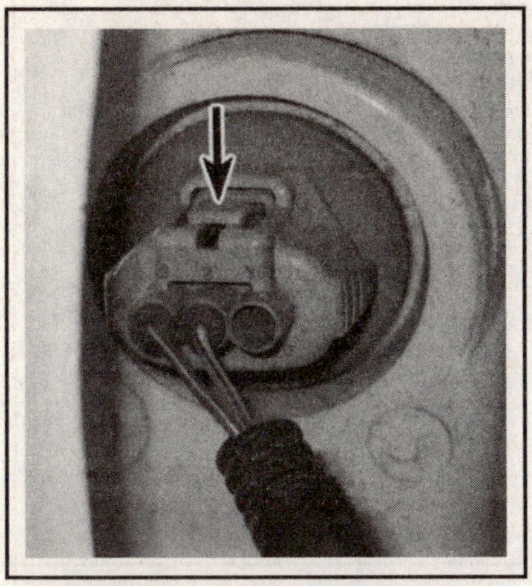

17.7 To disconnect the electrical connector from the turn signal bulb socket, depress this release tab and pull off the connector

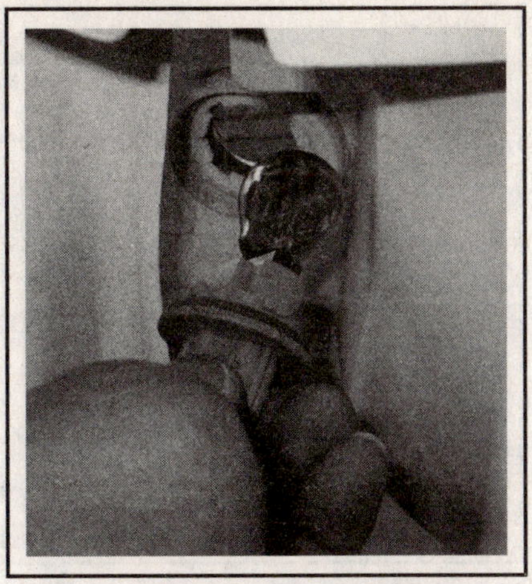

17.8 To remove the front turn signal bulb socket from the headlight housing, turn it counterclockwise and pull it out

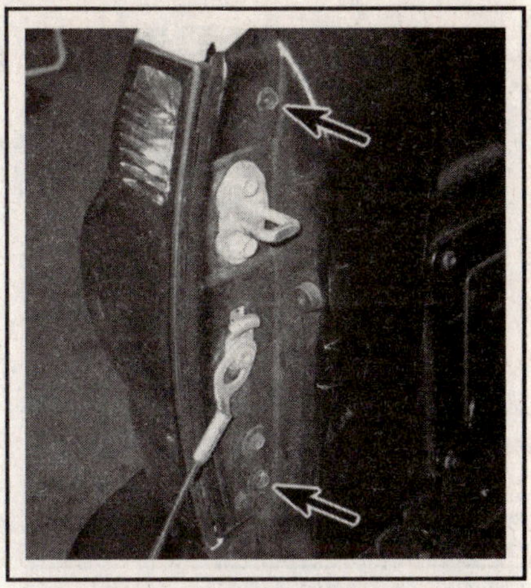

17.14 To detach the taillight housing, remove these two bolts

4 To remove the parking/front sidemarker bulb from its socket, pull it straight out of the socket. To install a new bulb in the socket, push it into the socket until it stops.

5 When installing the bulb socket, align the larger lug with the larger cutout in the headlight housing, then insert the socket into the mounting hole and turn it clockwise until it locks into place. Installation is otherwise the reverse of removal.

Front turn signal bulbs

➧ Refer to illustrations 17.7 and 17.8

6 The front turn signal bulb is located near the inner end of the headlight housing (see illustration 15.4).

7 Disconnect the electrical connector from the bulb socket (see illustration).

8 Turn the front turn signal bulb socket counterclockwise and pull it out of the headlight housing (see illustration). Note that there are four lugs on the socket and four cutouts in the headlight housing, and one of the lugs is slightly bigger than the other three.

9 To remove the turn signal bulb from the socket, pull it straight out of the socket. To install a new bulb in the socket, push it straight in until it stops.

10 When installing the bulb socket, align the larger lug with the larger cutout in the headlight housing, then insert the socket into the mounting hole and turn it clockwise until it locks into place. Installation is otherwise the reverse of removal.

Fog light bulbs

➡Note: The fog light bulbs, if equipped, are located in their own housings in the outer ends of the bumper cover. You can access them from the underside of the vehicle without raising it off the ground.

11 Depress the release tab and disconnect the electrical connector from the fog light bulb socket.

12 Turn the fog light bulb counterclockwise and pull it out of the fog light housing.

13 When installing the fog light, align the lugs on the bulb mounting flange with the cutouts in the fog light housing, then insert the bulb into the mounting hole in the fog light housing and turn it clockwise until it locks into place. Installation is otherwise the reverse of removal.

Rear turn signal, brake/taillight, rear sidemarker and back-up light bulbs

➧ Refer to illustrations 17.14, 17.16 and 17.17

14 Open the tailgate and remove the two taillight housing bolts (see illustration).

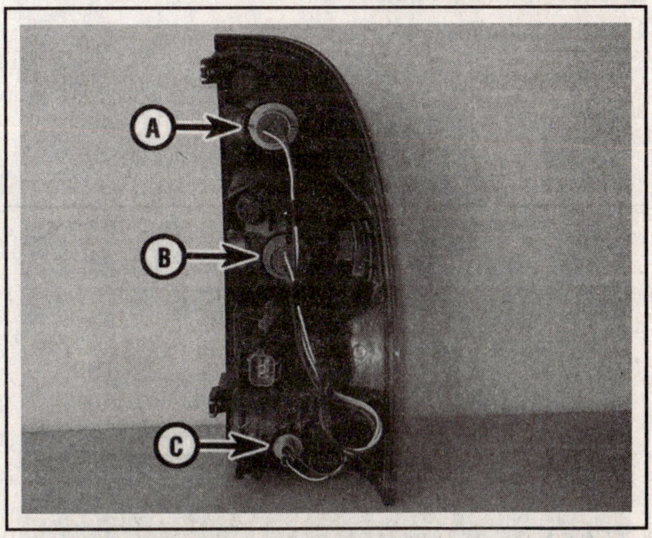

17.16 Taillight housing light bulbs:

A Rear turn signal light
B Brake/taillight and rear sidemarker light
C Back-up light

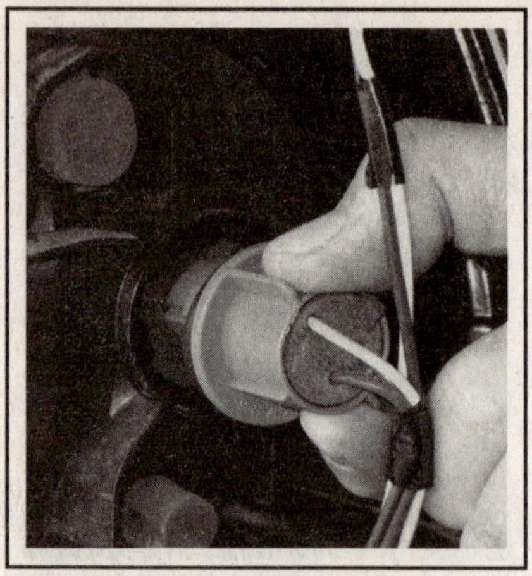

17.17 To remove a taillight bulb socket, turn it counterclockwise and pull it out of the taillight housing

17.21 To detach a license plate light housing, remove this screw and pull out the housing

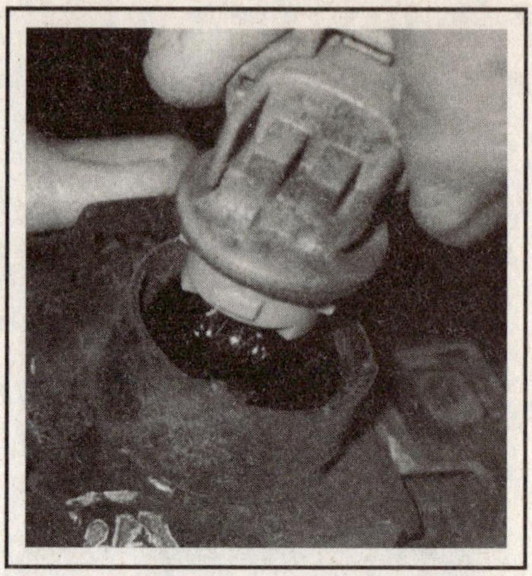

17.22 To remove the bulb socket from a license plate light housing, turn it counterclockwise and pull it out of the housing

15 Pull off the taillight housing for access to the bulbs.

16 There are three bulbs in the taillight housing (see illustration).

17 Remove a bulb socket by rotating it counterclockwise (see illustration), then pull it out of the taillight housing.

18 Remove the bulb from the socket by pulling it straight out of the socket. Install a new bulb by pushing it straight into the socket until it stops.

19 When installing a bulb socket, align the lugs on the socket with the cutouts in the taillight housing, insert the socket into the housing and turn the socket clockwise until it locks into place.

20 Installation is otherwise the reverse of removal.

License plate light bulbs

➧ Refer to illustrations 17.21 and 17.22

21 Remove the license plate light retaining screw (see illustration) and pull out the housing.

22 To remove a license plate light bulb socket, turn it counterclockwise and pull it out of the housing (see illustration).

23 To remove the bulb from the socket, pull it straight out. To install a new bulb, push it straight into the socket until it stops.

24 Installation is the reverse of removal.

Center high-mounted brake light bulb

➧ Refer to illustrations 17.25 and 17.26

25 Remove the center high-mounted brake light lens (see illustration).

26 Remove and inspect the O-ring type lens gasket (see illustration). If it's cracked, torn or deteriorated, replace it.

27 To remove a bulb from its socket in the housing, pull it straight out. To install a new bulb, push it straight into the socket until it stops.

28 Installation is the reverse of removal. Make sure that the O-ring gasket is properly seated before installing the lens.

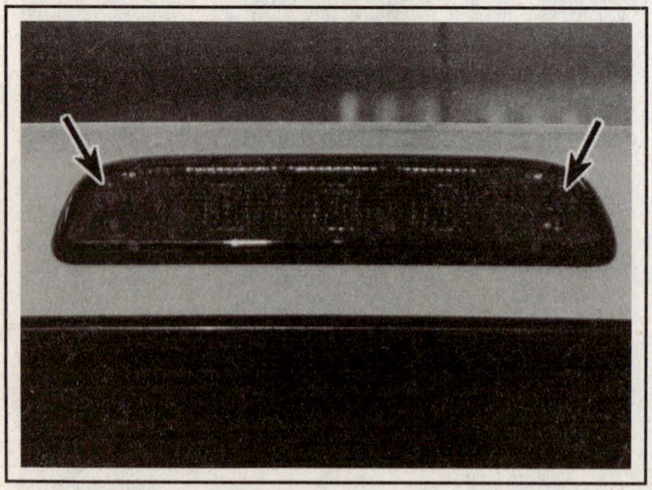

17.25 To remove the center high-mounted brake light lens, remove these two screws

17.26 Remove and inspect the O-ring type lens gasket. If it's damaged, replace it

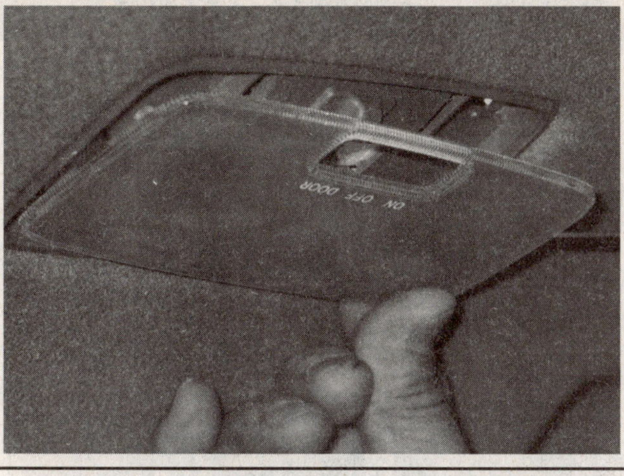

17.29 To remove a dome light lens, carefully pry it loose with a small screwdriver

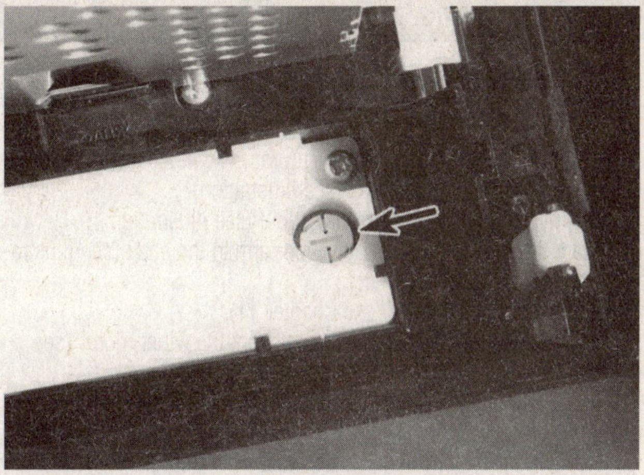

17.32 To remove the illumination bulb from the hazard flasher switch, turn it counterclockwise. To install a new bulb, turn it clockwise until it locks into place

INTERIOR LIGHT BULBS

Dome light bulbs

➧ Refer to illustration 17.29

29 Carefully pry off the dome light lens (see illustration).
30 Remove the old bulb and install a new one. Install the lens.

Hazard flasher switch illumination bulb

➧ Refer to illustration 17.32

31 Remove the hazard flasher switch (see Section 10).
32 To remove the old bulb, turn it counterclockwise and pull it out (see illustration).
33 When installing the new bulb, turn it clockwise until it stops.
34 Installation is the reverse of removal.

18 Horns - replacement

➧ Refer to illustrations 18.1 and 18.2

1 The horn or horns are located on the radiator support (see illustration). You have to remove the grille to access a horn (see Chapter 11).
2 Disconnect the electrical connector (see illustration).
3 Unscrew the nut and remove the horn from the mounting bracket.
4 Installation is the reverse of removal.

18.1 The horns are mounted on small brackets bolted to the radiator support (there's only one horn on this vehicle, but many vehicles have two horns)

18.2 To disconnect the electrical connector from the horn, depress the release tab (A) and pull off the connector. To detach the horn from the mounting bracket, remove this nut (B)

19 Wiper motor - replacement

➧ Refer to illustrations 19.1, 19.3, 19.5, 19.6, 19.8, 19.9 and 19.10

1 Carefully pry off the protective cap for the windshield wiper arm retaining nut from each wiper arm (see illustration).

2 Remove each wiper arm retaining nut (see illustration 19.1).

3 Mark the relationship of each wiper arm to the shaft (see illustration), then remove both wiper arms.

4 Remove the cowl cover (see Chapter 11).

5 Disconnect the electrical connector from the wiper motor (see illustration).

6 Remove the wiper motor/linkage assembly mounting bolts (see illustration).

7 Remove the wiper motor/linkage assembly from the cowl area and place it on a clean work bench space.

8 Detach the linkage arms from the crank arm (see illustration).

9 Remove the crank arm retaining nut and mark the relationship of the crank arm to the motor shaft (see illustration).

10 Remove the wiper motor mounting bolts (see illustration) and separate the motor from its mounting bracket.

11 Installation is the reverse of removal.

19.1 Carefully pry off the protective caps for the windshield wiper arm retaining nuts and remove the nuts

19.3 Before removing each windshield wiper arm, mark the relationship of the wiper arm to the shaft

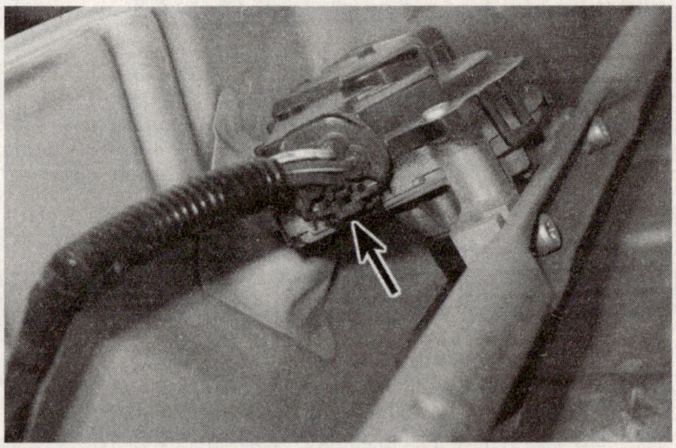

19.5 Depress the release tab and disconnect the electrical connector from the windshield wiper motor

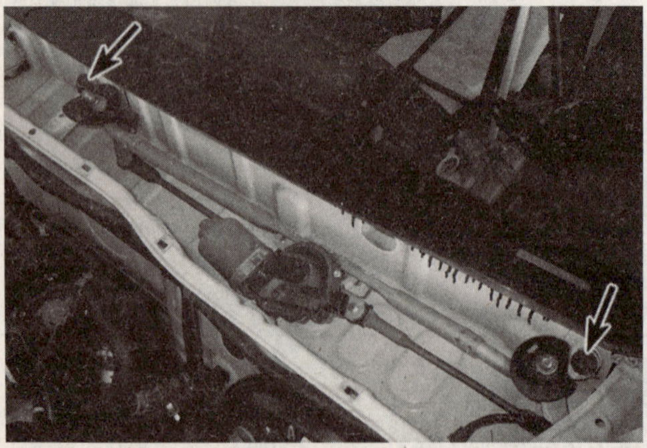

19.6 To detach the windshield wiper motor/linkage assembly from the cowl, remove these two bolts

19.8 To detach the linkage arms from the crank arm, pry them off, then remove the crank arm nut

19.9 Before removing the crank arm from the wiper motor shaft, be sure to mark the relationship of the crank arm to the shaft

19.10 To detach the wiper motor from its mounting bracket, remove these two bolts

20 Rear window defogger - check and repair

1 The rear window defogger consists of a number of horizontal elements baked onto the glass surface.

2 Small breaks in the element can be repaired without removing the rear window.

CHECK

➧ **Refer to illustrations 20.4, 20.5 and 20.7**

3 Turn the ignition switch and defogger system switches to the ON position. Using a voltmeter, place the positive probe against the defogger grid positive terminal and the negative lead against the ground terminal. If battery voltage is not indicated, check the fuse, defogger switch and related wiring.

4 When measuring voltage during the next two tests, wrap a piece of aluminum foil around the tip of the voltmeter positive probe and press the foil against the heating element with your finger (see illustration).

5 Check the voltage at the center of each heating element (see illustration). If the voltage is 6-volts, the element is okay (there is no break). If the voltage is 12-volts, the element is broken between the center of the element and the ground side. If the voltage is 0-volts the element is broken between the center of the element and positive side.

6 If none of the elements are broken, connect the negative lead to a good body ground. The voltage reading should stay the same; if it doesn't the ground connection is bad.

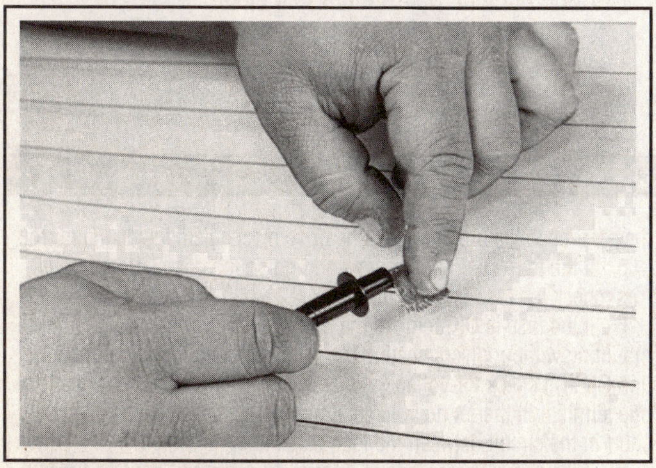

20.4 When measuring the voltage at the rear window defogger grid, wrap a piece of aluminum foil around the positive probe of the voltmeter and press the foil against the wire with your finger

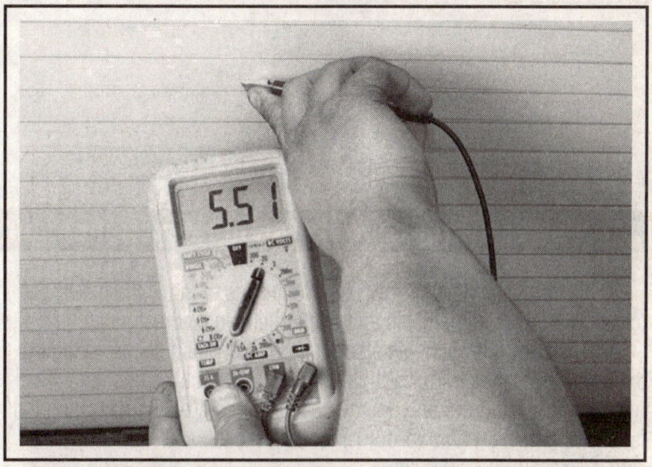

20.5 To determine if a heating element has broken, check the voltage at the center of each element; if the voltage is 5 or 6-volts, the element is unbroken, but if the voltage is 10 or 12-volts, the element is broken between the center and the ground side. If there is no voltage, the element is broken between the center and the positive side

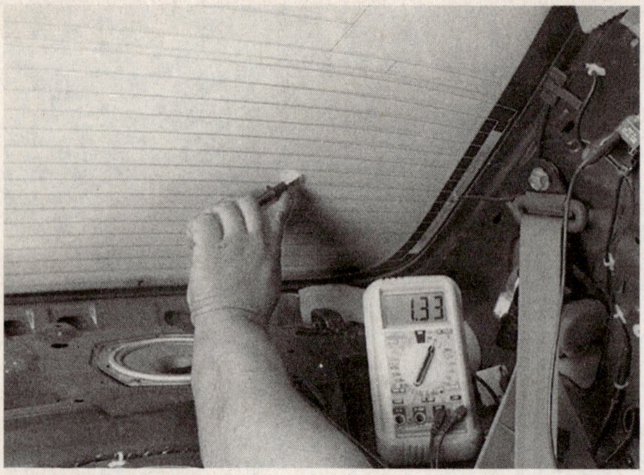

20.7 To find the break, place the voltmeter negative lead against the defogger ground terminal, place the voltmeter positive lead with the foil strip against the heating element at the positive terminal end and slide it toward the negative terminal end. The point at which the voltmeter reading changes abruptly is the point at which the element is broken

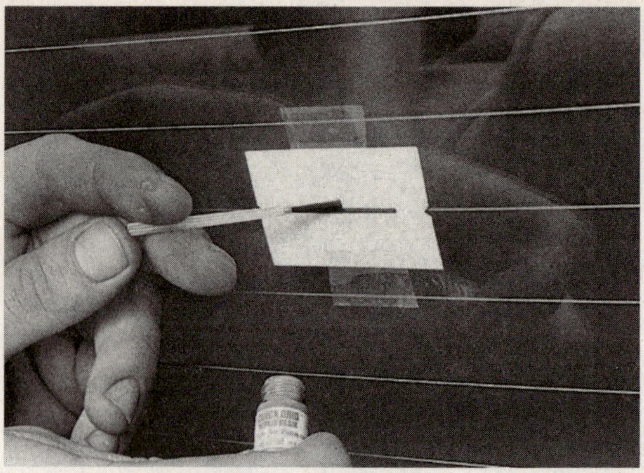

20.13 To use a defogger repair kit, apply masking tape to the inside of the window at the damaged area, then brush on the special conductive coating

7 To find the break, place the voltmeter negative lead against the defogger ground terminal. Place the voltmeter positive lead with the foil strip against the heating element at the positive terminal end and slide it toward the negative terminal end. The point at which the voltmeter deflects from several volts to zero is the point at which the heating element is broken (see illustration).

REPAIR

➧ Refer to illustration 20.13

8 Repair the break in the element using a repair kit specifically recommended for this purpose, such as DuPont paste No. 4817 (or equivalent). Included in this kit is plastic conductive epoxy.

9 Prior to repairing a break, turn off the system and allow it to cool off for a few minutes.

10 Lightly buff the element area with fine steel wool, then clean it thoroughly with rubbing alcohol.

11 Use masking tape to mask off the area being repaired.

12 Thoroughly mix the epoxy, following the instructions provided with the repair kit.

13 Apply the epoxy material to the slit in the masking tape, overlapping the undamaged area about 3/4-inch on either end (see illustration).

14 Allow the repair to cure for 24 hours before removing the tape and using the system.

21 Power mirror control system - description and check

1 Electric rear view mirrors use two motors to move the glass; one for up and down adjustments and one for left-right adjustments.

2 The control switch has a selector portion which sends voltage to the left or right side mirror. With the ignition ON but the engine OFF, roll down the windows and operate the mirror control switch through all functions (left-right and up-down) for both the left and right side mirrors.

3 Listen carefully for the sound of the electric motors running in the mirrors.

4 If the motors can be heard but the mirror glass doesn't move, there's probably a problem with the drive mechanism inside the mirror.

5 If the mirrors do not operate and no sound comes from the mirrors, check the fuse (see Section 3).

6 If the fuse is OK, remove the mirror control switch from its mounting without disconnecting the wires attached to it. Turn the ignition ON and check for voltage at the switch. There should be voltage at one terminal. If there's no voltage at the switch, check for an open or short in the circuit between the fuse panel and the switch.

7 If the mirror motor fails to operate as described, replace the mirror assembly (see Chapter 11).

22 Cruise control system - general information

1 All models have an electronically-controlled throttle body - there is no accelerator cable (and no cruise control cable either). When you select the speed that you want to maintain, the PCM controls vehicle speed by opening and closing the throttle plate by means of a computer-controlled solenoid (motor) inside the throttle body.

2 The diagnostic procedures for troubleshooting the cruise control system are beyond the scope of this manual, but if the system can't be set, or the set speed doesn't cancel when the brake pedal is depressed, check the fuses. Start with the fuses in the engine compartment fuse and relay box, then check the fuses in the under-dash fuse and relay box.

3 Other than checking the fuses, the diagnostic procedures for troubleshooting the cruise control system on these models are beyond the scope of this manual. A dealer service department should handle any further testing.

23 Power window system - description and check

1 The power window system operates electric motors, mounted in the doors, which lower and raise the windows. The system consists of the control switches, relays, the motors, regulators, glass mechanisms and associated wiring.

2 The power windows can be lowered and raised from the master control switch by the driver or by remote switches located at the individual windows. Each window has a separate motor which is reversible. The position of the control switch determines the polarity and therefore the direction of operation.

3 The circuit is protected by a fuse and a circuit breaker. Each motor is also equipped with an internal circuit breaker; this prevents one stuck window from disabling the whole system.

4 The power window system will only operate when the ignition switch is ON. In addition, many models have a window lockout switch at the master control switch which, when activated, disables the switches at the rear windows and, sometimes, the switch at the passenger's window also. Always check these items before troubleshooting a window problem.

5 These procedures are general in nature, so if you can't find the problem using them, take the vehicle to a dealer service department or other properly equipped repair facility.

6 If the power windows won't operate, always check the fuse and circuit breaker first.

7 If only the rear windows are inoperative, or if the windows only operate from the master control switch, check the rear window lockout switch for continuity in the unlocked position. Replace it if it doesn't have continuity.

8 Check the wiring between the switches and fuse panel for continuity. Repair the wiring, if necessary.

9 If only one window is inoperative from the master control switch, try the other control switch at the window.

➡Note: This doesn't apply to the driver's door window.

10 If voltage is reaching the motor, disconnect the glass from the regulator (see Chapter 11). Move the window up and down by hand while checking for binding and damage. Also check for binding and damage to the regulator. If the regulator is not damaged and the window moves up and down smoothly, replace the motor. If there's binding or damage, lubricate, repair or replace parts, as necessary.

11 If voltage isn't reaching the motor, check the wiring in the circuit for continuity between the switches and motors. You'll need to consult the wiring diagram for the vehicle. If the circuit is equipped with a relay, check that the relay is grounded properly and receiving voltage.

12 Test the windows after you are done to confirm proper repairs.

24 Power door lock system - description and check

1 A power door lock system operates the door lock actuators mounted in each door. The system consists of the switches, actuators, a control unit and associated wiring. Diagnosis can usually be limited to simple checks of the wiring connections and actuators for minor faults that can be easily repaired.

2 Power door lock systems are operated by bi-directional solenoids located in the doors. The lock switches have two operating positions: Lock and Unlock. When activated, the switch sends a ground signal to the door lock control unit to lock or unlock the doors. Depending on which way the switch is activated, the control unit reverses polarity to the solenoids, allowing the two sides of the circuit to be used alternately as the feed (positive) and ground side.

3 Some vehicles may have an anti-theft system incorporated into the power locks. If you are unable to locate the trouble using the following general Steps, consult a dealer service department or other qualified repair shop.

4 Always check the circuit protection first. Some vehicles use a combination of circuit breakers and fuses.

5 Operate the door lock switches in both directions (Lock and Unlock) with the engine off. Listen for the click of the solenoids operating.

6 Test the switches for continuity. Remove the switches and have them checked by a dealer service department or other qualified automobile repair facility.

7 Check the wiring between the switches, control unit and solenoids for continuity. Repair the wiring if there's no continuity.

8 Check for a bad ground at the switches or the control unit.

9 If all but one lock solenoids operate, remove the trim panel from the affected door (see Chapter 11) and check for voltage at the solenoid while the lock switch is operated. One of the wires should have voltage in the Lock position; the other should have voltage in the Unlock position.

10 If the inoperative solenoid is receiving voltage, replace the solenoid.

11 If the inoperative solenoid isn't receiving voltage, check the relay for an open or short in the wire between the lock solenoid and the control unit.

➡Note: It's common for wires to break in the portion of the harness between the body and door (opening and closing the door fatigues and eventually breaks the wires).

25 Daytime Running Lights (DRL) - general information

The Daytime Running Lights (DRL) system used on some models illuminates the headlights whenever the engine is running. The only exception is with the engine running and the parking brake engaged. Once the parking brake is released, the lights will remain on as long as the ignition switch is on, even if the parking brake is later applied.

The DRL system supplies reduced power to the headlights so they won't be too bright for daytime use, while prolonging headlight life.

26 Airbag system - general information

These models are equipped with a Supplemental Restraint System (SRS), which consists of a number of airbag modules. The SRS system is designed to protect the driver and the front seat passenger from serious injury in the event of a head-on or frontal collision. It consists of an airbag module in the center of the steering wheel and another airbag module on the right side of the instrument panel, two impact sensors (located near each side of the radiator), and a sensing/diagnostic module which is mounted in the center of the vehicle below the instrument panel. On some models, the SRS system also includes side airbags and curtain shield airbags designed to protect the occupants in a side impact. These models are also equipped with a pair of impact sensors that are located near the front seat belt retractors, and another pair of sensors located near the rear seat belt retractors.

These vehicles are also equipped with seatbelt pre-tensioners, also part of the airbag system. The pre-tensioners are pyrotechnic (explosive) devices designed to retract the seat belts in the event of a collision. Do not remove the front seat belt retractor assemblies. Problems with the pre-tensioners will turn on the SRS (airbag) warning light on the dash. If any pre-tensioner problems are suspected, take the vehicle to a dealer service department.

AIRBAG MODULE

Steering wheel-mounted

The airbag inflator module contains a housing incorporating the cushion (airbag) and inflator unit, mounted in the center of the steering wheel. The inflator assembly is mounted on the back of the housing over a hole through which gas is expelled, inflating the bag almost instantaneously when an electrical signal is sent from the system. A spiral cable assembly on the steering column under the module carries this signal to the module. This spiral cable assembly can transmit an electrical signal regardless of steering wheel position.

Instrument panel-mounted

The passenger side airbag is mounted above the glove compartment and designated by the letters SRS (Supplemental Restraint System). It consists of an inflator containing an igniter, a bag assembly, a reaction housing and a trim cover.

The passenger airbag is considerably larger than the steering wheel-mounted unit and is supported by the steel reaction housing. The trim cover has a molded seam which splits when the bag inflates.

Side and curtain airbags

The side airbags are mounted on the outer sides of the front seats The curtain shield airbag assemblies run along the interior of the roof from the front A-pillar to the rear of the passenger compartment.

SENSING AND DIAGNOSTIC MODULE

The sensing and diagnostic module supplies the current to the airbag system in the event of the collision, even if battery power is cut off. It checks this system every time the vehicle is started, causing the AIR BAG light to go on then off, if the system is operating properly. If there is a fault in the system, the light will go on and stay on, flash, or the dash will make a beeping sound. If this happens, the vehicle should be taken to your dealer immediately for service.

PRECAUTIONS

Disabling the SRS system

⁂ WARNING 1:

Failure to follow these precautions could result in accidental deployment of the airbag and personal injury.

⁂ WARNING 2:

Never install a memory-saver device, used to preserve PCM memory and radio station presets, when working on or around any of the airbag system components.

Whenever working in the vicinity of the steering wheel, instrument panel or any of the other SRS system components, the system must be disarmed. To disarm the system:

a) *Point the wheels straight ahead and turn the ignition key to the LOCK position.*
b) *Disconnect the cable from the negative terminal of the battery.*
c) *Wait at least two minutes for the back-up power supply capacitor to be depleted.*

Whenever handling an airbag module, always keep the airbag opening (trim side) pointed away from your body. Never place the airbag module on a bench or other surface with the airbag opening facing the surface. Always place the airbag module in a safe location with the airbag opening (trim side) facing up.

Never measure the resistance of any SRS component. An ohmmeter has a built-in battery supply that could accidentally deploy the airbag.

Never use electrical welding equipment on a vehicle equipped with an airbag without first disconnecting the negative battery cable.

Never dispose of a live airbag module. Return it to your dealer for safe deployment, using special equipment, and disposal.

27 Wiring diagrams - general information

Since it isn't possible to include all wiring diagrams for every year covered by this manual, the following diagrams are those that are typical and most commonly needed.

Prior to troubleshooting any circuits, check the fuse and circuit breakers (if equipped) to make sure they're in good condition. Make sure the battery is properly charged and check the cable connections (see Chapter 1).

When checking a circuit, make sure that all connectors are clean, with no broken or loose terminals. When unplugging a connector, do not pull on the wires. Pull only on the connector housings themselves.

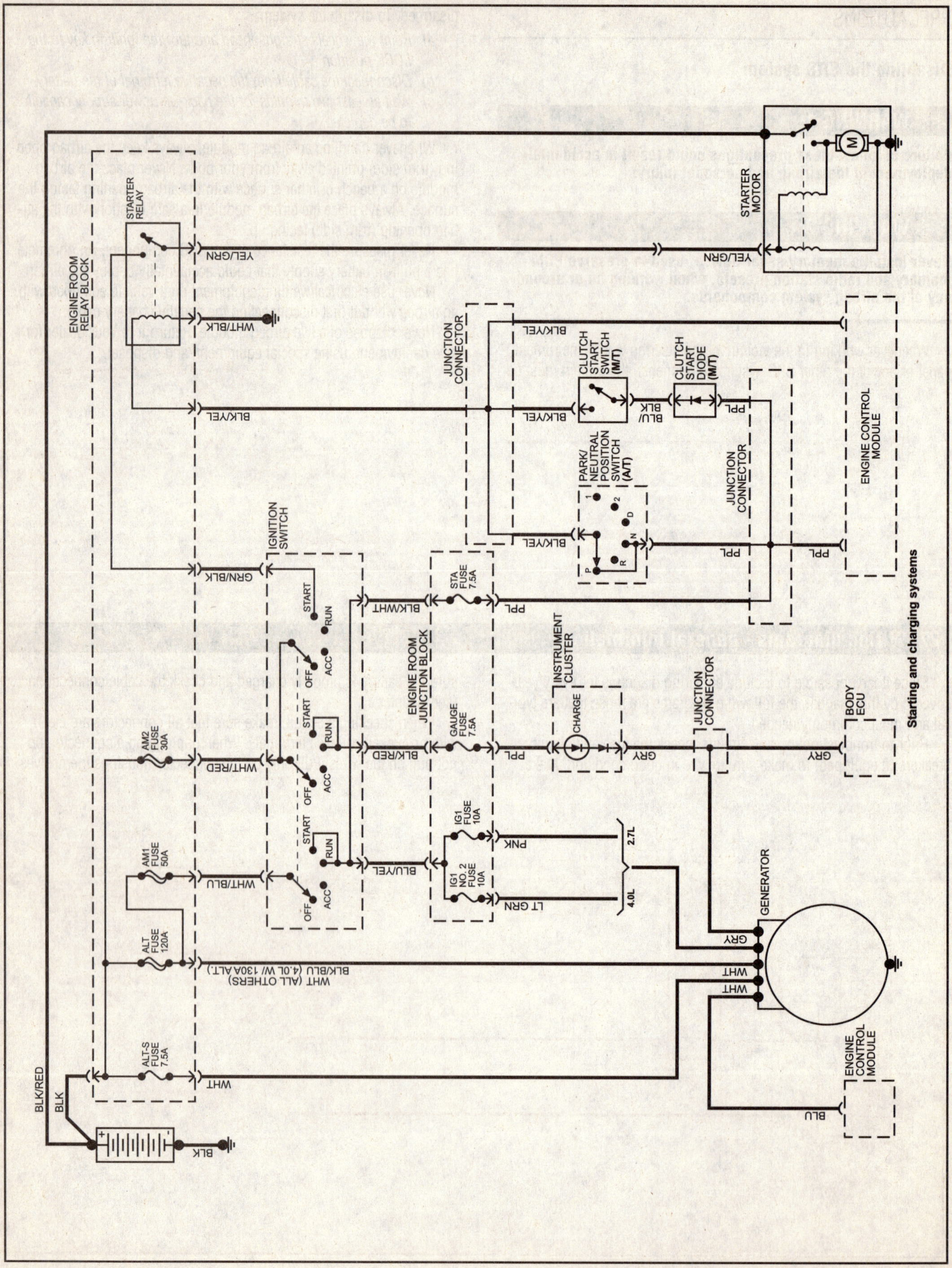

Starting and charging systems

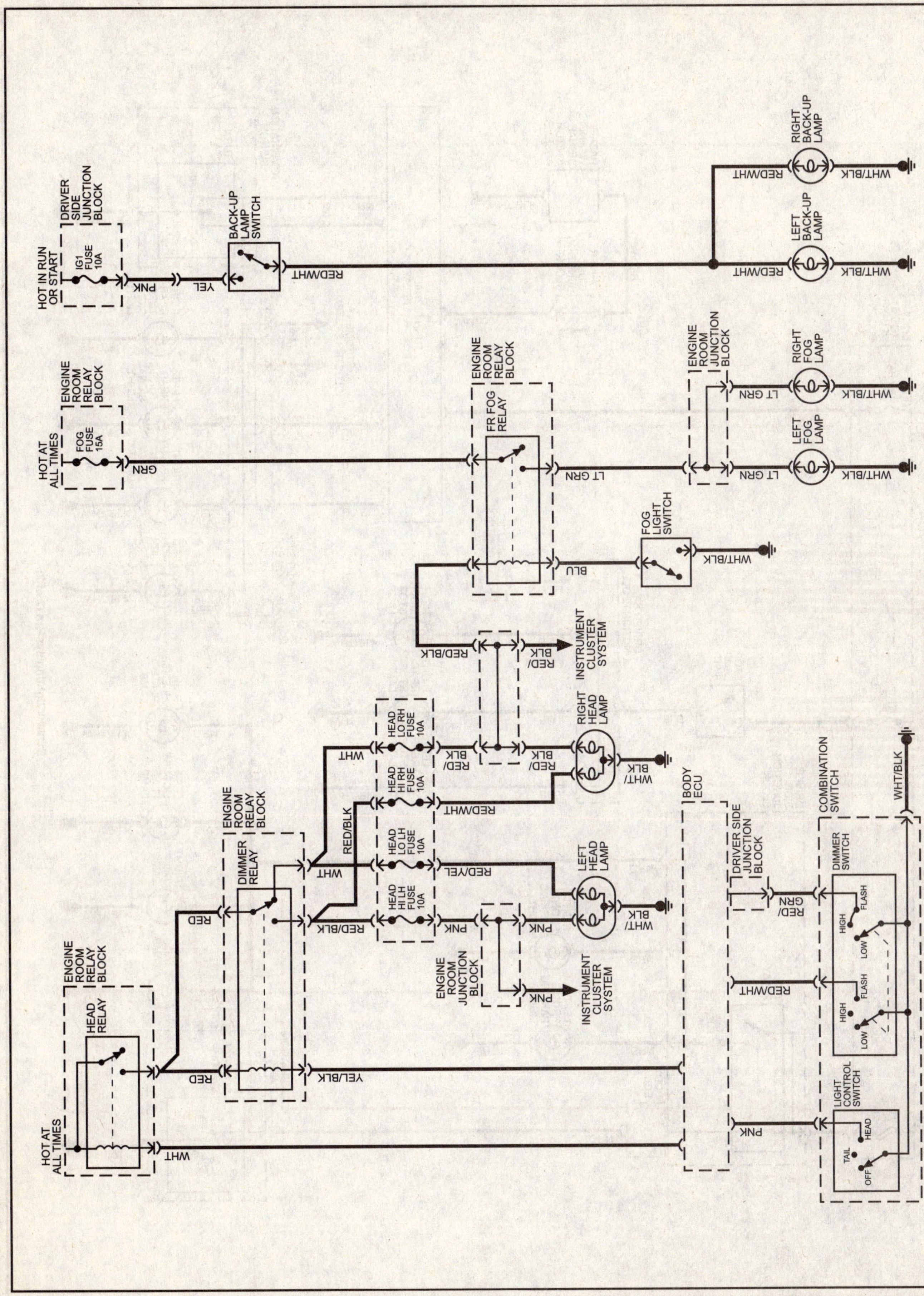

Exterior lighting system (1 of 2)

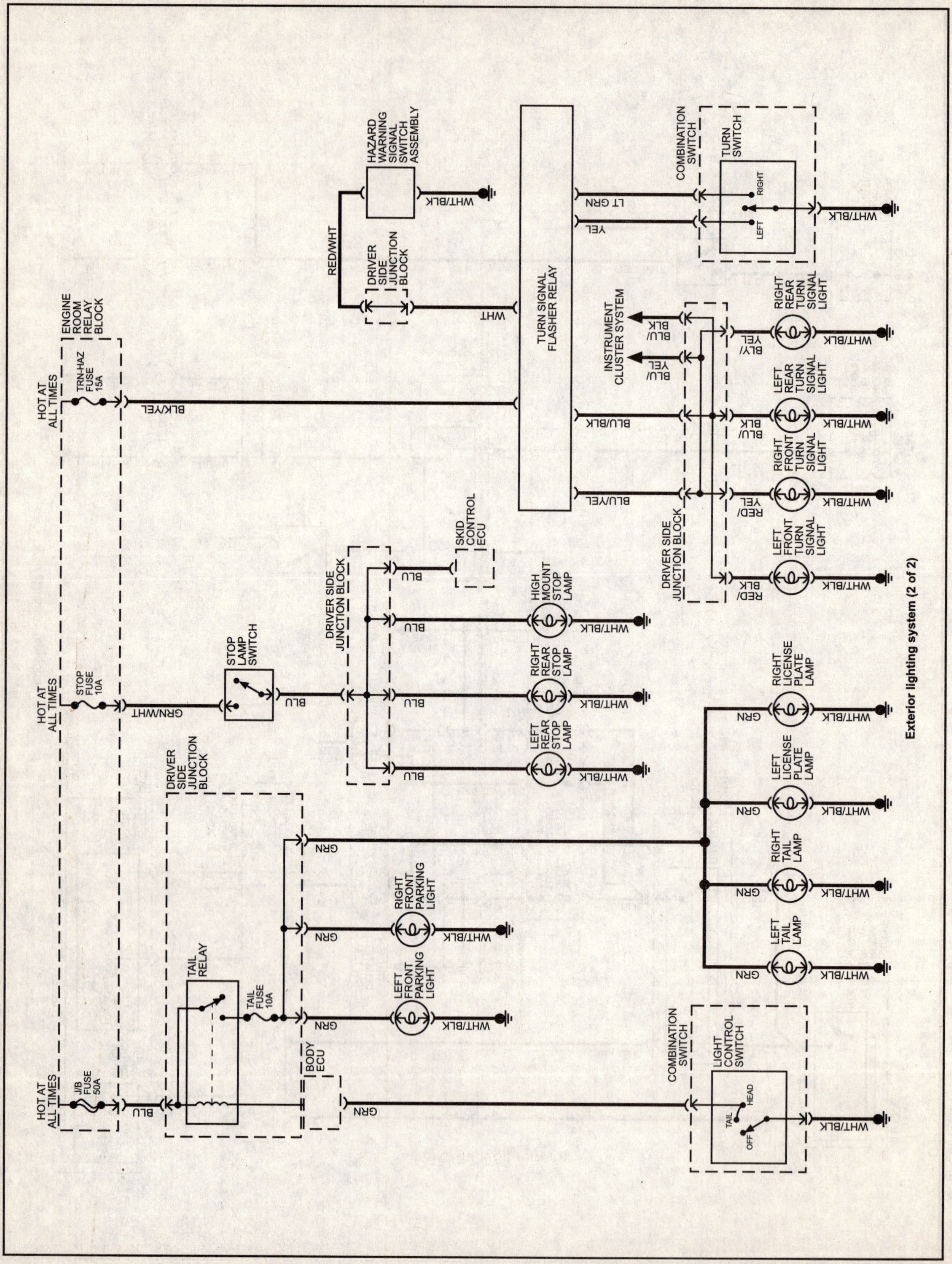

Exterior lighting system (2 of 2)

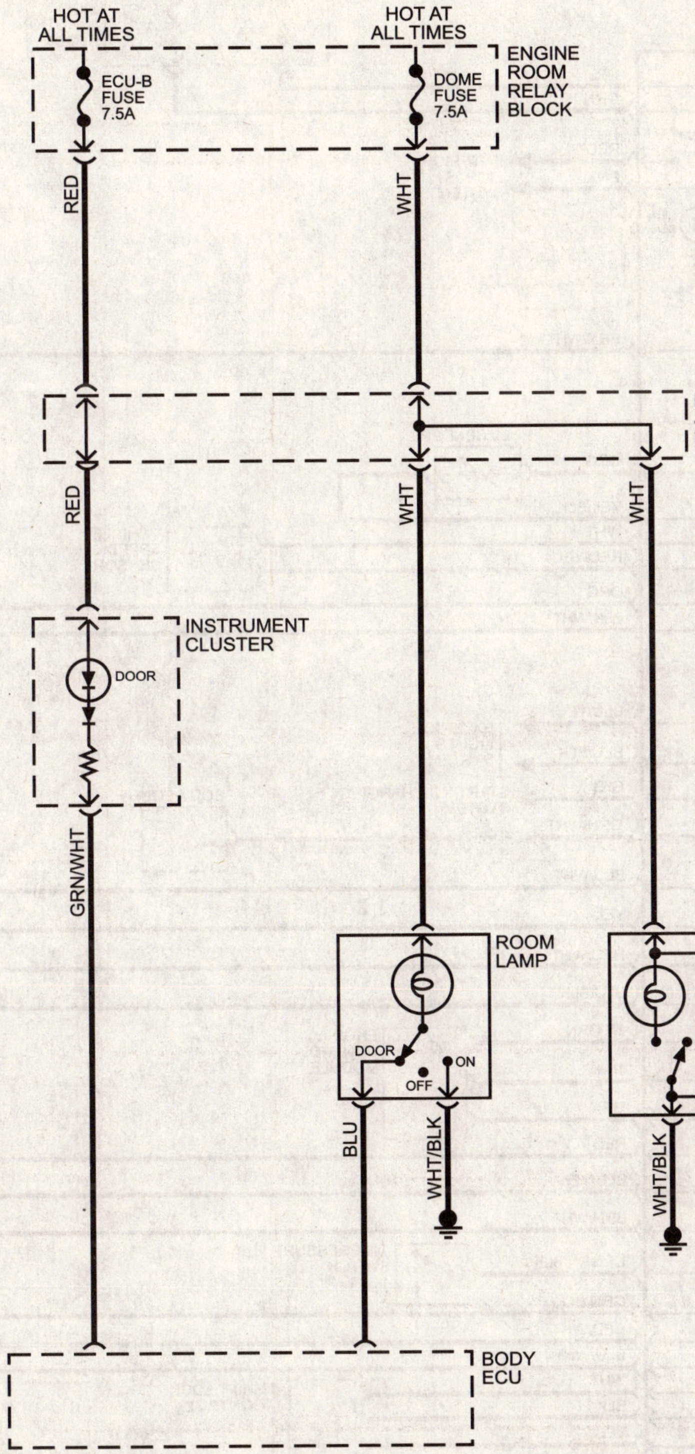

Interior lighting system

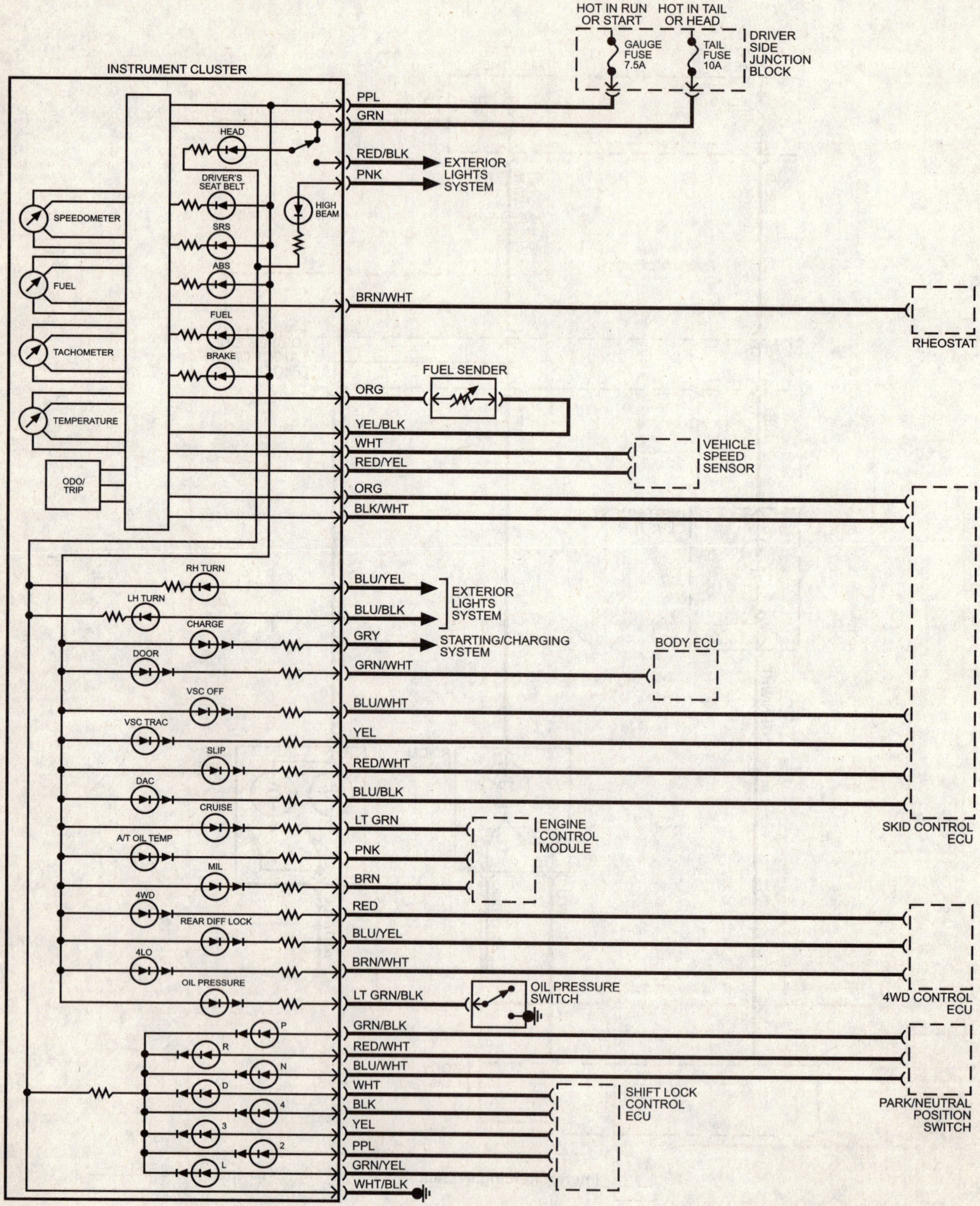

Instrument cluster warning system

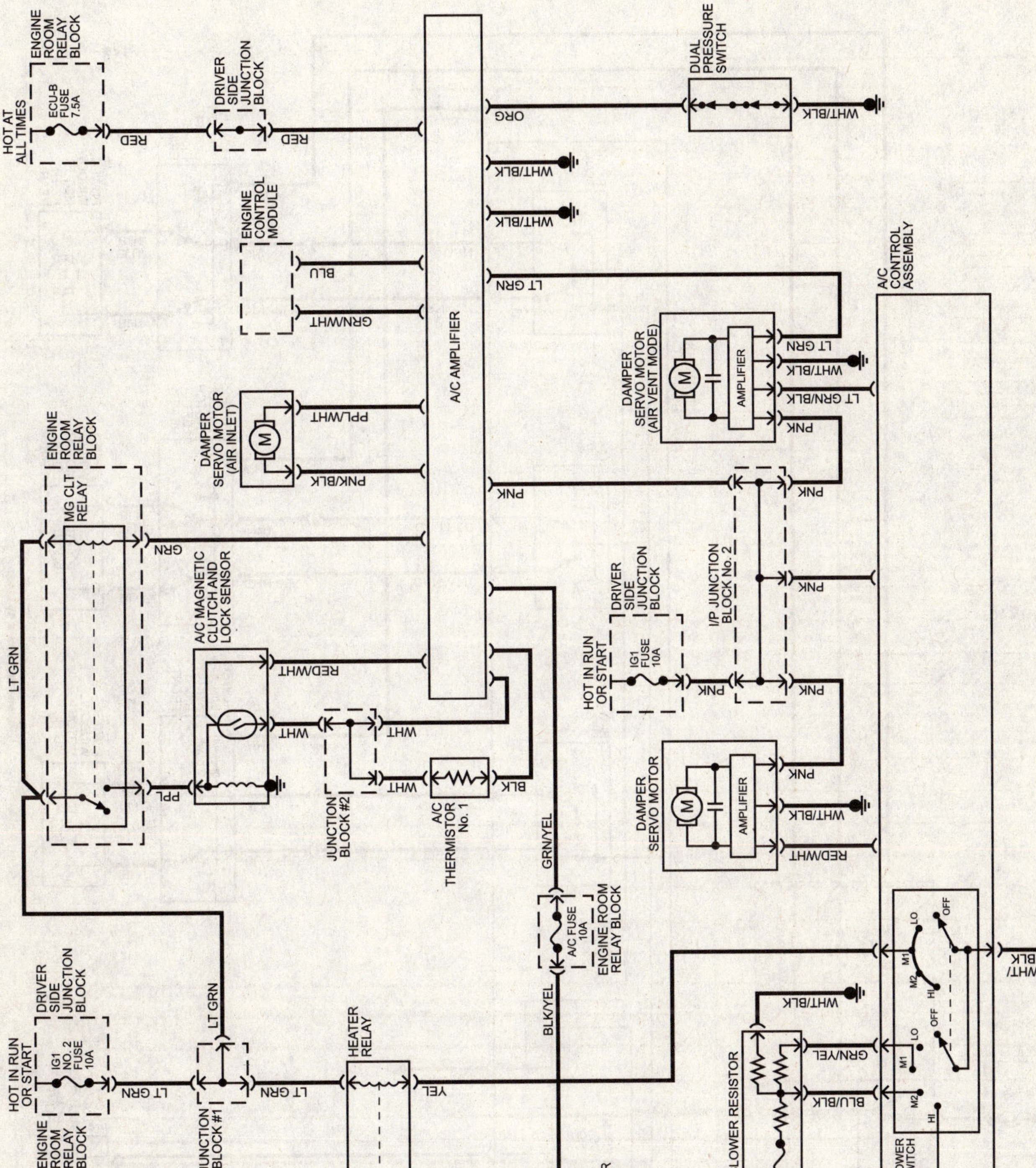

Air conditioning system

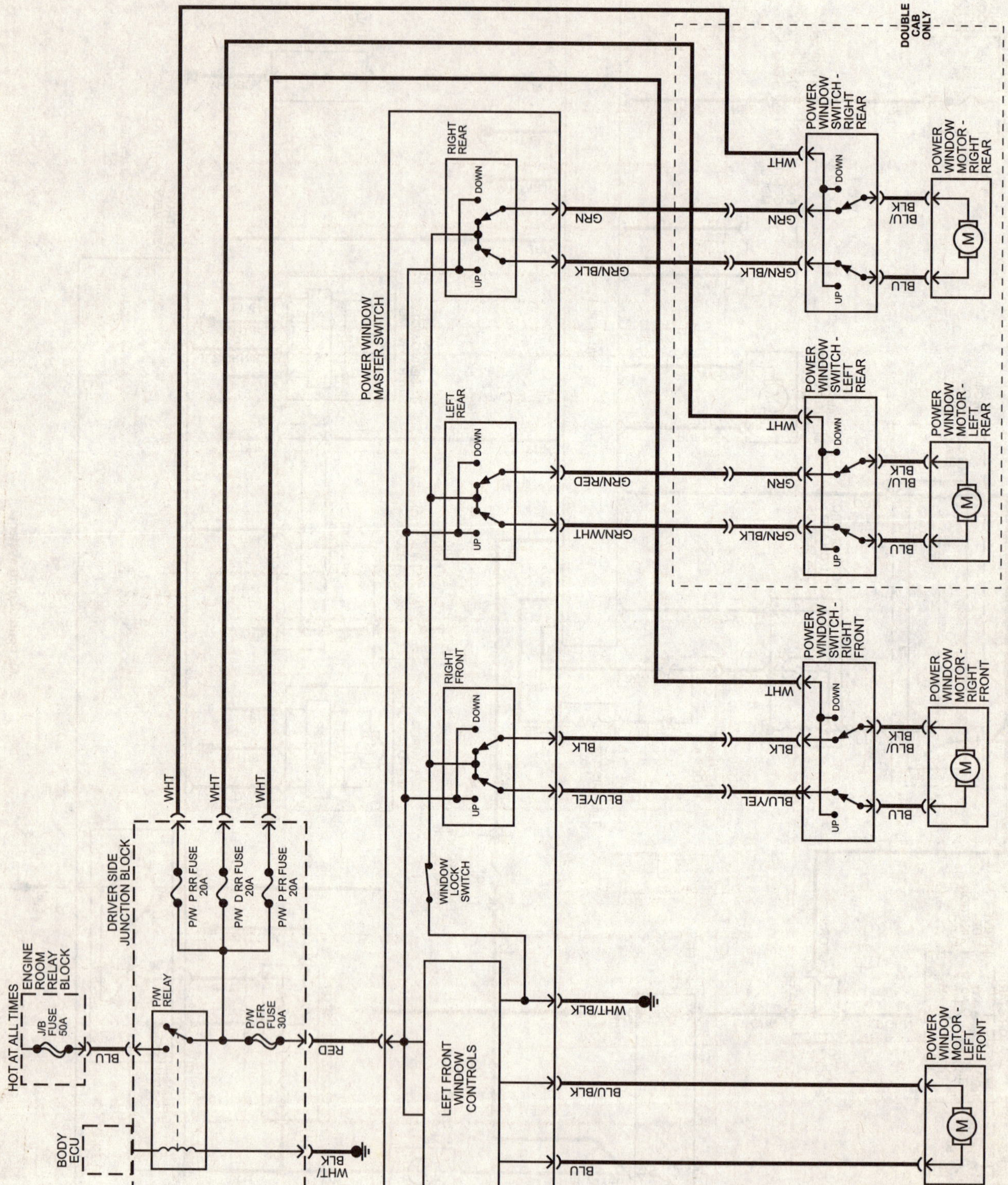

Power windows system

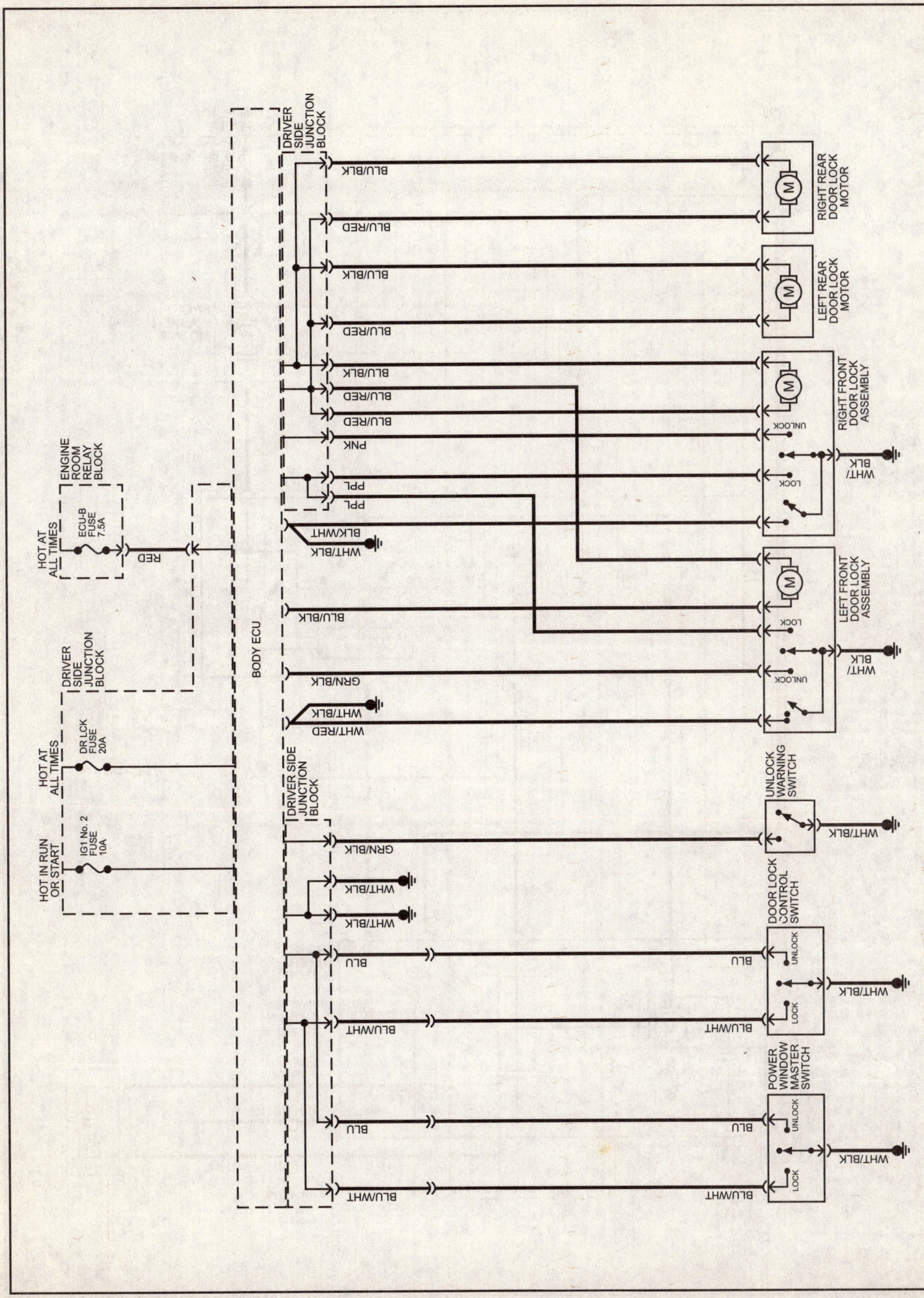

Power door locks system

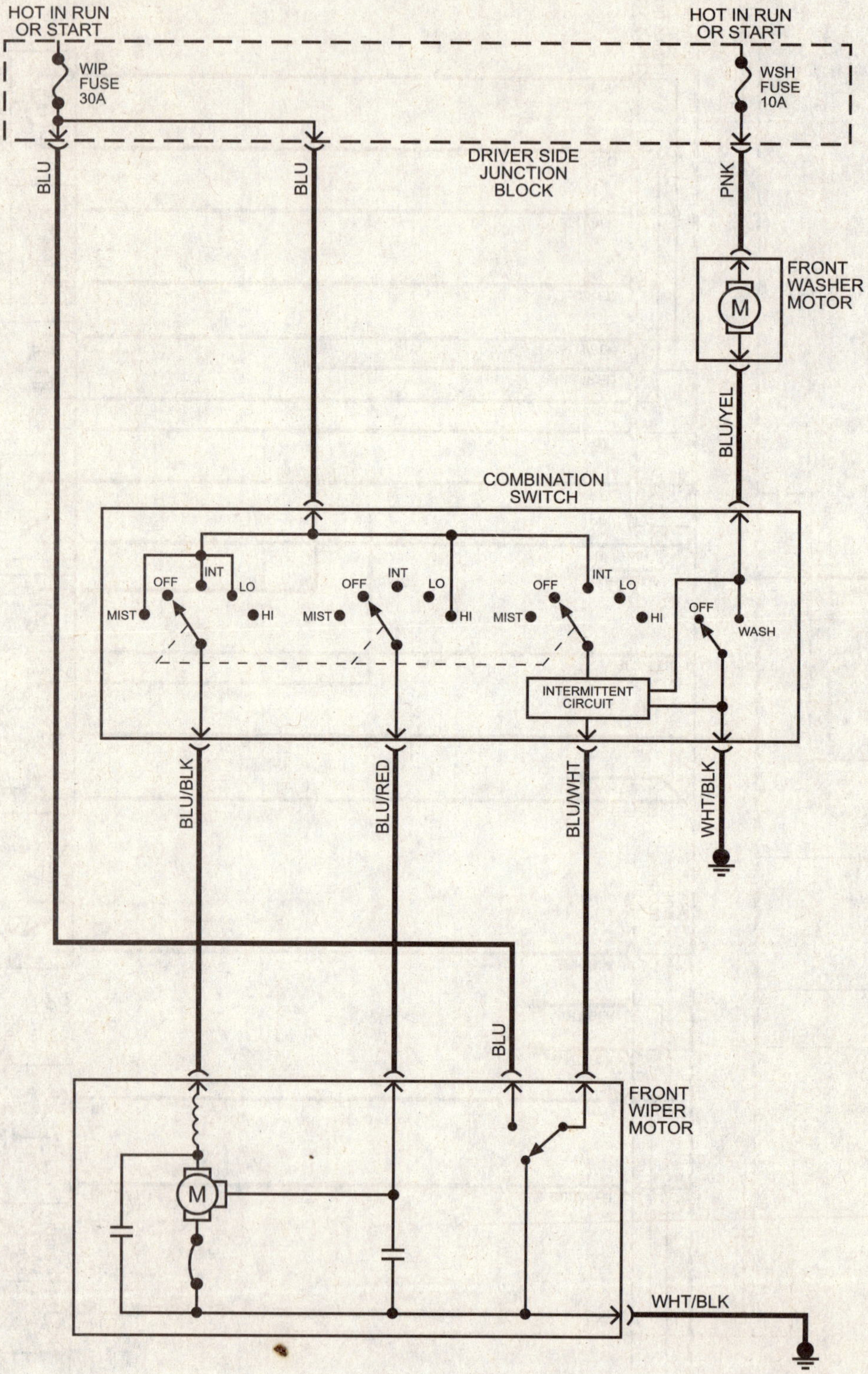

Windshield wiper and washer system (with INT)

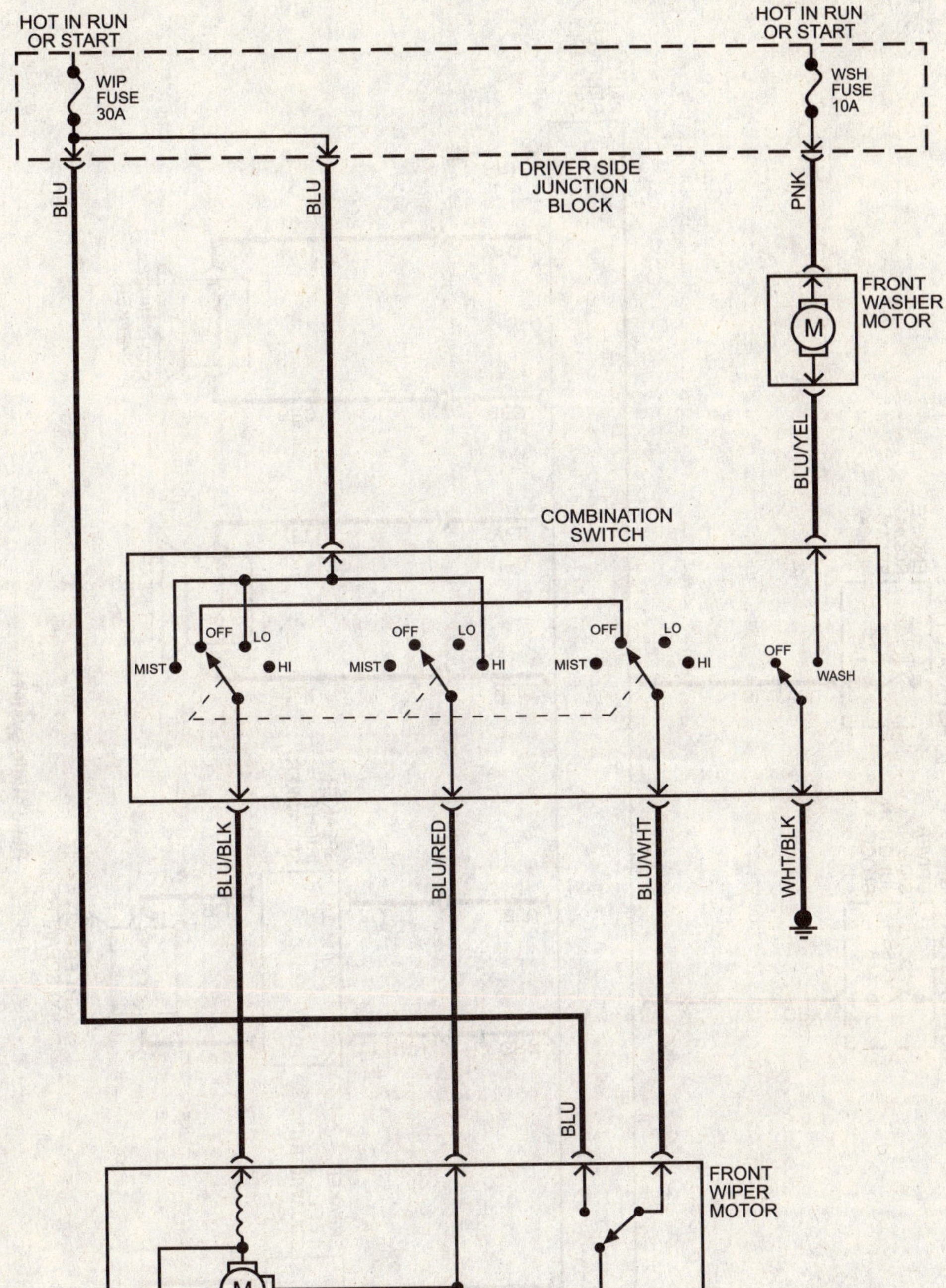

Windshield wiper and washer system (without INT)

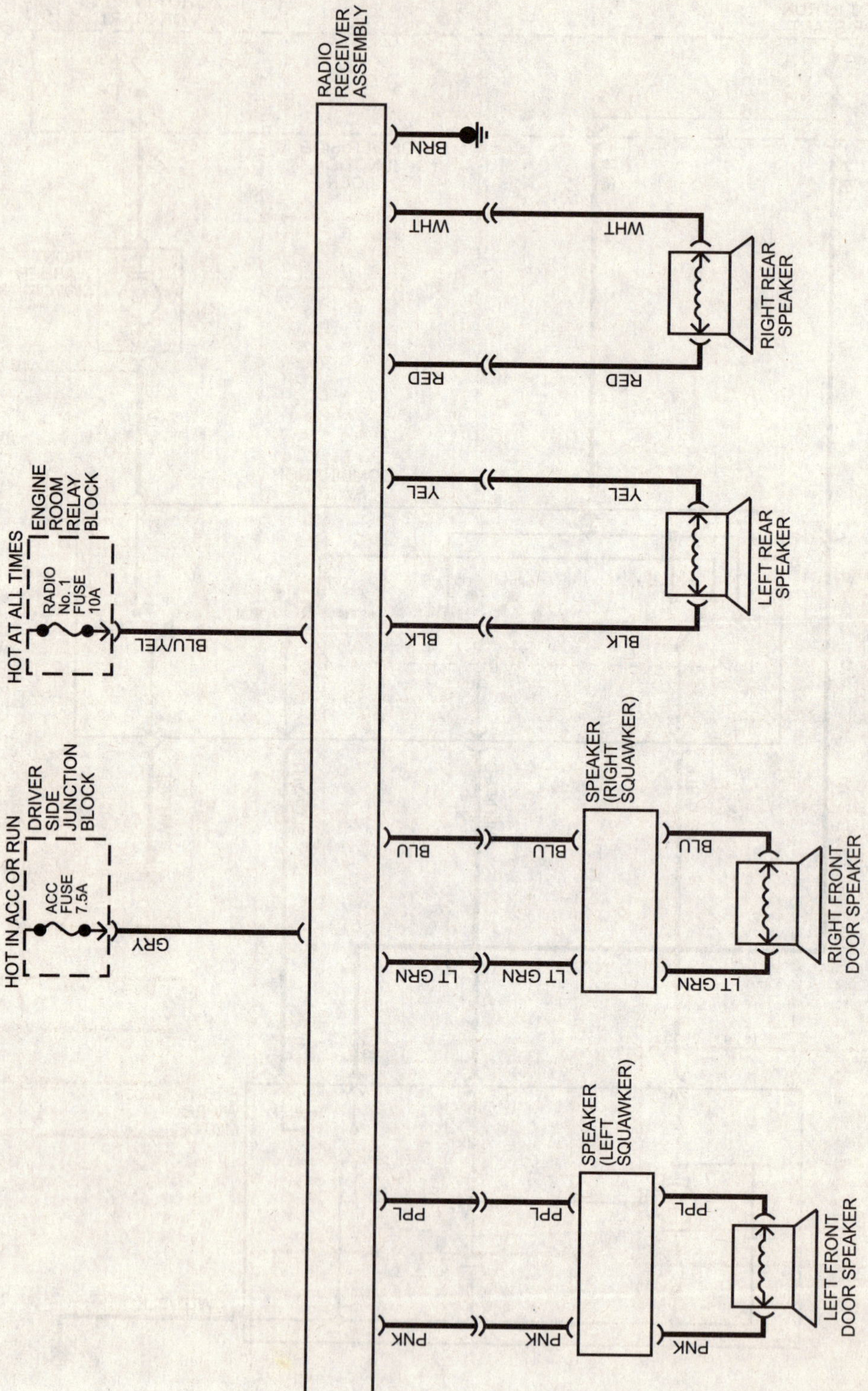

Base audio system

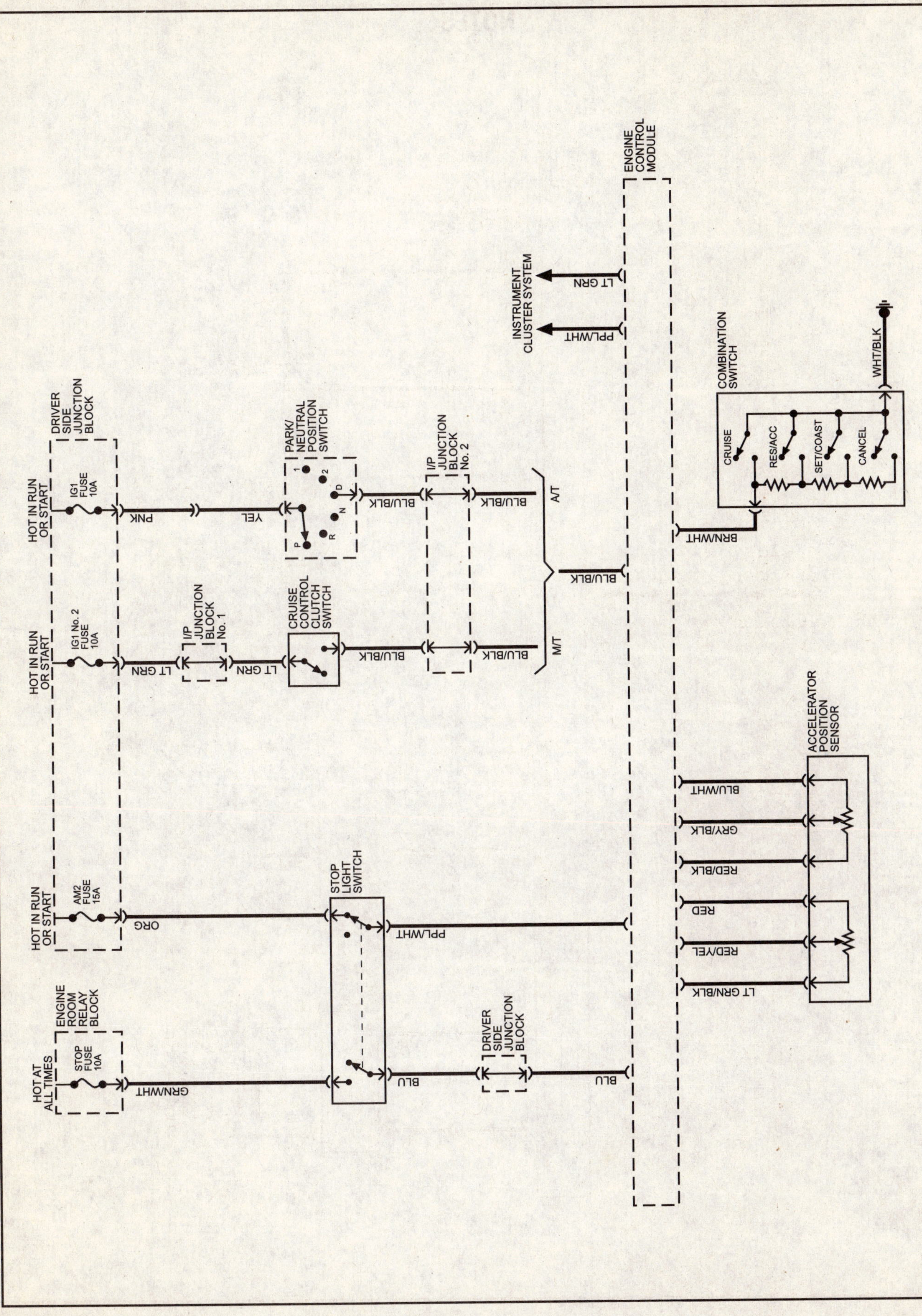

Cruise control system

NOTES

GLOSSARY

AIR/FUEL RATIO: The ratio of air-to-gasoline by weight in the fuel mixture drawn into the engine.

AIR INJECTION: One method of reducing harmful exhaust emissions by injecting air into each of the exhaust ports of an engine. The fresh air entering the hot exhaust manifold causes any remaining fuel to be burned before it can exit the tailpipe.

ALTERNATOR: A device used for converting mechanical energy into electrical energy.

AMMETER: An instrument, calibrated in amperes, used to measure the flow of an electrical current in a circuit. Ammeters are always connected in series with the circuit being tested.

AMPERE: The rate of flow of electrical current present when one volt of electrical pressure is applied against one ohm of electrical resistance.

ANALOG COMPUTER: Any microprocessor that uses similar (analogous) electrical signals to make its calculations.

ARMATURE: A laminated, soft iron core wrapped by a wire that converts electrical energy to mechanical energy as in a motor or relay. When rotated in a magnetic field, it changes mechanical energy into electrical energy as in a generator.

ATMOSPHERIC PRESSURE: The pressure on the Earth's surface caused by the weight of the air in the atmosphere. At sea level, this pressure is 14.7 psi at 32°F (101 kPa at 0°C).

ATOMIZATION: The breaking down of a liquid into a fine mist that can be suspended in air.

AXIAL PLAY: Movement parallel to a shaft or bearing bore.

BACKFIRE: The sudden combustion of gases in the intake or exhaust system that results in a loud explosion.

BACKLASH: The clearance or play between two parts, such as meshed gears.

BACKPRESSURE: Restrictions in the exhaust system that slow the exit of exhaust gases from the combustion chamber.

BAKELITE: A heat resistant, plastic insulator material commonly used in printed circuit boards and transistorized components.

BALL BEARING: A bearing made up of hardened inner and outer races between which hardened steel balls roll.

BALLAST RESISTOR: A resistor in the primary ignition circuit that lowers voltage after the engine is started to reduce wear on ignition components.

BEARING: A friction reducing, supportive device usually located between a stationary part and a moving part.

BIMETAL TEMPERATURE SENSOR: Any sensor or switch made of two dissimilar types of metal that bend when heated or cooled due to the different expansion rates of the alloys. These types of sensors usually function as an on/off switch.

BLOWBY: Combustion gases, composed of water vapor and unburned fuel, that leak past the piston rings into the crankcase during normal engine operation. These gases are removed by the PCV system to prevent the buildup of harmful acids in the crankcase.

BRAKE PAD: A brake shoe and lining assembly used with disc brakes.

BRAKE SHOE: The backing for the brake lining. The term is, however, usually applied to the assembly of the brake backing and lining.

BUSHING: A liner, usually removable, for a bearing; an anti-friction liner used in place of a bearing.

CALIPER: A hydraulically activated device in a disc brake system, which is mounted straddling the brake rotor (disc). The caliper contains at least one piston and two brake pads. Hydraulic pressure on the piston(s) forces the pads against the rotor.

CAMSHAFT: A shaft in the engine on which are the lobes (cams) which operate the valves. The camshaft is driven by the crankshaft, via a belt, chain or gears, at one half the crankshaft speed.

CAPACITOR: A device which stores an electrical charge.

CARBON MONOXIDE (CO): A colorless, odorless gas given off as a normal byproduct of combustion. It is poisonous and extremely dangerous in confined areas, building up slowly to toxic levels without warning if adequate ventilation is not available.

CARBURETOR: A device, usually mounted on the intake manifold of an engine, which mixes the air and fuel in the proper proportion to allow even combustion.

CATALYTIC CONVERTER: A device installed in the exhaust system, like a muffler, that converts harmful byproducts of combustion into carbon dioxide and water vapor by means of a heat-producing chemical reaction.

CENTRIFUGAL ADVANCE: A mechanical method of advancing the spark timing by using flyweights in the distributor that react to centrifugal force generated by the distributor shaft rotation.

CHECK VALVE: Any one-way valve installed to permit the flow of air, fuel or vacuum in one direction only.

CHOKE: A device, usually a moveable valve, placed in the intake path of a carburetor to restrict the flow of air.

CIRCUIT: Any unbroken path through which an electrical current can flow. Also used to describe fuel flow in some instances.

CIRCUIT BREAKER: A switch which protects an electrical circuit from overload by opening the circuit when the current flow exceeds a predetermined level. Some circuit breakers must be reset manually, while most reset automatically.

COIL (IGNITION): A transformer in the ignition circuit which steps up the voltage provided to the spark plugs.

COMBINATION MANIFOLD: An assembly which includes both the intake and exhaust manifolds in one casting.

COMBINATION VALVE: A device used in some fuel systems that routes fuel vapors to a charcoal storage canister instead of venting them into the atmosphere. The valve relieves fuel tank pressure and allows fresh air into the tank as the fuel level drops to prevent a vapor lock situation.

COMPRESSION RATIO: The comparison of the total volume of the cylinder and combustion chamber with the piston at BDC and the piston at TDC.

CONDENSER: 1. An electrical device which acts to store an electrical charge, preventing voltage surges. 2. A radiator-like device in the air conditioning system in which refrigerant gas condenses into a liquid, giving off heat.

CONDUCTOR: Any material through which an electrical current can be transmitted easily.

CONTINUITY: Continuous or complete circuit. Can be checked with an ohmmeter.

COUNTERSHAFT: An intermediate shaft which is rotated by a mainshaft and transmits, in turn, that rotation to a working part.

CRANKCASE: The lower part of an engine in which the crankshaft and related parts operate.

CRANKSHAFT: The main driving shaft of an engine which receives reciprocating motion from the pistons and converts it to rotary motion.

CYLINDER: In an engine, the round hole in the engine block in which the piston(s) ride.

CYLINDER BLOCK: The main structural member of an engine in which is found the cylinders, crankshaft and other principal parts.

CYLINDER HEAD: The detachable portion of the engine, usually fastened to the top of the cylinder block and containing all or most of the combustion chambers. On overhead valve engines, it contains the valves and their operating parts. On overhead cam engines, it contains the camshaft as well.

DEAD CENTER: The extreme top or bottom of the piston stroke.

DETONATION: An unwanted explosion of the air/fuel mixture in the combustion chamber caused by excess heat and compression, advanced timing, or an overly lean mixture. Also referred to as "ping".

DIAPHRAGM: A thin, flexible wall separating two cavities, such as in a vacuum advance unit.

DIESELING: A condition in which hot spots in the combustion chamber cause the engine to run on after the key is turned off.

DIFFERENTIAL: A geared assembly which allows the transmission of motion between drive axles, giving one axle the ability to turn faster than the other.

DIODE: An electrical device that will allow current to flow in one direction only.

DISC BRAKE: A hydraulic braking assembly consisting of a brake disc, or rotor, mounted on an axle, and a caliper assembly containing, usually two brake pads which are activated by hydraulic pressure. The pads are forced against the sides of the disc, creating friction which slows the vehicle.

DISTRIBUTOR: A mechanically driven device on an engine which is responsible for electrically firing the spark plug at a predetermined point of the piston stroke.

DOWEL PIN: A pin, inserted in mating holes in two different parts allowing those parts to maintain a fixed relationship.

DRUM BRAKE: A braking system which consists of two brake shoes and one or two wheel cylinders, mounted on a fixed backing plate, and a brake drum, mounted on an axle, which revolves around the assembly.

DWELL: The rate, measured in degrees of shaft rotation, at which an electrical circuit cycles on and off.

ELECTRONIC CONTROL UNIT (ECU): Ignition module, module, amplifier or igniter. See Module for definition.

ELECTRONIC IGNITION: A system in which the timing and firing of the spark plugs is controlled by an electronic control unit, usually called a module. These systems have no points or condenser.

END-PLAY: The measured amount of axial movement in a shaft.

ENGINE: A device that converts heat into mechanical energy.

EXHAUST MANIFOLD: A set of cast passages or pipes which conduct exhaust gases from the engine.

FEELER GAUGE: A blade, usually metal, or precisely predetermined thickness, used to measure the clearance between two parts.

FIRING ORDER: The order in which combustion occurs in the cylinders of an engine. Also the order in which spark is distributed to the plugs by the distributor.

FLOODING: The presence of too much fuel in the intake manifold and combustion chamber which prevents the air/fuel mixture from firing, thereby causing a no-start situation.

FLYWHEEL: A disc shaped part bolted to the rear end of the crankshaft. Around the outer perimeter is affixed the ring gear. The starter drive engages the ring gear, turning the flywheel, which rotates the crankshaft, imparting the initial starting motion to the engine.

FOOT POUND (ft. lbs. or sometimes, ft.lb.): The amount of energy or work needed to raise an item weighing one pound, a distance of one foot.

FUSE: A protective device in a circuit which prevents circuit overload by breaking the circuit when a specific amperage is present. The device is constructed around a strip or wire of a lower amperage rating than the circuit it is designed to protect. When an amperage higher than that stamped on the fuse is present in the circuit, the strip or wire melts, opening the circuit.

GEAR RATIO: The ratio between the number of teeth on meshing gears.

GENERATOR: A device which converts mechanical energy into electrical energy.

HEAT RANGE: The measure of a spark plug's ability to dissipate heat from its firing end. The higher the heat range, the hotter the plug fires.

HUB: The center part of a wheel or gear.

HYDROCARBON (HC): Any chemical compound made up of hydrogen and carbon. A major pollutant formed by the engine as a byproduct of combustion.

HYDROMETER: An instrument used to measure the specific gravity of a solution.

INCH POUND (inch lbs.; sometimes in.lb. or in. lbs.): One twelfth of a foot pound.

INDUCTION: A means of transferring electrical energy in the form of a magnetic field. Principle used in the ignition coil to increase voltage.

INJECTOR: A device which receives metered fuel under relatively low pressure and is activated to inject the fuel into the engine under relatively high pressure at a predetermined time.

INPUT SHAFT: The shaft to which torque is applied, usually carrying the driving gear or gears.

INTAKE MANIFOLD: A casting of passages or pipes used to conduct air or a fuel/air mixture to the cylinders.

JOURNAL: The bearing surface within which a shaft operates.

KEY: A small block usually fitted in a notch between a shaft and a hub to prevent slippage of the two parts.

MANIFOLD: A casting of passages or set of pipes which connect the cylinders to an inlet or outlet source.

MANIFOLD VACUUM: Low pressure in an engine intake manifold formed just below the throttle plates. Manifold vacuum is highest at idle and drops under acceleration.

MASTER CYLINDER: The primary fluid pressurizing device in a hydraulic system. In automotive use, it is found in brake and hydraulic clutch systems and is pedal activated, either directly or, in a power brake system, through the power booster.

MODULE: Electronic control unit, amplifier or igniter of solid state or integrated design which controls the current flow in the ignition primary circuit based on input from the pick-up coil. When the module opens the primary circuit, high secondary voltage is induced in the coil.

NEEDLE BEARING: A bearing which consists of a number (usually a large number) of long, thin rollers.

OHM: (Ω) The unit used to measure the resistance of conductor-to-electrical flow. One ohm is the amount of resistance that limits current flow to one ampere in a circuit with one volt of pressure.

OHMMETER: An instrument used for measuring the resistance, in ohms, in an electrical circuit.

OUTPUT SHAFT: The shaft which transmits torque from a device, such as a transmission.

OVERDRIVE: A gear assembly which produces more shaft revolutions than that transmitted to it.

OVERHEAD CAMSHAFT (OHC): An engine configuration in which the camshaft is mounted on top of the cylinder head and operates the valve either directly or by means of rocker arms.

OVERHEAD VALVE (OHV): An engine configuration in which all of the valves are located in the cylinder head and the camshaft is located in the cylinder block. The camshaft operates the valves via lifters and pushrods.

OXIDES OF NITROGEN (NOx): Chemical compounds of nitrogen produced as a byproduct of combustion. They combine with hydrocarbons to produce smog.

OXYGEN SENSOR: Use with the feedback system to sense the presence of oxygen in the exhaust gas and signal the computer which can reference the voltage signal to an air/fuel ratio.

PINION: The smaller of two meshing gears.

PISTON RING: An open-ended ring with fits into a groove on the outer diameter of the piston. Its chief function is to form a seal between the piston and cylinder wall. Most automotive pistons have three rings: two for compression sealing; one for oil sealing.

PRELOAD: A predetermined load placed on a bearing during assembly or by adjustment.

PRIMARY CIRCUIT: the low voltage side of the ignition system which consists of the ignition switch, ballast resistor or resistance wire, bypass, coil, electronic control unit and pick-up coil as well as the connecting wires and harnesses.

PRESS FIT: The mating of two parts under pressure, due to the inner diameter of one being smaller than the outer diameter of the other, or vice versa; an interference fit.

RACE: The surface on the inner or outer ring of a bearing on which the balls, needles or rollers move.

REGULATOR: A device which maintains the amperage and/or voltage levels of a circuit at predetermined values.

RELAY: A switch which automatically opens and/or closes a circuit.

RESISTANCE: The opposition to the flow of current through a circuit or electrical device, and is measured in ohms. Resistance is equal to the voltage divided by the amperage.

RESISTOR: A device, usually made of wire, which offers a preset amount of resistance in an electrical circuit.

RING GEAR: The name given to a ring-shaped gear attached to a differential case, or affixed to a flywheel or as part of a planetary gear set.

ROLLER BEARING: A bearing made up of hardened inner and outer races between which hardened steel rollers move.

ROTOR: 1. The disc-shaped part of a disc brake assembly, upon which the brake pads bear; also called, brake disc. 2. The device mounted atop the distributor shaft, which passes current to the distributor cap tower contacts.

SECONDARY CIRCUIT: The high voltage side of the ignition system, usually above 20,000 volts. The secondary includes the ignition coil, coil wire, distributor cap and rotor, spark plug wires and spark plugs.

SENDING UNIT: A mechanical, electrical, hydraulic or electromagnetic device which transmits information to a gauge.

SENSOR: Any device designed to measure engine operating conditions or ambient pressures and temperatures. Usually electronic in nature and designed to send a voltage signal to an on-board computer, some sensors may operate as a simple on/off switch or they may provide a variable voltage signal (like a potentiometer) as conditions or measured parameters change.

SHIM: Spacers of precise, predetermined thickness used between parts to establish a proper working relationship.

SLAVE CYLINDER: In automotive use, a device in the hydraulic clutch system which is activated by hydraulic force, disengaging the clutch.

SOLENOID: A coil used to produce a magnetic field, the effect of which is to produce work.

SPARK PLUG: A device screwed into the combustion chamber of a spark ignition engine. The basic construction is a conductive core inside of a ceramic insulator, mounted in an outer conductive base. An electrical charge from the spark plug wire travels along the conductive core and jumps a preset air gap to a grounding point or points at the end of the conductive base. The resultant spark ignites the fuel/air mixture in the combustion chamber.

SPLINES: Ridges machined or cast onto the outer diameter of a shaft or inner diameter of a bore to enable parts to mate without rotation.

TACHOMETER: A device used to measure the rotary speed of an engine, shaft, gear, etc., usually in rotations per minute.

THERMOSTAT: A valve, located in the cooling system of an engine, which is closed when cold and opens gradually in response to engine heating, controlling the temperature of the coolant and rate of coolant flow.

TOP DEAD CENTER (TDC): The point at which the piston reaches the top of its travel on the compression stroke.

TORQUE: The twisting force applied to an object.

TORQUE CONVERTER: A turbine used to transmit power from a driving member to a driven member via hydraulic action, providing changes in drive ratio and torque. In automotive use, it links the driveplate at the rear of the engine to the automatic transmission.

TRANSDUCER: A device used to change a force into an electrical signal.

TRANSISTOR: A semi-conductor component which can be actuated by a small voltage to perform an electrical switching function.

TUNE-UP: A regular maintenance function, usually associated with the replacement and adjustment of parts and components in the electrical and fuel systems of a vehicle for the purpose of attaining optimum performance.

TURBOCHARGER: An exhaust driven pump which compresses intake air and forces it into the combustion chambers at higher than atmospheric pressures. The increased air pressure allows more fuel to be burned and results in increased horsepower being produced.

VACUUM ADVANCE: A device which advances the ignition timing in response to increased engine vacuum.

VACUUM GAUGE: An instrument used to measure the presence of vacuum in a chamber.

VALVE: A device which control the pressure, direction of flow or rate of flow of a liquid or gas.

VALVE CLEARANCE: The measured gap between the end of the valve stem and the rocker arm, cam lobe or follower that activates the valve.

VISCOSITY: The rating of a liquid's internal resistance to flow.

VOLTMETER: An instrument used for measuring electrical force in units called volts. Voltmeters are always connected parallel with the circuit being tested.

WHEEL CYLINDER: Found in the automotive drum brake assembly, it is a device, actuated by hydraulic pressure, which, through internal pistons, pushes the brake shoes outward against the drums.

MASTER INDEX

A

D

E

F

S

T

U

V

W